Current Topics in Pathology

77

Managing Editors

C.L. Berry E. Grundmann

Morphological Tumor Markers

General Aspects and Diagnostic Relevance

Contributors

M. Altmannsberger · R. Berndt · P. Brandtzaeg
J. Caselitz · T.M. Chu · A.D. Cox · I. Damjanov
H. Denk · Ph. U. Heitz · R.V. Iozzo · R.C. Janzer
M. Kiessling · P. Kleihues · G. Klöppel · F.R. Korsrud
Th. Löning · K. Milde · P. Möller · R. Moll · M. Osborn
H.F. Otto · T.O. Rognum · K. Schwechheimer
P.S. Thrane · G.L. Wright Jr.

Editor

Gerhard Seifert

Springer-Verlag
Berlin Heidelberg NewYork
London Paris Tokyo

G. SEIFERT, Professor Dr., Institut für Pathologie
der Universität Hamburg, Martinistraße 52, UKE, D-2000 Hamburg 20

C.L. BERRY, Professor Dr., Department of Morbid Anatomy,
The London Hospital Medical College, Whitechapel,
London E1 1BB, Great Britain

E. GRUNDMANN, Professor Dr., Gerhard-Domagk-Institut für
Pathologie der Universität, Domagkstraße 17, D-4400 Münster

With 95 Figures, Some in Colour

Library of Congress Cataloging-in-Publication Data. Morphological tumor markers. (Current topics in pathology; 77) Includes bibliographies and index. 1. Tumor markers – Diagnostic use. 2. Tumor antigens – Diagnostic use. I. Altmannsberger, M. II. Seifert, Gerhard, 1921– . III. Series: Current topics in pathology; v. 77. [DNLM; 1. Antigens, Neoplasm – analysis. 2. Histocytochemistry. 3. Immunochemistry. 4. Immunologic Technics. 5. Neoplasms – diagnosis. 6. Receptors, Endogenous Substances. W1 CU821H v. 77/QZ 241 M871] RB1.E6 vol. 77 616.07 s 87-20709 [RC270.3.T84] [616.99′207′58]

ISBN-13: 978-3-642-71358-3 e-ISBN-13: 978-3-642-71356-9
DOI: 10.1007/978-3-642-71356-9

Softcover reprint of the hardcover 1st edition 1987

2122/3130-543210

List of Contributors

ALTMANNSBERGER, M., Prof. Dr.	Institut für Pathologie, Klinikum der Justus-Liebig-Universität Gießen, Langhansstraße 10, D-6300 Gießen
BERNDT, R., Dr.	Institut für Allgemeine Pathologie und pathologische Anatomie der Universität, Im Neuenheimer Feld 220-221, D-6900 Heidelberg
BRANDTZAEG, P., Prof. Dr.	Laboratory for Immunohistochemistry and Immunopathology, Rikshospitalet, N-0027 Oslo 1
CASELITZ, J., Dr.	Institut für Pathologie der Universität Hamburg, Martinistraße 52, UKE, D-2000 Hamburg 20
CHU, T.M., Prof. Dr.	Diagnostic Immunology Research and Biochemistry Department, Roswell Park Memorial Institute, Buffalo, NY 14263, USA
COX, A.D., Dr.	Department of Microbiology and Immunology, Eastern Virginia Medical School, PO Box 1980, Norfolk, VA 23501, USA
DAMJANOV, I., Prof. Dr.	Jefferson Medical College, Thomas Jefferson University, Philadelphia, PA 19107, USA
DENK, H., Univ.-Prof. Dr.	Institut für Pathologische Anatomie der Universität, Auenbruggerplatz 25, A-8036 Graz
HEITZ, Ph.U., Prof. Dr.	Institut für Pathologie der Universität, Schönbeinstraße 40, CH-4003 Basel

IOZZO, R.V., Dr. | Department of Pathology and Laboratory Medicine, University of Pennsylvania School of Medicine, 3400 Spruce Street G/1, Philadelphia, PA 19104 USA

JANZER, R.C., Dr. | Abteilung Neuropathologie, Institut für Pathologie, Universität Zürich, CH-8091 Zürich

KIESSLING, M., Dr. | Abteilung Neuropathologie, Pathologisches Institut, Universität Freiburg, D-7800 Freiburg

KLEIHUES, P., Prof. Dr. | Abteilung Neuropathologie, Institut für Pathologie, Universität Zürich, CH-8091 Zürich

KLÖPPEL, G., Prof. Dr. | Institut für Pathologie der Universität Hamburg, Martinistraße 52, UKE, D-2000 Hamburg 20

KORSRUD, F.R., Dr. | Laboratory for Immunohistochemistry and Immunopathology, Rikshospitalet, N-0027 Oslo 1

LÖNING, TH., Prof. Dr. | Institut für Pathologie der Universität Hamburg, Martinistraße 52, UKE, D-2000 Hamburg 20

MILDE, K., Dr. | Institut für Pathologie der Universität Hamburg, Martinistraße 52, UKE, D-2000 Hamburg 20

MÖLLER, P., Dr. | Institut für Allgemeine Pathologie und pathologische Anatomie der Universität, Im Neuenheimer Feld 220-221, D-6900 Heidelberg

MOLL, R., Dr. | Institut für Pathologie, Klinikum der Johannes Gutenberg-Universität, Langenbeckstraße 1, D-6500 Mainz

OSBORN, MARY, Dr. | Max-Planck-Institut für biophysikalische Chemie, Am Fassberg, D-3400 Göttingen

OTTO, H.F., Prof. Dr.	Institut für Allgemeine Pathologie und pathologische Anatomie der Universität, Im Neuenheimer Feld 220-221, D-6900 Heidelberg
ROGNUM, T.O., Dr.	Institute of Forensic Medicine and Laboratory for Immunohistochemistry and Immunopathology, Institute of Pathology, Rikshospitalet, N-0027 Oslo 1
SCHWECHHEIMER, K., Dr.	Institut für Allgemeine Pathologie und pathologische Anatomie der Universität, Im Neuenheimer Feld 220-221, D-6900 Heidelberg
THRANE, P.S., Dr.	Laboratory for Immunohistochemistry and Immunopathology, Institute of Pathology, Rikshospitalet, N-0027 Oslo 1, and Department of Oral Surgery and Oral Medicine, Dental Faculty, University of Oslo, Geitemyrsveien 71, N-0455 Oslo 1
WRIGHT Jr., G.L., Prof. Dr.	Department of Microbiology and Immunology, Eastern Virginia Medical School, PO Box 1980, Norfolk, VA 23501, USA

Preface

New methods open new insights and provide the basis for new concepts in cancer reseach. The discovery of monoclonal antibodies and the further development of immunocytochemistry and hybridization techniques has led to a "renaissance" of pathology in the field of tumor classification and prognosis as well as in oncotherapy. The application of the new marker methods may result in a more intensive integration of morphological findings into clinical practice.

In recent years an increasing battery of commercially available antibodies has made a progressively more detailed classification of tumors possible using a broader spectrum of immunocytochemical methods. All scientific journals publish papers about the results of these new applications in tumor diagnosis. Principally, we must differentiate between biological and morphological tumor markers.

In this monograph the importance of morphological tumor markers is the focus of the presentation by an international group. As an introduction, two chapters offer accounts of the application of monoclonal antibodies to human tumor antigens and the important role of biochemical markers in human cancer. A special chapter refers to the different immunocytochemical methods for the demonstration of tumor markers and their relevance in diagnostic pathology.

The so-called tumor markers are very often cell or tissue markers which provide data related to the problems of cytogenesis, cell function or cell differentiation. In contrast with the evidence from biological markers, the methods of immunocytochemistry or hybridization techniques permit the demonstration of tumor markers precisely, on the cell membrane, in the organelles of the cytoplasm or in the nucleus. In the very heterogeneous group of morphological tumor markers the intermediate filaments of the cytoskeleton, glycoconjugate antigens of the cell membrane and special proteins such as enzymes play the major role in routine diagnostic work.

From the histogenetic point of view, markers of different tissues can be distinguished. This fact is the organizing principle of the majority of chapters. The epithelial tumor markers include widely varying substances (cytokeratins, tissue polypeptide antigen, epithelial membrane antigen, oncofetal antigens, markers of glandular differentiation). Mesenchymal markers consist of some intermediate filaments (e.g. vimentin, desmin etc.), special proteins and enzymes. Other markers are suitable for demonstration of the intercellular matrix (proteoglycans, basal membrane antigens), blood group substances, lectins or neuroendocrine products. Further chapters demonstrate the importance of tumor markers in neuro-oncology and viral-associated tumors. The conclusion of the monograph is an overview on the theme of "cell, tissue and organ specific tumor markers".

Altogether the monograph is a review of methods and practical application of the different morphological tumor markers in the field of tumor diagnosis and prognosis. Organ specific diagnosis has a very great clinical relevance for treatment and prognosis. Therefore, this volume brings all pathologists and oncologically orientated clinicians up to the state of the art in cancer detection and diagnosis.

I would like to thank very much all contributors for their excellent cooperation in timing and arrangement of the chapters. I am also grateful for the outstanding work on this volume by Springer-Verlag, in particular Mrs. H. Herion and Mrs. U.S. Davis.

Hamburg G. SEIFERT

Contents

Indexed in ISR

Monoclonal Antibodies to Human Tumor Antigens

G.L. WRIGHT JR. and A.D. COX

1 Introduction

The introduction of monoclonal antibody (Mab) technology (KÖHLER and MILSTEIN 1975) revolutionized the serological and biochemical analysis of human cancer. By fusing spleen cells from immunized mice with myeloma cell partners, followed by cloning of the resulting hybrids, it is possible to produce monoclonal cell populations, each of which has acquired both the property of immortality and the ability to produce a specific antibody. The power of this technology results from the ability to generate large quantities of stable, sensitive, specific Mab probes which can detect but a single epitope on a complex molecule. Tumor markers may be qualitatively or quantitatively different from those expressed on normal cells; Mabs can detect such markers even if they are present only in very small quantity or differ only slightly from other molecules. Although human tumor-specific antigens have continued to prove elusive, and indeed may not exist, many tumor-associated antigens (TAA) from a wide variety of human malignancies have been identified and characterized using Mab probes. Tumor markers detected by Mabs to date include oncofetal antigens, differentiation antigens, growth factors and oncogene products, hormones, receptors, enzymes, and many novel antigens for which no function has yet been described.

1.1 Specificity of Monoclonal Antibodies

Although the class/subclass, affinity and biological effector functions of a Mab clearly affects its utility, the careful, thorough analysis of Mab specificity is the most essential aspect of monoclonal antibody characterization. Since heterogeneity within and among tumors is the rule rather than the exception (vide infra), the use of multiple assays (e.g. immunoperoxidase (IP), immunofluorescence (IF) and radioimmunoassay (RIA)) on frozen or fixed tissue sections and on live and fixed cells, as well as the use of multiple tumor and normal tissue sources is crucial to determining the true distribution of antigen specified by the Mab under study.

Although Mabs react with their antigens in a highly specific manner, the recognition of the same epitope borne on an unrelated antigen is a common source of lack of specificity, or cross-reactivity. In a typical example, Mabs directed against a glycolipid antigen found on prostate carcinoma were found to cross-react with red blood cells (LINDGREN et al. 1986). The findings suggested that the carbohydrate moiety was attached to ceramide in the prostate-associated antigen but to a protein core in the red blood cell.

1.2 Murine Versus Human Monoclonal Antibodies

A theoretically major advantage of human Mabs is the increased potential to detect subtle antigenic differences between cells of different human malignancies or between malignant and normal human cells.

Unfortunately, despite considerable effort, only limited progress has been made in generating useful human anti-tumor Mabs. Human monoclonal antibodies to tumors of such diverse origin as breast, lung, gastric, colorectal and urogenital carcinomas, as well as melanomas, gliomas and leukemias, have been produced by a variety of interspecies fusion, human/human fusion and Epstein-Barr virus transformation techniques (reviewed in STRELKAUSKAS 1985). Most of the Mabs are not yet well-characterized, but their quality has been largely disappointing. In general, most have been characterized by low affinity (particularly since many are IgM) and by low levels and unstable antibody production. Interestingly, and perhaps most disappointingly, none has detected a human tumor-specific antigen: most react with normal membrane or cytoplasmic components of the tumor cell. It is possible that many interesting TAA may not be immunogenic in humans.

A critical appraisal of current progress thus suggests that many technical problems need to be resolved before clinical application of human Mab becomes a reality. Some of these formidable problems may be more immediately amenable to solution than others. For example, in vitro immunization, perhaps using splenic lymphocytes, is unlikely to be more successful without improved understanding of the growth and differentiation factor requirements of antigen-responsive cells. However, recent advances such as electrofusion methodology (Lo et al. 1984) may improve fusion frequencies, while DNA transfer techniques (Takeda et al. 1985) may make fusion unnecessary altogether.

As an alternative to wholly human Mabs, several investigators have generated human/mouse chimeric antibodies, in which the antigen-binding region is from a murine Mab and the constant region is of human origin (Morrison et al. 1985; Takeda et al. 1985; Sun et al. 1986). In a marriage of hybridoma, recombinant DNA and gene transfer technologies, the genes of interest are cloned and transfected into a myeloma cell line for expression. Such a "transfectoma" approach permits the construction of antibodies with desired specificity, as can be produced in the mouse, but which are less immunogenic than murine Mabs and which possess human effector functions for use in human hosts. These attributes are desirable for the *in vivo* diagnostic and therapeutic use of anti-tumor Mabs, as discussed below.

1.3 Tumor Antigens, Growth Factors and Oncogenes

A wide range of antigens expressed in tumor cells have been detected and characterized by monoclonal antibodies (for reviews see Wright 1984; Sell and Reisfeld 1985; Schlom and Weeks 1985). Most Mab-defined TAA are generally not confined to tumors of one type, many are expressed on normal cells, and "preferential tumor expression" is often identified in quantitative terms. Properties of potential tumor markers which must be examined before strategies for their use can be developed are: 1) their biochemical nature, 2) number of epitopes, 3) cellular location, 4) stability, 5) distribution on normal/tumor tissue types (public vs. private specificity), 6) distribution within and among tumors (heterogeneity), and 7) biological function.

One class of TAA deserves special mention as a field of active inquiry to which Mabs are being newly applied. Aberrant expression of cellular oncogenes has been found in carcinomas, melanomas, sarcomas, neuroblastomas and hematopoietic malignancies. They appear to be associated with abnormalities of growth and differentiation, and some have been shown to encode cellular growth factors or their receptors. There is evidence for greater amplification and enhanced expression of oncogenes in the advanced stages of lung carcinomas, neuroblastomas, and leukemias/lymophomas (*myc*) (reviewed in Colb and Krontiris 1986), and (controversially, see below) in more aggressive forms of breast, colon and prostatic carcinoma (*ras*) (reviewed in Thor et al. 1986). If this is so, then simple methods to detect specific oncogene products would be of enormous clinical utility.

Mabs have now been produced to members of several families of oncogenes: *ras, myc, erbB* and *neu*, enabling the detection and characterization of oncogene products by antibody binding in immunoblotting or immunohistochemical techniques. For example, Mabs directed at the *erbB* product, i.e. the EGF receptor, have shown that overexpression of EGFR is a property of squamous, but not other, tumor types (OZANNE et al. 1986), and these Mabs have been shown to have anti-tumor effects (MASUI et al. 1986).

The apparent presence of activated or otherwise aberrant *ras* genes in approximately 20% of solid tumors is enticing, although the role of *ras* in malignancies is unclear at present. The p21 protein product of the *ras* family, which may be an obligate intermediate in the transduction of growth factor signals (GOUSTIN et al. 1986), has been extensively studied using anti-p21 Mabs. The Rap 1–5 series of Mabs (HORAN HAND et al. 1984) and the Y13-259 Mab (FURTH et al. 1982) detect both normal and activated forms of the *ras* proteins; at this writing, controversy exists over the significance of tissue binding by Rap-5, since IP studies (THOR et al. 1986) have shown a much higher presence of *ras* than can be accounted for by oncogene transcripts. It has been suggested that the additional reactivity to tissue sections seen in IP studies may be due to cross-reactivity of the anti-p21 Mabs with both normal and activated *ras*; however, the Mabs identified by CARNEY et al. (1986) detect only mutated p21 yet also show more cross-reactivity in IP studies than in vitro specificity would predict. Work by ROBINSON et al. (1986) indicates that Y13-259, but not Rap-5, can discriminate between cell lines with and without transforming *ras* genes. It is therefore possible that Mabs such as Rap-5 bind to cancer tissues at high levels at least in part through cross-reaction with an epitope also found on an unrelated antigen, as described above in Section 1.1. Further studies will have to be performed to settle the controversy. Such experiences are common when Mab technology is applied to any new area of tumor immunology, and illustrate the necessity to apply rigorous control systems to the specificity analysis of anti-TAA Mabs. Whatever the outcome of this particular case, Mabs directed against oncogenes and their products appear to show great promise in tracking the etiology, classification, and prognosis of human tumors.

1.4 Antigenic Heterogeneity

Heterogeneity of antigenic expression on tumor cells is a major potential problem for the development of strategies for the immunological diagnosis and therapy of cancer. The increased use of highly specific, non-crossreacting monoclonal antibodies has both highlighted the presence of antigenic heterogeneity and accentuated the problems due to it.

Immunohistochemical studies using Mabs have shown both qualitative and quantitative heterogeneity, such that (1) multiple antigenic phenotypes can exist in tumors of a given type, ranging from cells expressing all the antigens under study to cells expressing none; (2) variable phenotypic expression can occur within a given tumor, showing a patchwork distribution of antibody-binding and non-antibody-binding cells or areas of cells; (3) a particular antigen may be expressed in the cytoplasm of cells in one part of the tumor and in the membrane of cells in a different part; (4) different metastases from the same patient may show different antigenic phenotypes and (5) individual cells of the same phenotype may differ significantly in antigen density (reviewed in EDWARDS 1985).

The multiple genotypic and phenotypic causes of the heterogeneity described above have been reviewed in SCHLOM (1986). Most intriguing is the possibility that different antigenic expression may sometimes reflect biological behavior: certain Mab-defined TAAs have been shown to correlate with more (SUTER et al. 1985) or less (XIANG and KIMURA 1986, personal communication) aggressive behavior im melanomas. It might be possible to exploit such variation in antigen expression for diagnostic and prognostic purposes.

The problem of antigenic heterogeneity becomes clear and vexing when one considers that a primary advantage of using Mabs lies in their exquisite sensitivity and specificity and their ability to discriminate between morphologically similar cells. Unfortunately, cells which do not, for whatever reason, express the appropriate antigen are rendered "invisible" to the Mab. Since TAA-directed Mabs tend to react not with 100% of tumors or tumor cells, but rather with a subpopulation of those tumors or cells, the use of a single Mab thus inevitably will mean missed diagnoses, unrecognized recurrences and metastases, and tumor cells resistant to Mab-directed therapy.

One answer to the problem of variable tumor marker expression is therefore the use of panels of Mabs or defined mixtures (cocktails) of Mabs that react with distinct epitopes on the same or different antigens. Such panels have improved the detection of TAA in serum by RIA and ELISA, in urine by flow cytometry, in tissue sections by IP, and *in vivo* by radioscintigraphy, and are discussed in the appropriate sections below.

Another interesting approach is the enhancement of marker expression in a tumor population by the use of biologic response modifiers. For example, interferon has been shown to enhance the expression of TAA and HLA-DR antigens on the surface of human colon or breast carcinoma, but not normal, cells (GREINER et al. 1984; ROWLINSON et al. 1986). In the latter study, HLA-DR antigens were induced on the surface of 80% of formerly HLA-DR-negative breast carcinoma cells, where they then served as targets for radioimmunoimaging in a nude mouse xenograft model.

2 Diagnostic Applications of Monoclonal Antibodies to Human Tumor Antigens

2.1 Serum Assays

Tumor markers may be released into the serum by antigen secretion, antigen shedding or tumor cell death, where they may serve as markers of residual or recurrent disease, or aid in the assessment of response to therapy.

Although CEA and AFP will be covered in detail in other chapters, they are the prototypical seroassays for tumor-associated antigens; therefore a few salient points relating to monoclonal antibody technology will be discussed below.

2.1.1 *Alpha-Fetoprotein (AFP)*

AFP is an important marker for the detection and monitoring of primary hepatocellular carcinoma (HCC) and germ cell tumors. In the past its use, particularly in populations at high risk for HCC, has been plagued by cross-reactivity of conventional polyclonal antisera with degradation products of albumin. BELLET et al. (1984) produced 2 high affinity Mabs to different epitopes on the AFP molecule that are not shared with albumin, and used these Mabs in a double determinant RIA on a very large series of sera. This impressive study showed that 80% of patients with hepatitis B antigen-positive (HBsAg) HCC had strikingly high levels of AFP, >200 ng/ml, while all normal donors, and 99% of disease controls had AFP levels of less than 20 ng/ml. These encouraging results are substantially different from those seen with conventional polyclonal antisera, in which 31–40% of patients with hepatitis and 8–33% of patients with cirrhosis show elevated levels of AFP. The ability of the Mabs to specifically detect AFP-specific, and possibly tumor-associated, epitopes therefore appears to have greatly enhanced the potential of this assay for screening for AFP-producing tumors. The Mabs themselves may be potential candidates for immunotherapy regimens.

2.1.2 *Carcinoembryonic Antigen (CEA)*

Several studies have shown the advantage of panels of anti-CEA Mabs directed against CEA-specific epitopes. One of the major drawbacks of the use of anti-CEA Mabs in immunodiagnosis, monitoring and therapy has been their cross-reactivity with several normal adult tissues, particularly polymorphonuclear leukocytes (PMN). Since "CEA" appears to be a family of isoantigens with differing properties, investigators have attempted to improve the discrimination of their Mabs by developing sets or series of Mabs detecting different epitopes of CEA (PRIMUS et al. 1983; HERLYN et al. 1983). One series of Mabs, COL 1-15 (MURARO et al. 1985), selected early in the screening process for lack of reactivity to PMN, react with at least five epitopes on the CEA molecule and are highly selective for reactivity to colon carcinomas compared to normal tissues, other carcinomas and benign gastrointestinal disease. The use of such improved Mabs may thus enhance both the sensitivity and the specificity of seroassays for CEA and other TAAs.

2.1.3 *Mab 19-9*

Mab 19-9 is directed against the sialylated Lewis^a antigen, also called GICA, a large mucin-containing antigen found in many gastrointestinal cancers. In a longitudinal study of 100 colorectal cancer patients by serum RIA, persistent or rising antigen levels preceded recurrence of their disease by 3 to 18 months

(SEARS et al. 1982). Elevated levels of Mab 19-9 were also found to be associated with pancreatic, gastric and hepatic carcinomas (79%, 57% and 49% positive, respectively), but not with benign lesions (3%) or healthy individuals (0.5%) (MAGNANI et al. 1983). The correlation between elevated levels and disease states indicates that Mab 19-9 will be useful for the serum monitoring of tumor burden in gastrointestinal cancer patients. The inability of this Mab (like others) to detect all cases of known disease indicates that it may benefit from being included in a panel of Mabs to TAA known to be expressed on such tumors. A new Mab, Co29.11, isolated by the same workers (HERLYN et al. 1985), detects a different epitope on the same antigen as 19-9. A large prospective study is now underway using a panel of Mabs to detect the 19-9, Co29.11, LNFIII and CEA antigens in serum.

2.1.4 Mab OC125

A highly useful new Mab is OC125, which reacts with a glycoprotein antigen (CA125) found in most nonmucinous epithelial ovarian tumors. Several large studies have shown elevated levels of this antigen by RIA and ELISA in the serum of 85–96% of ovarian cancer patients, but only 1–7% of healthy controls (BAST et al. 1983). In two separate studies, CA125 levels were found to correlate well with progression or regression of ovarian cancer (42/45 patients (BAST et al. 1983) and 18/20 patients (RICOLLEAU et al. 1984)). Unfortunately, many patients with benign diseases also may show elevated levels of CA125; in one study, 24% of patients with benign pancreatic, biliary or hepatocellular disease had such levels (HAGLUND 1986). Finally, although pancreatic, colon and breast carcinomas have also been shown to shed CA125 antigen into serum (BAST et al. 1983), the CA125 assay appears to be of limited diagnostic or prognostic importance in such cases.

2.1.5 Other Anti-TAA Monoclonal Antibodies

Several other antitumor Mabs, although less extensively tested, also appear to have potential clinical application. The DU-PAN-2 Mab reacts to a large mucin-like molecule distinct from the 19-9 antigen. Elevated levels of the DU-PAN-2 antigen have been demonstrated in 89% of patients with pancreatic cancer (METZGAR et al. 1984). The DF3 Mab has detected elevated antigen levels in 76% of patients with metastatic breast cancer compared to 8% of normal controls (HAYES et al. 1985).

The B72.3 Mab detects yet another large, mucin-like molecule, designated TAG-72; TAG-72 antigen expression appears to be widespread in carcinomas. In a pilot study, elevated levels of TAG-72 were found in sera from some ovarian, advanced colorectal, pancreatic and lung carcinoma patients (PATERSON et al. 1986). TAG-72 was elevated in only 35% of these patients, perhaps indicating the existence of TAG-72-positive and -negative subpopulations of these tumors. Interestingly, although B72.3 was raised against a breast carcinoma and reacts with 50% of breast cancer tissues (NUTI et al. 1982), no shed TAG-

72 antigen could be detected in sera of breast cancer patients in the study cited above. The results of this study therefore emphasize once again the need to anticipate antigenic heterogeneity within and among tumor types.

2.2 Cytology of Body Fluids

2.2.1 Flow Cytometry

Patients with residual or recurrent bladder carcinoma shed malignant transitional epithelial cells (TCC) into their urine. DNA flow cytometry (FCM) can detect aneuploid (tumor) populations, but will give a falsely low value in low grade malignancies, which have a near diploid DNA content, or if a large number of inflammatory cells is present. In patients with cystitis, or after BCG treatment for TCC, the presence of inflammatory cells in the urine poses a serious obstacle to the accurate differential diagnosis of TCC. ALEXANDER et al. (1986) have used TCC-directed Mabs in indirect IF procedures and FCM to selectively stain TCC cells, but not inflammatory cells, collected from voided urine or bladder washings. The TAA-positive TCC cells could then be sorted for further purification and study. These investigators found that the immunoselection process improved the sensitivity of TCC detection, and also that the use of TCC-directed Mabs represents an improvement over tissue-specific markers such as cytokeratin and HMFG/EMA (ALEXANDER, personal communication).

Flow cytometric techniques are also highly useful for the subclassification of leukemic phenotypes according to the expression of CALLA and other tumor markers, and have been of major importance in determining prognosis and appropriate treatment strategies for leukemic patients (reviewed in BERNARD et al. 1984).

2.2.2 Immunocytochemistry

Malignant cells in serous effusions of pleura, pericardium and peritoneum are often exceptionally difficult to differentiate from reactive mesothelial cells; once identified, the source of the malignancy may remain obscure. The ability of Mabs to detect even a single cell expressing Mab-defined markers offers the possibility of improving the detection and classification of these cells. For example, the F36/22 Mab has been used to detect malignant adenocarcinoma, but not reactive mesothelial cells, in ascites from ovarian adenocarcinoma patients (CROGHAN et al. 1984). Additionally, a pool of breast cancer Mabs could distinguish breast, ovarian and lung carcinoma cells from reactive mesothelial cells in an IF assay (TAGLIABUE et al. 1986).

Despite the growing number of cases of malignant mesothelioma, there is currently no positive Mab marker available for this disease, and the best method for its diagnosis is controversial (reviewed in SHEIBANI et al. 1986). The immunological diagnosis of such tumors now largely rests on their failure to react with Mabs directed against lung, breast, ovarian or other carcinomas. The B72.3 (SZPAK et al. 1986) and HMFG Mabs (WRIGHT, unpublished observations) seem especially suited for this purpose. Again, because none of the Mabs react with 100% of carcinomas, the use of a panel of Mabs has the potential to improve the sensitivity of discrimination. Such Mabs might also be used for the immunocytochemical differential diagnosis of cells in sputum.

2.2.3 Immunohistopathology

Severe morphological distortions can make interpretation of fine needle aspiration biopsy material difficult. The use of Mabs such as B72.3 have been shown to be of value in distinguishing between adenocarcinomas and benign lesions of breast, lung and pancreas (Johnston et al. 1986). It is probable that the application of other TAA-directed Mabs will have similar potential to improve the accuracy of diagnoses based on fine needle aspiration biopsy material.

The most prominent use of Mabs to tumor markers today lies in the immunohistopathological detection of malignant cells in tissue sections. Recent reviews have described the use of Mabs to tumor markers for this purpose (Borowitz and Stein 1984; Wright 1984), and several of the chapters in this volume describe current work with both monoclonal and polyclonal antibodies in detail. The brief discussion below describes the use of Mabs in prostate cancer to exemplify 1) turning cross-reacting epitopes to advantage and 2) the use of a panel of Mabs to improve the detection of malignant cells in tissue sections.

Wahab and Wright (1985) have shown that the Leu-7 marker, although directed against natural killer cells, also reacts with an antigen present on cells of normal, benign and malignant prostate origin. Unlike other organ-specific markers such as prostate-specific antigen, the Leu-7 marker was not lost on any of the metastatic tumors examined. Therefore, anti-Leu 7 may be useful in the diagnosis and prognosis of prostate cancer.

Since no tumor-selective anti-prostate Mabs are available, the routine immunohistochemical diagnosis of prostate cancer currently relies on reactivity with the organ-specific prostate markers prostatic acid phosphatase (PAP) and prostate-specific antigen (PA). However, not all tumors express these antigens, particularly poorly differentiated tumors, and heterogeneity of staining is observed. Wahab (1985) examined a series of normal prostate, benign prostatic hyperplasia and primary and metastatic prostatic carcinomas, using a panel of 7 anti-prostate Mabs in IP studies on formalin-fixed, paraffin embedded tissues. The panel included PAP, PA, CEA, NCA, Leu-7, and two new anti-prostate Mabs designated TURP-27 and TURP-73 (Starling et al. 1986). No single Mab in the panel could detect all cases of prostate carcinoma; however, the collective use of TURP-27, Leu-7, anti-PAP and anti-PA permitted the correct identification of all specimens.

2.3 Radioimmunodetection (RID) of Cancer

The creation of site-specific radiopharmaceutical probes by coupling antibodies to radionuclides has the potential to improve the diagnostic capabilities of conventional imaging techniques in the detection of very small tumors, in determining the tissue origin of a mass, or in distinguishing tumor recurrence from post-operative changes. Further, there is the very real possibility of individually tailoring immunotherapy regimens, by using the same Mab to track a tumor and to treat it (see Sect. 3.2.3).

It has been clearly established that Mabs detecting human tumor markers can specifically localize in human tumor xenografts in model systems. Mabs to tumor markers found in carcinomas of the breast, colon and rectum, lung and ovary, as well as melanoma and osteogenic sarcoma have been intensively investigated and some have now gone on to clinical trials. Considerable clinical experience has been gained with Mabs HMFG-2, 791T/36, B72.3, 19-9, Col17-1A (specificities given elsewhere in text), T-101 (anti-T cell), and several anti-CEA and anti-melanoma Mabs (for review, see PRIMUS et al. 1984; LARSON 1985; BALDWIN and BYERS 1985).

In the few short years since MACH et al. (1981) successfully detected colorectal cancer sites in patients scanned with ^{131}I-anti-CEA Mabs, understanding of the complex interactions between Mab, antigen, tumor, host and imaging technology necessary to produce useful tumor imaging *in vivo* has advanced tremendously. As an example of the complexity involved, a recent multicenter study used $F(ab')_2$ fragments of the anti-melanoma Mab 225.28S in RID of 254 patients. These investigators confirmed the influence of tumor size, site, blood flow and antigen density, and route of Mab injection, and also demonstrated effects of the isotope ($99m^{Tc}$ vs. ^{111}In) and the clinical stage of the patients on the results of the scans (SICCARDI et al. 1986). This study also provides a typical illustration of both the power and the problems associated with the use of RID: occult lesions were visualized in many patients, and most of these were confirmed to be melanoma by other means; however, nonspecific radioactivity accumulated in bone marrow, spleen, liver and kidneys, in some cases limiting the detection of tumor sites.

Some problems which have yet to be satisfactorily resolved are 1) high blood background, 2) nonspecific uptake in normal organs, 3) low absolute levels of labeled Mab in the tumor, and 4) the fact that not all antigen-positive tumors can be imaged by a given Mab directed to that antigen. Solving the first two problems is probably a matter of technical improvements, for example the use of second antibody clearance (SHARKEY et al. 1984) to reduce blood background. There is not enough data at present to say whether the third problem reflects a true physiological limit. Unfortunately, the last problem has serious clinical consequences, implying that each patient must be tested individually, and should temper enthusiasm with the realization that not every patient will be able to be imaged and/or treated with any Mab directed against antigen(s) found on his or her tumor.

In a similar vein, various investigators using different systems regularly disagree about whether the optimal isotope to use is ^{111}In, ^{131}I, ^{125}I, ^{123}I or $99m^{Tc}$. The failure to agree on a single isotope is a reflection of the cardinal rule in RID: each Mab and tumor system configuration must be individually optimized.

In an exciting recent advance in instrumentation, the advent of a hand-held gamma probe has now made it possible to detect tumor cells intraoperatively, for example in the selective detection of ^{125}I-Co17-1A in subclinical tumor masses compared to surrounding normal tissues (MARTIN et al. 1986).

In an exciting advance in Mab and labeling technology, bifunctional Mabs have been generated which have one combining site specific for antigen and the other site specific for hapten. GOODWIN et al. (1986) have used such a bifunctional antibody to label Mab in vivo. By separately administering Mab and isotope, this new technique has the potential to decrease blood background and to allow the use of short-lived radionuclides such as $99m^{Tc}$ to be used with Mabs which take 24–96 hours or more to localize at the tumor site.

3 Immunotherapeutic Applications

One of the most difficult problems in conventional cancer therapy is how to eradicate malignant cells while sparing normal ones. The advent of highly specific monoclonal antibodies has brought with it the possibility of greatly improving the selective targeting of cytotoxic agents to cancer cells, and thus decreasing their systemic toxicity. Some of the issues in Mab-based immunotherapy which must be addressed are: Mab specificity, cross-reactivity, and immunogenicity; antigenic heterogeneity; target cell access; conjugate stability; delivery of the toxic effector(s) within appropriate range of the target, and the susceptibility of the target to cytolysis by the effector.

3.1 Cytotoxic Mabs

Many Mabs have the ability to kill targets expressing their antigen in vivo, either directly or by the activation of host effector mechanisms. In clinical trials, regression of metastatic malignant melanoma lesions were observed in 4 of 21 patients following therapy with a complement-fixing IgG3 Mab, R_{24}, which recognizes the GD3 ganglioside antigen (HOUGHTON 1986). Inflammation and complement deposition at tumor sites were seen (HOUGHTON et al. 1985). The R_{24} Mab also activates human effector cells, i.e. is active in antibody-dependent cell-mediated cytotoxicity (ADCC) and binds to a subpopulation of $T8^+$ lymphocytes (HERSEY et al. 1986). The cytodestruction of tumors in vivo by this Mab appears to result from all these activities, and might be enhanced by the addition of a second antibody directed at a distinct epitope on the same (GD3) or a closely related (GD2) antigen (HELLSTRÖM et al. 1983); such studies are underway.

Murine IgG3 and IgG2a Mabs are particularly efficient activators of ADCC (HERLYN et al. 1985). An IgG2a Mab, Co17-1A, directed against gastrointestinal adenocarcinomas, has been shown to mediate ADCC in vitro and in vivo by the activation of mouse/human monocytes and killer cells (HERLYN and KOPROWSKI 1981). Mab 17-1A has now been used in several early clinical trials (for review, see STEPLEWSKI 1986), and some patients have shown objective responses to treatment. WEINER et al. (1986) are currently pretreating colorectal and pancreatic carcinoma patients with gamma interferon, in an attempt to increase Fc receptor expression on peripheral blood monocytes and thereby potentiate the anti-tumor ADCC activity of Mab 17-1A.

These early clinical trials have clearly shown that, 1) while the administration of a single dose of murine Mab is safe, a significant fraction of patients develop human anti-murine antibodies (HAMA); 2) HAMA may be directed against the murine Fc portion, in which case chimeric or human Mabs would decrease the incidence of HAMA, or against the antigen-binding site (idiotype); 3) HAMA have the potential to cause severe allergic reactions, but some, especially anti-idiotypic antibodies, may bode well for the patient by mimicking the tumor antigen and provoking a beneficial immune response to the tumor (Herlyn et al. 1986); 4) only a very small percentage of administered Mab actually localizes to tumor deposits; 5) the number of binding sites on the tumor occupied by Mab correlates with the effectiveness of Mab-mediated tumor destruction; 6) at least partly due to poor Mab binding and to antigenic heterogeneity, the degree of tumor burden in these patients precludes the possibility of eradication of the tumor by Mab-based immunotherapy, and 7) it therefore seems unlikely that unmodified Mab alone will be successful in immunotherapy. When preliminary trials to establish the performance parameters of Mab alone are farther along, further trials incorporating Mab-based immunotherapy as part of a multimodality treatment regimen would seem to be in order, especially in patients with minimal residual disease.

3.2 Mab Conjugates

Most current immunotherapeutic strategies use Mabs as tumor targeters (warhead) only, letting the Mabs deliver other effectors (missiles) to the site of action. Most effectors now in use are drugs, toxins or radionuclides.

3.2.1 Mab-Toxin Conjugates

The use of toxins has proven to be very difficult to translate from theory into practice. Mabs conjugated with intact ricin and the subunit ricin A-chain are being intensively investigated, and have been shown to be active both in vitro and in animal models. However, at present their in vivo clinical use does not yet appear promising. Intact ricin-Mab conjugates are highly toxic to normal cells and there are problems with the in vivo stability and delivery into target cells of ricin-A chain-Mab conjugates. In contrast, the ex vivo use of immunotoxins for cleansing the bone marrow of tumor cells prior to autologous grafting or of mature T cells prior to allogeneic grafting seems very promising indeed (Saärinen et al. 1985).

3.2.2 Mab-Drug Conjugates

Many biologically active Mab-drug conjugates have been prepared, most notably with adriamycin, daunomycin, vindesine (VDS) and methotrexate (MTX), and these have been analyzed both in vitro and in vivo in nude mouse xenograft model systems. The anti-osteogenic sarcoma Mab 791T/36 has been extensively tested with different agents and shows great promise in model systems (Embleton et al. 1984). VDS conjugates with 791T/36, antimelanoma Mab 96.5 and anti-CEA 11.285.14 and 14.95.55 Mabs (Rowland et al. 1985), as well as MTX conjugated to an anti-prostatic acid phosphatase Mab (Deguchi et al. 1986), have been able to retard or suppress growth of the appropriate tumor xenografts. In the VDS conjugate study above, effective doses of free drug were highly toxic, whereas conjugation with Mab decreased drug toxicity but not effectiveness. The MTX conjugate study was hampered by conjugate uptake in normal liver and spleen. The results of these studies and others emphasize that: 1) delivery of drug into the target cell is crucial for most current anti-cancer drugs, but most Mabs bind to stable (non-internalized) surface antigens; 2) drug conjugates require very high levels of antigen binding to produce cytotoxic effects; 3) such high levels may often not be achievable due to differences in antigen expression; 4) preferential targeting of drug to tumor by specific Mab does not prevent conjugate uptake in normal organs and tissues; 5) antigenic heterogeneity will continue to pose a problem for tumor eradication, since only cells binding drug conjugates will be killed; and 6) drug resistant populations introduce another form of "antigenic" heterogeneity which must be dealt with. Nevertheless, many of these problems appear amenable to solution, perhaps with novel techniques such as hybrid Mab-toxin conjugates and drug potentiation (Webb et al. 1986).

3.2.3 Mab-Radionuclide Conjugates

Mab-radionuclide conjugates currently appear to have the most immediate immunotherapeutic application. The ability to deliver cytotoxic doses of radiation to cells without entering (internalizing) or even binding to them, i.e. the ability to produce radiocytotoxicity over several cell diameters, suggests that Mab-radionuclide conjugates may prove to be capable of dealing with antigenic heterogeneity. (Of course, it also suggests that non-specific uptake will be a critical limiting factor for this modality, as it is for RID, see Sect. 2.3.) Further, the prospects for using RID with the same Mab to determine 1) prior to therapy, the biodistribution of labeled Mab in that particular patient, and 2) post-therapy, the tumor response, suggest that radioimmunotherapy may be more amenable to individualization than other forms of treatment.

While tumor xenografts have been completely eradicated with radioimmunotherapy (Cheung et al. 1986; Lange et al. 1985), it is too early to properly evaluate clinical trials of Mab-radionuclide conjugates, as these have been conducted on only a small number of patients to date. Nevertheless, the results are encouraging, as objective responses have been seen in melanoma patients treated with ^{131}I-anti-p97 Mabs (Carrasquillo et al. 1984) and in lung, ovarian and breast cancer patients with effusions treated with ^{131}I-HMFG2 Mab (Pectasides et al. 1986), for example. Intracavitary Mab administration in the latter group of patients was not associated with toxicity, in contrast to bone marrow toxicity reported by the former group following intravenous administration of labeled Mab.

Table 1. Possible solutions to current problems with monoclonal antibody therapy

Problem	Possible solution
Tumor heterogeneity	1. Use a mixture of monoclonal antibodies reacting to different antigens or epitopes. 2. Use monoclonal antibodies conjugated to radionuclides that emit radiation beyond a single cell.
Antigenic modulation	1. Use monoclonal antibodies reacting to stable membrane antigens. 2. Use a mixture of monoclonal antibodies reacting to different antigens or epitopes.
Human anti-mouse antibodies (HAMA)	1. Induce tolerance with large doses of monoclonal antibody. 2. Prevent development of HAMA by use of human or chimeric human/mouse antibodies. 3. Remove HAMA with plasmapheresis and immunoabsorption.
Blocking antigen	1. Remove TAA and TAA-antibody complexes by plasmapheresis and immunoabsorption.
Unmodified monoclonal antibody lacks cytotoxicity	1. Use of monoclonal antibodies conjugated to radionuclides, drugs, enzymes, or toxins. 2. Class switch the monoclonal antibody to IgG_{2a} or IgG_3. 3. Make a chimera with an appropriate Fc region. 4. Make heterobifunctional monoclonal antibodies with cytotoxic properties and original antigen binding site.
Nonspecific monoclonal antibody uptake	1. Block with cold monoclonal antibody.

Extensive experience with ^{131}I-anti-ferritin polyclonal antibodies has shown dramatic clinical responses in 40–50% of patients with hepatoma and ferritin-positive Hodgkin's disease (ORDER 1985). Interestingly, these workers found monoclonal anti-ferritin to lack efficacy in their hands, due to dehalogenation of antibody. In this regard, beta-emitters such as ^{90}Y are being actively investigated, as they bind stably, can be administered on an outpatient basis, and are expected to reduce systemic toxicity.

Problem areas to be addressed and prospects for radioimmunotherapy have recently been discussed (COBB and HUMM 1986; EPENETOS et al. 1986). Possible approaches to improving current techniques in Mab-based immunotherapy in general, directed at specific problems encountered today are listed in Table 1. It will be noted that many of these approaches are currently under investigation, while others are impractical at the moment, given the current state-of-the-art.

4 Conclusions and Outlook

The increased sensitivity, specificity, and reliable, homogeneous binding properties of monoclonal antibodies have contributed to great improvements in many conventional, polyclonal antibody-based diagnostic techniques. In addition, the ability of Mabs to detect a single epitope on complex molecules has allowed the detection and characterization of many new and useful tumor markers which have served to define, track and manipulate tumor cells. There is every reason to expect that, with a better understanding of the mechanisms of growth, differentiation and carcinogenesis, a new generation of Mabs defining more accurate tumor markers will become available. Meanwhile, heterogeneity in the expression of many of the presently defined tumor markers remains a difficult problem to overcome, but the use of panels of Mabs of differing specificity is a promising approach.

In the near future, the ability to manipulate Mab structure in the production of bifunctional Mabs may allow the generation of families of "all-purpose reagent" Mabs in which the antigenic specificity of the Mab remains the same, but different reagents are linked to it for use in different techniques or to create an additive effect. Examples might include a radionuclide for in vivo diagnosis and an enzyme with a colored substrate for in vitro (IP) diagnosis; or, in a form of "multimodality" immunotherapy, the same Mab could carry both drug and toxin to the cell.

Advances in gene cloning and transfer techniques have already allowed the production of chimeric antibodies, and promise to lead to increasingly engineered forms of Mab reagents. Within the constraints of gene expression, protein folding and antigenic specificity, it should be possible to create sets of Mab probes with differing specificity and function. These statements are made with the caveat that it is important to temper one's expectations, in order to avoid the roller coaster of too-high expectations followed by deep disappointments that has long plagued tumor immunology. The roller coaster effect results from the overly optimistic belief that all things theoretically possible are in fact practically possible, and has proved difficult to resist in the past.

Nevertheless, despite these caveats and the known limitations of today's Mabs, monoclonal antibodies to tumor antigens have opened up new vistas; in all probability new uses for them have yet to be discovered.

References

Alexander JP, Schellhammer PF, Konchuba AM et al. (1986) DNA analysis of bladder tumor cells following immunoselection with monoclonal antibodies against tumor-associated antigens. Submitted for publication

Baldwin RW, Byers VS (eds) (1985) Monoclonal antibodies for cancer detection and therapy. Academic Press, London, pp 1–395

Bast RC, Klug TL, St. John E et al. (1983) A radioimmunoassay using a monoclonal antibody to monitor the course of epithelial ovarian cancer. N Engl J Med 309:883–887

Bellet DH, Wands JR, Isselbacher KJ et al. (1984) Serum alpha-fetoprotein levels in human disease: Perspective from a highly specific monoclonal radioimmunoassay. Proc Natl Acad Sci USA 81:3869–3873

Bernard A, Boumsell L, Daussett J et al. (eds) (1984) Leucocyte typing. Springer, Berlin Heidelberg New York, pp 1–314

Borowitz MJ, Stein RB (1984) Diagnostic applications of monoclonal antibodies to human cancer. Arch Pathol Lab Med 108:101–105

Carney WP, Hamer P, Petit D et al. (1986) A monoclonal antibody reactive with an activated ras protein expressing valine at position 12. J Cell Biochem 32:207–214

Carrasquillo JA, Krohn KA, Beaumier P et al. (1984) Diagnosis of and therapy of solid tumors with radiolabeled antibodies and immune fragments. Cancer Treat Rep 68:317–328
Cheung N-KV, Landmeier B, Neely J et al. (1986) Complete tumor ablation with iodine 131-radiolabeled disialoganglioside GD2-specific monoclonal antibody against human neuroblastoma xenografted in nude mice. JNCI 77:739–745
Cobb LM, Humm JL (1986) Radioimmunotherapy of malignancy using antibody targeted radionuclides. Br J Cancer 54:863–870
Colb M, Krontiris TG (1986) Oncogenes. Adv Int Med 31:47–70
Croghan GA, Wingate MB, Gammarra M et al. (1984) Reactivity of monoclonal antibody F36/22 with human ovarian adenocarcinomas. Cancer Res 44:1954–1962
Deguchi T, Chu TM, Leong SS et al. (1986) Effect of methotrexate-monoclonal antiprostate acid phosphatase antibody conjugate on human prostate tumor. Cancer Res 46:3751–3755
Edwards PAW (1985) Heterogeneous expression of cell-surface antigens in normal epithelia and their tumours, revealed by monoclonal antibodies. Br J Cancer 51:149–160
Embleton MJ, Garnett, Jacobs E et al. (1984) Antigenicity and drug susceptibility of human osteogenic sarcoma cells "escaping" a cytotoxic methotrexate-albumin-monoclonal antibody conjugate. Br J Cancer 49:559–565
Epenetos AA, Britton KE, Mather J et al. (1982) Targeting of iodine 123-labeled tumor associated monoclonal antibodies to ovarian, breast and gastrointestinal tumors. Lancet II:1000–1004
Epenetos AA, Snook D, Durbin H et al. (1986) Limitations of radiolabeled monoclonal antibodies for localization of human neoplasms. Cancer Res 46:3183–3191
Furth ME, Davis LJ, Fleurdelys B et al. (1982) Monoclonal antibodies to the p21 products of the transforming gene of Harvey murine sarcoma virus and of the cellular ras gene family. J Virol 43:294–304
Goodwin DA, Meares CF, McTigue M et al. (1986) Rapid localization of hapten in sites containing previously administered antibody for immunoscintigraphy with short half-life tracers. J Nucl Med 27:9599
Goustin AS, Leof EB, Shipley GD et al. (1986) Growth factors and cancer. Cancer Res 46:1015–1029
Greiner JW, Horan Hand P, Noguchi P et al. (1984) Enhanced expression of surface tumor associated antigens on human breast and colon tumor cells after recombinant leukocyte alpha-interferon treatment. Cancer Res 44:3208–3214
Haglund C (1986) Tumour marker antigen CA125 in pancreatic cancer: A comparison with CA19-9 and CEA. Br J Cancer 54:897–901
Hayes DF, Sekine H, Ohno T et al. (1985) Use of a murine monoclonal antibody for detection of circulating plasma DF3 antigen levels in breast cancer patients. J Clin Invest 75:1671–1678
Hellström I, Brown J, Hellström KE (1983) Monoclonal antibodies to two determinants of melanoma-antigen p97 act synergistically in complement-dependent cytotoxicity. J Immunol 127:157–160
Herlyn D, Koprowski H (1981) IgG2 monoclonal antibodies inhibit human tumor growth through interaction with effector cells. Proc Natl Acad Sci USA 79:4761–4765
Herlyn D, Ross AH, Koprowski H (1986) Anti-idiotype antibodies bear the internal image of a human tumor antigen. Science 232:100–102
Herlyn M, Blaszczyk M, Sears HF et al. (1983) Detection of a carcinoembryonic antigen and related antigens in sera of patients with gastrointestinal tumors using monoclonal antibodies in double determinant radioimmunoassays. Hybridoma 2:329–339
Herlyn M, Blaszczyk M, Bennicelli J et al. (1985) Selection of monoclonal antibodies detecting serodiagnostic human tumor markers. J Immunol Methods 80:107–116
Hersey P, Schibeci SD, Townsend P et al. (1986) Potentiation of lymphocyte responses by monoclonal antibodies to the ganglioside GD3. Cancer Res 46:6083–6090
Horan Hand P, Thor A, Wunderlich D et al. (1984) Monoclonal antibodies of predefined specificity detect activated ras gene expression in human mammary and colon carcinomas. Proc Natl Acad Sci (USA) 81:5227–5231
Houghton AN (1986) Immunotherapy of solid tumors. Hybridoma 5:173–174
Houghton AN, Mintzer D, Cordon-Cardo C et al. (1985) Mouse monoclonal antibody detecting GD3 ganglioside: A phase I trial in patients with malignant melanoma. Proc Natl Acad Sci USA 82:1242–1246
Johnston WW, Szpak CA, Lottich SC et al. (1986) Use of a monoclonal antibody (B72.3) as a novel immunohistochemical adjunct for the diagnosis of carcinoma in fine needle aspiration biopsy specimens. Hum Pathol 17:502–513

Köhler G, Milstein C (1975) Continuous cultures of fused cells secreting antibody of predefined specificity. Nature 256:495–496
Kufe DW, Nadler L, Sargent L et al. (1983) Biological behavior of human breast carcinoma-associated antigens expressed during cellular proliferation. Cancer Res 43:851–857
Lange PH, Vessella RL, Chiou R-K et al. (1985) Monoclonal antibodies in human renal cell carcinoma and their use in radioimmune localization and therapy of tumor xenografts. Surgery 98:143–150
Larson SM (1985) Radiolabeled monoclonal antitumor antibodies in diagnosis and therapy. J Nucl Med 26:538–545
Lindgren J, Blaszczyk M, Atkinson B et al. (1986) Monoclonal antibody-defined antigens of human prostate cancer cell line PC3. Cancer Immunol Immunother 22:1–7
Lo MMS, Tsony TY, Conrad MK et al. (1984) Monoclonal antibody production by receptor-mediated electrically induced cell fusion. Nature 310:792
Mach J-P, Buchegger F, Forni M et al. (1981) Use of radiolabeled monoclonal anti-CEA antibodies for the detection of human carcinomas by external photoscanning and tomoscintigraphy. Immunol Today 2:239–249
Magnani JL, Steplewski Z, Koprowski H et al. (1983) Identification of the gastrointestinal and pancreatic cancer-associated antigen detected by monoclonal antibody 19-9 in the sera of patients as a mucin. Cancer Res 43:5489–5492
Martin EW Jr, Tuttle SE, Rousseau M et al. (1986) Radioimmunoguided surgery: Intraoperative use of monoclonal antibody 17-1A in colorectal cancer. Hybridoma (Suppl 1) 5:S97–S108
Masui H, Moroyama T, Mendelsohn J (1986) Mechanism of antitumor activity in mice for anti-epidermal growth factor receptor monoclonal antibodies with different isotypes. Cancer Res 46:5592–5598
Metzgar RS, Rodriquez N, Finn O et al. (1984) Detection of a pancreatic cancer-associated antigen (DU-PAN-2 antigen) in serum and ascites of patients with adenocarcinoma. Proc Natl Acad Sci USA 81:5242–5246
Morrison S (1985) Transfectomas provide novel chimeric antibodies. Science 229:1202–1207
Muraro R, Wunderlich D, Thor A et al. (1985) Definition by monoclonal antibodies of a repertoire of epitopes on carcinoembryonic antigen differentially expressed in human colon carcinomas versus normal adult tissue. Cancer Res 45:5769–5780
Nuti M, Teramoto YA, Mariani-Costanti R et al. (1982) A monoclonal antibody (B 72.3) defines patterns of distribution of a novel tumor associated antigen in human mammary carcinoma cell populations. Int J Cancer 29:539–545
Order SE (1985) Analysis, results, and future prospective of the therapeutic use of radiolabeled antibody in cancer therapy. In: Baldwin RW, Byers VS (eds) Monoclonal antibodies for cancer detection and therapy. Academic Press, London, pp 303–316
Ozanne B, Richards CS, Hendler F et al. (1986) Overexpression of the EGF receptor is a hallmark of squamous cell carcinomas. J Pathol 149:9–14
Paterson AJ, Schlom J, Sears HF et al. (1986) A radioimmunoassay for detection of a human tumor-associated glycoprotein (TAG-72) using monoclonal antibody B72.3. Int J Cancer 37:659–666
Pectasides D, Stewart S, Courtenay-Luck N et al. (1986) Antibody-guided irradiation of malignant pleural and pericardial effusions. Br J Cancer 53:727–732
Primus FJ, Newell KD, Blue N et al. (1983) Immunological heterogeneity of carcinoembryonic antigen: antigenic determinants on carcinoembryonic antigen distinguished by monoclonal antibodies. Cancer Res 43:686–692
Primus FJ, Deland FH, Goldenberg DM (1984) Monoclonal antibodies for radioimmunodetection of cancer. In: Wright GL Jr (ed) Monoclonal antibodies and cancer. Marcel-Dekker, New York, pp 305–323
Ricolleau G, Chatal J-F, Fumoleau P et al. (1984) Radioimmunoassay of the CA125 antigen in ovarian carcinomas: Advantages compared with CA19-9 and CEA. Tumour Biology 5:151–159
Robinson A, Williams ARW, Piris J et al. (1986) Evaluation of a monoclonal antibody to ras peptide, RAP-5, claimed to bind preferentially to cells of infiltrating carcinomas. Br J Cancer 54:877–883
Rowland GF, Axton CA, Baldwin RW et al. (1985) Antitumor properties of vindesine-monoclonal antibody conjugates. Cancer Immunol Immunother 19:1–7
Rowlinson G, Balkwill F, Snook D et al. (1986) Enhancement by gamma-interferon of *in vivo* tumor radiolocalization by a monoclonal antibody against HLA-DR antigen. Cancer Res 46:6413–6417
Saärinen UM, Coccia PF, Gerson SL (1985) Eradication of neuroblastoma cells in vitro by monoclon-

al antibody and human complement: Method for purging autologous bone marrow. Cancer Res 45:5969–5975
Schlom J (1986) Basic principles and applications of monoclonal antibodies in the management of carcinomas: The Richard and Hinda Rosenthal Foundation Award Lecture. Cancer Res 46:3225–3238
Schlom J, Weeks MO (1985) Potential clinical utility of monoclonal antibodies in the management of human carcinomas. In: DeVita VJ Jr, Hellman S, Rosenberg SH (eds) Important advances in oncology. Lippincott, Philadelphia, pp 170–192
Sears HF, Herlyn M, Del Villano B et al. (1982) Monoclonal antibody detection of a circulating tumor-associated antigen. II. A longitudinal evaluation of patients with colorectal cancer. Clin Immunol 2(2):141–149
Sears HF, Herlyn D, Steplewski Z et al. (1985) Phase II clinical trial of a murine monoclonal antibody cytotoxic for gastrointestinal adenocarcinoma. Cancer Res 45:5910–5913
Sell S, Reisfeld R (eds) (1985) Monoclonal antibodies in cancer. Contemporary biomedicine. Humana, Clifton, NJ, pp 1–485
Sharkey RM, Primus FJ, Goldenberg DM (1984) Second antibody clearance of radiolabeled antibody in cancer radioimmunodetection. Proc Natl Acad Sci USA 81:2843–2846
Sheibani K, Battifora H, Burke JS (1986) Antigenic phenotype of malignant mesotheliomas and pulmonary adenocarcinomas. AJP 123(1):212–219
Siccardi AG, Buraggi GL, Callegaro L et al. (1986) Multicenter study of immunoscintigraphy with radiolabeled monoclonal antibodies in patients with melanoma. Cancer Res 46:4817–4822
Starling JJ, Sieg SM, Beckett ML et al. (1986) Human prostate tissue antigens defined by murine monoclonal antibodies. Cancer Res 46:367–374
Steplewski Z (ed) (1986) Wistar Symposium on Immunodiagnosis and Immunotherapy with CO17-1A Mab in Gastrointestinal Cancer. Hybridoma (Suppl 1) 5:S1–S187
Strelkauskaus A (ed) (1985) Human hybridomas: diagnostic and therapeutic application. Marcel-Dekker, New York, pp 1–297
Sun LK, Curtis P, Rakowicz-Szulczynska E et al. (1986) Chimeric antibodies with 17-1A derived variable and human constant regions. Hybridoma (Suppl 1) 5:17–20
Suter L, Brüggen J, Bröcker EB et al. (1985) A tumor-associated antigen expressed in melanoma cells with lower malignant potential. Int J Cancer 35:787–791
Szpak CA, Johnston WW, Roggli V et al. (1986) The diagnostic distinction between malignant mesothelioma of the pleura and adenocarcinoma of the lung as defined by a monoclonal antibody (B72.3). Am J Pathol 122:252–260
Tagliabue E, Porro G, Barbanti P et al. (1986) Improvement of tumor cell detection using a pool of monoclonal antibodies. Hybridoma 5(2):107–115
Takeda S, Naito T, Hama K et al. (1985) Construction of chimaeric processed immunoglobulin genes containing mouse variable and human constant region sequences. Nature 314:446–449
Thor A, Ohuchi N, Horan Hand P et al. (1986) Biology of Disease. Ras gene alterations and enhanced levels of ras p21 expression in a spectrum of benign and malignant human mammary tissues. Lab Invest 55(6):603–615
Wahab ZA (1985) Immunohistochemical markers for human prostate carcinoma. Doctoral dissertation, Eastern Virginia Medical School, Norfolk, VA, USA
Wahab ZA, Wright GL Jr (1985) Monoclonal antibody (anti-Leu 7) directed against natural killer cells reacts with normal, benign and malignant prostate tissues. Int J Cancer 36:677–683
Webb KS, Liberman SN, Ware JL et al. (1986) *In vitro* synergism between hybrid immunotoxins and chemotherapeutic drugs: relevance to immunotherapy of prostate carcinoma. Cancer Immunol Immunother 21:100–106
Weiner LM, Steplewski Z, Koprowski H et al. (1986) Biologic effects of gamma interferon pretreatment followed by monoclonal antibody 17-1A administration in patients with gastrointestinal carcinoma. Hybridoma (Suppl 1) 5:S65–S77
Wright GL Jr (1984) Application of monoclonal antibodies to the study of human cancer. In: Wright GL Jr (ed) Monoclonal antibodies and cancer. Marcel-Dekker, New York, pp 1–29

Biochemical Markers for Human Cancer

T.M. CHU

1 Introduction

Development of biochemical markers for human cancer in recent years that have been shown to be of some clinical value in management of patients will be presented. Previous approaches of using conventional immunochemical techniques to the measurement of target markers still dominate the area of laboratory investigation and clinical application. However, in addition to radioimmunoassay and enzyme linked immunosorbent techniques, immunohistochemistry, flow cytometry and new molecular biology procedures are increasingly being utilized. Also, tissue and body fluid other than serum are acquired as the specimens for measurement of these markers.

Tumor markers are ideally to be as the indicator for the presence of preneoplastic and neoplastic diseases. Cancer is described as a disorder of cell proliferation. Neoplastic cells basically are different from their normal counterparts in its failure to respond to growth control. At the present stage of development no qualitative tumor markers, so called tumor-specific markers, have been identified. Rather, quantitative difference between normal and neoplastic cells is used as the index.

Almost without exception, monoclonal antibody is the choice of reagent in immunochemical techniques, although it should be kept in mind that polyclonal antiserum reagent of proved value and quality can be equally effective. Typical advantages of monoclonal antibody are standardization of test reagent and consistency of quality control. At present no marker for general screening of cancer is yet in existence. Most of the clinical utilizations are in patient management of post-diagnosis, with a few that are potentially applicable in early detection of cancer for high risk population. During recent years, with the use of monoclonal antibodies, many new tumor markers have also been recognized, characterized and evaluated for clinical applications.

This chapter aims primarily to discuss and emphasize recent development of tumor markers in in vitro clinical applications.

2 Tumor Associated Antigens

2.1 Carcinoembryonic Antigen

Although carcinoembryonic antigen (CEA) was first reported more than twenty years ago (Gold and Freedman 1965), it is still the most extensively investigated tumor marker with the largest number of publications. Recent review articles are readily available (Roger 1983; Steele and Zamcheck 1985; Beard and Haskell 1986; Fletcher 1986), only a few recent aspects will be discussed.

Complete chemical structure of CEA is still unknown. Using conventional biochemical and chemical techniques, CEA molecule has been characterized partially. Elucidation of the total amino acid sequence of the peptide portion has been tried but to no avail. Modern techniques of molecular biology to sequence the nucleotides encoded for the amino acids would be a more effective approach. The microheterogeneity in carbohydrate portion of the molecule still would be a formidable problem. A recent interesting piece of information regarding the molecular structure of CEA is that CEA is composed of two identical subunits (Lisowska et al. 1983). Using sodium dodecyl sulfate-polyacrylamide gel electrophoresis and cross-linking experiments, CEA has been shown to be a dimer consisting of two identical or closely similar non-covalently bound subunits. CEA is dissociated upon heating in the presence of sodium dodecyl sulfate, and the dissociation is reversible upon the removal of the detergent. Subunits of CEA are not linked by way of the disulfide bonds, but rather by the hydrophobic interactions between the subunits and among peptide chains.

Should this result be confirmed it may simplify the work on characterization of the entire CEA molecular structure. In spite of the extensive clinical evaluation of CEA, biological function, if any, of CEA still remains totally unknown.

More than twenty years since CEA was first reported, clinical usefulness and limitation of CEA have been mostly assessed with very few outstanding clinical issues. The role of CEA in second-look surgery is one that awaits definitive information. Although information is available concerning this critical subject, with some describing favorable data and some expressing reservation; none has been really systematically and statistically analyzed. One report describing an extensive investigation just became available recently (Minton et al. 1985). This study was initiated by the Society of Surgical Oncology in 1978. It involved 400 patients in a nonrandomized prospective study to evaluate the usefulness of CEA as an indicator of recurrent colon and rectum cancer, as conducted by 37 surgical oncologists in 31 institutes across the USA. Patients were evaluated post-operatively with customary follow-up procedures and serial CEA determinations. During follow-up, 130 patients had recurrence (21 Dukes' B, 8 B_2, 31 C_1, 58 C_2 and 12 unknown), and 75 were reported on, with 43 reoperations were CEA-directed and 32 clinically directed. Two of these 75 patients died within one month after the second surgery. Twenty-two second-look patients remained disease-free five years after their second operation. The highest resectability of recurrent cancer occurred in patients with a CEA level less than 11 ng/ml, in whom the CEA was assayed at intervals of one to two months.

Four important guidelines for follow-up management were proposed from this study. For Dukes' A carcinoma, there may be no justification for a CEA follow-up when the tumor is resected, since no patient in this group (a total of 17 patients) developed recurrent disease. CEA follow-up is recommended for patients with Dukes' stages B_1 to C_2 patients, since about 25% of B_1 (21/91), B_2 (18/31) and C_1 (31/119) carcinomas and 50% of C_2 patients (58/122) developed recurrent disease. CEA determinations should be performed at one- or two-month intervals, since this practice offers the best opportunity for early discovery of resectable recurrent cancer during the first and second years of post-operative follow-up. Reoperation should be performed before the CEA exceeds 11 ng/ml, since patients with CEA less than this level have shown the greatest resectability rate (63%) and the best five-year disease-free survival results (68%).

Of interest to note is the observation that there is no significant difference in overall cumulative surviving between CEA directed and clinically directed second-look surgeries. A previous report has described similar observation regarding disease-free interval and survival (Hine and Dykes 1984). This may be due to the fact that there is still no effective therapy for colon/rectum cancer. Therefore, the most effective use of CEA in management of patients will be in those where potentially effective therapeutic manipulation is available, such as small cell carcinoma of the lung.

Investigations on other subjects of CEA, such as assays, screening, diagnosis, prognosis and monitoring, including that of in vivo radioimmunodetection, have provided no dramatically new information in recent years. With the use of monoclonal antibodies, at least five different antigenic determinants (epitopes) have been identified in CEA molecule (Muraro et al. 1985). Whether these CEA-reactive monoclonal antibodies could improve the clinical effectiveness of CEA assay remains to be determined.

A refined CEA assay was reported a few years ago that showed an improvement in the specificity of serum CEA for carcinomas (Hedin et al. 1983). Using two monoclonal antibodies specific for two separate CEA epitopes, an enzyme-linked immunosorbent procedure was developed and shown to be quite specific in detecting CEA levels in serum specimens from patients with solid tumors.

So far, no further reports on a large scale specimens including patients with early cancer have been available.

2.2 Alpha-Fetoprotein

In comparison with CEA, we know much more about α-fetoprotein (AFP) (Hirai 1982). The primary structures including the complete sequence of 590 amino acids of human AFP have been elucidated (Morinaga et al. 1983; Sakai et al. 1985). Its clinical use in primary hepatocellular carcinoma and in testicular tumors of germ-cell origin (along with human chorionic gonadotrophin) is well known (Norgaard-Pedersen 1979). AFP also is a potential prenatal diagnostic tool for neural-tube defects. For reasons that are not scientifically justified, full diagnostic application of this clinically useful marker unfortunately has not been achieved (Sell 1981).

The most important clinical utilization of biochemical marker for cancer is its use as a screening test for cancer, either in general populations or in a high risk population. No such marker is yet available. The best candidate marker potentially for such a use in fact is AFP. An elevated serum AFP in apparently healthy adult (non-pregnant) is the single most discriminating marker indicative of cancer (Sell 1981). AFP has been a well known diagnostic marker associated with primary hepatocellular carcinoma for over twenty years. Simple laboratory tests are available with the use of either polyclonal or monoclonal antibody reagent.

Several studies using AFP as a screening tool for high risk population have been conducted in China, Southeast Asia and Africa, where the incidence rate of primary hepatocellular carcinoma is highest in the world (Sell 1981). When a tumor marker is used for cancer screening, the factor of cancer prevalence as well as specificity and sensitivity of the assay should be considered (Wagener 1984). As far as AFP is concerned, infection with hepatitis B virus also needs to be taken into account, since highest rates of hepatitis B infection and primary hepatocellular carcinoma are commonly found in the same populations. Therefore, should an AFP test be of screening value, it should be able to discriminate, at least quantitatively, primary hepatocellular carcinoma from patients with hepatitis B infection and other non-malignant hepatic diseases, such as cirrhosis.

In one study conducted in China, where almost a half million asymptomatic individuals were screened with AFP test, 57 were identified to have a serum AFP of greater than 500 ng/ml. All 57 of these apparently healthy individuals were subsequently diagnosed to have liver cancer. In a later study involving 115,000 Chinese, the results were not so drastically definitive. Although 20 individuals having high AFP levels were identified to have primary hepatocellular carcinoma, another 270 individuals had slightly elevated levels of AFP and none was found to have evidence of cancer. It is not clear whether these two studies were conducted in the same region or different pocket areas where incidence rate of liver cancer could be different.

As indicated previously, when AFP is used as a screening cancer test, the prevalence of cancer is a critical factor. This can be best illustrated by a study conducted in Germany, where liver cancer is not common. A highly elevated serum AFP level, greater than 2,000 ng/ml, was found in 10 of 18 patients with primary hepatocellular carcinoma, who also had cirrhosis of the liver. In addition, among 18 patients with primary hepatocellular carcinoma, 13 were found to have an AFP of greater than 400 ng/ml; but also were 14 of the 385 patients who had cirrhosis of the liver without liver cancer (AFP ranged 400–2,000 ng/ml) (Lehmann and Wegener 1979).

In a recent report screening an entirely different population for early detection of primary hepatocellular carcinoma, the results were quite encouraging (HEYWARD et al. 1985). The subjects of screened population were Alaskan Natives who had high rates of primary hepatocellular carcinoma and hepatitis B virus infection. In addition to screening, it was also to determine if primary hepatocellular carcinoma could be detected at an early stage at the time when tumor could be surgically removed. Serum AFP was tested semiannually by an ELISA (AUSZYME II) among 1,394 Alaskan Natives who were infected with hepatitis B virus tested by an RIA (AUSRIA) for hepatitis B surface antigen (HBsAg). A total of 3,387 AFP tests were performed among these natives during a 26-month period (1982–1984). Of 126 persons with elevated AFP (>25 ng/ml), nine males were found to have primary hepatocellular carcinoma and all had an AFP of >350 ng/ml. Six of these nine persons were asymptomatic and four had small tumors (<6 cm) that were surgically resected.

It is interesting to note that of all 602 HBsAg-positive females, no cases of primary hepatocellular carcinoma were found in 107 persons who had at least an elevated AFP test. Most elevations of AFP could be attributed to pregnancy. Of the 792 HBsAg-positive males screened, 19 (2.4%) had at least an elevation of AFP, and 9 were found to have primary hepatocellular carcinoma with 6 asymptomatic at the time of the blood test as noted already. Three of these six were detected in a prospective manner, i.e. all had normal AFP 6 to 12 months before their first elevation of AFP. In all six patients, the AFP level increased very rapidly, doubling every 1 to 3 months.

This is by far the most well controlled study indicating that AFP can be used as a screening test for primary hepatocellular carcinoma in a high risk population with hepatitis B virus infection. It also reveals that early stage of surgically resectable tumor can be detected by a simple AFP test.

As illustrated from the reports presented above, should a tumor marker be used as an aid for diagnosis of cancer, it should be able to provide a parameter to differentiate malignant neoplasm from benign disease, and also to identify an early resectable tumor which can be successfully treated. In these respects, AFP certainly is a clinically useful diagnostic marker for primary hepatocellular carcinoma.

In addition to primary hepatocellular carcinoma, AFP in combination with human chorionic gonadotrophin (HCG) can be used effectively for clinical diagnosis of testicular tumors of germ cell origin. Circulating AFP and HCG have been reported to correlate with histology of testicular tumor. It has been well documented that elevated serum AFP is associated with tumors containing endodermal sinus, yolk sac elements and HCG with choriocarcinomatous elements or syncytial giant cells. In fact, these markers are so well studied that recommendations have been made that every patient with a testicular tumor should have a preoperative determination of serum AFP and HCG, then at least twice weekly in the first six weeks after operation (NORGAARD-PEDERSEN 1979). This also may apply when a germ-cell tumor of ovary is suspected. It should be noted, though, that combined AFP and HCG tests still would miss a small portion of patients with germ cell tumors of the testes. On the patients who were found to exhibit measurable and elevated AFP and/or HCG, the levels and rates of change of AFP and HCG in the postoperative follow-up, and during and after chemotherapy, can provide additional information on the course of disease.

Both CEA and AFP are well known biochemical tumor markers and tumor associated antigens. The term antigen is more of an operational use than that of classical immunological terminology. Since, as far as human hosts are concerned, the presence of anti-CEA or anti-AFP antibody has not been unequivocally documented. In recent years with the availability of an enormous amount of monoclonal antibodies generated, many previously unrecognized tumor associated antigens and components have been reported. Some of them have been used in cancer management especially patient monitoring. Only those that have been reported extensively will be presented here.

2.3 CA 19-9 Antigen

CA 19-9 is a monoclonal antibody-defined tumor marker representing a carbohydrate antigenic determinant of glycoprotein and glycolipid substances (DEL VILLANO et al. 1983). Like other tumor associated antigen markers, low concentration of CA 19-9 is found in healthy individuals, but frequently increased in sera from patients with gastrointestinal adenocarcinomas, especially those of the pancreas. Similar to that of CEA, CA 19-9 is detected in fetal, normal and neoplastic gastrointestinal tissues.

Serum CA 19-9 measurement has been shown to be of more clinical value, i.e. sensitivity and specificity, than CEA in pancreatic cancer (STEINBERG et al. 1986). Serum levels of CA 19-9 also has been reported to assist in the differentiation of chronic pancreatitis from pancreas cancer. Although it has been claimed that early localized pancreatic adenocarcinoma can be detected by the measurement of CA 19-9, this interesting and clinical application remains to be documented with further study (GUPTA et al. 1985). The test kit of CA 19-9 is commercially available. So far, no tumor marker has been available that can identify patients with early and resectable pancreatic tumor.

2.4 CA 125 Antigen

CA 125 is an ovarian cancer associated mucin-like antigen as recognized by a murine monoclonal antibody (BAST et al. 1981). Its clinical use in ovarian cancer is by a radioimmunoassay test, and the kit is available commercially. In the initial clinical report (BAST et al. 1983), only 1% of apparently healthy blood donor controls and 6% of benign disease patients were detected to have an elevated serum CA 125 level. Significantly, 82% of patients with ovarian cancer exhibited an elevated level, some with extremely high levels. Serum CA 125 levels appear to correlate with tumor burden and stages of disease. Twenty-nine percent of patients with non-gynecological cancers also exhibited an elevated CA 125 level, especially those with pancreatic cancer (59% of 29 patients).

Like most other tumor markers, serum CA 125 is more useful in monitoring the course of epithelial ovarian cancer. CA 125 has been reported to be effective in monitoring the progression or regression of disease in more than 90% of ovarian cancer patients. For common epithelial ovarian cancers, serum CA 125 has been reported to be the most valuable marker for disease management, better than CEA and CA 19-9. As discussed previously, AFP and HCG are useful for ovarian germ cell tumor similar to testicular germ cell tumors, while choriocarcinoma may be managed by HCG monitoring.

A murine monoclonal antibody, F36/22, has been shown to identify 100% of epithelial ovarian cancers as examined by tissue immunostain techniques (CROGHAN et al. 1983). Our unpublished data reveal that more than 95% of patients with active epithelial ovarian cancers exhibit an elevated serum F36/22 antigenic components, while patients without evidence of disease have normal range of antigen level. Pregnant women and patients with non-epithelial ovarian cancers, benign ovarian and benign gynecological conditions all have normal blood levels. Preliminary study involving serial specimens also reveals that F36/22 may be an effective marker for monitoring therapy and recurrent disease in epithelial ovarian cancer patients. Neither CA 125 nor F36/22 has been shown to be of clinical value for the early detection of common epithelial ovarian carcinomas.

3 Organ-Site and Cell-Type Specific Antigen

AFP, CEA, CA 19-9 and CA 125 are tumor associated antigens and clinically useful in the sense that their quantitative expression is utilized as an aid to differentiate neoplastic disease from normal or benign conditions. There is another class of substance that is organ-site and cell-type specific and has been used effectively as tumor marker. The best example is a human prostate-specific antigen, commonly abbreviated as PA or PSA (WANG et al. 1979). In fact, PA has been approved by the Food and Drug Administration (USA); the first tumor marker to receive such an approval since CEA was approved in the early 1970s.

Unlike most of tumor associated antigens, PA is present at an almost equal concentration in normal, benign hypertrophic and cancerous human prostatic tissues (WANG et al. 1982). It is not present in other normal or tumor tissues. This tissue specificity is confirmed by both biochemical and immunocytochemical techniques, and by both polyclonal and monoclonal antibody reagents. In the prostate, PA is located in the ductal epithelial cells and not in other cellular components. PA is a glycoprotein of Mr 33,000 with microheterogeneity in carbohydrate contents. The complete amino acid sequence of peptide portion of PA has been elucidated (WATT et al. 1986). Despite the well-characterized physico-chemical properties, the biological function of PA is not known, although proteolytic activities have been shown to be associated with PA (BAN et al. 1984). Test kits for serum PA assay are available commercially, and that manufactured by Hybritech Incorp. has been approved by the FDA for clinical use as an aid in management of patients.

Prostatic acid phosphatase (PAP), a well known tumor marker, has been used for over 50 years to assist in diagnosis and monitoring of prostate cancer

(GUTMAN and GUTMAN 1938). From data available to date, PA is superior to prostatic acid phosphatase; and this is primarily due to the fact that PA is absolutely specific for the prostate. Although serum assay of PA alone cannot be used for early detection of prostate cancer, because of the PA levels of early stage of prostatic cancer patients are often overlapped with those with benign prostatic hypertrophy, its role in monitoring therapy and predicting recurrent disease has been firmly established. Further, combination assay with prostatic acid phosphatase may be of use in early diagnosis of prostate cancer.

Simultaneous serum assays of PA and PAP have detected more patients with prostate cancer than either assay alone (KURIYAMA et al. 1982). Combination assays have revealed a discordance between these two markers, reflecting the expression of these two biochemically and immunologically distinct prostate glycoproteins may represent different aspects on the biology of prostate cancer. Since patients with benign prostatic hypertrophy also exhibited slightly elevated levels of both markers, which can be set as the operation cut-off points for a combination test and thus decrease the false positive rate. Using this approach, data are available supporting that both assays are of additive value and suggesting that diagnosis of early stage of prostate cancer is feasible in certain circumstances.

Single assay of serum PA alone has been shown to be of prognostic and monitoring applications of prostate cancer. Data obtained from large-scale and long-term clinical evaluations have been analyzed and reported (KILLIAN et al. 1985). In patients with regionally confined prostate cancer who had received curative therapies, levels of PA are effective in predicting early recurrence of disease and are of prognostic value in disease-free survival time, irrespective of stage and therapy of patients. A single elevated PA level also has been shown to be of predictive value for recurrence of disease. For patients with early stages (B_2 to D_1) of prostate cancer with regionally confined disease, an association of increasing serum levels of PA to increasing tumor burden and decreasing time to disease recurrence is statistically established. Therapeutic manipulation based upon the PA level may improve the patients' survival in a prospective manner.

For patients with advanced prostate cancer who were receiving treatments, the levels of serum PA were correspondent with clinical courses of patients. Using patient as his own reference, PA levels in general increase as disease progresses and decrease as disease regresses. In addition for those patients with disease progression, a quantitative and significant association between the risk of progression or death and a PA level exceeding a certain elevated value has been statistically established (KILLIAN et al. 1985). Thus, the prognostic reliability of PA is firmly established.

Certainly, should a biochemical tumor marker be clinically useful and practically applicable, it should be better than existing assays for management of patients. In this respect, PA has been compared with other commonly used markers for prostate cancer and shown to be the best marker in its reliability for predicting disease progression (KILLIAN et al. 1986). Acid phosphatase, prostatic acid phosphatase, total alkaline phosphatase and its bone isoenzyme all have been shown previously to be useful in varying degrees for management of prostate cancer (KILLIAN et al. 1981). Comprehensive statistical analyses on a large series of patients have revealed that either PA or PAP is of value in detection of disease progression in monitoring early stages of prostate cancer. Progression of disease in patients with advanced prostatic cancer can be monitored reliabily by PA assay alone.

In summary, as far as serum PA in monitoring of prostate cancer is concerned, it is reasonable to state that what CEA can do for colon-rectum cancer, PA can do for prostate cancer, and may be better. Practical use of PA as an immunohistochemical marker for prostatic neoplasms is detailed in Sect. 8.

4 Steroid Hormone Receptor

An important clinical value of biochemical tumor marker is its use in staging of disease. Although CEA has been reported to be of some value in staging surgically resectable patients in lung cancer (VINCENT et al. 1979), the best marker for such a clinical utility is estrogen receptor. The importance of estrogen receptor protein (estrophilin) analysis in staging of patients with estrogen-related tumors is well known as a guide to therapy and prognosis. Biochemical assays, based upon the interaction between a radioactive estrogen ligand and the receptor protein at the receptor binding site, are commonly used. It requires sufficient amount of fresh tumor tissue and tedious labor-intensive exercise with careful quality control to preserve biologically active binding site (WITTLIFF et al. 1982). False estrogen receptor-poor results is the most common problem, which often deprives a patient from the benefit of hormonal therapy.

Monoclonal antibodies directed against distinct determinants of human estrogen receptor protein have been generated (GREENE et al. 1980), and immunoradiometric or enzyme-linked immunosorbent assays also have been developed (GREENE and JENSEN 1982). The estrogen receptor contents measured by monoclonal antibody-based assay are found to be indistinguishable from those as quantified by conventional biochemical technique, such as sucrose gradient ultracentrifugation.

A test kit of enzyme immunoassay (EIA) for estrogen receptor measurement became available recently commercially (Abbott Laboratories). Two reports evaluating this kit presented somewhat different results. In comparison with the results obtained from classical dextran-coated charcoal method, high estrogen receptor levels were detected by the EIA method in one report (RAAM and VRABEL 1986). However, another report indicated that the EIA method is valid in measuring estrogen receptor contents as compared with those as detected by established radioligand-based assays, dextran-coated charcoal and hydroxylapatite (HOLT and BOLANOS 1986). Estrogen receptor in breast, ovarian and uterine carcinomas can be accurately measured. Regardless of these results, at the present stage of development monoclonal antibody-based immunoassays will be the method of choice in the foreseeable future.

This notion has been fortified by most recent reports involving multicenter evaluation of EIA method in both Europe and the USA (LECLERCQ et al. 1986;

JORDAN et al. 1986). ER-EIA monoclonal immunoassay (Abbott) was evaluated in eleven laboratories. Results suggest that the EIA method is comparable in measuring estrogen receptor contents with conventional steroid binding assays. Routine use of this newly developed immunoassay can be anticipated in the not too distant future.

Theoretically, progesterone receptor can be quantified similarly. As of this date, no such test is yet available.

It cannot be overemphasized the clinical importance of steroid hormone receptors. It also should be noted that quality control assurance of the assay is of utmost importance in obtaining accurate laboratory results. Proficiency-testing specimens are readily available and should be performed. Availability of specific monoclonal antibody for estrogen receptor also provided a useful tool for immunocytochemical assessment of tumor tissues (MCCARTY et al. 1986). Potential clinical significance of this approach is to be discussed later in Sect. 8.

5 Potential Markers for Early Cancer Diagnosis

5.1 Thyrocalcitonin

For early detection of cancer, AFP is by far the best candidate as already discussed. There is another potential marker which also has been shown to be effective, i.e. thyrocalcitonin in medullary thyroid carcinoma (COOPER 1976). Calcitonin is a 32 amino acid peptide hormone produced in non-follicular thyroidal C-cells. Physiologically it plays an important role in calcium utilization and bone resorption. It is a hypocalcemic and hypophosphatemic hormone. Data available suggest the presence of a closed-loop feedback system between the gut (gastrin) and thyroid (calcitonin) (COOPER 1976), and gastrin-calcitonin relationship in controlling the level of calcium during intestinal absorption.

The application of calcium or more conveniently of gastrin for a provocative test in early diagnosis of medullary thyroid carcinoma in high risk family members of the patients has been reported. Suspected patients with medullary thyroid carcinoma would respond to the injection of the synthetic gastrin analogue, pentagastrin, to produce a rapid and highly elevated serum level of calcitonin. Subclinical or questionable cases of this C-cell tumor have been identified by this simple manipulation which involves a provocative test of pentagastrin injection and evaluation of blood calcitonin by immunoassay. Pentagastrin injection is more commonly practiced than calcium infusion. Urine calcitonin level also has been suggested to be a reliable screening test (SILVA et al. 1979).

5.2 Fibrinogen Degradation Products

Detection of urinary fibrinogen degradation products could be a simple and potential test for diagnosis of bladder cancer. The association of urinary bladder carcinoma and elevation of urinary fibrinogen degradation products has been known for some time (Wajsman et al. 1975). Recently, a monoclonal antibody-based enzyme linked immunosorbent assay of fibrinogen degradation products was reported (McCabe et al. 1984). The specificity of the test was reported to be 96% with a false-positive rate of 5%. Although these results are not acceptable for screening use of urinary bladder cancer, it is a reasonable expectation that a better assay could be developed in the near future.

It would be worthwhile to make some comments regarding serological use of biochemical tumor markers at this point. At the present stage of scientific and technical development, the most difficult and yet the most rewarding clinical application is cancer screening for general population or asymptomatic subjects in high-risk populations. Markers of absolute specificity and sensitivity are required. The former is a dream rather than a goal; the latter is an achievable goal, as examplified by AFP for primary hepatocellular carcinoma. More realistical is the use of biochemical tumor markers in early detection of cancer. Since target patients are those already with signs or symptoms suspected or suggestive of cancer, the requirement of specificity is differentiation of malignant from benign disease, and that of sensitivity is identification of early curable lesion from advanced metastatic tumor (Herberman 1982). AFP, HCG, PA and calcitonin are potential tumor markers of this category.

Another two clinical applications of biochemical markers for cancer are their use in staging of disease, and monitoring of therapeutic efficacy and disease recurrence. Steroid hormone receptors is a good example of the former, and many tumor markers, those discussed here and others published, are proved parameters for the purposes of the latter. Therefore, at present only biochemical markers suitable for these last two clinical applications are mostly in practical use.

6 Enzymes and Isoenzymes

Enzyme has been traditionally used as a biochemical marker for cancer. Abnormality in enzyme activity is a major biochemical characteristic of uncontrolled tumor cell growth. This phenomenon has been used as a laboratory parameter in differentiation of tumor cell from its normal counterpart, and used as a target in developing therapeutic agents against cancer. The use of enzyme as tumor marker is based upon the concept that alteration in the gene expression during malignant transformation can be detected at the level of end product. A qualitative difference in enzyme expression would be an ideal tumor marker. No such cancer specific enzyme has been identified and clinically utilized. Enzymatic inbalance in tumor cells has been used instead for clinical evaluation of cancer.

Biochemical activity of the enzyme is commonly used as a parameter for tumor marker. With recent availability of immunologic reagents, mass of enzyme protein or specific antigenic determinant also has been measured. However, the measurement of an enzyme alone still cannot clinically detect a small tumor mass, since the secreted enzyme is vastly diluted or actively metabolized in serum, urine or other extracellular fluids. Despite these inherent problems, enzyme and isoenzyme are useful probes as an aid for management of cancer patients. Some recent developments using enzymes and isoenzymes as markers for diagnosis and treatment monitoring are briefly discussed.

6.1 Prostatic Acid Phosphatase

Acid phosphatase of the prostate is the first enzyme marker for cancer. Serum acid phosphatase activity by conventional biochemical assays has been used for over 50 years as a laboratory aid for metastatic prostate cancer (Gutman and Gutman 1938). Prostatic acid phosphatase (PAP) is one of a few enzyme markers exhibiting relative organ and cell-type specificity. Although acid phosphatase of the prostate is the most well known and investigated enzyme marker for cancer, its biology in prostate or prostate cancer remains totally unknown. Information available recently has shown that PAP is related to hydrolysis of tyrosine-phosphate in phosphoprotein molecules (Li et al. 1984). Its role in oncogene and malignant transformation thus has been speculated, since tyrosinekinase is a well known product of oncogene.

PAP is biochemically a glycoprotein of 100,000 molecular weight, consisting of two dimers and approximately 10% carbohydrate and 90% peptide with microheterogeneity (i.e. multiple isoelectric points) (Chu et al. 1982; Lin et al. 1983a). N-terminal amino acid is lysine and partial sequence is also known (Lin et al. 1983b). Reagents for biochemical and immunological assays are readily available from commercial sources.

It should be noted that PAP is prostate specific rather than prostate-tumor specific. Thus, the assay alone is of limited use in distinguishing early prostate cancer from benign prostatic hypertrophy, although its clinical utilization can be enhanced by other prostate tumor markers. PAP is not suitable as a screening modality for prostate cancer.

Clinically, PAP is useful in identifying metastatic prostate cancer, and in monitoring disease status and therapeutic response of prostate cancer (Chu et al. 1982). There are some recent well-controlled reports describing the role of PAP in monitoring prostate cancer with similar though not identical results. One report found that the PAP level in some patients did decrease to normal range after successful treatment, but the incidence of decrease in the PAP level was less with increasing severity of the disease (Cooper et al. 1983). Serial longitudinal follow-ups of PAP in patients who had been treated with hormonal and chemotherapy are available. The results revealed that PAP assay is a valuable indicator for treatment response and disease progression. One study reported that although serum PAP level decreased in patients responding to hormonal therapy, only in a small number of patients responding to chemotherapy did the serum PAP decrease (Zweig and Ihde 1985). These data suggest that PAP is more senitive in indicating disease progression than disease regression, and that PAP assay in general can be considered to be of value in confirming clinical assessment of patients (Killian et al. 1986).

As discussed already, PAP is a useful tool in detecting disease recurrence in patients with early stage of prostate cancer who have received apparently curative therapy (Killian et al. 1985). It can be used also as a prognosticator for predicting disease-free survival time. Although not as reliable as prostate-specific antigen, PAP still is a reasonably good marker for monitoring disease progression in advanced prostate cancer (Killian et al. 1986). Since prostate-specific antigen is relatively new and may not be available readily, PAP then will be the choice of biochemical marker. When combined with alkaline phosphatase and its bone isoenzyme, the clinical effectiveness of PAP can be greatly enhanced (Killian et al. 1981).

6.2 Neuron-Specific Enolase

Neuron-specific enolase (NSE) is another important enzyme marker of cancer. Enolase is a key enzyme involving the synthesis of ATP in glycolysis. Clinical significance of NSE is based upon its presence only in the cells associated with amine precursor uptake and decarboxylation (APUD) in the neurons of the diffuse neuroendocrine system. This unique feature provides the basis for NSE to serve as an enzyme marker for tumors of APUD or neuroendocrine origin. Highly specific antibody has been used for assay development. It should be emphasized also that NSE is not a tumor specific enzyme.

Small cell carcinoma of the lung was the first tumor shown to be associated with elevated NSE level (Carney et al. 1982). Even patients with limited disease (40%) exhibited mildly elevated serum NSE levels. NSE level was related to extent of disease. Virtually all patients with metastases at 3 or more sites were shown to have a highly elevated NSE. Treatment response was shown to be effectively monitored by serial NSE measurements in patients receiving combination chemotherapy. Furthermore, the source of serum NSE was identified to be originating from patients' tumor cells, as documented from continuous cell lines established from patients. These initial data suggest some interesting points. A new enzyme tumor marker has been identified. A simple serum NSE assay can be used effectively in monitoring clinical status of small cell carcinoma of the lung, one of the few cancers that can be treated effectively, and thus may contribute to overall survival of these patients. This is one exciting development in recent years using enzymes as tumor markers.

Serum NSE also has been shown to be of prognostic value in metastatic neuroblastoma. More than 95% of these patients had an elevated NSE. Patients exhibiting highly elevated NSE levels (>100 ng/ml) were generally associated with poor prognosis, especially in infants of less than 1 year old at the time of diagnosis.

Serum NSE, when combined with pancreatic polypeptide and HCG, also could be useful biochemical marker for determining prognosis and monitoring therapy response in pancreatic islet cell tumors and carcinoid tumors of the gastrointestinal tract. NSE is usually elevated in nonfunctioning pancreatic islet cell tumors, but less frequently in glucagonomas and intestinal islet cell tumors.

Additional clinical use of NSE is its diagnostic application in immunohistopathology of these tumors (KLOPPEL et al. 1983).

6.3 Galactosyltransferase

During the past ten years, the potential importance of glycosyltransferases in cancer has been extensively investigated. Data available indicate that their role in tumor metastasis is still largely speculative and that their role as tumor markers has been shown to be of some value. Glycosyltransferases are involved in the transfer and addition of a specific carbohydrate from a donor to a specific protein or glycoprotein acceptor. Galactosyltransferase, sialyltransferase and fucosyltransferase are more investigated than others in clinical oncology. These three enzymes catalyze the addition of galactose, sialic acid and fucose, respectively, to their specific acceptors.

A variety of cancer patients have been shown to exhibit elevated serum total galactosyltransferase activity (SCHWARTZ 1982). Like many other nonspecific enzyme markers, more patients with advanced disease have elevated value than patients with early staged disease. Its primary use is in monitoring and prognosis (DAVEY et al. 1986). Also like other enzymes, serum galactosyltransferase activity is the sum of many isoenzymes. One isoenzyme, designated galactosyltransferase II (GT-II), has been identified to be primarily associated with cancer. An initial paper indicated that serum GT-II was elevated in 85/117 patients with colorectal cancer, 18/23 breast cancer, 13/20 lung cancer, 15/18 pancreas cancer and 12/16 gastric cancer, along with some patients in many other cancers, but no elevation was found in melanoma or osteosarcoma (PODOLSKY et al. 1981).

A prospective study on GT-II isoenzyme involving 270 patients in detecting pancreatic cancer and comparing it with other biochemical (CEA, AFP, ferritin, C1q binding and RNAase) and physical diagnostic modalities has been reported (PODOLSKY et al. 1981). GT-II isoenzyme was shown to be the most sensitive (67%) and specific (98%) for differentiating malignant pancreatic cancer from benign disease. GT-II was as sensitive as ultrasound and computerized body tomography were, but was less sensitive than endoscopic retrograde cholangiopancreatography. Combining with GT-II and radiologic or endoscopic test resulted in greater sensitivity than any single test alone. However, combination with other serologic markers indicated above yielded no significant increase in detecting pancreatic cancer. In this prospective study, CA 19-9 was not included since it was not yet available at that time. Serum RNAase detected only 30% of pancreas cancer and in 14% of patients with benign disease. Apparently RNAase was not of any clinical value in pancreatic cancer.

Although a simple ELISA using specific monoclonal antibody to assay GT-II isoenzyme has been reported, no further clinical data to support the initial promising study on pancreatic cancer were available. It appeared that this monoclonal antibody-based assay was more useful for ovarian cancer.

It should be mentioned again at this point that as far as serologic markers and pancreatic cancer are concerned, there is no reliable assay available to date. CA 19-9, DU PAN-2 (METZGAR et al. 1984), and perhaps GT-II, pancreas cancer associated antigen and pancreas specific antigen have been shown to be of value in various degrees (LOOR et al. 1984). Biochemical marker for pancreas cancer represents the most difficult area of diagnostic oncology.

6.4 Sialyltransferase

Serum sialyltransferase activity has been reported to be associated with cancer patients for a long time. Although a significant number of cancer patients exhibited abnormal sialyltransferase activity, a variety of non-neoplastic and benign disorders also were associated with elevated levels (SCHWARTZ 1982). Its clinical value is limited, except perhaps in monitoring disease status and therapy response.

6.5 Fucosyltransferase

At least three different fucosyltransferases have been described. Elevated levels of fucosyltransferase activity have been reported in patients with solid tumors and in AML and non-Hodgkin's lymphoma (SCHWARTZ 1982). An elevated enzyme level generally is associated with proliferative and secretory processes of both normal and neoplastic cells. Therefore, its clinical value in cancer diagnosis is extremely limited.

6.6 Terminal Deoxynucleotidyl Transferase

Most of the enzyme markers so far discussed are identified with solid tumors, very few enzymes have been shown to be associated with nonsolid tumors. Among this very small group of enzymes, terminal deoxynucleotidyl transferase (TdT) is the most extensively studied and is a potential marker for some forms of leukemia (SRIVASTAVA 1982). TdT is a DNA-polymerizing enzyme by catalyzing the addition of deoxynucleotide to the 3′-hydroxyl groups of a primer polydeoxynucleotide without the requirement of a template polynucleotide. TdT was found more than twenty years ago first in calf thymus, and reported to be present in ALL cells and T-ALL cells about ten years later. Antibody specific for TdT without cross-reactivity with α, β or γ DNA polymerase is available. Biochemical assay is still widely used as an analytical tool.

Highly elevated TdT is more commonly found in patients with ALL. Some patients with CML, AML and non-Hodgkin's also exhibited an elevated level. The enzyme level in lymphocytes from normal peripheral blood generally is very low. In CML, TdT positive cells usually increase during lymphoblastic crisis.

Most clinical application of TdT is its use in prognosis for adult patients with leukemia. TdT activity is generally correlated with frequency and duration of remission, and survival in patients with ALL, ANLL and malignant non-Hodgkin's lymphoma. It is of less value for young patients. Combining with different levels of expression by adenosine deaminase, TdT also can be used for subclassification of ALL; e.g. T-ALL, TdT positive and high adenosine deaminase; non-T and non-B ALL, TdT positive and moderate adenosine deaminase; B-ALL, TdT negative and low adenosine deaminase (MERTELSMANN 1982).

6.7 Lysozyme

Lysozyme is another enzyme reported to be of some clinical value in leukemia. Lysozyme is found primarily in monocytes and neutrophils. An elevated level is commonly associated with granulocyte destruction. Elevated serum lysozyme levels are associated with acute granulocytic leukemia, acute myelomonocytic leukemia, and AML. Patients with ALL rarely have an elevated level. Therefore, an elevated lysozyme can be used to exclude the presence of ALL. Also, serial measurement of lysozyme value has been shown to be of value in following the disease course and chemotherapy (Schwartz 1982).

6.8 Other Enzyme and Isoenzyme Markers

There is an extensive list of other enzymes and isoenzymes that have been reported to be associated with cancer and are of some clinical value. Only a few more common markers are briefly mentioned. γ-Glutamyltranspeptidase (γ-GTP) is well known in its association with preneoplastic liver and hepatoma in experimental animals. γ-GTP has been shown in human cancer to be associated predominantly with secondary hepatic metastasis and can be useful in combination with CEA or other assay (Chu and Douglass 1986). Another enzyme marker reported to be useful in detecting hepatic metastasis and primary hepatocellular carcinoma is 5′-nucleotide phosphodiesterase isoenzyme V (Chu 1982). It is an exonuclease and different from 5′-nucleotidase.

Adenosine deaminase, in addition to that already mentioned with TdT, is shown to be a biochemical marker for CML, since an elevated level is found in almost 90% of patients in the accelerated phase. Thymidine kinase isoenzyme I is reported to be a valuable marker for identifying sub-group of adult non-Hodgkin's lymphoma who respond poorly to therapy and require a new therapeutic approach.

Isoenzymes of alkaline phosphatase have been extensively investigated. Several isoenzymes have been identified in cancer patients' sera: Nagao isoenzyme in ovarian cancer, liver isoenzyme in hepatic metastasis and bone isoenzyme in bone metastasis (Chu 1982; Kottel and Fishman 1982). Most clinical application is in combination with other markers. The combination of liver isoenzyme with CEA is an economical screen method for liver metastasis. The combination of bone isoenzyme with PAP is effective for monitoring advanced prostate cancer (Killian et al. 1981).

Aldolase activity was recently shown to be associated with cancer. There are two primary aldolase isoenzymes A and B. During the process of malignant transformation, the aldolase isoenzyme pattern is changed from B-dominant, liver or adult type, to A-dominant, muscle or fetal type. The aldolase B level in the serum is known to increase in patients with liver cancer and hepatic disease due to its high content in the liver and is released from the damaged hepatic tissue (Asaka and Alpert 1983). Whether it would be a better parameter than existing hepatic markers already discussed remains to be seen.

An increase in enzyme activity or protein mass so far has been discussed as a parameter for manifestation of cancer. In contrast to all these observations, the decrease in an enzyme activity, α-L-fucosidase, has been reported to be associated with some ovarian cancer patients (BARLOW et al. 1981). Tumor burden, disease stage, histologic type or grade of differentiation do not appear to be correlated with decreased level of enzyme activity. Familial study suggested that the deficiency of enzyme may be heritable and may be associated with an increased risk of developing ovarian cancer. Further studies certainly are needed to confirm this interesting report.

7 Other Biochemical Markers for Cancer

7.1 Ferritin

Ferritin is a well known protein related to iron storage and metabolism, and is present abundantly in the liver, spleen and bone marrow. Biochemically it exists in isomeric forms and is generally termed as isoferritins. The change in isoferritin molecules is associated with embryological development and pathological/malignant transformation. A carcinofetal form of ferritin has been identified (ALPERT 1982).

Serum ferritin is elevated mostly in patients with cancer related to liver and bone marrow cells. Highly elevated levels are found in patients with AML and other myeloproliferative diseases. Successful chemotherapy results in decrease of serum ferritin level to normal range, as leukemic cells and Hodgkin's lymphomas synthesize ferritin at a faster rate than that of normal lymphocytes. Patients with liver carcinoma and with acute or chronic liver inflammation all exhibit elevated serum ferritin levels. Ferritin is also elevated in patients with teratoblastoma and breast cancer. The major use of serum ferritin level is in evaluating prognosis and therapy response. It should also be noted that ferritin synthesized by tumor cells is different biochemically and immunologically from that of normal tissues.

7.2 Polyamines

The association of polyamines (spermidine, spermine and putrescine) with cancer was first reported about fifteen years ago (RUSSELL 1982). They have been unequivocally established as biochemical markers for normal and malignant cell growth. Although recent investigations have concentrated on basic work related to pharmacology and biochemistry of tumor cells, their use as biochemical tumor markers should not be overlooked. In fact, data available have shown that urinary and plasma levels of spermidine are related to tumor cell loss, and that those of putrescine are associated with both the number of tumor cells in cell cycle and tumor cell loss. These findings provide the basis for polyamines to serve as effective markers for monitoring cancer patients during therapeutic manipulation. Putrescine can also be useful in determining whether the chemotherapy affects a subsequent phase of cell proliferation.

Measurement of polyamines are commonly performed by tedious and labor-intensive high pressure liquid chromatography. That is one major reason why these markers have not been routinely used. Since polyamines are relatively small molecules in size, reliable and simple immunochemical assays are more difficult to establish than other protein tumor markers.

7.3 Immunoglobulins

Abnormal immunoglobulins are well-known tumor markers. In fact, Bence Jones protein is perhaps the first tumor marker used effectively in clinical diagnosis of cancer patients (Bence Jones 1847). The abnormal feature of immunoglobulins have been well known for patients with multiple myeloma and Waldenstrom's macroglobulinemia. Monoclonal gammapathies are most commonly observed in cancer patients involving plasma cells where a single malignant clone of precursor cell outgrow other cells and resulting in the production of a homogeneous monoclonal immunoglobulin. Biclonal gammapathies are rare but do occur. Detection of abnormal serum or urine immunoglobulins can be easily accomplished by routine electrophoresis with appropriate and specific antibody reagents that are readily available.

7.4 Peptide Hormones

Peptide hormones ectopically produced by non-endocrine tumors are well-recognized tumor markers. They are produced by tumor cells that are normally not regarded as being engaged in the synthesis and secretion of the hormones. So far only peptide hormones, their precursors or fragments have been documented. In addition, ectopic production of multiple hormones is another common feature.

The hormones commonly found in ectopic tumor production are adrenocorticotrophic hormone and related molecules, parathyroid hormone, gonadotrophin, growth hormones, prolactin, calcitonin, arginine vasopression, vasoactive intestinal peptide, and somatomedin, among others. Production of growth hormone-releasing activity or corticotrophin-releasing activity also has been observed. Oat cell carcinoma of the lung and carcinoid tumors of the liver, stomach, colon and thymus are the most common cancers producing ectopic hormones. Their use as biochemical markers in diagnosis, monitoring and localization to cancer has been reported. Apart from serving as biochemical markers which may be a secondary effect, ectopic hormone production by tumor can provide clues for investigating biologic significance and pathological implication of these tumors (Neville 1982).

7.5 Immune Complexes

Elevated circulating antigen-antibody immune complexes have been used as parameters for a wide variety of malignant diseases, from lung, breast, colon-rectum cancer, thyroid cancer, to malignant melanoma, osteosarcoma, and leukemia/lymphoma. Unfortunately, a variety of non-malignant disorders also exhibit elevated immune complexes. Furthermore, all the assay procedures used for detection and measurement of circulating immune complexes are non-specific, thus diminishing the clinical value of immune complexes (CHU et al. 1983; SALINAS and HANNA 1985).

Theoretically the formation of antigen (cancer)-antibody (host) immune complexes is a result of the patient's natural immunological defense function. Measurement of such parameters should be of great value in clinical oncology, provided that antigen-specific assay is available. To this date, no such assay or methodology is yet available. Therefore, the isolation and characterization of the antigen components of immune complexes should be the focus for cancer immunodiagnosis. This information also is critical for the elucidation of immune complexes in human pathology and the development of new concepts in pathogenesis of immune complex-associated diseases.

8 Immunohistochemical Markers

In addition to serologic assays, biochemical markers can be used in the tissue and cell stain for immunopathologic assessment of tumors. Conventional morphological examination is enhanced by immunohistochemical stain for markers expressed by tumors. Further classification, phenotyping and differential diagnosis of tumor can be achieved at either light or electron microscopic level. Immunoperoxidase stain and immunofluorescence technique are most commonly used. Techniques and other related subjects are described in detail elsewhere in this book. A few biochemical markers that have been described already in this chapter and shown to be of practical application in tumor pathology are discussed.

One of the best examples in the use of biochemical markers in in vitro cell and tissue stain is the localization of estrogen receptor protein in breast cancer and other estrogen-sensitive tumors. Estrogen receptor has been traditionally recognized to include two forms, cytosol and nuclear, and that these parameters have been used in elucidation of action mechanism of estrogen and providing the basis for hormonal therapy (WITTLIFF et al. 1982).

Using monoclonal antibodies specific for estrogen receptor protein, recent reports have indicated that estrogen receptor is exclusively localized in nucleus of the cells (KING and GREENE 1984). The presence or absence of nuclear immunostain was significantly associated with the cytosol level of estrogen receptor as determined by conventional radioligand binding assay. These results suggest that previously observed cytosol and nuclear forms of the estrogen receptor may be present only in the nuclear compartment in both steroid-binding and nonbinding forms. Data available also suggest the unoccupied estrogen receptor recovered in low salt cytosol preparation from tissue homogenate and determined by steroid binding assay represents the portion of estrogen receptor that is loosely bound to nuclear fraction (GREENE and JENSEN 1982).

The availability of monoclonal antibodies specific for estrogen receptor protein provides another tool for estrogen receptor measurement by simple immunohistochemical technique. Comparison of results obtained from immunohistochemical assay for estrogen receptor in breast tumors with those of traditional steroid-binding assay has resulted in highest degree of correlation (KING et al. 1985). A clear and significant association was shown between the presence or absence of nuclear staining for estrogen receptor and the level of cytosolic receptor measured in tumor extracts. All tumor specimens with estrogen-receptor negative immunostain were found to have insignificant levels of cytosolic estrogen receptor. Of clinical interest is that those tumors expressing none to minimum levels (0 to 636 fmol/g) of cytosolic estrogen receptor indeed contain decreased degree of immunostaining in both the cellularity and the percentage of stained tumor cells. These were the tumors from those patients generally presumed to be hormone unresponsive. Specimens with less than 300 fmol/g of cytosolic estrogen receptor show only focal nuclear staining.

The defined technique using estrogen receptor as an immunohistochemical marker also can identify sub-populations of hormone-unresponsive cells in patients, due to the heterogeneity of tumor cells. Additional consideration of therapeutic manipulation thus can be made in these patients. With further refinement, immunohistochemical procedure hopefully will provide valuable information regarding patients' prognosis and treatment responsiveness. Immunohistochemical reagent kit for estrogen receptor is now available commercially.

Detection of early metastasis or micrometastasis, and differential diagnosis of metastatic tumor are of critical importance in management of patients. For these purposes, it is not necessarily required to use tumor-specific markers. Antibody reagents to organ-site or cell-type specific antigen markers, and to tumor associated antigens have been extensively utilized for these purposes. Again, only a few examples using biochemical markers already discussed in this chapter will be illustrated.

Using conventionally prepared formalin-fixed paraffin-embedded tissue sections, PA has been shown to be an effective immunohistochemical marker for detection of secondary metastatic prostate tumors and for differential diagnosis (NADJI et al. 1981). Since PA is prostate specific, it is expressed also in metastatic tumors of prostate origin. With our reagents prepared in-house, we have been able to detect all secondary metastatic prostate tumors examined, including those in the regional lymph node and bone marrow. Similar results have been reported by other investigators. This test is most useful in identification of distant metastasis and in differential diagnosis of tumor involving prostate and adjacent tissues, especially the differentiation of poorly differentiated transitional cell carcinoma of the bladder from prostate carcinoma.

Another clinical application of histochemical markers is on cytopathologic assessment of tumor cells in serous effusions or extracellular fluid. It is well known that one of the most difficult diagnostic problems in tumor cytology is to differentiate malignant cells from mesothelial cells. A couple of monoclonal antibodies have been shown to be of great value in the use of this very area.

Monoclonal antibody B72.3, recognizing TAG72 antigen, has been shown to stain adenocarcinoma cells in effusion specimens. In one study using large numbers of specimens involving both malignant and benign effusions, greater than 90% of adenocarcinomas were identified without staining any mesothelial cells (SZPAK et al. 1984). Another murine monoclonal antibody F36/22, recognizing a human ductal carcinoma antigen, also exhibits such a quality, especially for identification of ascitic epithelial ovarian carcinomas (CROGHAN et al. 1984).

9 Oncogene Products as Biochemical Cancer Markers

Recent advancement in molecular biology has identified the expression of more than twenty genes that appear to be associated with the conversion of normal cells to neoplastic cells, so called oncogenes (GARRETT 1986). Proteins encoded by some of these oncogenes also have been identified. Monoclonal antibodies directed against specific fragment (synthetic peptide) of oncogene products are available and some have been applied in cancer detection.

Since the presence of the oncogene products, if any, in circulation will be extremely small in amount, so far all reports concerning their expression have been obtained from examination of tissue sections. The oncogene coded protein most extensively studied is *ras* oncogene protein p21, a protein product encoded by the members of the *ras* oncogene, the major neoplastic transformation-inducing genes of human solid tumors.

In one study involving its expression in prostate cancer, all normal prostate and benign prostate tissues were shown to be devoid of p21 expression (VIOLA et al. 1986). In 29 patients with prostate cancer, p21 was detected in 2 of 6 prostate with Gleason Grade I carcinoma, 4 of 6 with Grade II, and all of 17 tumors with higher grades (III to V). Semiquantitatively, the expression of p21 in prostate tumor tissues was correlated strongly with nuclear anaplasia and inversely related to the degree of glandular differentiation. These results appear to suggest that *ras* oncogene p21 is a "prostate tumor-specific" marker, and that increased expression is associated with high grade, presumably more malignant tumors. Certainly, this interesting and provocative finding requires further confirmation with additional studies.

Interesting data on the expression of *ras* oncogene p21 also has been obtained from human breast tissues (OHUCHI et al. 1986). Enhanced expression was found in invasive carcinoma, while decreasing expression was detected in carcinoma in situ, atypical hyperplasia and nonatypical hyperplasia, respectively. Carcinomas from postmenopausal patients were generally associated with higher levels of p21 expression than that from premenopausal patients. No correlation in p21 expression was found with estrogen receptor status. Additionally, normal mammary epithelium in terminal duct lobular units from patients with hyperplasia demonstrated higher level of p21 expression than did epithelium in large ducts. This finding appears to be consistent with the notion that these are the areas preferentially undergoing malignant transformation. Taking all these results together, it would suggest that an enhanced expression of *ras* p21 may be involved in the early malignant development of human breast tumor. This exciting observation again needs to be confirmed.

Without question, the reports concerning oncogene product and cancer diagnosis will be forthcoming at a greater rate in the next few years. The use of oncogene related products in tumor therapy would be another area of clinical interest.

Several oncogene probes for cancer markers including that mentioned above and those for leukemia/lymphoma and others are already available from commercial sources.

10 Biochemical Markers for Carcinogen Testing

Another potential application of biochemical tumor markers is their use in carcinogen testing. In fact, this is an area of investigation that could provide an enormous scientific return. The time-consuming and expensive animal bioassay in rodents is still the best available method and standard technique currently officially acceptable for identifying and evaluating chemical carcinogens. Although a large series of animals are generally required, ambiguous statistical results are often obtained. Therefore, a biochemical marker capable of indicating the cause of malignant transformation or correlating the progress of carcinogenesis by a carcinogen would offer a simple and objective criterion for the identification of chemical carcinogen. Qualitative or quantitative biochemical markers of carcinogenicity will enhance the validity of existing short-term bioassays for chemical carcinogens.

In addition to the determination of maximum tolerated dose, animal bioassay for carcinogens primarily depend upon evaluation of morphologic changes in affected organs following carcinogen exposure. Should a biochemical marker be identified or available to indicate morphologic changes associated with carcinogen exposure, this marker would be of great value in carcinogen testing. Several biochemical markers are effective in indicating such changes.

Serum AFP has been shown to be a very reliable marker for determining exposure to liver carcinogens in animals, especially in rats, regardless of the nature of neoplastic lesions (SELL et al. 1983). Early prolonged elevations of serum AFP or sudden drastic elevations of AFP can be used as conclusive evidence of activity by strong hepatocarcinogens. Some carcinogens may not produce such patterns of serum AFP elevation even after potentiation with choline-deficient diet and may result in a possibly false negative result. In mice that are exposed to chemical hepatocarcinogens or develop spontaneous primary hepatocellular carcinoma, an elevation of serum AFP may not be observed in early stage. If an AFP elevation is produced in animals upon exposure to carcinogens, immunostain also will provide another probe to delineate the cellular mechanisms involved.

Gamma-glutamyltranspeptidase (γ-GTP) is another well known biochemical marker for preneoplastic hepatocyte foci and of hepatocellular carcinoma in animals (RICHARDS 1983). Although its function and biologic significance in normal cells and in malignant transformation is still unclear, γ-GTP has been used increasingly as an enzyme marker for chemical hepatocarcinogens. Simple histochemical stain allows easy identification of putative preneoplastic hepatocytes. It should be noted, though, that γ-GTP is not a tumor marker per se, it is a marker of cell differentiation, cell aging and reduced cell proliferation rather than of elevated cell proliferation. Furthermore, some rat and human hepatocellular carcinomas and most mouse hepatomas are devoid of γ-GTP expression.

Polyamines are well known for their important role in cell growth, and have been discussed as potential biochemical tumor markers. Ornithine decarboxylase is a critical enzyme involved in metabolism of polyamines, especially in catalyzing the formation of putrescine (diamine) from ornithine, serving the precursor for spermidine (tertiary amine) and spermine (quarternary amine). Induction of ornithine decarboxylase during tumor promotion and in transformed cells as a result of chemical carcinogenesis or other mechanisms has been suggested to be a potential universal marker of neoplastic expression (HADDOX and GREENFIELD 1983). This notion represents a basis for its use in carcinogen testing but requires further investigation.

Biochemical markers, such as aryl hydrocarbon hydroxylase, epoxide hydrolase, DT-diaphorase, and others, have been shown to be of some value as enzyme markers for chemical carcinogenesis.

The use of DNA-carcinogen adducts for possible monitoring exposure to chemical carcinogen is a relatively new and extremely important area of investigation. Considerable characterization of DNA component structurally modified by chemical carcinogen has been reported. Several biochemical and immunologic methods are available which detect and quantify human DNA modified by carcinogens, even to the level of accuracy in defined restriction fragments of DNA molecules modified to less than one carcinogen per DNA molecule (BUSTIN et al. 1983). This new approach represents a great potential research and scientific reward for biochemical markers for cancer.

11 Summary

At this stage of development, there is no tumor specific biochemical marker. Although sensitive and specific procedures are available for measurements of cancer-associated biochemical markers, none has been shown to be of value as a general screen for cancer. Biochemical markers do have potential as screening tools for high-risk populations and symptomatic patients, e.g. α-fetoprotein for primary hepatocellular carcinoma, thyrocalcitonin for medullary thyroid carcinoma, and perhaps fibrinogen degradation products for carcinoma of the urinary bladder. Also, biochemical markers, such as estrogen receptor protein, are clinically useful for staging of breast cancer and other estrogen-sensitive tumors. Most of the biochemical markers available to date are of clinical utility as a means of monitoring the response to treatment and/or of detecting disease recurrence at an early stage. At present, there are only two markers, carcinoembryonic antigen and prostate specific antigen, approved by the Food and Drug Administration. Conventional radioimmune assay and biochemical assay, including enzyme-linked immunosorbent assay, are still commonly applied in detection of biochemical markers. Other new approaches or technologies, such as immunohistochemistry, flow cytometry and molecular biology, are being increasingly utilized. Another potential application of biochemical tumor markers is their use in carcinogen testing. Biochemical marker can serve as a simple and objective criterion correlating morphological changes in malignant transformation. Finally, recent advancement in molecular biology has identified oncogene encoded protein as a potential biochemical marker for human cancer.

References

Alpert E (1982) Isoferritins: alteration in human malignancy and clinical significance. In: Chu TM (ed) Biochemical markers for cancer. Marcel Dekker, New York, pp 229–239

Asaka M, Alpert E (1983) Subunit-specific radioimmunoassay for aldolase A, B and C subunits. Ann NY Acad Sci 417:359–367

Ban Y, Wang MC, Watt KW, Loor RM, Chu TM (1984) The proteolytic activity of human prostate-specific antigen. Biochem Biophys Res Commun 123:482–488

Barlow JJ, DiCoccio RA, Dillard PH, Blumenson LE, Matta KL (1981) Frequency of an allele for low activity of fucosidase in sera. J Natl Cancer Inst 67:1005–1009

Bast RC, Feeney M, Lazarus H, Nadler LM, Colvin RB, Knapp RC (1981) Reactivity of a monoclonal antibody with human ovarian cancer. J Clin Invest 68:1331–1337

Bast RC, Klug TL, St. John E, Jenison E, Niloff JM, Lazarus H, Berkolwitz RS, Leavitt T, Griffiths T, Parker L, Zurawski VR, Knapp RC (1983) Radioimmunoassay using a monoclonal antibody to monitor the course of epithelial ovarian cancer. N Engl J Med 309:883–887

Beard DB, Haskell CM (1986) Carcinoembryonic antigen in breast cancer. Clinical review. Am J Med 80:241–245

Bence Jones H (1847) Papers on chemical pathology, Lecture III. Lancet 2:88–92

Bustin M, Kurth PD, Slor H, Seidman M (1983) Immunological studies in the influence of chromatin structure on the binding of a chemical carcinogen to the genome. In: Millman HA, Sell S (eds) Application of biological markers to carcinogen testing. Plenum Press, New York, pp 349–372

Carney DN, Ihde DC, Cohen MH, Marangos J, Bunn PA, Minna JD, Gazdar AF (1982) Serum neuron-specific enolase. Lancet I:583–585

Chu TM (1982) The detection of liver metastasis by laboratory tests. In: Weiss L, Gilbert HA (eds) Liver metastasis. G.K. Hall Medical Publishers, Boston, pp 255–265

Chu TM, Douglass HO (1986) Laboratory tests in diagnosis of liver disease. In: Bengmark S (ed) Liver surgery. Churchill Livingstone, London, pp 8–18

Chu TM, Wang MC, Lee CL, Killian CS, Murphy GP (1982) Prostatic acid phosphatase in human prostate cancer. In: Chu TM (ed) Biochemical markers for cancer. Marcel Dekker, New York, pp 117–136

Chu TM, Maidment BW, Koestler TP, Papsidero LD, Inaji H, Croghan G, Killian CS, Loor RM, Douglass HO, Berjian R, Nemoto T (1983) Immune complexes and cancer. Immunodiagnostics 8:259–268

Cooper CW (1976) Recent advances with thyrocalcitonin. Ann Clin Lab Sci 6:119–129

Cooper EH, Pidcock NB, Daponte D, Glashan RW, Rowe E, Robinson MRG (1983) Monitoring of prostatic cancer by an enzyme-linked immunosorbent assay of prostatic acid phosphatase. Eur Urol 9:17–23

Croghan GA, Papsidero LD, Valenzulea LA, Nemoto T, Penetrante R, Chu TM (1983) Tissue distribution of an epithelial and tumor-associated determinant recognized by monoclonal antibody F36/22. Cancer Res 43:4980–4998

Croghan GA, Wingate MB, Gamarra M, Johnson E, Chu TM, Allen H, Valenzuela L, Tsukada Y, Papsidero LD (1984) Reactivity of monoclonal antibody F36/22 with human ovarian adenocarcinomas. Cancer Res 44:1954–1962

Davey RA, Milliken ST, Harvie RM, Cahill EJ, Morgan LJ, Kerestes Z, Levi HA (1986) Serum galactosyltransferase as a prognostic marker in patients with solid tumors. Cancer Res 46:5973–5975

Davies SN, Giffiths JC (1982) Prostate specific acid phosphatase. Further studies with immunologic technique. Clin Chim Acta 122:29–38

Del Villano BC, Brennen S, Brook P, Bucher C, Lice V, McClure M, Rake B, Space S, Westrick B, Schoemaker H, Zurawski VR (1983) Radioimmunometric assay for a monoclonal antibody-defined tumor marker. Clin Chem 29:549–552

Fletcher RH (1986) Carcinoembryonic antigen. Ann Int Med 104:66–73

Garrett CT (1986) Oncogenes. Clin Chim Acta 156:1–40

Gold P, Freedman SO (1965) Demonstration of tumor-specific antigens in human colonic carcinomata by immunological tolerance and absorption techniques. J Exp Med 121:439–459

Greene GL, Fitch FW, Jensen EV (1980) Monoclonal antibodies to estrophilin. Proc Natl Acad Sci USA 77:157–161

Greene G, Jensen EV (1982) Monoclonal antibodies as probes for estrogen receptor detection and characterization. J Steroid Biochem 16:353–359

Gupta MK, Arciaga R, Bocci Tubbs R, Bukowski R, Deodhar SD (1985) Measurement of a monoclonal antibody-defined antigen (CA 19-9) in the sera of patients with malignant and nonmalignant disease. Cancer 56:277–283

Gutman AB, Gutman ED (1938) An "acid" phosphatase occurring in the serum of patients with metastazing carcinoma of the prostate gland. J Clin Invest 17:473–478

Haddox MK, Greenfield RL (1983) Ornithine decarboxylase: alterations in carcinogenesis. In: Millman HA, Sell S (eds) Application of biological markers to carcinogen testing. Plenum Press, New York, pp 221–228

Hedin A, Carlsson L, Berglund A, Hammarstrom S (1983) A monoclonal antibody-enzyme immunoassay for serum carcinoembryonic antigen with increased specificity for carcinomas. Proc Natl Acad Sci USA 80:3470–3474

Herberman RB (1982) Immunological approaches to the biochemical markers for cancer. In: Chu TM (ed) Biochemical markers for cancer. Marcel Dekker, New York, pp 1–23

Heyward WL, Lanier AP, McMahon BJ, Fitzgerald MA, Kilkenny S, Paprocki TR (1985) Early detection of primary hepatocellular carcinoma. J Am Med Assoc 254:3052–3054

Hine KR, Dykes PW (1984) Prospective randomized trial of early cytotoxic therapy for recurrent colorectal carcinoma detected by serum CEA. Gut 25:682–688

Hirai H (1982) Alpha-fetoprotein. In: Chu TM (ed) Biochemical markers for cancer. Marcel Dekker, New York, pp 25–60

Holt JA, Bolanos J (1986) Enzyme-linked immunochemical measurement of estrogen receptor in gynecologic tumors, and an overview of steroid receptors in ovarian carcinomas. Clin Chem 32:1836–1843

Jordan VC, Jacobson HI, Keenan EJ (1986) Determination of estrogen receptor in breast cancer using monoclonal antibody technology: results of a multicenter study in the United States. Cancer Res 46:4237s–4240s

Killian CS, Vargas FP, Pontes JE, Beckley S, Slack NH, Murphy GP, Chu TM (1981) The use of serum isoenzymes of alkaline and acid phosphatases as possible quantitative markers of tumor load in prostate cancer. Prostate 2:187–206

Killian CS, Yang N, Kuriyama M, Vargas FP, Emrich LJ, Slack NH, Wang MC, Papsidero LD, Murphy GP, Chu TM (1985) Prognostic importance of prostate-specific antigen for monitoring patients with localized prostate cancer. Cancer Res 45:886–891

Killian CS, Emrich LJ, Vargas FP, Yang N, Wang MC, Priore RL, Murphy GP, Chu TM (1986) Relative reliability of five serially measured markers for prognosis of progression in prostate cancer. J Natl Cancer Inst 76:179–185

King WJ, Greene GL (1984) Monoclonal antibodies localize estrogen receptors in the nuclei of target cells. Nature 307:745–747

King WJ, DeSombre ER, Jensen EV, Greene GL (1985) Comparison of immunocytochemical and steroid-binding assays for estrogen receptor in human breast tumors. Cancer Res 45:293–304

Kloppel G, Girard J, Polak JM, Vaitukaitis JL, Kasper M, Heitz PU (1983) Alpha-human chorionic gonadotrophin and neuron-specific enolase as markers for malignancy and neuroendocrine nature of pancreatic endocrine tumors. Cancer Detect Prev 6:161–166

Kottel RH, Fishman WH (1982) Developmental alkaline phosphatases as biochemical tumor markers. In: Chu TM (ed) Biochemical markers for cancer. Marcel Dekker, New York, pp 93–115

Kuriyama M, Wang MC, Lee CL, Killian CS, Papsidero LD, Inaji H, Loor RM, Lin MF, Nishiura T, Slack NH, Murphy GP, Chu TM (1982) Multiple marker evaluation in prostate cancer using tissue specific antigens. J Natl Cancer Inst 68:99–105

Leclerq G, Bojar H, Goussard J, Nicholson RI, Pichon MF, Piffanelli A, Pousette A, Thorpe S, Lonsdorfer M (1986) Abbott monoclonal enzyme immunoassay measurement of estrogen receptors in human breast cancer: a European multicenter study. Cancer Res 46:4233s–4236s

Lehmann FG, Wegener T (1979) Etiology of human liver cancer. J Toxicol Environ Health 5:281–299

Li HC, Chernoff J, Chen LB, Kirschonbaum A (1984) A phosphotyrosyl-protein phosphatase activity associated with acid phosphatase from human prostate. Eur J Biochem 138:45–51

Lin MF, Lee CL, Li SL, Chu TM (1983a) Purification and characterization of a new human prostatic acid phosphatase isoenzyme. Biochem 22:1055–1062

Lin MF, Lee CL, Sahrief FS, Li SL, Chu TM (1983b) Glycoprotein exhibiting immunological and enzymatic activities of human prostatic acid phosphatase. Cancer Res 43:3841–3846

Lisowska E, Krop WA, Sedlaczek P (1983) The dimeric structure of carcinoembryonic antigen. Biochem Biophys Res Commun 115:206–211

Loor RM, Kuriyama M, Manzo Bodziak ML, Inaji H, Douglass HO, Berjian R, Nicolai JJ, Tytgat GN, Chu TM (1984) Simultaneous evaluation of a pancreas-specific antigen and a pancreas cancer-associated antigen in pancreas carcinoma. Cancer Res 44:3604–3607

McCabe RP, Lamm DL, Haspel MV, Pomato N, Smith KO, Thompson E, Hanna MG (1984) A diagnostic-prognostic test for bladder cancer using monoclonal antibody-based enzyme linked immunoassay for detection of urinary fibrinogen degradation products. Cancer Res 44:5886–5893

McCarty KS Jr, Szabo E, Flowers JL, Cox EB, Leight GS, Miller L, Konrath J, Soper JT, Budwit DA, Creasman WT, Seigler HF, McCarty KS Sr (1986) Use of monoclonal anti-estrogen receptor antibody in the immunohistochemical evaluation of human tumors. Cancer Res 46:4244s–4248s

Mertelsmann R (1982) The prognostic significance of terminal deoxynucleotidyl transferase in patients with leukemia and malignant lymphomas. Adv Exp Med Biol 145:259–277

Metzgar RS, Rodriguez M, Finn OJ, Lan MS, Dausch VN, Fernsten PD, Meyers WC, Sindelar WF, Sandler RS, Seigler HF (1984) Detection of a pancreatic cancer-associated antigen (DUPAN-2 antigen) in serum and ascites of patients with adenocarcinoma. Proc Natl Acad Sci USA 81:5242–5246

Minton JP, Hoehn JL, Gerber DM, Horsley JS, Connolly DP, Salwan F, Fletcher WS, Cruz AB, Gatchell FG, Oviedo M, Meyer KK, Leffall LD, Berk RS, Stewart PA, Kurucz SE (1985) Results of a 400-patient carcinoembryonic antigen second-look colorectal cancer study. Cancer 55:1284–1290

Morinaga T, Sakai M, Wegmann TG, Tamaoki T (1983) Primary structure of human α-fetoprotein and its mRNA. Proc Natl Acad Sci USA 80:4604–4608

Muraro R, Wunderlich D, Thor A, Lundy J, Noguchi P, Cunnigham R, Schlom J (1985) Definition of monoclonal antibodies of a repetoire of epitopes on carcinoembryonic antigen differentially expressed by human colon carcinomas versus normal adult tissues. Cancer Res 45:5769–5780

Nadji M, Tabei SZ, Castro A, Chu TM, Wang MC, Murphy GP, Morales AR (1981) Prostatic specific antigen: An immunohistologic marker for prostatic neoplasms. Cancer 48:1229–1232

Neville AM (1982) Ectopic hormone production by tumors. In: Chu TM (ed) Biochemical markers for cancer. Marcel Dekker, New York, pp 267–300

Norgaard-Pedersen B (1979) Clinical use of AFP and HCG in testicular tumors of germ-cell origin. Lancet II:1042

Ohuchi N, Thor A, Page D, Horan Hand P, Halter SA, Schlom J (1986) Expression of the 21,000 molecular weight ras protein in a spectrum of benign and malignant human mammary tumor tissues. Cancer Res 46:2511–2519

Podolsky DK, McPhee MS, Alpert E, Warshaw AL, Isselbacher KJ (1981) Galactosyltransferase isoenzyme II in the detection of pancreatic cancer. N Engl J Med 304:1313–1318

Raam S, Vrabel DM (1986) Evaluation of an enzyme immunoassay kit for estrogen receptor measurements. Clin Chem 32:1496–1502

Richards WL (1983) Gamma-glutamyl transpeptidase expression in neoplasia and development. In: Millan HA, Sell S (eds) Application of biological markers to carcinogen testing. Plenum Press, New York, pp 199–220

Roger GT (1983) Carcinoembryonic antigen and related glycoproteins. Biochim Biophys Acta 695:227–249

Russell DH (1982) Polyamines as biochemical markers for tumor growth parameters. In: Chu TM (ed) Biochemical markers for cancer. Marcel Dekker, New York, pp 241–245
Sakai M, Morinaga T, Uranos Y, Watanabe K, Wegmann TG, Tamaoki T (1985) The human α-fetoprotein gene. J Biol Chem 260:5055–5060
Salinas FA, Hanna MG (eds) (1985) Immune complexes and human cancer. Plenum Pub Corp, New York
Schwartz MK (1982) Enzymes as tumor markers. In: Chu TM (ed) Biochemical markers for cancer. Marcel Dekker, New York, pp 81–92
Sell S (1981) Diagnostic application of alpha-fetoprotein. Hum Pathol 12:959–963
Sell S, Becker F, Leffert H, Osborn K, Salman J, Lombardi B, Shinozuka H, Reddy J, Ruoshahti E, Sala-Trepat J (1983) Alphafetoprotein as a marker for early event and carcinoma development during chemical hepatocarcinogenesis. In: Millman H, Sell S (eds) Application of biological markers to carcinogen testing. Plenum Press, New York, pp 271–294
Silva OL, Snider RH, Moore CF, Becker KL (1979) Urine calcitonin as a test for medullary thyroid cancer. Ann Surg 189:269–274
Srivastava BIS (1982) Biochemical markers for the differential diagnosis of leukemia. In: Chu TM (ed) Biochemical markers for cancer. Marcel Dekker, New York, pp 137–181
Steele G, Zamcheck N (1985) The use of carcinoembryonic antigen in the clinical management of patients with colorectal cancer. Cancer Detect Prev 8:421–427
Steinberg WM, Gelfand R, Anderson KK, Glenn J, Kurtzman SH, Sindelar WF, Toskes PP (1986) Comparisons of the sensitivity and specificity of the CA 19-9 and carcinoembryonic antigen assays in detecting cancer of the pancreas. Gastroenterol 90:343–349
Szpak CA, Johnston WW, Lottich SC, Kufe D, Thor A, Schlom J (1984) Patterns of reactivity of four novel monoclonal antibodies with cells in human malignant and benign effusions. Acta Cytol 28:356–367
Vincent RG, Chu TM, Lane WW (1979) The value of carcinoembryonic antigen with carcinoma of the lung. Cancer 44:685–691
Viola MV, Fromowitz F, Oravez S, Deb S, Finkel G, Lundy J, Hand P, Thor A, Schlom J (1986) Expression of ras oncogene p21 in prostatic cancer. N Engl J Med 314:133–137
Wagener C (1984) Diagnostic sensitivity, diagnostic specificity and predictive value of the determination of tumor markers. J Clin Chem Clin Biochem 22:969–979
Wang MC, Valenzuela LA, Murphy GP, Chu TM (1979) Purification of a human prostate specific antigen. Invest Urol 17:159–163
Wang MC, Kuriyama M, Papsidero LD, Loor RM, Valenzuela LA, Murphy GP, Chu TM (1982) Prostate antigen of human cancer patients. In: Busch H, Yeoman LC (eds) Methods in cancer research XIX. Academic Press, New York, pp 179–197
Wajsman Z, Merrin C, Chu TM, Moore RH, Murphy GP (1975) Evaluation of biological markers in bladder cancer. J Urol 114:879–883
Watt KWK, Lee PJ, M'Timkulu T, Chan WP, Loor RM (1986) Human prostate specific antigen: structure and functional similarity with serine proteases. Proc Natl Acad Sci USA 83:3166–3170
Wittliff JL, Mehta RG, Lewko WM, Park DC (1982) Steroid-receptor interactions in normal and neoplastic mammary tissues. In: Chu TM (ed) Biochemical markers for cancer. Marcel Dekker, New York, pp 183–227
Zweig NH, Ihde DC (1985) Assessment of serum radioimmune and enzymatic prostatic acid phosphatase and radioimmune creatine kinase BB for monitoring response to therapy in metastatic prostatic carcinoma. Cancer Res 45:3945–3950

Immunohistochemical Methods for the Demonstration of Tumor Markers

H. DENK

1 Introduction

Immunhistochemistry plays a major role in biological and biomedical research as well as in diagnostic histopathology. In pathology, it has not only contributed significantly to our understanding of basic pathogenetic principles but also to the precision of diagnosis (see TAYLOR and KLEDZIK 1981; DE LELLIS et al. 1979; for further information). An accurate classification of neoplasms is an important prerequisite of the newly developed specific and highly efficient therapeutic approaches in oncology. A major problem in histopathology is the identification and classification of tumors on the basis of light microscopic appearance in routinely stained and processed sections. This task is relatively easily accomplished in well differentiated neoplasms with (more or less) close resemblance to the respective non-neoplastic tissue, but may be difficult or even impossible in undifferentiated tumors and in small tissue fragments. In these situations, the evaluation of tumor cell constituents or products (tumor markers) is a valuable supplement for refinement of diagnosis. The family of tumor markers includes various immunoglobulin classes in tumors or atypical proliferative conditions of lymphoreticular origin, hormones, particularly in endocrine tumors, tumor cell constituents, such as membrane and cytoskeletal components, and oncofetal substances (for further information see HEYDERMAN 1983 and this volume).

As a promising tool, immunohistochemistry started with the first description of antibody labeling with a fluorescent dye by COONS et al. (1941), and has been considerably improved with the introduction of enzyme labels and the unlabeled antibody methods (STERNBERGER 1979; VAN NOORDEN and POLAK 1983; DENK and PLATZER 1982; TAYLOR 1980; BOSMAN 1983). Its superior sensitivity and specificity now allows its application to fixed and paraffin-embedded material, the material of choice referred to the surgical pathologist. Numerous modifications of the original procedures have been described with the aim increasing sensitivity, specificity and applicability further. A significant improvement, particularly with regard to specificity, has been achieved by the introduction of the monoclonal antibody technology (for review and further information see DIAMOND and SCHARFF 1982). Moreover, immunochemical analyses of antibodies and antigens (for example, by immunoblotting procedures) allows a more precise definition of antigenic determinants recognized by the antibodies. However, the still expanding range of application of immunohistochemical techniques raises the risk of pitfalls and misinterpretations of results and makes proper evaluation of the methodology, particularly with respect to its specificity, imperative. Moreover in diagnostic pathology, immunohistochemistry should be guided by conventional histopathological evaluation and cannot replace the experience and skill of the pathologist. Diagnostic errors resulting from uncritical use of these costly and time-consuming procedures must be avoided.

The present review will focus on the various methods applied today in diagnostic pathology in general and in oncology, for tumor marker detection in particular.

2 Principles of Immunohistochemistry

Immunohistochemistry allows in situ-identification of antigenic cell or tissue components, including secretion products, on the basis of specific antigen-antibody interactions which are visualized by suitable labels at the light and electron microscopic level. Besides pure immunological methods, depending solely on the specific interaction of antibodies (or antigen) with the tissue-bound antigen (or antibodies), non-immunological fixation of substances in combination with immunological methods (mixed methods), or non-immunological binding alone can be successfully used for visualization of diverse cell and tissue components (for recent review see, for example, VAN NOORDEN and POLAK 1983; TAYLOR

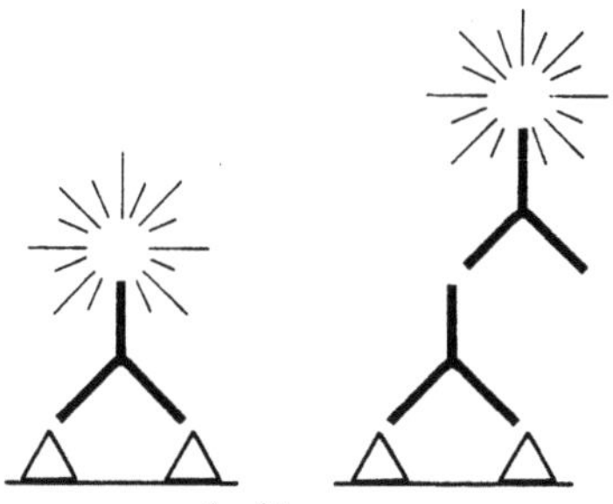

Fig. 1. In direct immunofluorescence (*left*) the fluorochrome-labeled antibody reacts with the tissue bound antigen. In indirect immunofluorescence (*right*) the labeled secondary antibody marks the position of the first antibody bound to the antigen

1983; BOSMAN 1983). Labels used for visualization of binding are either fluorescent dyes (immunofluorescence), enzymes (immunoenzymatic methods), electron dense substances and particles for light and electron microscopic immunocytochemistry (e.g. colloidal gold, ferritin), and other particles (e.g. erythrocytes).

2.1 Immunofluorescence Microscopy

Immunofluorescence microscopy (Fig. 1), first introduced by COONS et al. (1941), is based on labeling of the antibodies with fluorescent dyes (fluorochromes). Several fluorochromes are suitable for antibody labeling, including fluorescein isothiocyanate (FITC), tetramethylrhodamine isothiocyanate (TRITC), lissamine rhodamine B 200 sulphonyl chloride (RB 200SC) and Texas Red^R sulfonyl chloride. An optimal fluorochrome-protein ratio is required for low signal-to-noise ratio, that is the intensity of specific staining in relation to nonspecific background staining. Since conjugation of antibody with label involves a chemical reaction it may to some degree affect antibody structure and immunoreactivity adversely. The disadvantages of immunofluorescence microscopy are the impermanence of staining (although this disadvantage can now largely be overcome by embedding the stained frozen section in special media such as Moviol^R), the need for special and expensive equipment (fluorescence microscope), the need, in many instances, for frozen or specially treated tissues, and difficulties in the analysis of structural details in frozen material subjected to immunofluorescence staining.

2.2 Enzymes as Labels

Attempts to overcome disadvantages inherent in immunofluorescence techniques resulted in the widespread use of enzymes as markers (Fig. 2) (STERNBERGER 1979; NAKANE and PIERCE 1966; VAN NOORDEN and POLAK 1983). Horseradish peroxidase and alkaline phosphatase, glucose oxidase, and beta-galactosidase are used as marker enzymes which, for visualization, are developed histochemically with various substrates (chromogens) yielding differently coloured end-products (CORDELL et al. 1984; MASON et al. 1983; VAN NOORDEN

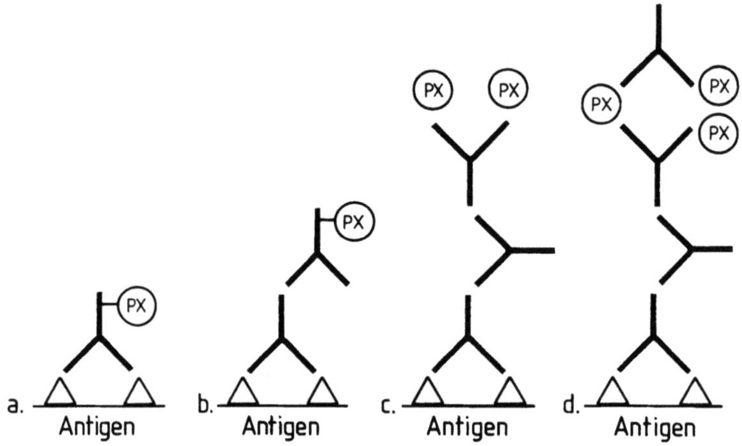

Fig. 2. The coupled enzyme (peroxidase) methods (**a**, **b**) resemble the immunofluorescence methods (direct and indirect immunoperoxidase method). In the unlabeled methods the enzyme marker (peroxidase) is fixed to the site of the antigen by an antibody sequence (the final antibody directed to the enzyme). In the antibody bridge method a peroxidase antibody is bound to a primary antibody of the same species by a bridging antibody (**c**). The enzyme is added as the fourth layer. In the peroxidase-anti-peroxidase (PAP) method a peroxidase-anti-peroxidase PAP complex is bound to the primary antibody by a bridging antibody (**d**)

and POLAK 1983; for review of recent literature). Glucose oxidase has the advantage that no endogenous enzyme activity exists in mammalian tissues (see CLARK et al. 1982; GOWN et al. 1986 for further references), in contrast to alkaline phosphatase and peroxidase, which has to be blocked in order to avoid false positive results (peroxidase by methanol-H_2O_2 or periodate oxidation; alkaline phosphatase by levamisole, acetic acid, or periodate oxidation). These procedures occasionally result in impairment of antigenicity (for review and references see MASON et al. 1983). A sensitive method involves the combination of glucose oxidase and horseradish peroxidase (KUHLMANN and PESCHKE 1986) which are coimmobilized onto the same cellular sites and act as components of a reaction sequence. The H_2O_2 generated during the glucose oxidase reaction is the substrate of horseradish peroxidase and is utilized for the oxidation of chromogens. This method is superior to conventional immunoperoxidase labeling for the localization of antigens in tissues rich in endogenous peroxidase activity.

The chromogen used most frequently in the peroxidase methods is 3,3′-diaminobenzidine (DAB) in H_2O_2, yielding a brown color which can be further intensified by treatment with osmium tetroxide or $NiCl_2$, $CoCl_2$ or $CuSo_4$ (ADAMS 1981; HSU and SOBAN 1982; TROJANOWSKI et al. 1983; DE JONG et al. 1985; see also section 6 on sensitivity). Alternative substrates are 4-chloro-1-naphthol (blue color, but not alcohol resistant), 3-amino-9-ethylcarbazol (red color), and p-phenylenediamine-HCl/pyrocatechol (black color) (see MASON et al. 1983; VAN NOORDEN and POLAK 1983). The 1-naphthol-basic dye technique can be used as alternative to DAB in immunoperoxidase techniques (MAURO et al. 1985). The oxidation product of 1-naphthol is capable of binding basic dyes resulting in precipitates with characteristic color, e.g. blue-green with azur A, toluidine blue, cresyl violet, and Giemsa stain, dark green with methyl-green, blue-violet with crystal violet, red-violet with safranin, etc., which are alcohol-resistant. In alkaline phosphatase immunohistochemistry the enzyme label is revealed with naphthol AS-MX (or naphthol AS-BI) plus Fast Red (red color), hexazotized New Fuchsin (red color) or Fast blue (blue color) (CORDELL et al. 1984; DE JONG et al. 1985). Glucose oxidase activity is detected with a substrate medium containing B-D glucose, nitro blue tetrazolium and phenozine methosulfate (KUHLMANN and AVRAMEAS 1971; GOWN et al. 1986).

Horseradish peroxidase was originally selected as marker enzyme because it could easily be coupled to antibodies and yielded a stable color reaction. However, chemical coupling procedures have several short-comings, particularly by affecting adversely antibody as well as enzyme reactivities (STERNBERGER 1979; VAN NOORDEN and POLAK 1983). Therefore, unconjugated enzyme-antibody methods, such as the peroxidase-antiperoxidase-method (STERNBERGER 1979), have been developed in which the marker enzyme is fixed to the site of the antigen by several steps of antibody binding instead of chemical coupling. The superior sensitivity of these methods allows the detection of antigens, including tumor markers, in routinely fixed and processed material making retrospective studies on a large scale possible. In addition, the excellent structural preservation of fixed material and counterstaining allow a combination of immunohistochemistry with traditional light microscopy and thus facilitate interpretation (TAYLOR 1980; BOSMAN 1983; VAN NOORDEN and POLAK 1983).

2.3 Colloidal Gold and Other Particles as Labels

The stable and rapid binding of colloidal gold particles to proteins, including antibodies, enzymes, protein A, and lectins, without affecting biological activity is exploited in light as well as in electron microscopic immunocytochemistry (including scanning electron microscopy) (Fig. 3). With colloidal gold – coupled probes neither substrates (as in enzyme methods) nor expensive equipment (as in immunofluorescence microscopy) are required for visualization (DE MEY 1983b). Immunogold procedures can be combined with other staining methods, e.g. enzyme cytochemistry, for correlation of morphological with functional criteria (DE WAELE et al. 1983). For visualization of the gold label under the light microscope (red staining) a relatively high local concentration of immunogold reagent is necessary. Sensitivity is considerably increased and, hence, less reagent is required when immunogold staining is combined with silver precipitation (HOLGATE et al. 1983). Metallic gold catalyzes the reduction of silver ions (derived, for example, from silver lactate) to metallic silver in the presence of a reducing agent (e.g. hydroquinone) resulting in an enlarging shell of silver grains around the gold particle which is then easily visible at the light microscopic level (black staining). Several modifications of this method have been designed (SPRINGALL et al. 1984; FRITZ et al. 1986; DANSCHER and NÖRGAARD 1983 for resin-embedded tissue). Advantages are the high sensitivity (see below), the lack of diffusion of the reaction product, low background staining, and easy combination with routine staining procedures, e.g. haematoxilin-eosin staining.

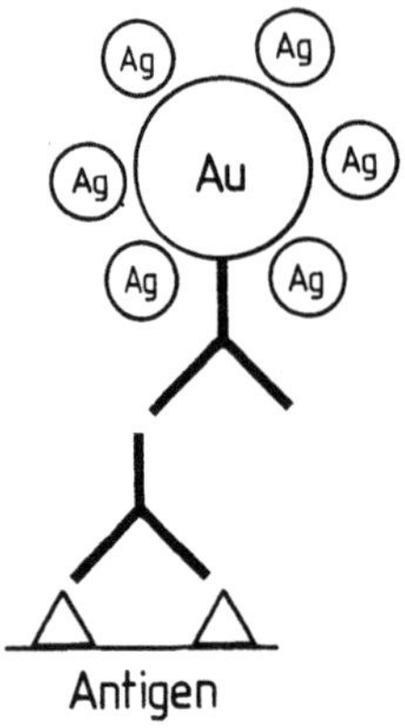

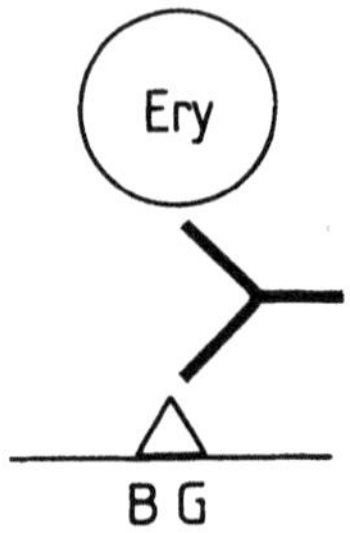

Fig. 4. In the erythrocyte adherence reaction erythrocytes (*Ery*) containing blood group antigens are fixed by a bridging antibody to blood group antigens (*BG*) present in the tissue. The site of the antigen is indicated by erythrocyte adherence

◁ **Fig. 3.** The secondary antibody labeled with colloidal gold binds to the unlabeled primary antibody which reacts with the tissue-bound antigen. The color reaction is intensified by silver precipitation

Particles containing a specific antigenic structure, either artificially coupled to its surface or present in their membranes, e.g. blood group antigens in erythrocytes, can be used as probes by fixation to identical antigens present in tissues or cells by a bridging antibody. An example is the erythrocyte adherence reaction for the demonstration of tissue-bound blood group antigens (Fig. 4; see DENK and PLATZER 1982 for review), originally described by DAVIDSON 1972.

3 Immunohistochemical Methods: Types and Mechanisms

3.1 Direct Labeled Antibody Method

The labeled antibody identifies the tissue-bound antigen by specifically binding to the antigenic sites (Fig. 1, Fig. 2). This method is less sensitive and requires a vast library of labeled antibodies depending on the antigens to be studied.

3.2 Indirect Labeled Antibody Method

An unlabeled (primary) antibody bound to its specific antigen in the tissue is revealed by a labeled secondary antibody raised in another animal species against the primary (Fig. 1, Fig. 2). This method exceeds the direct method in sensitivity. Since primary antibodies are produced in a limited number of animal species, preferentially in rabbits, sheep, goats, pigs, guinea pigs, only relatively few different labeled antibodies are required to detect a diversity of antigens.

3.3 Unlabeled Antibody Enzyme Bridge Method

A bridging antibody fixes an anti-enzyme antibody (derived from the same species as the first antibody) to the unlabeled first antibody directed against

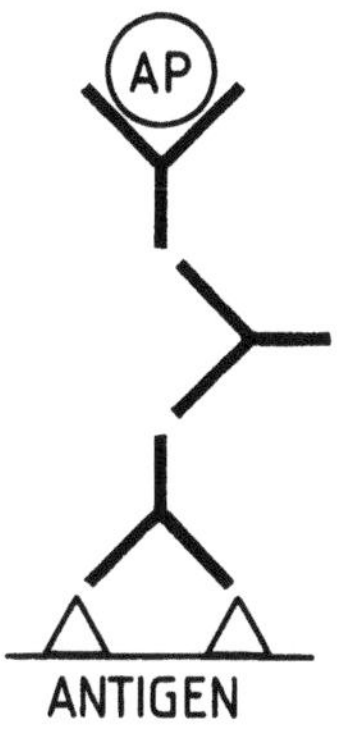

Fig. 5. An alkaline-anti-alkaline phosphatase complex (APAAP) is fixed to the primary antibody by a bridging antibody. When the primary antibody and the alkaline phosphatase antibody are monoclonal (mouse) an anti-mouse IgG is used as bridging antibody

the antigen (STERNBERGER et al. 1970; STERNBERGER 1979; VAN NOORDEN and POLAK 1983). The enzyme is added and immunologically bound in a final step and visualized histochemically (Fig. 2c). Since chemical coupling of antibody and enzyme is avoided maximum reactivity of both is guaranteed. In the three step enzyme-anti-enzyme complex method a preformed antibody-enzyme complex (e.g. peroxidase-antiperoxidase complex, PAP; alkaline phosphatase-antialkaline phosphatase complex, APAAP; CORDELL et al. 1984) is applied (Fig. 2d, Fig. 5). The concentration of enzyme molecules at the antigenic site, and hence sensitivity, can be increased by repeating the sequences (multistep methods). It has been shown that a linear relationship exists between the number of incubation cycles and the amount of enzyme, e.g. peroxidase, bound. Consequently, a high enzyme (e.g. peroxidase)-to-antigen ratio is achieved and the primary antibody can be used highly diluted giving the advantage of low non-specific background staining (LANSDORP et al. 1984). Several variants of the unlabeled antibody-enzyme method have been described, such as the double PAP technique (ORDRONNEAU et al. 1981). Limitations may ensue from dissociation of anti-immunoglobulin from immunoglobulin, dissociation of anti-peroxidase from peroxidase, and impaired accessibility of substrate within large complexes (LANSDORP et al. 1984).

For monoclonal primary antibodies (rat or mouse) monoclonal peroxidase-antiperoxidase (PAP) immunocytochemical reagents have recently been developed (MASON et al. 1982; CUELLO et al. 1984). Since monoclonal antibodies recognize only one specific type of antigenic determinant, monoclonal PAP complexes have lower molecular weights than polyclonal PAP complexes and probably contain only two molecules of horseradish peroxidase per antibody molecule which facilitates penetration (MASON et al. 1982). The use of monoclonal PAP complexes results in very low background staining and intense specific staining (LANSDORP et al. 1984). In "hybrid-hybridoma" immunocytochemistry "bi-specific" monoclonal antibodies which recognize both peroxidase and another antigen are introduced (MILSTEIN and CUELLO 1983).

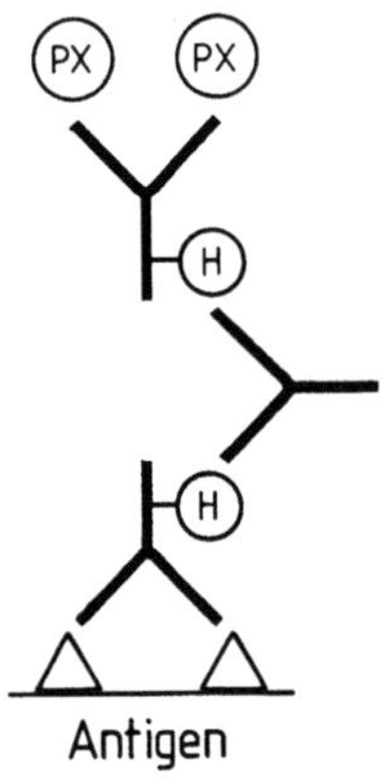

Fig. 6. Uncoupled hapten bridge method. Primary antibodies as well as enzyme antibodies are haptenated and bridged by an antibody directed to the hapten

3.4 Hapten-Labeled Antibody Method

Labeling of a primary antibody with a hapten (e.g. arsanilic acid, dinitrophenol, p-amino-benzoylglycine, p-aminobenzoylglutamic acid), which is then recognized by a labeled hapten antibody, is the principle of this reaction. Since large numbers of haptens can be coupled to the antibody its antigenicity towards the second antibody is increased and sensitivity augmented (CAMMISULI and WOFSI 1976; JASANI et al. 1981; MASON et al. 1983). In a modified version of this method an unlabeled anti-hapten antibody (reacting with the haptenated primary antibody) is used as the second and a hapten-labeled enzyme antibody or PAP complex as the third layer (Fig. 6).

3.5 Labeled Antigen Method

This method is highly specific and is based on the binding of a labeled antigen to its specific antibody before this antibody reacts with its other binding site with the tissue-bound antigen (MASON and SAMMONS 1979). Enzymes, radioactive isotopes, or gold particles can serve as labels. The necessity of labeling individual antigens is a disadvantage of this method (see VAN NOORDEN and POLAK 1983; BOSMAN 1983 for further information).

3.6 Biotin-Avidin Method

Avidin is an egg white glycoprotein (molecular weight 68,000) with pronounced affinity to the vitamin biotin (see HSU and RAINE 1984 for review). In the labeled avidin-biotin technique (GUESDON et al. 1979; HSU et al. 1981; BOORSMA et al. 1986) biotin-labeled antibodies and enzyme-labeled avidin are applied in sequence. In addition to horseradish peroxidase alkaline phosphatase can be used as marker enzyme. With biotin-labeled secondary antibodies (3-step technique) only few biotinylated antibodies are required to detect a wide range of primary antibodies. However, biotin-labeled primary antibodies (2-step tech-

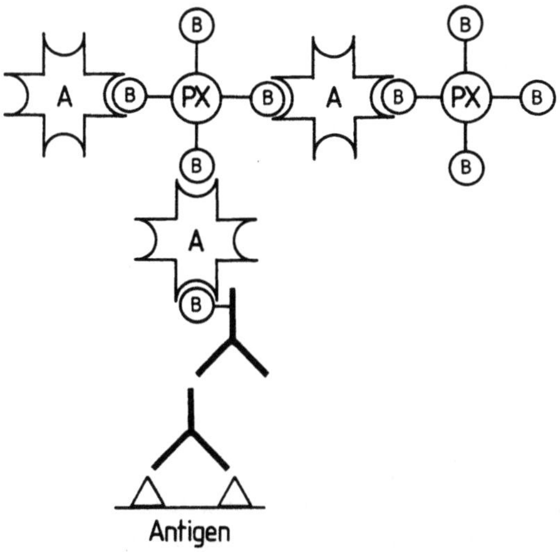

Fig. 7. A biotinylated antibody (either primary or secondary antibody) is visualized by an avidin (*A*)-biotin (*B*)-peroxidase (*PX*)-complex. Since several complexes can bind, several peroxidase molecules are fixed to the place of the tissue bound antigen increasing the sensitivity of the reaction

nique) have the benefit of lower background staining and thus yield a high signal-to-noise ratio, resulting in a distinct staining reaction. In the direct avidin-biotin bridge method avidin bridges the biotin-labeled primary antibody and the biotin-labeled enzyme. In the indirect bridged avidin-biotin technique (Hsu et al. 1981) the biotin-labeled secondary antibody and the biotin-labeled enzyme are combined by avidin. The reagents of the different steps are either applied sequentially or a preformed avidin-biotin-peroxidase complex (ABC) is used (Fig. 7). The high sensitivity of the indirect bridged avidin-biotin technique may ensue from the following facts: (1) Biotin is bound to the antibody or the enzyme (peroxidase) through amino groups. Since multiple biotin moieties can be introduced into immunoglobulin or peroxidase, the number of binding sites for avidin is increased; (2) avidin has four active binding sites for biotin, not all of which react with antibody-associated biotin but which do react with biotin molecules coupled to peroxidase leading to amplification of the staining reaction by formation of a large complex containing multiple peroxidase molecules; (3) penetration of the reagents is facilitated by their smaller size (Hsu and Raine 1984). In order to further enhance antibody penetration biotin labeled Fab fragments can be used for light as well as for electron microscopy.

The quality of the ABC reagent also determines sensitivity. An excess of free avidin molecules decreases sensitivity by competition with the avidin-biotin-peroxidase complex for the biotin molecules coupled to the antibody. On the other hand, with biotin-peroxidase in excess the biotin binding sites of the avidin molecules in the complex may become saturated and the complex is then devoid of free biotin-binding sites impeding its reactivity with the biotinylated (secondary) antibody.

Non-specific binding of ABC is observed with high concentrations and can be diminished by appropriate dilutions (for other pitfalls see section 9).

Further possible pitfalls of avidin-based systems may result from the fact that avidin at neutral pH is highly positively charged (isoelectric point around

10) and, therefore, non-specifically attaches to negatively charged molecules, including nucleic acids and phospholipids (SAVAGE 1986).

Avidin can be replaced by streptavidin, which is a protein from Streptomyces avidinii and has four high-affinity binding sites for biotin. It can be conjugated with labels without loss of biotin-binding characteristics. Labels used in the streptavidin system include fluorescein, Texas red, enzymes (peroxidase, alkaline phosphatase, β-galactosidose), colloidal gold, and 125 I. Non-specific binding of streptavidin to negatively charged molecules and to lectins is avoided by its lack of carbohydrate residues and by an isoelectric point around neutral pH (see SAVAGE 1986 for further information). Amplification of the signal can be achieved by biotinylation of the second antibody at multiple sites or by using streptavidin containing several label molecules per molecule.

3.7 Labeled Protein A Method

Protein A is a highly resistant cell wall constituent present in most strains of Staphylococcus aureus and consists of a single polypeptide with a molecular weight of 42,000 and an isoelectric point value of 5.1. It shows a high affinity to the Fc region of immunoglobulins, particularly of IgG, which makes it a suitable tool for antibody recognition (see ROTH 1984 for additional information). Differences, however, exist in its binding affinities to IgG subclasses derived from different species (see ROTH 1984 for information). Since it binds strongly to rabbit and guinea pig but rather weakly to goat and sheep immunoglobulins, reagents derived from the former species should be preferred. Protein A can be labeled with several substances, including fluorescein isothiocyanate or colloidal gold particles, and is usable in both light and electron microscopic immunocytochemistry (see also BENDAYAN and DUHR 1986 for modifications).

3.8 Lectins

Lectins are sugar binding proteins or glycoproteins (derived from plants, bacteria, fungi, invertebrates, and vertebrates) which non-covalently bind to specific carbohydrate residues of polysaccharides and glycoproteins, and are able to agglutinate cells containing complementary saccharides (PONDAR 1983; ALROY et al. 1984). Lectins coupled to labels, including fluorescent dyes, enzymes and electron-dense particles, have been used as probes in carbohydrate-directed histochemical studies.

Lectin histochemistry has contributed to our knowledge of the distribution of sugar residues, including blood group-related antigens, on cell surfaces and secretions during maturation, differentiation, but also malignant transformation (for different lectin and sugar specificities see ALROY et al. 1984). Evaluation of the presence, deletion, or distribution of lectin receptors may help to define neoplastic changes and, consequently, the biological behaviour of tumor cells, and may, therefore, contribute to our understanding of pathological processes, in addition to providing diagnostic and prognostic hints.

In addition to direct methods with labeled lectins, lectin binding sites are also revealed by the PAP as well as the ABC method in formalin-fixed and paraffin-embedded material. In the PAP method, the tissue sections are first incubated with a lectin followed by sequential incubations with lectin antibodies, secondary antibodies, and a corresponding PAP complex. In the ABC method, a biotinylated lectin is revealed by ABC. Controls include blocking of the lectin by preincubation with the specific sugar.

4 Comparison of Different Immunohistochemical Techniques

4.1 Immunofluorescence Versus Immunoenzymatic Techniques

Important factors determining the reliability of immunohistochemistry are the sensitivity of the technique as well as the detection efficiency reflected by the signal-to-noise ratio. A valid comparison of different methods in this respect requires a defined test system (VALNES et al. 1984). VALNES and BRANDTZAEG (1984) concluded from their studies using artificial substrates (human immunoglobulins IgA and IgG incorporated) that only minor differences in sensitivity and efficiency exist between the direct and indirect immunofluorescence, the indirect immunoperoxidase conjugate and the unlabeled PAP methods. According to their experience, masking of antigenicity in formaldehyde-fixed substrates played some role in immunofluorescence but less in immunoenzymatic techniques and could be overcome by raising conjugate concentrations, prolongation of incubation time, and protease pretreatment. However, in the opinion of most authors enzyme bridge, PAP and avidin-biotin methods surpass the direct and indirect methods in sensitivity (e.g. BERGROTH 1983; see also VAN NOORDEN and POLAK 1983).

4.2 PAP Versus ABC Method

By using the ABC method staining intensity is increased in comparison with the PAP technique, thus allowing considerably higher dilution of primary antibodies (HSU et al. 1981). However, other authors (STERNBERGER and STERNBERGER 1986) disagree on the basis of their studies using a new method of quantification by image analysis of optical densities. STERNBERGER and STERNBERGER (1986) obtained stronger staining intensities, and hence sensitivities (signal-to-background ratios), with the PAP than with the ABC method. In their hands, the PAP method was ten times more sensitive than the ABC method within a certain range of antibody dilutions.

4.3 Immunogold-Silver Staining Versus PAP Method

Immunogold-silver staining excells the PAP technique in sensitivity. Increased sensitivity (up to 200 fold) is particularly obtained by the silver precipitation step (see also SPRINGALL et al. 1984), but inclusion of detergent (0.5% Tween 80 or 0.2% Triton X-100) and higher sodium chloride concentrations into the buffers, higher pH of the immunogold reagent and modifications of the silver development procedures (SPRINGALL et al. 1984) may also contribute to enhancement of staining. Small-sized gold particles adsorbed to affinity-purified antibodies facilitate penetration of the labeled antibodies.

5 Simultaneous Detection of Two or Several Antigens

5.1 Double Immunofluorescence

Differently labeled secondary antibodies (fluorescein isothiocyanate, rhodamine isothiocyanate, Texas Red) are used as a mixture or in sequence to detect primary antibodies derived from different species (indirect method). Primary antibodies conjugated with different fluorochromes can also be used (direct method). Immunofluorescence may also be combined with other methods, e.g. with the immunogold (WANG and LARSSON 1985) or with the PAP method (GAUDINO et al. 1982).

5.2 Immunoenzymatic Techniques

Simultaneous demonstration of multiple antigens by enzyme immunohistochemistry may be achieved by using one type of enzyme with different substrates or different enzymes with or without elution of the preceding sequences (for information and references see WANG and LARSSON 1985; MASON et al. 1983). If one enzyme is used with different substrates for labeling of two different antigens the procedure has to be performed sequentially with the disadvantages of prolongation of the immunohistochemical procedure. With two different enzymes, peroxidase is usually applied in the first sequence. Alkaline phosphatase, as the second enzyme, can be developed with different substrates revealing reaction products with exellent contrast to the DAB reaction product as well as to haematoxilin as counter stain (MASON et al. 1983; VALNES and BRANDTZAEG 1984). For double labeling a combination of avidin-biotin-glucose oxidase in conjunction with avidin-biotin-peroxidase method can be used (GOWN et al. 1986): the avidin-biotin-peroxidase with DAB development is applied first, followed by an elution step (5% acetic acid or 10% oxalic acid) and/or incubation with Triton X-100 (0.1%), or Tween 20 (0.1%), and then by avidin-biotin-glucose oxidase system with nitro blue tetrazolium as chromogen. For maximum staining intensity in paraffin-embedded tissue approximately ten-fold higher concentrations of primary antibodies were required in the avidin-biotin-glucose oxidase than in the avidin-biotin-peroxidase system, but this varied with different antibodies and antigens (GOWN et al. 1986).

In the elution techniques, the antibodies of the first reaction are eluted after color development and the section is then incubated for demonstration of the second antigen using an immunoenzymatic procedure with a differently colored reaction product either by changing the enzyme or the substrate (MASON et al. 1983; WANG and LARSSON 1985). Difficulties of removal of the first sequence antibodies may arise with antibodies with high binding avidity, and some of the methods efficient in this respect may adversely affect antigenicity (TRAMU et al. 1978; STERNBERGER 1979; see also WANG and LARSSON 1985). Removal of the immunoreagents of the first sequence without affecting the DAB reaction product before application of the second sequence for staining of the second antigen can be achieved by acidification (for further references see STERNBERGER and JOSEPH 1979) or by dimethyl formamide treatment (VANDESANDE and DIERICKX 1975) by which most antigen-antibody complexes are dissociated.

Elution procedures may be avoided by modifications of the techniques, for example by a mixed PAP-immunofluorescence staining sequence (LARSSON et al. 1979). In other non-elution procedures differently labeled primary antibodies, primary antibodies derived from different species, labeled antigen and protein A are used.

According to STERNBERGER and JOSEPH (1979) in the unlabeled antibody enzyme (PAP) method, performed under standard conditions, color mixing did not occur despite the fact that primary antibodies with differing specificities in the first and second sequence were from the same animal species, provided that the first sequence primary antibodies were sufficiently highly concentrated. Apparently, the DAB oxidation product is able to mask antigenic and catalytic sites of the first sequence of immunoreagents thus obviating elution prior to application of the second sequence for double antigen staining. Another possibility is that excess linking antibody in the first sequence blocks all antigenic determinants of the primary antibody and that the PAP complex of the first sequence blocks all free combining sites of the linking antibody.

Another non-elution technique introduced by WANG and LARSSON (1985) is based on the observation that formaldehyde vapour selectively destroys the antigen – combining sites of the second layer anti-IgG antibodies. Under these conditions antibodies of different specificities raised in the same species can be used in sequence for the demonstration of several antigens in indirect immunohistochemical staining. The prerequisite of the suitability of this technique, however, is the stability of the antigens towards formaldehyde fixation. Conditions, which have to be established in pilot studies, include optimal dilutions of primary antibodies and optimal (minimal) time of formaldehyde vapour treatment. Good results have been obtained with several peptide hormones (WANG and LARSSON 1985).

In the hapten sandwich double labeling methods the haptenated primary antibody directed to the first antigen is linked to the haptenated third stage (e.g. PAP) by a antihapten (e.g. to dinitrophenol) bridging antibody. The second antigen is revealed by a second sandwich with another hapten (e.g. p-aminobenzoylglutamic acid or p-aminobenzoylglycine; for further information see MASON et al. 1983).

If alkaline phosphatase is conjugated with (periodate oxidized) horseradish peroxidase a violet reaction product is revealed upon sequential alkaline phosphatase (Naphthol AS-MX/Fast blue) and horseradish peroxidase (3-amino-9-ethylcarbazole) development. This double enzyme conjugate can be coupled either to an antigen (for determination of antibody producing cells) or to antibodies and allows (in combination with pure alkaline phosphatase and horseradish peroxidase immunohistochemistry) the simultaneous detection of three different antigens or antibodies in a single section after a one-step incubation and a two-step standard cytochemical procedure (CLAASSEN et al. 1986).

5.3 Other Techniques

Double staining can also be achieved by sequential application of immunogold reagents, the first staining sequence intensified with silver precipitation reaction (black reaction product) and the second without (red color of colloidal gold) as described by Manigley and Roth 1985, or by a combination of immunogold-silver staining with the ABC method (Sako et al. 1986).

6 Sensitivity

Sensitivity can be defined in terms of the smallest amount of antigen to be detected and, therefore, depends greatly on the ratio of specific to nonspecific (background) staining. Sensitivity depends on the quality of the antibody, the chemical structure of the antigenic determinant to be detected, and on the methodology (as described above).

Increasing the incubation time with the different layers can improve specific staining, and, in addition, allows further dilution of the antibodies resulting in decreased background reaction. Repetition of layers, e.g. of the primary antibody layer after washing or of the second layer after application of the PAP complex, can increase the amount of label concentrated at the antigenic site, and thus the signal-to-noise ratio. Combination of immunogold techniques with the PAP method with development of peroxidase (with DAB) and subsequent silver intensification results in further increase in sensitivity (Scopsi and Larsson 1985).

Reduction of non-specific background staining can be achieved by using highly specific antibodies, such as affinity – purified or monoclonal antibodies. In immunofluorescence microscopy, background staining may result from binding of unconjugated fluorochrome (e.g., fluorescein isothiocyanate), which can be removed, by Sephadex G-25 chromatography (see Van Noorden and Polak 1983 for further information). Nonspecific adsorption of antibodies can be prevented by pretreatment of antisera with tissue powders. Nonspecific reactive sites in the tissue section are blocked by normal serum applied prior to the immune reaction. This is, of course, not applicable if protein A is used as a component of the immunohistochemical procedure because of its affinity to the Fc portion of immunoglobulins. Moreover, non-specific binding of antibodies can also be minimized or prevented by detergents such as Triton X-100 or Tween 20 included in the washing buffers (Van Noorden and Polak 1983; Juhl et al. 1984). Pretreatment of dewaxed formalin-fixed paraffin sections with (2%) Tween 20 and dilution of the additional layer (first and second antibody layer in indirect peroxidase immunohistochemistry) with the detergent (0.05–2%) significantly reduces background staining as evaluated with blood group antigen A in human urothelium (Juhl et al. 1984). Antibody penetration may also be facilitated by detergent action.

Increased sensitivity can also be obtained by changing the chromogen protocol in peroxidase immunohistochemistry (Trojanowski et al. 1983). According to TROJANOWSKI et al. (1983), with neuro-and glial filaments as antigens DAB in combination with imidazole exceeded all other protocols (amino-ethylcarbazole, 0-tolidine, paraphenylenediamine-pyprocatechol, tetramethylbenzidine, diaminobenzidine, and diaminobenzidine at neutral pH with or without intensification with cobalt chloride and at pH 5.1) in sensitivity. 0-tolidine was the least sensitive chromogen in the hands of these authors although the results may vary with other systems. The positive effect of imidazol added to the DAB incubation medium is apparently due to acceleration of DAB oxidation especially at neutral pH (STRAUS 1982).

Further important determinants of sensitivity are the chemical structure and, particularly, the stability of the antigens in question, especially in fixed and embedded material. Denaturation of antigenic determinants occurs with aldehyde fixatives in addition to masking of antigenic determinants by formation of intermolecular linkages (BRANDTZAEG and ROGNUM 1983, 1984a, b; see VALNES et al. 1984 for further information). Masking may also result from cross-linking with other proteins as shown by VALNES et al. (1984) and this can be prevented by protease pretreatment (see RADASZKIEWICZ and DENK 1982 for further information).

7 Fixation and Embedding Procedures

Fixation and embedding may destroy and/or mask antigenic determinants. On the other hand, however, formaldehyde fixation diminishes the net positive charge of tissues by blocking amino groups and, consequently, inhibits electrostatic (nonspecific) attraction of antibodies. In immunohistochemistry, adequate fixation causes a rapid and total immobilization of the antigens along with a sufficient preservation of antigenicity and maintenance of accessibility of the antigens to the immunochemical reagents. However, excellent tissue preservation and antigenicity are often mutually exclusive and therefore compromises have to be made.

Particularly detrimental to antigenicity may be the denaturing effect of aldehyde fixatives (e.g. formaldehyde) at low antigen concentrations. With IgA and IgG as antigens, BRANDTZAEG and ROGNUM (1983, 1984a, b) found that the concentration of IgG detectable in formaldehyde-fixed specimens had to be 8 times higher than in ethanol-fixed ones. The difference was less pronounced with IgA. In Bouin – and Susa – fixed material more than 150 times higher antigen concentrations were required for the detection of IgG, but only 3–8 times higher for IgA (BRANDTZAEG and ROGNUM 1984a, b) indicating the influence of chemical structure, substrate matrix, and other variables.

Antigenic masking by aldehyde fixatives not only depends on the antigen concentration but also on environmental proteins surrounding the antigen which may prevent antigen – antibody interaction by steric hindrance (see BRANDTZAEG and ROGNUM 1983).

However, even fixatives which cross-link proteins and which are commonly used in pathology, like buffered formaldehyde solution, paraformaldehyde or

Bouin's fixative can yield satisfactory results in immunohistochemistry (RADASZKIEWICZ and DENK 1979; BRANDTZAEG and ROGNUM 1983, 1984a, b; NATHRATH et al. 1985; HOPWOOD 1985; VAN EWIJK et al. 1984; BEROD et al. 1981; CURRAN and GREGORY 1980; MEPHAN 1982; JUDD and BRITTEN 1982). Glutaraldehyde mixed in low concentrations with formaldehyde or paraformaldehyde often gives positive results, but destroys most of the antigenic determinants if applied in higher concentrations (see RADASZKIEWICZ and DENK 1982 for references and further information). Alternative fixatives are dimethylsuperimidate (reacting specifically with the ε-amino-groups of lysine) and periodate-lysine-paraformaldehyde (MCLEAN and NAKANE 1979; see also RADASZKIEWICZ and DENK 1982) which acts by oxidation of membrane carbohydrates to aldehydes (by periodate) and cross-linking via the amino groups of lysine. This fixation procedure, therefore, stabilizes carbohydrate moieties with preservation of antigenic sites (MCLEAN and NAKANE 1979) and has been found to be suitable for a variety of immunohistochemical reactions, including those performed with monoclonal antibodies on paraffin-embedded material (BRENES et al. 1986; HANCOCK et al. 1982). Fixation time is usually not critical, but prolonged fixation (several days) may adversely affect antigenicity (BRENES et al. 1986 for further information).

Alteration of antigenicity may be prevented by using dehydrating substances for fixation instead of aldehyde fixatives, such as ethanol, methanol, ethanol-acetic acid, chloroform or ethyl ether which act by protein precipitation. Ethanol, however, does not immobilize all protein antigens equally well in order to resist prolonged incubation and washing procedures during immunohistochemical staining (BRANDTZAEG and ROGNUM 1983).

During routine paraffin embedding procedures temperature-dependent destruction of antigenic determinants may occur which can be prevented by using paraffins with lower melting points instead of the conventional paraffin. The paraffin embedding procedure designed by SAINTE-MARIE (1962) after alcohol fixation and dehydration at low temperature (4° C) leads to excellent antigen preservation but is less practicable in routine pathology. An expanded range of antigens can also be detected in paraffin-embedded tissue after freeze-drying (JUDD and BRITTEN 1982; STEIN et al. 1984) or acetone fixation (JUDD and BRITTEN 1982; TANAKA et al. 1984). Several antigens can be demonstrated in plastic-embedded tissue sections after previous fixation in (1%) paraformaldehyde and digestion with trypsin (BECKSTEAD 1985) or in undecalcified methylmethacrylate – embedded formalin-fixed bone biopsies (SCHRÖDER and DELLING 1986).

In conclusion, therefore, it does not seem to be possible to develop a fixative of general applicability but the techniques should be modified depending on the antigens, the antibodies, and the methods to be applied.

Immunhistochemistry is complicated if material requiring decalcification has to be processed (MATHEWS and MASSON 1984). MATHEWS and MASSON (1984) demonstrated that treatment of tissues with neutral EDTA, formic, or acetic acid after formalin fixation combined with trypsin treatment did not adversely affect immunoreactivity of a variety of antigens, including immunoglobulins, factor VIII-related antigen, lysozyme and keratin. However, with decalcifying agents containing mineral acids unreliable results were obtained. Under these conditions, trypsin concentration and duration of treatment had to be reduced in most cases after extended periods of decalcification, although with some decalcifying agents the tissue was refractory to trypsin action (for detailed information consult MATHEWS and MASSON 1984).

MULLINK et al. (1985) regard fixation in a mercuric chloride – formaldehyde mixture followed by decalcification in acetic acid-formaldehyde-saline as the best procedure for antigen preservation (including CEA and lysozyme) and protection of morphological details in paraffin-embedded human bone marrow biopsies. Trypsinization was usually not needed except for the demonstration of factor VIII-related antigen in megakaryocytes. To date it appears that not all markers specific for a certain cell type (say, lymphocyte subsets) can be detected under these circumstances. Frozen sections are often required.

8 Proteases and Other Pretreatments

Specific immunostaining of several (but not all) antigens can be enhanced by pretreatment of dewaxed sections with proteases, such as pronase and trypsin (see DENK et al. 1977; RADASZKIEWICZ and DENK 1982). Among the various enzymes tested, including pronase, papain, pepsin, trypsin, hyaluronidase (for demonstration of fibronectin; HOLUND and CLEMMENSEN 1982; see also DENK et al. 1977; PINKUS et al. 1985; CURRAN and GREGORY 1978; HUANG et al. 1976; MEPHAM et al. 1979) trypsin seems to be the most suitable in the hands of most authors since its action is easily controlled and structural details of the section are better preserved. Trypsin hydrolyzes arginine-lysine bonds. Attention to the type of trypsin as well as to incubation times is essential and the results may vary accordingly (see PINKUS et al. 1985; HAJDU 1983). The mechanism of the beneficial effect of protease pretreatment, however, is not fully elucidated. According to HUANG et al. (1976) certain immunoreactive sites are unmasked, due eventually to release of protein cross-links, and more antigenic sites become available to interaction with the antibody (see also RADASZKIEWICZ and DENK 1982 for further information). However, individual treatment is needed and optimal regimens should be assessed in pilot studies for different fixation conditions and antigens. Common fixatives, such as formalin-mercury, Bouin and Carnoy, formalin-acetic acid, fix immunoglobulins at least in such a way that antigenicity is preserved and demonstrable without protease pretreatment. However, in tissues fixed with neutral (buffered) formalin, the demonstration of certain antigens (immunoglobulins) requires pretreatment with proteolytic enzymes (MEPHAM 1982). A number of fixatives were also tested, particularly with respect to their effects on immunoglobulins, by CURRAN and GREGORY (1980). Sections of tonsil fixed with buffered (4%) formaldehyde revealed positive immunostaining (PAP method) only after trypsinization, whereas enzyme pretreatment was not required with tissues fixed with Bouin's or Carnoy's fluid, with solutions containing mercuric chloride, and particularly with formol-saline containing (2–10%) acetic acid, as well as with a combination of acetic acid (10%) – formol-saline and formol sublimate (CURRAN and GREGORY 1980).

9 Specificity and Pitfalls

The specificity of the immunohistochemical staining reaction has to be thoroughly established and should be evaluated with respect to antibody and method specificity (see Van Leeuwen 1986).

Affinity purification of antibodies on Sepharose-4B-CNBr columns to which the antigen is coupled or on gels containing antigens separated by SDS-polyacrylamide gel electrophoresis is a practical method to reveal monospecific antibodies (see De Mey 1983 for review). Considerable progress regarding specificity has been made by the development of the hybridoma technology for production of monoclonal antibodies. Moreover, this technique creates the potential for obtaining large amounts of antibodies against a specific antigenic determinant even if impure antigens are used for immunization, and is thus important in oncology, for example to detect tumor-specific antigens (see Diamond and Scharff 1982; Bhan 1984). Disadvantages are the often poor avidity and stability of monoclonal antibodies and the possibility of cross-reactivity (Diamond and Scharff 1982).

The identification and characterization of the antigen recognized by the antibodies is another basis for specific immunohistochemistry. This can be achieved by immunoprecipitation, immunoblotting procedures and radioimmunoassay (Van Noorden and Pollak 1983; De Mey 1983 for review). Method specificity can be checked by absorption of primary antibodies with their respective antigens or omission of one of the reagents of the staining sequences.

False positive results may be caused by peroxidase-like activity in haeme (in erythrocytes) which is blocked by methanol/H_2O_2. Endogenous peroxidase (catalase in peroxisomes and cytochrome c in mitochondria) is capable of reacting with DAB-H_2O_2 but is inhibited by methanol, nitroferricyanide or phenylhydrazine, as well as (10%) methanol and (3%) H_2O_2. Free reactive groups in the tissue may lead to non-specific binding of IgG (particularly after aldehyde fixation) and this is prevented by pretreatment with normal serum (not recognized by the immunoreagents), borohydride, serum albumin, or buffers containing lysine, glycine, or TRIS. Charged groups within proteins can be responsible for ionic interactions with antibodies which can be minimized with buffers with increased ionic strength (e.g. 0.5 M NaCl). Moreover, incubation of tissue sections at pH 8.6 also minimizes the influence of ionic charges since IgG has an isoelectric point at 8.6. Hydrophobic interactions of antibodies and labels with hydrophobic tissue components can be prevented by addition of Triton X-100 to the media. Immunoglobulin aggregates (eliminated by dilution or high speed centrifugation) may adhere nonspecifically to hydrophobic tissue components (or embedding media). Natural and contaminating antibodies (due to impurities of the antigen preparation used for immunization) should also be considered as source of false-positive results. In frozen sections, unwanted non-specific background staining by the ABC method may occur in liver, pancreas and kidney, possibly due to the presence of biotin or related substances reacting with avidin. This non-specific binding can be abolished by pretreatment of the section with avidin followed by biotin (Hsu and Raine 1984). Non-specific avidin binding can also be induced by the presence of endogenous lectins (for example, in nerve fibers) which react with mannose-containing glycoproteins, such as avidin. This unwanted reactivity can be blocked by pretreatment with alpha-methyl-D-mannoside (Hsu and Raine 1984). The reaction of mast cell granules with ABC in frozen as well as paraffin sections can be prevented by using the ABC solution at pH 9.4 (Bussolati and Gugliotta 1983).

False negative results may be due to modification of antigenic sites by fixation, dehydration and embedding procedures. Antigen-antibody interaction may be hampered by impaired antibody penetration, which is then facilitated by non-ionic detergents (Triton X-100, 0.1–0.5%). Steric hindrance exerted either by embedding medium or associated proteins may also impede interaction of antibodies with the antigen. In the unlabeled antibody-enzyme method, a high concentration of antigen within the tissue section and insufficient dilution of the primary antibodies may prevent bridge formation by the second antibody (see Van Leeuwen 1986 for references).

10 Outlook

Immunohistochemistry has contributed considerably to our understanding of basic biological processes but also to refinement of histopathologic evaluation. It plays a particularly valuable role in tumor histopathology, provided strict controls and critical interpretation are used. The immunohistochemical demonstration of a given substance depends on several factors, including quantity and quality of the antigen to be determined, antibody reactivity and the reliability and efficiency of the immunohistochemical procedure employed. Commercial institutions now play a major role in provision of reagents as well as in specificity testing. This should, however, not obscure the fact that valuable information in histopathology can only be obtained by those who understand the basic principles and pitfalls of the reactions and who are experienced in histopathological diagnosis.

References

Adams JC (1981) Heavy metal intensification of DAB-based HRP reaction product. J Histochem Cytochem 29:775

Alroy J, Ucci AA, Pereira MEA (1984) Lectins: histochemical probes for specific carbohydrate residues. In: De Lellis RA (ed) Advances in immunohistochemistry. Masson Publ. USA, New York Paris Barcelona Milan Mexico City Rio de Janeiro, pp 67–88

Beckstead JH (1985) Optimal antigen localization in human tissues using aldehyde-fixed plastic embedded sections. J Histochem Cytochem 33:954–958

Bendayan M, Duhr MA (1986) Modification of protein A-gold immunocytochemical technique for the enhancement of its efficiency. J Histochem Cytochem 34:569–575

Bergroth V (1983) Comparison of various immunohistochemical methods. Demonstration of extracellular antigens in cryostat sections. Histochem 77:177–184

Berod A, Hartman BK, Pujol JF (1981) Importance of fixation in immunohistochemistry: use of formaldehyde solutions at variable pH for the localization of tyrosine hydroxylase. J Histochem Cytochem 29:844–850

Bhan AK (1984) Application of monoclonal antibodies to tissue diagnosis. In: DeLellis RA (ed) Advances in immunohistochemistry. Masson Publ. USA, New York Paris Barcelona Milan Mexico City Rio de Janeiro, pp 1–29
Boorsma DM, Van Bommel J, Vanden Heuvel J (1986) Avidin HRP conjugates in biotin avidin immunoenzyme cytochemistry. Histochem 84:333–337
Bosman FT (1983) Some recent developments in immunocytochemistry. Histochem J 15:189–200
Brandtzaeg P, Rognum TO (1983) Evaluation of tissue preparation methods and paired immunofluorescence staining for immunocytochemistry of lymphomas. Histochem J 15:655–689
Brandtzaeg P, Rognum TO (1984a) Evaluation of nine different fixatives. 1. Preservation of immunoglobulin isotypes. J chain and secretory component in human tissues. Path Res Pract 179:250–266
Brandtzaeg P, Rognum TO (1984b) Evaluation of nine different fixatives. 2. Preservation of IgG, IgA and secretory component in an artificial immunohistochemical test substrate. Histochem 81:213–219
Brenes F, Harris S, Paz M, Petrovic LM, Scheuer PJ (1986) PLP fixation for combined routine histology and immunocytochemistry of liver biopsies. J Clin Pathol 39:459–463
Bussolati G, Gugliotta P (1983) Nonspecific staining of mast cells by avidin-biotin-peroxidase complexes (ABC). J Histochem Cytochem 31:1419–1421
Cammisuli S, Wofsy L (1976) Hapten-sandwich labelling. III. Bifunctional reagents for immunospecific labelling of surface antigens. J Immunol 117:1695–1704
Claassen E, Boorsma DM, Kors N, Van Rooijen N (1986) Double-enzyme conjugates, producing an intermediate color, for simultaneous and direct detection of three different intracellular immunoglobulin determinants with only two enzymes. J Histochem Cytochem 34:423–428
Clark CA, Downs EC, Primus JF (1982) An unlabelled antibody method using glucose oxidase-antiglucose oxidase complexes (GAG): a sensitive alternative to immunoperoxidase for the detection of tissue antigens. J Histochem Cytochem 30:27–34
Coons AH, Creech HJ, Jones RN (1941) Immunological properties of an antibody containing a fluorescent group. Proc Soc Exp Biol Med 47:200–202
Cordell JL, Falini B, Erber WN, Ghosh AK, Abdulaziz Z, MacDonald S, Pulford KAF, Mason DY (1984) Immunoenzymatic labelling of monoclonal antibodies using immune complexes of alkaline phosphatase and monoclonal anti-alkaline phosphatase (APAAP complexes). J Histochem Cytochem 32:219–229
Cuello AC, Milstein C, Wright B, Braunwell S, Priestley JV, Jarvis J (1984) Development and application of a monoclonal rat peroxidase antiperoxidase (PAP) immunocytochemical reagent. Histochem 80:257–261
Curran RC, Gregory J (1977) Masking of antigens in paraffin sections of tissue by trypsin. Experientia 33:1400–1401
Curran RC, Gregory J (1980) Effects of fixation and processing on immunohistochemical demonstration of immunoglobulin in paraffin sections of tonsil and bone marrow. J Clin Pathol 33:1047–1057
Danscher G, Nörgaard JOR (1983) Light microscopic visualization of colloidal gold on resin-embedded tissue. J Histochem Cytochem 31:1394–1398
Davidsohn I (1972) Early immunologic diagnosis and prognosis of carcinoma. Am J Clin Pathol 57:715–730
De Jong ASH, Van Kessel-Van Vark M, Raap AK (1985) Sensitivity of various visualization methods for peroxidase and alkaline phosphatase activity in immunoenzyme histochemistry. Histochem J 17:1119–1130
De Mey J (1983a) Raising and testing antibodies for immunocytochemistry. In: Polak JM, Van Noorden S (eds) Immunocytochemistry. Practical applications in pathology and biology. Wright, Bristol London Boston, pp 43–52
De Mey J (1983b) Colloidal gold probes in immunocytochemistry. In: Polak JM, Van Noorden S (eds) Immunocytochemistry. Practical applications in pathology and biology. Wright, Bristol London Boston, pp 82–112
De Lellis RA, Sternberger LA, Mann RB, Banks PM, Nakane PK (1979) Immunoperoxidase technics in diagnostic pathology. Report of a workshop sponsored by the National Cancer Institute. Am J Clin Pathol 71:483–488
De Waele M, De Mey J, Moeremans M, De Brabander M, Van Camp B (1983) Immunogold staining method for the light microscopic detection of leukocyte cell surface antigens with mono-

clonal antibodies. Its application to the emuneration of lymphocyte subpopulations. J Histochem Cytochem 31:375–381

Denk H, Radaszkiewicz T, Weirich E (1977) Pronase pretreatment of tissue sections enhances sensitivity of the unlabelled antibody enzyme (PAP) technique. J Immunol Methods 15:163–167 (1977)

Denk H, Platzer B (1982) Immunofluorescence demonstration of tumor-associated antigens. In: Wick G, Trail KN, Schauenstein K (eds) Immunofluorescence technology. Selected theoretical and clinical aspects. Elsevier Biomedical Press, pp 325–348

Diamond B, Scharff MD (1982) Monoclonal antibodies. JAMA 248:3165–3169

Fritz P, Hoenes J, Schenk J, Mischlinski A, Gran A, Saal JG, Tuczek HV, Multhaupt H, Pfleiderer G (1986) Colour development of immunogold-labelled antibodies for light microscopy. Histochem 85:209–214

Gaudino G, Fasolo A, Accossato LP (1982) Double immunohistochemical methods. A procedure employing a PAP-indirect immunofluorescence sequence. Bas Appl Histochem 26:47–51

Gown AM, Garcia R, Ferguson M, Yamanaka E, Tippens D (1986) Avidin-biotin-immunoglucose oxidase: use in single and double labeling procedures. J Histochem Cytochem 34:403–409

Guesdon JL, Ternynck T, Avrameas S (1979) The use of avidin-biotin interaction, in immunoenzymatic techniques. J Histochem Cytochem 27:1131–1139

Hajdu I (1983) The immunohistochemical detection of J chain in lymphoid cells in tissue sections: the necessity of trypsin digestion. Cell Immunol 79:157–163

Hancock WW, Becker GJ, Atkins RC (1982) A comparison of fixatives and immunohistochemical technics for use with monoclonal antibodies to cell surface antigens. Am J Clin Pathol 78:825–831

Heyderman E (1983) Tumor markers. In: Polak JM, Van Noorden S (eds) Immunocytochemistry. Practical applications in pathology and biology. Wright, Bristol London Boston, pp 274–294

Holgate CS, Jackson P, Cowen PN, Bird CC (1983) Immunogold-silver staining. New method of immunostaining with enhanced sensitivity. J Histochem Cytochem 31:938–944

Holund B, Clemmensen I (1982) The value of hyaluranidase treatment of different tissues before demonstration of fibronectin by indirect immunoperoxidase technique. Histochem 76:517–525

Hopwood D (1985) Cell and tissue fixation, 1972–1982. Histochem J 17:389–442

Hsu SM, Soban E (1982) Color modification of diaminobenzidine (DAB) precipitation by metallic ions and its application for double immunohistochemistry. J Histochem Cytochem 30:1079–1082

Hsu SM, Raine L (1984) The use of avidin-biotin-peroxidase complex (ABC) in diagnostic and research pathology. In: De Lellis RA (ed) Advances in immunohistochemistry. Masson Publ. USA, New York Paris Barcelona Milan Mexico City Rio de Janeiro, pp 31–42

Hsu SM, Raine L, Fanger H (1981) Use of avidin-biotin-peroxidase complex (ABC) in immunoperoxidase techniques: A comparison between ABC and unlabeled antibody (PAP) procedures. J Histochem Cytochem 29:577–580

Huang SN, Munassian H, More JD (1976) Application of immunofluorescent staining on paraffin sections improved by trypsin digestion. Lab Invest 35:383–390

Jasani B, Wynford Thomas D, Williams ED (1981) Use of monoclonal anti-hapten antibodies for immunolocalization of tissue antigens. J Clin Pathol 34:1000–1002

Judd MA, Britten KJM (1982) Tissue preparation for the demonstration of surface antigens by immunoperoxidase techniques. Histochem J 14:747–753

Juhl B, Norgaard T, Bjerrum OJ (1984) The effect of Tween 20 on indirect immunoperoxidase staining of blood group antigen A in human urothelium. J Histochem Cytochem 32:935–941

Kuhlmann WD, Avrameas S (1971) Glucose oxidase as an antigen marker for light and electron microscopic studies. J Histochem Cytochem 19:361–368

Kuhlmann WD, Peschke P (1986) Glucose oxidase as label in histological immunoassays with enzyme-amplification in a two-step technique: coimmobilized horseradish peroxidase as secondary system enzyme for chromogen oxidation. Histochem 85:13–17

Landsdorp PM, Van der Kwast TH, De Boer M, Zeijlemaker WP (1984) Stepwise amplified immunoperoxidase (PAP) staining. I. Cellular morphology in relation to membrane markers. J Histochem Cytochem 32:172–178

Larsson LI, Childers S, Snyder SH (1979) Met- and leu-enkephalin immunoreactivity in separate neurons. Nature 282:407–410

Manigley C, Roth J (1985) Applications of immunocolloids in light microscopy. IV. Use of photochemical silver staining in a simple and efficient double-staining technique. J Histochem Cytochem 33:1247–1251

Mason DY, Sammons RE (1979) The labeled antigen method for immunoenzymatic staining. J Histochem Cytochem 27:832–840

Mason DY, Cordell JL, Abdulaziz Z, Naiem M, Bordenave G (1982) Preparation of peroxidase-antiperoxidase (PAP) complexes for immunohistological labeling of monoclonal antibodies. J Histochem Cytochem 30:1114–1122

Mason DY, Abdulaziz Z, Falini B, Stein H (1983) Double immunoenzymatic labelling. In: Polak JM, van Noorden S (eds) Immunocytochemistry. Practical applications in pathology and biology. Wright, Bristol London Boston, pp 113–128

Mathews JB, Mason GI (1984) Influence of decalcifying agents on immunoreactivity of formalin-fixed, paraffin-embedded tissue. Histochem J 16:771–787

Mauro A, Germano I, Giaccone G, Giordona MT, Schiffer D (1985) 1-naphthol basic dye (1-NBD). An alternative to diaminobenzidine (DAB) in immunoperoxidase techniques. Histochem 83:97–102

McLean IW, Nakane PK (1979) Periodate-lysine-paraformaldehyde fixative: a new fixative for immunoelectron microscopy. J Histochem Cytochem 27:1131–1139

Mepham BL (1982) Influence of fixatives on the immunoreactivity of paraffin sections. Histochem J 14:731–737

Mepham BL, Frater W, Mitchell BS (1979) The use of proteolytic enzymes to improve immunoglobulin staining by the PAP technique. Histochem J 11:345–358

Milstein C, Cuello AC (1983) Hybrid hybridomas and their use in immunohistochemistry. Nature 305:537–540

Mullink H, Henzen-Logmans SC, Tadema TM, Mol JJ, Meiyer CJLM (1985) Influence of fixation and decalcification on the immunohistochemical staining of cell-specific markers in paraffin embedded human bone biopsies. J Histochem Cytochem 33:1103–1109

Nakane PK, Pierce GB (1966) Enzyme labeled antibodies: preparation and application for the localization of antigens. J Histochem Cytochem 14:929–931

Nathrath WBJ, Heidenkummer P, Björklund V, Björklund B (1985) Distribution of tissue polypeptide antigen (TPA) in normal human tissues. Immunohistochemical study on unfixed, methanol-, ethanol-, and formalin-fixed tissues. J Histochem Cytochem 33:99–109

Ordroneau P, Lindström PBM, Petrusz P (1981) Four unlabeled antibody bridge techniques: a comparison. J Histochem Cytochem 29:1397–1404

Pinkus GS, O'Connor EM, Etheridge CL, Corson JM (1985) Optimal immunoreactivity of keratin proteins in formalin-fixed, paraffin-embedded tissue requires preliminary trypsinization. An immunoperoxidase study of various tumors using polyclonal and monoclonal antibodies. J Histochem Cytochem 33:465–473

Ponder BAJ (1983) Lectin histochemistry. In: Polak JM, Van Noorden S (eds) Immunocytochemistry. Practical applications in pathology and biology. Wright, Bristol London Boston, pp 129–142

Radaszkiewicz T, Denk H (1982) Immunofluorescence microscopy on paraffin sections. In: Wick G, Trail KN, Schauenstein K (eds) Immunofluorescence technology. Selected theoretical and clinical aspects. Elsevier Biomedical Press, pp 317–324

Roth J (1984) Light- and electronmicroscopic localization of antigens with the protein A-gold (pAg) technique. In: De Lellis RA (ed) Advances in immunohistochemistry. Masson Publ. USA, New York Paris Barcelona Milan Mexico City Rio de Janeiro, pp 43–69

Sainte-Marie G (1962) A paraffin embedding technique for studies employing immunofluorescence. J Histochem Cytochem 10:250–256

Sako H, Nakane Y, Okino K, Nishihara K, Kodama M, Kawata M, Yamada H (1986) Simultaneous detection of B-cells and T-cells by a double immunohistochemical technique using immunogold-silver staining and the avidin-biotin-peroxidase complex method. Histochem 86:1–4

Savage B (1986) The biotin-streptavidin amplification system. ICPR May/June: 34–39

Schröder S, Delling G (1986) Bone metastases of differentiated and medullary thyroid gland carcinomas. Usefulness and limitations of immunohistology performed on undecalcified plastic-embedded tissue specimens. Virchows Arch (Pathol Anat) 409:767–776

Scopsi L, Larsson LI (1985) Increased sensitivity in immunocytochemistry. Effects of double application of antibodies and of silver intensification of immunogold and peroxidase-antiperoxidase staining techniques. Histochem 82:321–329

Springall DR, Hacker GW, Grimelius L, Polak JM (1984) The potential of the immunogold-silver staining method for paraffin sections. Histochem 81:603–608

Stein H, Gatter KC, Heryet A, Mason DY (1984) Freeze-dried paraffin-embedded human tissue for antigen labelling with monoclonal antibodies. Lancet 2:71

Sternberger CA, Hardy PH, Cuculis JJ, Meyer HG (1970) The unlabeled antibody enzyme method of immunohistochemistry. Preparation and properties of soluble antigen-antibody complex (horseradish peroxidase-antihorseradish peroxidase) and its use in identification of spirochetes. J Histochem Cytochem 18:315–333

Sternberger LA (1979) Immunocytochemistry, 2nd edn. Wiley, New York

Sternberger LA, Joseph SA (1979) The unlabeled antibody method. Contrasting color staining of paired pituitary hormones without antibody removal. J Histochem Cytochem 27:1424–1429

Sternberger LA, Sternberger NH (1986) The unlabeled antibody method. Comparison of peroxidase-antiperoxidase with avidin-biotin complex by a new method of quantification. J Histochem Cytochem 34:599–605

Straus W (1982) Imidazol increases the sensitivity of the cytochemical reaction for peroxidase with diaminobenzidine at neutral pH. J Histochem Cytochem 30:491–493

Tanaka M, Tanaka H, Ishikawa E (1984) Immunohistochemical demonstration of surface antigen of human lymphocytes with monoclonal antibody in acetone-fixed paraffin-embedded sections. J Histochem Cytochem 32:452–454

Taylor CR (1980) Immunohistologic studies of lymphoma. Past, present, and future. J Histochem Cytochem 28:777–787

Taylor CR, Kledzik G (1981) Immunohistologic techniques in surgical pathology – a spectrum of "new" special stains. Hum Pathol 12:590–596

Trauni G, Pillez A, Leonardelli J (1978) An efficient method of antibody elution for the successive or simultaneous localization of two antigens by immunocytochemistry. J Histochem Cytochem 26:322–324

Trojanowski JQ, Obrocka MH, Lee VMY (1983) A comparison of eight different chromogen protocols for the demonstration of immunoreactive neurofilaments or glial filaments in rat cerebellum using the peroxidase-antiperoxidase method and monoclonal antibodies. J Histochem Cytochem 31:1217–1223

Valnes K, Brandtzaeg P (1984) Paired indirect immunoenzyme staining with primary antibodies from the same species. Application of horseradish peroxidase and alkaline phosphatase as sequential labels. Histochem J 16:477–487

Valnes K, Brandtzaeg P, Rognum TO (1984) Sensitivity and efficiency of four immunohistochemical methods as defined by staining of artificial sections. Histochem 81:313–319

Van Ewijk W, Van Soest PL, Verkerk A, Jongkind JF (1984) Loss of antibody binding to prefixed cells: fixation parameters for immunocytochemistry. Histochem J 16:179–193

Van Leeuwen F (1986) Pitfalls in immunocytochemistry with special references to the specificity problems in the localization of neuropeptides. Am J Anat 175:363–377

Van Noorden S, Polak JM (1983) Immunocytochemistry today. Techniques and practice. In: Polak JM, Van Noorden S (eds) Immunocytochemistry. Practical applications in pathology and biology. Wright, Bristol London Boston, pp 11–42

Vandesande F, Dierickx K (1975) Identification of the vasopressin producing and of the oxytocin producing neurons in the hypothalamic magnocellular neurosecretory system of the rat. Cell Tissue Res 164:153–162

Wang BL, Larsson LI (1985) Simultaneous demonstration of multiple antigens by indirect immunofluorescence or immunogold staining. Novel light and electron microscopical double and triple staining method employing primary antibodies from the same species. Histochem 83:47–56

Epithelial Tumor Markers: Cytokeratins and Tissue Polypeptide Antigen (TPA)

R. Moll

1 Introduction

The concept of morphological tumor markers is derived from the fact that tumor cells express a specific, albeit limited, program of differentiation that usually represents at least a part of the differentiation program of the putative cell of origin. Epithelial tumor markers, which are discussed in this and the following contributions, should facilitate the identification and characterization of poorly differentiated and/or metastatic epithelial tumors in tissue sections, even though the morphological appearance of such neoplasms may be uncharacteristic.

The epithelial cell system is highly complex. Thus, the process of epithelial differentiation does not simply involve a single cell type but rather a variety of cell forms of different histogenetic origin that differ widely with respect to their structure and function. However, normal epithelial cells have several morphological features in common, most notable the presence of desmosomes (maculae adhaerentes) and tonofilaments. Malignant epithelial tumors may lose these features to varying degrees, resulting in the absence of a recognizable epithelial morphology in certain anaplastic epithelium-derived tumors, even at the electron microscope level (Ghadially 1980). Recently, in diagnostic histopathology, molecular components that are specifically expressed in epithelial cells have acquired great importance as epithelial tumor markers or, more correctly, epithelial differentiation markers. Such markers have two main applications, i.e., in distinguishing epithelial from nonepithelial tumors, and in distinguishing different types of epithelial tumors. The present chapter is concerned with cytokeratins, a family of cytoskeletal proteins that have been convincingly demonstrated to be very reliable epithelial markers. Moreover, there is an increasing body of evidence indicating that the individual proteins of this family can be used for the subtyping of different epithelia and different epithelial tumors.

2 Cytokeratin as a General Marker of Epithelia

2.1 Cytokeratin-type Intermediate Filaments

Intermediate-sized filaments (intermediate filaments) are a ubiquitous cytoplasmic filament system (filament diameter, about 10 nm) and are part of the cytoskeleton. These filaments can be divided into five classes (cytokeratin, vimentin, desmin, glial filaments, and neurofilaments) on the basis of their protein composition, which is differentiation specific (for reviews see Lazarides 1980; Franke et al. 1982; Osborn and Weber 1983).

Cytokeratin filaments are expressed in epithelial cells. At the electron microscope level, they are indistinguishable from other types of intermediate filaments, being arranged in loose bundles or, in stratified squamous epithelia, occuring as tonofilaments that form densely packed bundles (tonofibrils). A typical feature of cytokeratin filaments is their association with the cytoplasmic plaques of desmosomes. In certain exceptional cases, they may be present in the form of paranuclear globular aggregates, also referred to as fibrous bodies (Racadot et al. 1964; see Section 2.6).

2.2 Distribution of Cytokeratin Filaments in Normal Tissues

In 1978, using immunofluorescence microscopy, it was found that antibodies raised against prekeratin (α-keratin) obtained from epidermis react with filament arrays in cultured cells of diverse origin, including simple epithelia of the mammary gland and kidneys as well as in a number of epithelial tissues, even though keratinization is completely absent in these cell types (Franke et al. 1978a, b; Sun and Green 1978). Since then α-keratin related proteins have been detected in all types of epithelial cells (Franke et al. 1979a, 1981a–d; Sun et al. 1979) and have been collectively termed cytokeratins or keratins. As these polypeptides are related to each other (see Section 3.1), they have certain antigenic determinants in common. Therefore, certain polyclonal (Franke et al. 1981a) and monoclonal (Table 1) antibodies are capable of detecting the cytokeratin

Table 1. Broad-spectrum cytokeratin antibodies

Antibody	Cytokeratin polypeptides recognized	References
K_G 8.13	CKs nos. 1, 5, 6, 7, 8, 18	GIGI et al. 1982
AE 3	CKs nos. 1–8	TSENG et al. 1982; COOPER et al. 1985
Clone 80	Basic CKs (including CKs nos. 1 and 5)	VAN MUIJEN et al. 1984
LP 34	Several CKs	LANE et al. 1985
lu-5	Most CKs	VON OVERBECK et al. 1985; FRANKE WW, unpublished
KL-1[a]	Epidermal CKs (55–57 KD polypeptide group)[b]	VIAC et al. 1983

[a] Basal layer of stratified squamous epithelia negative
[b] No biochemical data for non-epidermal CKs available

filaments present in all epithelial cells; these antibodies can be regarded as being broad-spectrum cytokeratin antibodies (Table 1).

In normal tissues, the cytokeratin filaments recognized by such broad-spectrum cytokeratin antibodies have been found (usually arranged in fibrillar cytoplasmic arrays) in all cell types that, according to the classical definition, are considered to be epithelial cells. Thus, cytokeratin filaments are not only expressed in the epidermis – the classical tissue of keratin research – but also in all types of noncornifying stratified squamous epithelia, pseudostratified epithelia, myoepithelial cells, and even simple epithelia (FRANKE et al. 1978a, 1979a, 1980, 1981a; SUN et al. 1979; ALTMANNSBERGER et al. 1981b; OSBORN and WEBER 1983), including epithelia of mesodermal origin, e.g., renal tubular cells (BACHMANN et al. 1983; RAMAEKERS et al. 1983c; HOLTHÖFER et al. 1984), epithelial cells of the upper male and female genital tract (CZERNOBILSKY et al. 1985; ACHTSTÄTTER et al. 1985), and mesothelial cells lacking any connection with other epithelial surfaces of the body. Cytokeratin filaments are also present in most endocrine cells (HOEFLER et al. 1984a; RAMAEKERS et al. 1983c; MIETTINEN et al. 1984a, 1985a–c; VAN MUIJEN et al. 1984a), including cells of the dispersed neuroendocrine system (HOEFLER and DENK 1984) such as Merkel cells of the skin (MOLL et al. 1984), which thus are also true epithelial cells in spite of their simultaneous expression of neural features.

Numerous studies have shown that nonepithelial cells lack cytokeratin filaments. These cytokeratin-negative cells include the various types of mesenchymal cells, myogenic cells, neuronal cells, glial cells, and melanocytes which, in turn, are characterized by the presence of other types of intermediate filaments (for a review see OSBORN and WEBER 1983). Certain epithelioid tissues, e.g., the so-called lens epithelium (RAMAEKERS et al. 1980), the pigment epithelium of the retina (at least in chicken), Sertoli cells (FRANKE et al. 1979b; for the detection of occasional cytokeratin-positive cells in human seminiferous tubules, see MIETTINEN et al. 1985d), glomerular podocytes (BACHMANN et al. 1983; RAMAEKERS et al. 1983c; HOLTHÖFER et al. 1984), and vascular endothelium (FRANKE et al. 1979c) are also cytokeratin negative, whereas they do express vimentin filaments. These cell types cannot, however, be regarded as being truly epithelial because they contain neither desmosomes nor tonofilaments. Immunohistochemical staining of certain muscle cells for cytoceratins has been reported by HUITFELDT and BRANDTZAEG (1985).

Thus, cytokeratin (as recognized by broad-spectrum antibodies) can be regarded as being a general, a specific, and, perhaps, the best marker of epithelial differentiation.

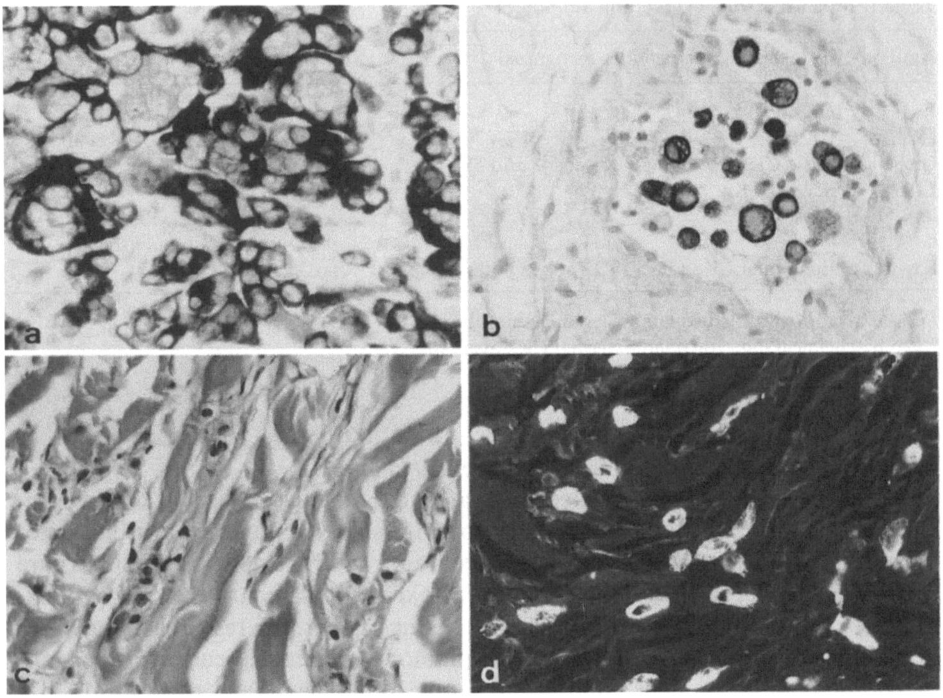

Fig. 1a–d. Immunocytochemical detection of cytokeratin in epithelial tumors, using broad-spectrum cytokeratin antibodies. **a** Anaplastic carcinoma, peritoneal metastasis (primary tumor probably in ovary), strongly positive with cytokeratin antibody AE1/AE3 (paraffin section; ABC-peroxidase). **b** Lymphangiosis carcinomatosa derived from an adenocarcinoma of the stomach, showing strong staining of individual tumor cells in a lymphatic vessel close to a lymph node with cytokeratin antibody AE1/AE3 (paraffin section; ABC-peroxidase). **c, d** Disparately growing tumor cells in a skin metastasis of a breast carcinoma, which are hard to recognize in HE-staining (**c**), are clearly identified by the cytokeratin antibody (**d**; antibody AE1/AE3; paraffin section; immunofluorescence microscopy). **a, b** ×240; **c, d** ×190

2.3 Expression of Cytokeratin Filaments in Malignant Tumors

From the observation that cultured carcinoma cells such as HeLa cells (a cell line derived from a cervical adenocarcinoma) contain intermediate filaments of the cytokeratin type (sometimes together with vimentin filaments), Franke et al. (1978a, 1979d) proposed the use of intermediate-filament antibodies for demonstrating whether a given tumor is of epithelial origin. Using available broad-spectrum cytokeratin antibodies, it has since been found that malignant epithelial tumors, including their metastases, consistently maintain the expression of cytokeratin-type intermediate filaments (Fig. 1a; Bannasch et al. 1980; Battifora et al. 1980; Löning et al. 1980; Schlegel et al. 1980; Altmannsberger et al. 1981a; Caselitz et al. 1981; Gabbiani et al. 1981; Sieinski et al. 1981). Numerous recent studies have further confirmed and demonstrated that the expression of cytokeratin filaments is a constant feature of all carcinomas,

Table 2. Cytokeratin-positive tumors

Tumor type	References
Squamous cell carcinomas (skin, upper digestive and respiratory tract, lung, cervix uteri)	1
Nasopharyngeal carcinomas (including lymphoepitheliomas)	2
Thymomas	3
Transitional cell carcinomas	1
Adenocarcinomas (lung, breast, gastrointestinal tract, female genital tract, prostate)	1
Choroid plexus carcinomas	4
Renal cell carcinomas[a]	5
Adenomatoid tumors	6
Malignant mesotheliomas (epithelial, fibrous)	7
Chordomas	8
Adamantinomas	9
Synovial sarcomas	10
Epithelioid sarcomas	11
Thyroid carcinomas	12
Adrenal cortical carcinomas[b]	13
Epithelial neuroendocrine tumors (Merkel cell carcinomas, carcinoid tumors, islet cell tumors, medullary carcinomas of thyroid, small cell carcinomas of lung)	14
Embryonal carcinomas, teratocarcinomas, choriocarcinomas, endodermal sinus tumors	15
Undifferentiated/anaplastic carcinomas	1, 16

[a] Rare cases cytokeratin-negative but vimentin-positive
[b] 13 of 25 cases were cytokeratin-positive

References: 1 Battifora et al. 1980; Schlegel et al. 1980; Bannasch et al. 1980; Gabbiani et al. 1981; Altmannsberger et al. 1981a, 1982; Sieinski et al. 1981; Caselitz et al. 1981; Espinoza and Azar 1982; Nagle et al. 1983; Saba et al. 1983; Ramaekers et al. 1983b, c; Makin et al. 1984; Pinkus et al. 1986; *2* Miettinen et al. 1982b; Madri and Barwick 1982; *3* Battifora et al. 1980; Loening et al. 1981; *4* Miettinen et al. 1986; *5* Holthöfer et al. 1983; Herman et al. 1983; *6* Barwick and Madri 1982; Said et al. 1982; *7* Corson and Pinkus 1982; Holden and Churg 1984; La Rocca and Rheinwald 1984; Churg 1985; Blobel et al. 1985b; *8* Miettinen et al. 1983a; Raju et al. 1983; *9* Rosai and Pinkus 1982; *10* Miettinen et al. 1982a, 1983b; see also pp. 155–178; *11* Chase et al. 1984; see also pp. 155–178; *12* Miettinen et al. 1984a; *13* Miettinen et al. 1985c; *14* For references, see Section 2.6; *15* Kahn et al. 1983; Battifora et al. 1984; Miettinen et al. 1985d; Ramaekers et al. 1985a; *16* Lauder et al. 1984; Kahn et al. 1984

irrespective of their degree of differentiation (for reviews, see Osborn and Weber 1983; Ramaekers et al. 1983c; Miettinen et al. 1984b; Erlandson 1984; Corson 1986). The various types of malignant tumors which have been found to be cytokeratin positive are listed in Table 2. In contrast to other epithelial markers such as epithelial membrane antigen (EMA; Pinkus and Kurtin 1985; Pinkus et al. 1986), cytokeratin filaments are usually uniformly distributed among carcinoma cells. In view of the steadily accumulating data concerning the heterogeneity of tumor cell populations (Fidler et al. 1978; Nicolson 1984), the degree of stability of cytokeratin expression in tumors is remarkably high. Therefore, cytokeratin is also a reliable marker for undifferentiated and anaplastic carcinomas (Fig. 1a), for disparately growing infiltrating carcinoma cells (Fig. 1c, d), for metastasizing single carcinoma cells in lymph nodes and the bone marrow (Fig. 1b), and for isolated carcinoma cells in suspension, e.g., in effusions and other body fluids.

Tumors containing spindle cells of a sarcomatoid appearance, e.g., mesotheliomas of the fibrous type, monophasic synovial sarcomas, and some carcinomas of the kidneys, skin (CORSON 1986), and other sites, are cytokeratin positive (Table 2) and are thus of an epithelial nature (for the co-expression of vimentin, see Section 2.5); this is in good agreement with the presence of epithelial features such as desmosomes in these tumors, as revealed by ultrastructural investigations (GHADIALLY 1980; ERLANDSON 1981). These findings are of particular importance for differential diagnosis, for they make it possible to distinguish such tumors from true sarcomas. Cytokeratin is also present in the epithelial subgroup of neuroendocrine tumors, sometimes occurring simultaneously with neurofilaments (see Section 2.6).

In certain rare tumors, the expression of cytokeratin appears to be confined to a subpopulation of cells. This has been reported for adrenal cortical carcinomas, some of which may also exhibit the complete absence of cytokeratin immunoreactivity (MIETTINEN et al. 1985b), suggesting that the extent of epithelial differentiation in these tumors is very low and variable. Certain tumors consisting mainly of vimentin-positive tumor cells, including granulosa cell tumors of the ovaries (CZERNOBILSKY et al. 1987), skeletal and extraskeletal Ewing's sarcomas (MOLL et al. 1987), and seminomas (MIETTINEN et al. 1985d; RAMAEKERS et al. 1985a; DENK et al. 1986) contain sparsely distributed, scattered cytokeratin-positive tumor cells. These tumors are also characterized by the presence of variable numbers of desmoplakin-positive true desmosomes (for the distribution of desmoplakin in human tumors, see MOLL et al. 1986a).

Nonepithelial tumors, including sarcomas, malignant fibrous histiocytomas, lymphomas, malignant melanomas, Schwannomas, gliomas, neuroblastomas, and pheochromocytomas are negative for cytokeratin proteins, but express other types of intermediate filaments (for references, see OSBORN and WEBER 1983; RAMAEKERS et al. 1983c; MIETTINEN et al. 1984b; CORSON 1986). With respect to diagnosis, this is especially important when nonepithelial tumors exhibit a solid or epithelioid growth pattern as is the case in, for example, some sarcomas, lymphomas, and melanomas that only express vimentin filaments.

2.4 Detection of Cytokeratin Filaments in Formalin-fixed, Paraffin-embedded Tissue

The reliable immunocytochemical detection of cytokeratin filaments is possible in frozen tissue, as well as in tissue fixed with alcohol (ALTMANNSBERGER et al. 1981b) or Methacarn (GOWN and VOGEL 1984) and embedded in paraffin.

When applied to formalin-fixed, paraffin-embedded material, many cytokeratin antibodies yield false-negative or misleadingly irregular results. Unfortunately, in many cases, only routinely fixed material may be available. This has prompted the search for conditions which allow the detection of cytokeratins in such material (see also pp. 47–69). The broad-spectrum cytokeratin antibodies that are reactive with formalin-fixed, paraffin-embedded tissue are listed in Table 3; among these, the antibody lu-5 appears to have the broadest spectrum of reactivity. Limited proteolytic digestion of sections is essential before incubation with these antibodies. The use of pepsin (MIETTINEN et al. 1983a; our own findings), trypsin (PINKUS et al. 1985; BATTIFORA and SILVA 1986), and pronase (GOWN and VOGEL 1985; our own data) has produced good or, at least, fairly reliable results (Fig. 1). However, even when such procedures are applied, negative results obtained in formalin-fixed material should be interpreted with caution, since small amounts of cytokeratin filaments may escape detection, or the antigens may be damaged during tissue processing.

Table 3. Monoclonal cytokeratin antibodies with broad or extended spectrum reactive on formalin-fixed, paraffin-embedded tissue

Antibody	Cell type spectrum	References
lu-5	Broad	von Overbeck et al. 1985
AE1	Extended (secretory simple epithelia[a] and suprabasal layers of epidermis negative)	Tseng et al. 1982; Spagnolo et al. 1985; Pinkus et al. 1986
KL-1	Extended (basal layer of stratified squamous epithelia negative)	Viac et al. 1983
PKK 1	Extended (epidermis negative)	Miettinen et al. 1985d; Virtanen et al. 1985
CAM 5.2	Extended (most hepatocytes and all stratified squamous epithelia negative)	Makin et al. 1984

[a] Hepatocytes, pancreatic acinar cells

2.5 Co-expression of Cytokeratin and Vimentin Filaments

In epithelial tumors, cytokeratin filaments are usually the only intermediate-filament class present; however, as has been described for cultured epithelial and carcinomatous cells (Franke et al. 1978a, 1979d, e, 1981b; Virtanen et al. 1981), certain types of epithelial tumors also express (in the majority of cases) vimentin filaments (Table 4; Fig. 2). This phenomenon was first observed in cells of pleomorphic adenomas of the parotid gland (Caselitz et al. 1981; Krepler et al. 1982). The co-expression of cytokeratin and vimentin filaments is particularly striking in cells exhibiting a clearly epithelial structure and tumors derived therefrom, e.g., certain cells of the endometrial glands and adenocarcinomas of the endometrium (McNutt et al. 1985), and the follicle epithelium (B. Dockhorn-Dworniczak and W.W. Franke, personal communication) and follicular and papillary carcinomas of the thyroid gland (Miettinen et al. 1984a). In malignant mesotheliomas, the co-expression of cytokeratin and vimentin occurs most consistently in fibrous tumors and areas consisting of spindleshaped, fibroblast-like tumor cells, but may also be observed in epithelial-like structures of these tumors (Blobel et al. 1985b). Here again, tumors can be seen to maintain a feature of the corresponding normal cells, i.e., pleural mesothelium cells (LaRocca et al. 1984). Normal granulosa cells of the ovary (Czernobilsky et al. 1985) as well as granulosa cell tumors (Czernobilsky et al. 1987) co-express both intermediate-filament types. Whereas most renal cell carcinomas co-express cytokeratin and vimentin (Fig. 2; Herman et al. 1983; Holthöfer et al. 1983; Waldherr and Schwechheimer 1985; Pitz et al. 1987) the normal adult kidney tubules exclusively express cytokeratin filaments, thus suggesting that, in these tumors, vimentin expression is switched on during tumorigenesis. A rare renal cell carcinoma subtype, the chromophobe cell renal carcinoma (Thoenes et al. 1985), as well as renal oncocytomas can be distinguished

Table 4. Co-expression of cytokeratin and vimentin filaments in malignant tumors

Tumor type	References
Adenoid-cystic carcinomas of salivary gland	1
Carcinomas of thyroid gland (follicular, papillary, anaplastic)	2
Adenocarcinomas of lung[a]	3, 4
Large cell carcinomas of lung[a]	4
Renal cell carcinomas (except chromophobe type)	5
Nephroblastomas[b]	6
Malignant rhabdoid tumors of kidney	7
Adenocarcinomas of endometrium	8, 9
Carcinomas of ovary[a]	8
Granulosa cell tumors of ovary[c]	10
Adenocarcinomas of prostate	11
Malignant mesotheliomas	12
Epithelioid sarcomas	13
Chordomas	14

[a] Only a minority of tumors revealed vimentin in addition to cytokeratins
[b] Co-expression of cytokeratins and vimentin was found in blastema cells
[c] These tumors express predominantly vimentin filaments, as well as desmoplakin
References: 1 CASELITZ et al. 1984; *2* MIETTINEN et al. 1984a; DROESE et al. 1984; SCHRÖDER et al. 1986; *3* JASANI et al. 1985; *4* UPTON et al. 1986; *5* HERMAN et al. 1983; HOLTHÖFER et al. 1983; PITZ et al. 1987; *6* ALTMANNSBERGER et al. 1984; DENK et al. 1985; *7* VOGEL et al. 1984; *8* MCNUTT et al. 1985; *9* DABBS et al. 1986; *10* CZERNOBILSKY et al. 1987; *11* WERNERT et al. 1986; *12* LA ROCCA and RHEINWALD 1984; CHURG 1985; BLOBEL et al. 1985b; *13* MIETTINEN and DAMJANOV 1985; see also pp. 155–178; *14* MIETTINEN et al. 1983; GOWN and VOGEL 1985; COINDRE et al. 1986; ABENOZA and SIBLEY 1986

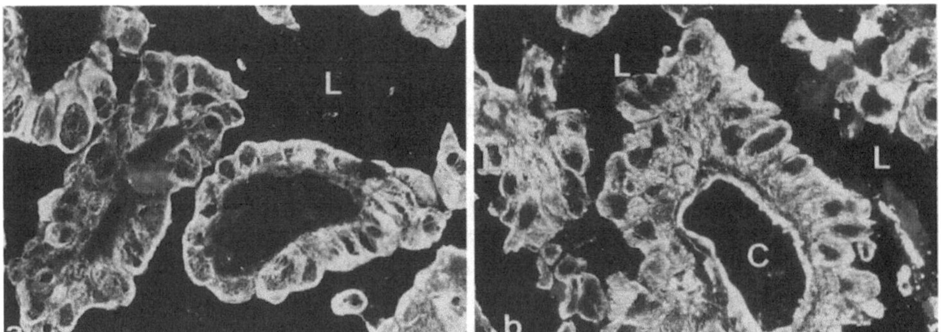

Fig. 2. Co-expression of cytokeratin filaments (**a**; monoclonal antibody against CK no. 18; frozen sections; immunofluorescence microscopy) and vimentin (**b**; antibody VIM-9) by the tumor cells of a renal cell carcinoma (eosinophilic-granular; tubulo-papillary). Note that the vimentin antibody (**b**) also stains the endothelium of a capillary (*C*). *L*, lumen. ×260

from the common types of renal cell tumors due to their exclusive expression of cytokeratin filaments (PITZ et al. 1987).

It is of interest that many of the tumors that exhibit the co-expression of vimentin and cytokeratin are derived from cells of mesodermal origin and, therefore, must be histogenetically related to (vimentin-expressing) mesenchymal cells. It is also important to note that the distribution of the expression of cytokeratin filaments in most of these tumor types is uniform, underlining their true epithelial character, regardless of the co-expression of vimentin. Clearly, testing for the presence or absence of cytokeratin-vimentin co-expression is of value for differential diagnosis, as this feature appears to be restricted to certain carcinoma types (Table 4; see also GOWN and VOGEL 1985; MCNUTT et al. 1985).

2.6 Co-expression of Cytokeratin and Neurofilaments

Recently, another type of intermediate-filament co-expression has been reported, i.e., the co-expression of cytokeratin and neurofilaments. This phenomenon appears to be a fairly characteristic feature of certain types of epithelial neuroendocrine tumors (Table 5).

The co-expression of cytokeratins (of the simple-epithelial type; see Section 3) and neurofilaments is most consistently found in neuroendocrine (Merkel cell) carcinomas of the skin (Fig. 3). In a series of 12 patients with such carcinomas, both intermediate-filament types were found in frozen tumor samples of all of the patients (MOLL et al. 1986b; see also HOEFLER et al. 1984b, 1985). In formalin-fixed samples of such tumors, neurofilaments have been detected with less frequency (HOEFLER et al. 1984b; SIBLEY and DAHL 1985). In such tumors, some of the filaments are arranged in fibrous bodies (see Section 2.1), a feature which is of diagnostic significance. Other types of neuroendocrine tumors, including carcinoid tumors of the bronchus and medullary carcinomas of the thyroid gland, also exhibit combined cytokeratin and neurofilament expression in some cases (Table 5); moreover, medullary carcinomas sometimes express vimentin filaments too. In contrast to earlier reports (LEHTO et al. 1983), several recent studies have shown that all small cell carcinomas of the lung contain cytokeratin intermediate filaments, and that most, if not all, of them

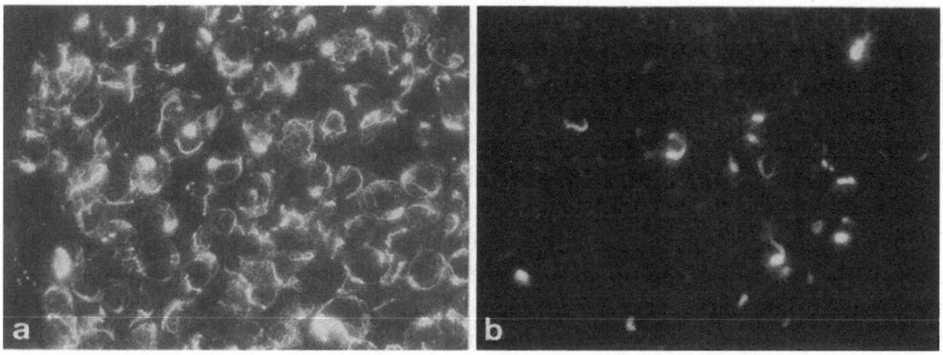

Fig. 3. Co-expression of cytokeratin (**a**; antibody PKK1; frozen sections; immunofluorescence microscopy) and neurofilaments (**b**; antibody 2F11) in a Merkel cell carcinoma of the skin. Note uniform positivity for cytokeratin (**a**), whereas neurofilaments are expressed only by some tumor cells, and are mostly arranged in globular aggregates (**b**). × 380

Table 5. Expression of cytokeratin and neurofilaments in epithelial neuroendocrine tumors

Tumor type	Cytokeratin	Neurofilament	References
Neuroendocrine (Merkel cell) carcinomas of skin	+	+	1
Carcinoid tumors of bronchus	+	+/−	2
Parathyroid adenomas	+	+/−	3
Islet cell tumors of pancreas	+	+/−	4, 5
Medullary carcinomas of thyroid gland[a]	+	+/−	6
Carcinoid tumors of small intestine	+	−/+	4, 7
Small cell carcinomas of lung	+	−	8

[a] These tumors may, in addition, express vimentin

References: 1 HOEFLER et al. 1984b; GOULD et al. 1985; SIBLEY and DAHL 1985; MOLL et al. 1986b; *2* LEHTO et al. 1985; BLOBEL et al. 1985a; *3* MIETTINEN et al. 1985a; *4* MIETTINEN et al. 1985b; *5* MOLL et al. 1986a; *6* DROESE et al. 1984; SCHRÖDER et al. 1986; *7* R. MOLL, unpublished results; *8* See text for references

are exclusively cytokeratin positive (VAN MUIJEN et al. 1984b; BLOBEL et al. 1985a; BROERS et al. 1985b; GATTER et al. 1985; MOSS et al. 1986). However, some 'variant cell lines' cultured from small cell carcinomas of the lung have been reported to be neurofilament positive and apparently lack cytokeratin filaments (BROERS et al. 1985a). It remains to be seen whether any in situ tumors have this phenotype (for the finding of neurofilaments in one case of a poorly differentiated squamous cell carcinoma of the lung, see VAN MUIJEN et al. 1984b).

Since normal islet cells of the pancreas (MIETTINEN et al. 1985b) and Merkel cells of the skin (MOLL et al. 1984) are exclusively cytokeratin positive, the corresponding tumors appear to acquire neurofilament expression during tumorigenesis. Obviously, the expression of neurofilaments is an optional part of the neuroendocrine differentiation program, which also involves the expression of other neuronal markers, such as neuron-specific enolase (NSE) and neurosecretory granules containing chromogranin and neuropeptide hormones. With regard to diagnosis, the co-expression of cytokeratin and neurofilaments can be used, for example, to distinguish a Merkel cell tumor from a malignant lymphoma or from a metastasis of a small cell carcinoma of the lung (MOLL et al. 1986b; BATTIFORA and SILVA 1986).

2.7 Co-expression of Cytokeratin and Glial Filaments

This rare type of co-expression has been observed in some normal myoepithelial cells of the parotid gland as well as in some cells present in pleomorphic adenomas (ACHTSTÄTTER et al. 1986), the latter of which also contain vimentin filaments. Cytokeratin, glial, and vimentin filaments may also be present together in choroid plexus carcinomas (MIETTINEN et al. 1986), but it has yet to be shown that they co-exist in individual tumors cells.

3 Diversity and Cell-Type Specificity of Cytokeratin Polypeptides

3.1 The Cytokeratin Family of Proteins

In contrast to all other classes of intermediate filaments, cytokeratin filaments are made up of a highly complex multigene family of proteins. In studies of cytokeratin polypeptides in human tissues, it has been shown that even epidermal keratinocytes contain several cytokeratin polypeptides (FUCHS and GREEN 1978, 1979; SUN and GREEN 1978). Nineteen distinct cytokeratin (CK) polypeptides have been found in the various types of human epithelial cells (CKs nos. 1–19) ranging in apparent molecular weight from 40,000 to 68,000 (Table 6; WINTER et al. 1980; DORAN et al. 1980; FRANKE et al. 1981c; MOLL et al. 1982; TSENG et al. 1982; WU et al. 1982; for reviews, see SUN et al. 1984; QUINLAN et al. 1985; COOPER et al. 1985; for the occurrence of a distinct set of eight keratin polypeptides in hair-forming cells, which are not discussed here, see HEID et al. 1986).

These polypeptides, all of which contain a central α-helical rod piece with an apparent molecular weight of 38,000, are related to each other to varying degrees with respect to their amino acid sequences (QUINLAN et al. 1985; FUCHS et al. 1985; MAGIN et al. 1986). According to their sequence relationships, they can be subdivided into a basic (B) subfamily (type II) comprising CKs nos. 1–8, and an acidic (A) subfamily (type I) comprising CKs nos. 9–19 (Table 6; FUCHS et al. 1981; MOLL et al. 1982; SCHILLER et al. 1982; TSENG et al. 1982; HANUKOGLU and FUCHS 1983; KIM et al. 1983; WEBER and GEISLER 1984; HATZFELD and FRANKE 1985; STEINERT et al. 1985; COOPER et al. 1985). At least one basic and one acidic polypeptide are present in each filament, forming heterotypic tetramer complexes (B_2A_2) that represent the actual subunits of the cytokeratin intermediate filaments (for a review, see QUINLAN et al. 1985). Accordingly, the cytokeratin polypeptides within a given cell are always expressed in pairs and in a coordinated fashion, i.e., a basic (type II) cytokeratin is always co-expressed with an acidic (type I) partner molecule.

With respect to the use of cytokeratins as differentiation markers, it is of particular importance that cytokeratin pairs are expressed in a differentiation-specific manner; thus, cytokeratin expression is correlated with the various epithelial cell types, so that different cytokeratin polypeptide patterns are observed in the various types of epithelia (Table 6).

3.2 Cytokeratin Polypeptide Patterns in Normal Epithelia

The cytokeratins can be roughly divided, according to their expression spectrum, into stratified-epithelium and simple-epithelium-type components. The *stratified-epithelium-type cytokeratins* comprise the higher-molecular-weight polypeptides of each subfamily, i.e., CKs nos. 1–6 of the basic group, and CKs nos. 9–17 of the acidic subfamily. Of these, CK no. 5, which is co-expressed with CK no. 14 and/or CK no. 17, is found in most stratified squamous epithelia as well as (most probably) in the basal cells of pseudostratified epithelia and in myoepithelial cells (Table 6; MOLL et al. 1982, 1983a, b; TSENG et al. 1982; NELSON and SUN 1983; BLOBEL et al. 1984; COOPER et al. 1985; NAGLE et al. 1985b; MOLL and FRANKE 1986); therefore, the cytokeratin pairs, 5/14 and

Table 6. Cytokeratin polypeptide patterns of normal epithelia and epithelial tumors, obtained using two-dimensional gel electrophoresis (MOLL et al. 1982; MOLL and FRANKE 1986). For each tumor, a typical and representative example is shown. *Large dot*, high relative proportion; *small dot*, low relative proportion of the respective polypeptide

	Stratified-epithelium type												Simple-epithelium type			
	BASIC					ACIDIC							BASIC		ACIDIC	
Cytokeratin No.[a]	1	3	4	5	6	10 11	12	13	14	15	16	17	7	8	18	19
Molecular weight (×10⁻³)	68	63	59	58	56	56	55	54	50	50	48	46	54	52.5	45	40
1. Normal epithelia																
Stratified squamous epithelia																
Epidermis	●			●		●			●	•						
Non-cornifying strat. squam. epithelium			●	●	•			●	•	•	•	•				•
Corneal epithelium		●		●			●									
Hair follicle outer root sheath				●	●				●	•	●	●				•
Simple epithelia																
Secretory epithelial cells [b]														●	●	
Intestinal epithelium														●	●	●
Ductal epithelium [c]													●	●	●	●
Urothelium			•	•				●					●	●	●	●
Mammary gland ducts				●					●	•		●	●	●	•	●
Respiratory epithelium				●	•					●		●	•	●	●	●
2. Epithelial tumors																
Squamous cell carcinomas																
Skin	•			●	●	•			●		●	•				
Tongue				●	●				●		●	●				•
Lung				●	●			•		•		●		●	•	●
Cervix (moderately differentiated)				●	●			●	●		•	●				•
Cervix (poorly differentiated)				●	●			●		•	•	●	•	●	•	●
Hepatocellular carcinoma														●	●	
Adenocarcinomas																
Colon														●	●	●
Pancreas												•	●	●	•	●
Endometrium [d]				•									•	●	●	●
Breast													●	●	●	●
Lung													●	●	●	●
Pleural mesothelioma (epithelial)				•									•	●	●	●
Small cell carcinoma of lung													•	●	●	●
Merkel cell carcinoma of skin														●	●	•
Transitional cell carcinoma (G II)				•				•				•	●	●	•	●

[a] The cytokeratin polypeptides no. 2 (molecular weight 65,500; expressed in epidermis) and no. 9 (molecular weight 64,000; expressed in palmar and plantar epidermis) are not included in this table

[b] E.g., hepatocytes, pancreatic acinar cells. Proximal and distal tubular cells of the kidney show the same pattern

[c] E.g., bile ducts, pancreatic ducts, renal collecting ducts

[d] CK no. 5 present in this special case but not in all cases

5/17, can be regarded as being the principal cytokeratin components of stratified epithelia. The remaining cytokeratin components of such epithelia are more selectively expressed, their presence being correlated with special forms of stratified squamous differentiation, and as shown by immunocytochemical study (see Table 7), they are usually expressed in more differentiated, suprabasal cells. These components include CKs nos. 1, 2, 9, 10, and 11 (typical of epidermis), the CK pair 3/12 (specific for corneal epithelium), and the CK pair 4/13 (typically expressed in noncornifying stratified squamous epithelia; Table 6; MOLL

Table 7. Selective single-polypeptide-specific cytokeratin antibodies

	Antibody	References
1. Antibodies against CK no. 10/11	K_K 8.60	HUSZAR et al. 1986
	K 92	PULFORD et al. 1985; GATTER et al. 1982, 1984
2. Antibodies against CK no. 9	Guinea pig antibodies	KNAPP et al. 1986
3. Antibodies against CK no. 3	A E 5	SCHERMER et al. 1986; COOPER et al. 1985
4. Antibodies against CK no. 4	6 B 10	VAN MUIJEN et al. 1986
5. Antibodies against CK no. 13	1 C 7	VAN MUIJEN et al. 1986
	2 D 7	VAN MUIJEN et al. 1986
6. Antibodies against CK no. 8	Troma-1	BRULET et al. 1980; TÖLLE et al. 1985
7. Antibodies against CK no. 18	CK 1–4	DEBUS et al. 1982
	RGE 53	RAMAEKERS et al. 1983a
	LE 61	LANE et al. 1982, 1985
	PKK 3	VIRTANEN et al. 1985
8. Antibodies against CK no. 7	CK 7	TÖLLE et al. 1985
9. Antibodies against CK no. 19	A 53/B-A2	KARSTEN et al. 1985
	BA 16, BA 17	BARTEK et al. 1985
	LP 2 K	LANE et al. 1985
	K_S 4.62	GIGI-LEITNER et al. 1986

et al. 1982; TSENG et al. 1982; for reviews, see COOPER et al. 1985; QUINLAN et al. 1985).

The *simple-epithelium-type cytokeratins* comprise the lower-molecular-weight polypeptides of each subfamily, i.e., CKs nos. 7 and 8 of the basic group, and CKs nos. 18 and 19 of the acidic subfamily (Table 6). These can also be divided into principal components (the CK pair 8/18) that are expressed in all simple epithelia, and more specific components; the latter include CK no. 19, which is found in ductal as well as in intestinal epithelium, and CK no. 7, which exhibits a much more restricted distribution (Table 6; MOLL et al. 1982; TSENG et al. 1982).

In complex epithelia such as the mammary gland epithelium (consisting of luminal and myoepithelial cells) and the respiratory epithelium, the different cell types present often express different cytokeratin polypeptides, resulting in complex patterns when pieces of whole epithelium are analyzed biochemically (Table 6; MOLL et al. 1982). Usually, the luminal cells of these complex epithelia express simple-epithelial-type components, while the basal or myoepithelial cells most likely contain stratified-epithelial-type components (mainly the CK pairs 5/14 or 5/17). Certain cytokeratins are heterogeneously distributed even within a particular epithelial cell layer, e.g., CKs nos. 4 and 13, which are expressed in single cells or in small groups of cells in the glandular epithelium of the prostate gland (VAN MUIJEN et al. 1986).

3.3 Selective Cytokeratin Antibodies

The biochemical diversity of cytokeratin polypeptides is reflected by the bewildering complexity and number of cytokeratin antibodies with different specificities that have been and are being developed. Selective cytokeratin antibodies are able to resolve expression differences within single cells, and the application of such antibodies is essential when cytokeratin typing is used for tumor diagnosis (Section 3.4). These antibodies, most of which are monoclonal antibodies, have been characterized to varying extents using immunocytochemical screening and biochemical/immunochemical tests. Selective cytokeratin antibodies can be divided into two main classes:

1. Group-specific Antibodies. These recognize two or more different cytokeratin polypeptides but, immunocytochemically, exhibit selectivity for certain types of epithelia.

Antibodies with a relatively extensive cell-type spectrum (being unreactive only with certain secretory epithelia such as hepatocytes and pancreatic acinar cells) fall into this class, e.g., some polyclonal antibodies against keratins from bovine hoof and muzzle epidermis (FRANKE et al. 1978b, 1979a) and human stratum corneum (SUN et al. 1979), as well as the monoclonal antibodies 34βE12 (GOWN and VOGEL 1982, 1984, 1985), clone 77 (VAN MUIJEN et al. 1984a), PKK2 (VIRTANEN et al. 1985), EKH4 (recognizing CK no. 14 and other cytokeratins; ETO et al. 1985), AE1 (recognizing many cytokeratins of the acidic subfamily; TSENG et al. 1982; COOPER et al. 1985), KA4 (NAGLE et al. 1985a, b), and LICR-LON-29b (KNIGHT et al. 1985). Other monoclonal antibodies decorate stratified squamous epithelia and myoepithelial/basal cells, e.g., KA1 (NAGLE et al. 1985b), CK B1 (CASELITZ et al. 1986a, b), and LICR-LON-16a (KNIGHT et al. 1985).

Monoclonal antibodies that selectively or preferentially stain stratified squamous epithelia include clone 78 (VAN MUIJEN et al. 1984a), 34ßB4 (GOWN and VOGEL 1984), ZK 61 and ZK 99 (LANG et al. 1986), and $K_S8.12$ (directed against CKs nos. 13 and 16; HUSZAR et al. 1986). Some monoclonal antibodies mainly react with the epidermis, e.g., K20 (GATTER et al. 1982, 1984; PULFORD et al. 1985), AE2 (against CKs nos. 1, 2, and 10; TSENG et al. 1982; COOPER et al. 1985), RKSE 60 (RAMAEKERS et al. 1985b), and KA5 (NAGLE et al. 1985a, b); of these antibodies, the last three do not decorate basal epidermal cells. Other antibodies react preferentially with simple epithelia and recognize at least CKs nos. 8, 18, and 19, e.g., PKK1 (HOLTHÖFER et al. 1983; MIETTINEN et al. 1985a; VIRTANEN et al. 1985).

Only for some of these group-specific antibodies has the exact spectrum of cytokeratin polypeptides that they recognize been determined.

2. Polypeptide-specific Antibodies. These are reactive with only a single cytokeratin polypeptide. At present, antibodies are available for at least nine of the cytokeratin polypeptides (Table 7). Antibodies of this class have been most important for elucidating the correlation between the pattern of cytokeratin expression and cellular differentiation. The most frequently used antibodies of this type are those directed against CK no. 18, which recognize simple epithelia and distinguish them from stratified squamous epithelia. The application of these antibodies for the subtyping of carcinomas (see Fig. 4) will be discussed in Section 3.4.

A negative immunocytochemical reaction with selective cytokeratin antibodies does not necessarily indicate the absence of the respective cytokeratin polypeptide(s), since the antigenic determinants may be masked (see also COOPER et al. 1985). Most polypeptide-specific cytokeratin antibodies do not react with formalin-fixed, paraffin-embedded tissue.

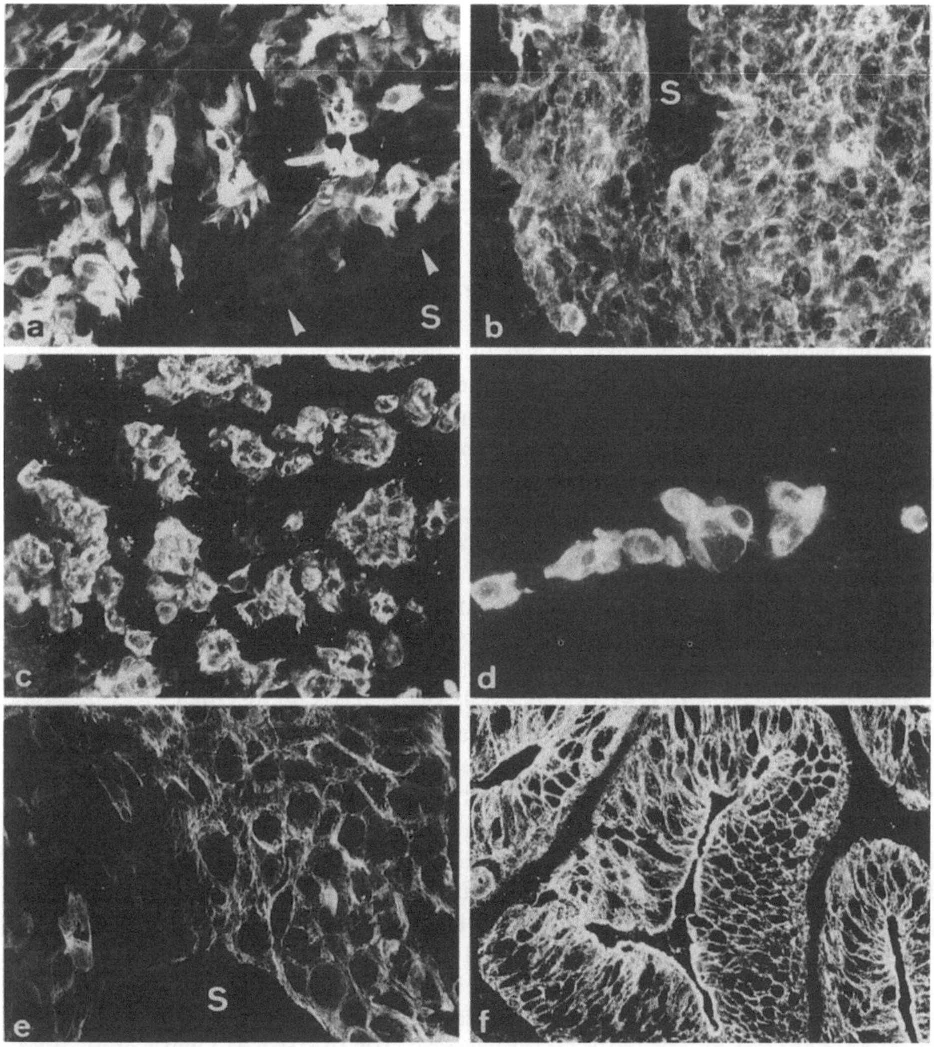

Fig. 4a–f. Detection of individual cytokeratin polypeptides in carcinomas using selective cytokeratin antibodies (frozen sections; immunofluorescence microscopy). **a** In a cornifying squamous cell carcinoma of the skin, antibody $K_K8.60$ against CKs nos. 10/11 stains a prominent proportion of tumor cells with central location within the tumor cell nodule (*arrowheads* denote the tumor-stroma(*S*)-interface). **b** Poorly differentiated squamous cell carcinoma of the palatine tonsil positive for antibody 1C7 against CK no. 13. **c, d** Lymphnode metastasis of a transitional cell carcinoma of the urinary bladder, uniformly positive with an antibody ($K_S18.18$) against CK no. 18 (**c**), and heterogeneously positive with antibody 1C7 against CK no. 13; the latter polypeptide is characteristic of transitional cell (and squamous cell) carcinomas. **e** Non-cornifying squamous cell carcinoma of the lung showing heterogeneous staining with antibody CK-2 against the simple-epithelial-type CK no. 18. *S*, stroma. **f** Adenocarcinoma of the colon, uniformly positive with antibody CK-2 against CK no. 18. **a, b, d, f** ×240, **c** ×190, **e** ×300

3.4 Differential Expression of Cytokeratin Polypeptides in Carcinomas

The finding of the cell-type specificity of cytokeratin polypeptide patterns has prompted the question as to whether analysis of the individual cytokeratin polypeptides present in carcinomas and their metastases may reveal the type of carcinoma and the epithelium of origin, as has been suggested by FRANKE et al. (1981c). Table 6 shows the typical cytokeratin polypeptide patterns of various types of epithelial tumors as determined using gel electrophoresis (MOLL et al. 1982, 1983a; QUINLAN et al. 1985; MOLL and FRANKE 1986). Complementary immunocytochemical data have been obtained by several investigators using selective cytokeratin antibodies.

Squamous cell carcinomas of various localizations are primarily characterized by the predominance of stratified-epithelium-type cytokeratins, a feature which distinguishes them from nonsquamous cell carcinomas (MOLL et al. 1982, 1983a, b; NELSON et al. 1984). *Squamous cell carcinomas of the skin* usually additionally express small amounts of the epidermis-typical CK pair 1 and 10/11, indicating a limited amount of terminal epidermal differentiation; antibodies directed against these cytokeratins stain a variable proportion of the cells of such tumors (Fig. 4a; LOENING et al. 1980; VIAC et al. 1982; GATTER et al. 1982; THOMAS et al. 1984; HUSZAR et al. 1986). CK no. 1 has not been detected in *squamous cell carcinomas of the oral cavity and esophagus*, which exhibit a relatively primitive, predominantly stratified-epithelium-type pattern (Table 6; Fig. 4b; MOLL et al. 1982, 1983a; NELSON et al. 1984; FUCHS et al. 1984); rather, they may contain small amounts of the simple-epithelium-type CK no. 19, which is also found in normal stratified squamous epithelia of the mucosae in the basal cell layer (BARTEK et al. 1986; FRANKE et al. 1986).

In contrast, *squamous cell carcinomas of the lung* and *cervix* are characterized by very complex cytokeratin patterns, in which stratified-epithelium-type cytokeratins are predominant, but simple-epithelium-type cytokeratins are also conspicuous (Table 6; MOLL et al. 1983b; BLOBEL et al. 1984; FUCHS et al. 1984; MOLL and FRANKE 1986; see also SAID et al. 1983; BANKS-SCHLEGEL et al. 1984). Accordingly, these tumors are positive, albeit heterogeneously (i.e., only in some cells), for antibodies against the simple-epithelium-type CKs nos. 8 and 18 (Fig. 4e; MOLL et al. 1983b; RAMAEKERS et al. 1983a, 1985b; DEBUS et al. 1984; GATTER et al. 1985). RAMAEKERS et al. (1985b) found that more than 90% of the squamous cell carcinomas of the lung that they examined were positively stained with an antibody against CK no. 18. Interestingly, squamous cell carcinomas of both lung and cervix usually arise (via squamous metaplasia) from simple or pseudostratified epithelia. It should be noted that the degree of differentiation may modify the cytokeratin pattern of carcinomas to a certain, although usually limited, extent (Table 6). As to whether differences in the cytokeratin pattern of squamous cell carcinomas of different origin is of any histodiagnostic value is a question that must remain open until larger series of patients have been investigated.

Carcinomas of the gastrointestinal tract exhibit a strikingly high level of conservation of cytokeratin polypeptide expression (MOLL et al. 1982, 1983a; MOLL and FRANKE 1986). Thus, *hepatocellular carcinomas* express the same primitive CK pattern (nos. 8:18) as normal hepatocytes (Table 6; DENK et al. 1982; MOLL et al. 1982). The absence of CKs nos. 7 and 19, which has recently also been demonstrated at the immunocytochemical level using polypeptide-specific antibodies (OSBORN et al. 1986), could probably be used to distinguish such tumors from various gastrointestinal adenocarcinomas, including cholangiocellular carcinomas (OSBORN et al. 1986; MOLL and FRANKE 1986). Hepatocellular carcinomas have also been found to be negative for the relatively-broad-spectrum antibodies AE1 (SPAGNOLO et al. 1985) and 34βE12 (GOWN and VOGEL 1985), this being in accordance with the fact that these tumors express only CKs nos. 8 and 18.

Adenocarcinomas of the colon also exhibit a very characteristic cytokeratin pattern that is identical to that of normal enterocytes, i.e., the presence of CKs nos. 8, 18, and 19, but with CK no. 7 being consistently absent (Table 6; MOLL et al. 1982). The absence of CK no. 7 is especially important for their differential diagnosis with respect to *adenocarcinomas of the pancreas* and *cholangiocellular carcinomas*, which both consistently express CK no. 7 (MOLL et al. 1982, 1983a; OSBORN et al. 1986). This CK can now be specifically detected using immunocytochemical procedures (OSBORN et al. 1986).

Adenocarcinomas of the endometrium and *ovary* often express the full complement of simple-epithelium-type CKs (nos. 7, 8, 18, and 19), but some lack CK no. 7 (Table 6; MOLL et al. 1983b; MOLL and FRANKE 1986; for the co-expression of vimentin, see Section 2.5). The finding of small amounts of the stratified-epithelium-type CK no. 5 in an endometrial adenocarcinoma (MOLL and FRANKE 1986) may indicate a correlation with squamous metaplastic processes. In endometrial adenocarcinomas with squamous metaplasia, antibodies against epidermal cytokeratins stain squamous foci much more strongly than glandular areas (BONAZZI DEL POGGETTO et al. 1983; WARHOL et al. 1984).

Most *invasive ductal and lobular carcinomas of the breast* express all of the simple-epithelium cytokeratins, i.e., nos. 7, 8, 18, and 19, as revealed in biochemical (Table 6; MOLL and FRANKE 1986) and immunocytochemical studies (ALTMANNSBERGER et al. 1986) and are thus related to the luminal rather than the myoepithelial cells of normal breast tissue. Most ductal carcinomas of the breast are strongly positive for the antibody 34βE12 (GOWN and VOGEL 1985). However, in another study, two well-differentiated breast carcinomas differed from all other types of adenocarcinomas tested due to their positive reaction with antibody clone 78, which is relatively specific for stratified squamous epithelial cells (VAN MUIJEN et al. 1984a). Interestingly, in one study, 1 of the 12 ductal carcinomas tested exhibited a positive reaction with antibody CK B1, which is specific for basal (squamous)/myoepithelial cells (ALTMANNSBERGER et al. 1986). Similar observations have been made by NAGLE et al. (1986). In biochemical investigations, certain rare breast carcinomas have been found to contain small amounts of stratified-epithelium-type CKs (MOLL et al. 1982; MOLL and FRANKE 1986). Thus, there appears to be a minor subtype of ductal carcinoma of the breast that expresses some cytokeratins typical of stratified epithelia (and myoepithelia) but whose morphology is not detectably different from that of the common breast carcinoma types (ALTMANNSBERGER et al. 1986). Intraepidermal Paget cells of mammary and extramammary *Paget's disease* express simple-epithelium-type cytokeratins (Moll and Moll 1985; Kariniemi et al. 1985; Nagle et al. 1985a) like breast carcinoma and secretory sweat gland cells and thus can easily be distinguished, using antibodies against such cytokeratins, from surrounding epidermal keratinocytes, as well as from normal and malignant melanocytes.

Interesting differences in cytokeratin polypeptide expression have also been found in *thyroid gland carcinomas*, some of which also express vimentin (Section 2.5). It has been found that, whereas all types of such carcinomas, including follicular carcinomas, exhibit staining for simple-epithelium-type cytokeratins, only papillary carcinomas are positive for antibodies against epidermal cytokeratins (Miettinen et al. 1984). The additional cytokeratin polypeptides of papillary carcinomas responsible for this immunoreactivity have yet to be determined.

Renal cell carcinomas exclusively contain simple-epithelium-type cytokeratins in varying patterns, in correlation with the histological tumor type (Pitz et al. 1987). While clear cell carcinomas express predominantly CKs nos. 8/18, eosinophilic-granular and basophilic (small cell cuboidal) carcinomas often also reveal CKs nos. 7 and 19 (for the simultaneous expression of vimentin, see Section 2.5). Chromophobe cell carcinomas typically contain CKs nos. 8 and 18 and, in some cases, no. 7 (Pitz et al. 1987).

Adenocarcinomas of the lung exhibit all four simple-epithelium-type cytokeratins, but they appear to lack stratified-epithelium-type cytokeratins (Blobel et al. 1984). Histologically, these tumors may be difficult to distinguish from malignant *pleural mesotheliomas* of the epithelial type. Attempts have been made to distinguish between these tumors on the basis of their cytokeratin expression (for a recent review, see Corson 1986). Epithelial mesotheliomas have been found to react positively with antibodies against epidermal (callus) keratins, while pulmonary adenocarcinomas are negative or react only weakly with such antibodies (Schlegel et al. 1980; Corson and Pinkus 1982). This is explained by the results of a recent biochemical study, in which certain stratified-epithelium-type cytokeratins, notably CK no. 5, were present in all five epithelial mesotheliomas tested and in four of the five biphasic pleural mesotheliomas examined (Table 6; Blobel et al. 1985b; see also Walts et al. 1983). Therefore, the use of antibodies specific for stratified-epithelium-type cytokeratins may prove to be very helpful in tackling this difficult differential-diagnostic problem. The expression of vimentin in the various types of mesotheliomas has been discussed above (Section 2.5).

Small cell carcinomas of the lung, which are characterized by the presence of certain neuroendocrine features such as neurosecretory granules, express cytokeratins (for references, see Section 2.6) of the simple-epithelium-type, of which only CKs nos. 8 and 18 are present in sizeable amounts, these being arranged in very delicate fibrils in the oat cell type of these tumors (Table 6; Blobel et al. 1985a; Moll and Franke 1986). Therefore, formalin-fixed tissue sections of such tumors are not usually stained by antibodies against epidermal keratins (Gusterson et al. 1982). *Neuroendocrine (Merkel cell) carcinomas of the skin*, whose histology closely resembles that of metastatic small cell carcinomas, can be distinguished from the latter by their fairly consistent (although sometimes rather weak) expression of neurofilaments (see Section 2.6). Merkel cell carcinomas also express simple-epithelium-type cytokeratins (nos. 8, 18 and, sometimes, 19; MOLL and FRANKE 1985; HOEFLER et al. 1985; MOLL et al. 1986). In addition, a cytoskeletal protein with an apparent molecular weight of 46,000 (IT protein) has been found in Merkel cell carcinomas but appears to be absent in other types of neuroendocrine carcinomas, including small cell carcinomas of the lung (MOLL and FRANKE 1985). This protein, which appears to be related to cytokeratins, is also a characteristic feature of intestinal epithelium and colonic adenocarcinomas (MOLL and FRANKE 1985; MOLL and FRANKE, unpublished results) and would appear to be useful for the differential diagnosis of these tumors.

The normal urothelium exhibits a unique and complex cytokeratin pattern (Table 6; MOLL et al. 1983a; ACHTSTÄTTER et al. 1985). Some antibodies against CK no. 18 only stain the superficial (umbrella) cells of normal urothelium and of low-grade papillary carcinomas, whereas they stain all cell layers in high-grade papillary carcinomas and invasive *transitional cell carcinomas* (RAMAEKERS et al. 1985b). Biochemically, many transitional cell carcinomas show cytokeratin patterns resembling that found in normal urothelium, including the expression of CK no. 13 (Table 6; ACHTSTÄTTER et al. 1985; MOLL and FRANKE 1986; see also Fig. 4c, d) but some grade-III tumors appear to have lost some of these polypeptides.

All of these data clearly show that there are profound differences in the patterns of cytokeratin polypeptide expression in different types of carcinomas. In our own studies, we have not found significant differences between the cytokeratin patterns of primary tumors and their metastases (MOLL and FRANKE 1986). It has to be borne in mind that some cytokeratin polypeptides may be heterogeneously expressed in the cells of a particular tumor, and that the degree of differentiation may also influence the pattern of cytokeratin expression. The expression of cytokeratins can also be influenced by experimental conditions, e.g., alterations in the concentration of vitamin A modify the expression of certain cytokeratin polypeptides, whereas others remain constant (FUCHS and GREEN 1981; ECKERT and GREEN 1984). It seems likely that the stability of expression of some cytokeratins is higher than that of others. In spite of this (limited) variability, it is evident from the data discussed here that many carcinoma types exhibit characteristic cytokeratin polypeptide phenotypes. Many of these differences can be detected using available polypeptide-specific antibodies, and it appears that, for a number of problems of differential diagnosis which cannot be satisfactorily resolved by morphological procedures alone, cytokeratin typing will prove to be very useful diagnostic adjunct.

4 Tissue Polypeptide Antigen (TPA) and Its Relationship to Cytokeratins

Tissue polypeptide antigen (TPA), which was originally prepared from the insoluble tissue residue of pooled carcinomas, was one of the first serum tumor markers (BJÖRKLUND and BJÖRKLUND 1957). The serum TPA level is often

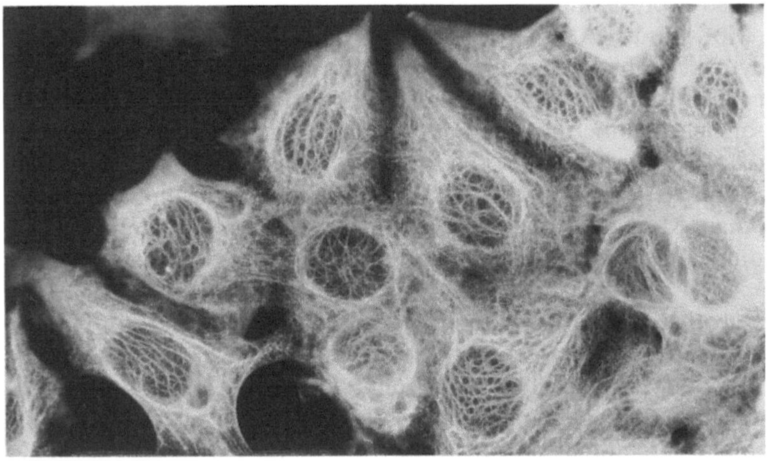

Fig. 5. Immunofluorescence microscopy of cultured cells of the human mammary carcinoma-derived cell line MCF-7, using rabbit antibodies against TPA (Weber et al. 1984). Note the decoration of a fibrillar cytoplasmic network as is also seen with antibodies against cytokeratins. × 600 (courtesy of Dr. M. Osborn, Göttingen)

elevated in patients with carcinomas and appears to be correlated with the proliferative activity of carcinomas (Björklund 1980). TPA has therefore been regarded as being a marker of proliferation (Björklund 1980).

TPA can be localized in normal and malignant cells using immunohistochemical techniques. In cultured epithelial cells, antibodies against TPA stain a fibrillar cytoplasmic network exhibiting a perinuclear concentration (Fig. 5; Kirsch et al. 1983; Weber et al. 1984), which is very reminiscent of the pattern of cytokeratin filament staining (Franke et al. 1979d). At immunoelectron microscopy, antibodies against TPA can be seen to bind to cytokeratin filaments (Zimmer et al. 1985). In a detailed immunohistochemical study, Nathrath et al. (1985) detected TPA in most types of epithelial cells, there being only a few exceptions, e.g., the epidermis, hepatocytes, and suprabasal layers of noncornifying stratified squamous epithelia (see also Loening et al. 1983). This cell-type distribution of TPA is almost identical with that of CK no. 19 (Bartek et al. 1986; Franke et al. 1986).

Using immunohistochemical techniques, TPA has also been detected in carcinomas of the breast (Björklund et al. 1982), in salivary gland tumors (Seifert and Caselitz 1983; Caselitz et al. 1983), squamous cell carcinomas of the esophagus, and various gastrointestinal carcinomas (Beham et al. 1986), exhibiting intracellular staining patterns similar to those of cytokeratins (Beham et al. 1986).

Biochemically, TPA purified according to the procedures of Björklund and co-workers (for a review, see Björklund 1980) consists of several polypeptides with a main component (TPA:B_1) having an apparent molecular weight of 43,000. Recently, in immunoblotting experiments, an antibody against TPA was found to bind specifically to the simple-epithelium-type CK polypeptides nos. 8, 18, and 19 (Weber et al. 1984). Similar observations have been reported by Beham et al. (1986). When the amino acid sequences determined for fragments of TPA subunits (Redelius et al.

1980; LÜNING et al. 1980; LÜNING and NILSSON 1983) are compared with the recently established amino acid sequences of simple-epithelium-type cytokeratins, a peptide fragment from the TPA subunit C reveals identity with respect to 48 out of 58 amino acids with a sequence in the α-helical rod domain of bovine CK no. 8 (MAGIN et al. 1986). Furthermore, complete identity with respect to amino acid sequence has now been shown for the TPA subunits C and E1 and human CK no. 8 (LEUBE et al. 1986).

The relationship between TPA and cytokeratins thus appears to be somewhat clarified. TPA is probably a mixture of proteolytic fragments containing the relatively stable α-helical rod domains of simple-epithelium-type cytokeratins. These fragments are probably released into the serum during necrosis and the lysis of carcinoma cells. Therefore, the TPA serum level is only indirectly correlated with proliferation, and TPA should be regarded as being a broad-spectrum epithelial tumor marker. These findings in no way compromise the clinical usefulness of TPA assays in tumor management, but together with our increasing knowledge of the molecular structure of cytokeratins, may prompt the search for more specific probes for the serological analysis of fragments of the various defined cytokeratin and intermediate filament polypeptides that might provide information concerning the nature of particular tumors.

5 Conclusion

In this article, the molecular and cell-biological features of cytokeratins as well as the various areas of their application in histopathological diagnosis have been discussed. It is evident that the expression of cytokeratins is closely correlated with epithelial differentiation and remaining stable in malignant tumors. Although of value, other epithelial markers such as epithelial membrane antigen (EMA) exhibit expression patterns in tumors that are less reliably correlated with the epithelial phenotype (PINKUS et al. 1986). In addition to the detectable pattern of cytokeratins being useful for the establishment of the epithelial nature of a primary or metastatic tumor, other intermediate-filament types co-expressed with cytokeratins may be highly relevant for diagnosis. Furthermore, cytokeratin subtyping, which is now feasible using immunohistochemical procedures, makes it possible to resolve a spectrum of different cytokeratin phenotypes, thus allowing carcinomas to be precisely classified into subgroups according to their type and origin.

The information obtained by the analysis of cytokeratin and other intermediate-filament proteins is, in principle, independent of morphological features. While it is true that the majority of tumors can be diagnosed using conventional morphological techniques, these markers are very valuable diagnostic adjuncts in problem cases exhibiting an equivocal morphological appearance. The correlation between the expression of the intermediate-filament proteins and previously known morphological data concerning normal histology, normal development, and tumor pathology is strikingly good. Thus, the examination of the expression of intermediate filaments provides a complement to established morphological procedures. It should be emphasized that all histological tumor markers have to be applied critically, and that morphological investigation retains a decisive role. However, in problem cases, the necessity of intermediate-filament and cytokeratin analysis for diagnosis is indisputable.

Acknowledgements. The author thanks Dr. W.W. Franke, Heidelberg, and Dr. W. Thoenes, Mainz, for helpful discussions, H. Breitbach and K. Essling for photographic and graphic work, A. Cowin for correcting, and C. Bürkner for typing of the manuscript, and the *Deutsche Forschungsgemeinschaft* for financial support.

References

Abenoza P, Sibley RK (1986) Chordoma: An immunohistologic study. Hum Pathol 17:744–747

Achtstätter T, Moll R, Moore B, Franke WW (1985) Cytokeratin polypeptide patterns of different epithelia of the human male urogenital tract. J Histochem Cytochem 33:415–426

Achtstätter T, Moll R, Anderson A, Kuhn C, Pitz S, Schwechheimer K, Franke WW (1986) Expression of glial filament protein (GFP) in nerve sheaths and non-neural cells re-examined using monoclonal antibodies, with special emphasis on the co-expression of GFP and cytokeratins in epithelial cells of human salivary gland and pleomorphic adenomas. Differentiation 31:206–227

Altmannsberger M, Osborn M, Hölscher A, Schauer A, Weber K (1981a) The distribution of keratin type intermediate filaments in human breast cancer: an immunohistological study. Virchows Arch (Cell Pathol) 37:277–284

Altmannsberger M, Osborn M, Hölscher A, Schauer A, Weber K (1981b) Antibodies to different intermediate filament proteins: cell type-specific markers on paraffin-embedded human tissues. Lab Invest 45:427–434

Altmannsberger M, Weber K, Hölscher A, Schauer A, Osborn M (1982) Antibodies to intermediate filaments as diagnostic tools: human gastrointestinal carcinomas express keratin. Lab Invest 46:520–526

Altmannsberger M, Osborn M, Schäfer H, Schauer A, Weber K (1984) Distinction of nephroblastomas from other childhood tumors using antibodies to intermediate filaments. Virchows Arch (Cell Pathol) 45:113–124

Altmannsberger M, Dirk T, Droese M, Weber K, Osborn M (1986) Keratin polypeptide distribution in benign and malignant breast tumors: subdivision of ductal carcinomas using monoclonal antibodies. Virchows Arch (Cell Pathol) 51:265–275

Bachmann S, Kriz W, Kuhn C, Franke WW (1983) Differentiation of cell types in the mammalian kidney by immunofluorescence microscopy using antibodies to intermediate filament proteins and desmoplakins. Histochemistry 77:365–394

Banks-Schlegel SP, McDowell EM, Wilson TS, Trump BF, Harris CC (1984) Keratin proteins in human lung carcinomas: combined use of morphology, keratin immunocytochemistry, and keratin immunoprecipitation. Am J Pathol 114:273–286

Bannasch P, Zerban H, Schmid E, Franke WW (1980) Liver tumors distinguished by immunofluorescence microscopy with antibodies to proteins of intermediate filaments. Proc Natl Acad Sci USA 77:4948–4952

Bartek J, Taylor-Papadimitriou J, Miller N, Millis R (1985) Patterns of expression of keratin 19 as detected with monoclonal antibodies in human breast tissues and tumours. Int J Cancer 36:299–306

Bartek J, Bartkova J, Taylor-Papadimitriou J, Rejthar A, Kovarik J, Lukas Z, Vojtesek B (1986) Differential expression of keratin 19 in normal human epithelial tissues revealed by monospecific monoclonal antibodies. Histochem J 18:565–575

Barwick K, Madri JA (1982) An immunohistochemical study of adenomatoid tumors utilizing keratin and Factor VIII antibodies. Evidence for a mesothelial origin. Lab Invest 47:276–280

Battifora H, Silva EG (1986) The use of antikeratin antibodies in the immunohistochemical distinction between neuroendocrine (Merkel cell) carcinoma of the skin, lymphoma, and oat cell carcinoma. Cancer 58:1040–1046

Battifora H, Sun T-T, Bahu RM, Rao S (1980) The use of antikeratin antiserum as a diagnostic tool: Thymoma versus lymphoma. Hum Pathol 11:635–641
Battifora H, Sheibani K, Tubbs RR, Kopinski MI, Sun T-T (1984) Antikeratin antibodies in tumor diagnosis. Distinction between seminoma and embryonal carcinoma. Cancer 54:843–848
Beham A, Weybora W, Lackinger E, Denk H, Björklund V, Björklund B (1986) Distribution of TPA and cytokeratins in gastrointestinal carcinomas as revealed by immunohistochemistry. Virchows Arch (Pathol Anat) 409:641–655
Björklund B (1980) On the nature and clinical use of Tissue Polypeptide Antigen (TPA). Tumor Diagnostik 1:9–20
Björklund B, Björklund V (1957) Antigenicity of pooled human malignant and normal tissues by cytoimmunological techniques: presence of an insoluble, heat-labile tumor antigen. Int Arch Allergy 10:153–184
Björklund V, Björklund B, Wittekind C, von Kleist S (1982) Immunohistochemical localization of Tissue Polypeptide Antigen (TPA) and carcinoembryonic antigen (CEA) in breast cancer. Acta Pathol Scand A6:471–476
Blobel GA, Moll R, Franke WW, Vogt-Moykopf I (1984) Cytokeratins in normal lung and lung carcinomas. I. Adenocarcinomas, squamous cell carcinomas and cultured cell lines. Virchows Arch (Cell Pathol) 45:407–429
Blobel GA, Gould VE, Moll R, Lee I, Huszar M, Geiger B, Franke WW (1985a) Coexpression of neuroendocrine markers and epithelial cytoskeletal proteins in bronchopulmonary neuroendocrine neoplasms. Lab Invest 52:39–51
Blobel GA, Moll R, Franke WW, Kayser KW, Gould VE (1985b) The intermediate filament cytoskeleton of malignant mesotheliomas and its diagnostic significance. Am J Pathol 121:235–247
Bonazzi del Poggetto C, Virtanen I, Lehto V-P, Wahlström T, Saksela E (1983) Expression of intermediate filaments in ovarian and uterine tumors. Int J Gynecol Pathol 1:359–366
Broers JLV, Carney DN, de Ley L, Vooijs GP, Ramaekers FCS (1985a) Differential expression of intermediate filament proteins distinguishes classic from variant small-cell lung cancer cell lines. Proc Natl Acad Sci USA 82:4409–4413
Broers J, Huysmans A, Moesker O, Vooijs P, Ramaekers F, Wagenaar S (1985b) Small cell lung cancers contain intermediate filaments of the cytokeratin type (letter). Lab Invest 52:113–115
Brulet P, Babinet C, Kemler R, Jacob F (1980) Monoclonal antibodies against trophectoderm-specific markers during mouse blastocyst formation. Proc Natl Acad Sci USA 77:4113–4117
Caselitz J, Osborn M, Seifert G, Weber K (1981) Intermediate-sized filament proteins (prekeratin, vimentin, desmin) in the normal parotid gland and parotid gland tumors. Immunofluorescence study. Virchows Arch (Pathol Anat) 393:273–281
Caselitz J, Seifert G, Björklund B, Björklund V (1983) Detection of Tissue Polypeptide Antigen in salivary gland tumors. Appl Pathol 1:115–120
Caselitz J, Becker J, Seifert G, Weber K, Osborn M (1984) Coexpression of keratin and vimentin filaments in adenoid cystic carcinomas of salivary glands. Virchows Arch (Pathol Anat) 403:337–344
Caselitz J, Osborn M, Hamper K, Wustrow J, Rauchfuß A, Weber K (1986a) Pleomorphic adenomas, adenoid cystic carcinomas and adenolymphomas of salivary glands analysed by a monoclonal antibody against myoepithelial/basal cells. An immunohistochemical study. Virchows Arch (Pathol Anat) 409:805–816
Caselitz J, Walther B, Wustrow J, Seifert G, Weber K, Osborn M (1986b) A monoclonal antibody that detects myoepithelial cells in exocrine glands, basal cells in other epithelia and basal and suprabasal cells in certain hyperplastic tissues. Virchows Arch (Pathol Anat) 409:725–738
Chase D, Enzinger FM, Weiss SW, Langloss JM (1984) Keratin in epithelioid sarcoma: An immunohistochemical study. Am J Surg Pathol 8:435–441
Churg A (1985) Immunohistochemical staining for vimentin and keratin in malignant mesothelioma. Am J Surg Pathol 9:360–365
Coindre J-M, Rivel J, Trojani M, de Mascarel I, de Mascarel A (1986) Immunohistological study in chordomas. J Pathol 150:61–63
Cooper D, Schermer A, Sun T-T (1985) Biology of disease. Classification of human epithelia and their neoplasms using monoclonal antibodies to keratins: strategies, applications, and limitations. Lab Invest 52:243–256
Corson JM (1986) Keratin protein immunohistochemistry in surgical pathology practice. In: Sommers

SC, Rosen PP, Fechner RE (eds) Pathology annual, vol 21, part 2. Appleton-Century-Crofts, Norwalk, Connecticut, pp 47–81

Corson JM, Pinkus GS (1982) Mesothelioma: Profile of keratin proteins and carcinoembryonic antigen. An immunoperoxidase study of 20 cases and comparison with pulmonary adenocarcinomas. Am J Pathol 108:80–87

Czernobilsky B, Moll R, Levy R, Franke WW (1985) Co-expression of cytokeratin and vimentin filaments in mesothelial, granulosa and rete ovarii cells of the human ovary. Eur J Cell Biol 37:175–190

Czernobilsky B, Moll R, Leppien G, Schweikhart G, Franke WW (1987) Desmosomal plaque-associated vimentin filaments in human ovarian granulosa cell tumors of various histologic patterns. Am J Pathol 126:476–486

Dabbs DJ, Geisinger KR, Norris HT (1986) Intermediate filaments in endometrial and endocervical carcinomas: The diagnostic utility of vimentin patterns. Am J Surg Pathol 10:568–576

Debus E, Weber K, Osborn M (1982) Monoclonal cytokeratin antibodies that distinguish simple from stratified squamous epithelia: Characterization on human tissues. EMBO J 1:1641–1647

Debus E, Moll R, Franke WW, Weber K, Osborn M (1984) Immunohistochemical distinction of human carcinomas by cytokeratin typing with monoclonal antibodies. Am J Pathol 114:121–130

Denk H, Krepler R, Lackinger E, Artlieb U, Franke WW (1982) Biochemical and immunocytochemical analysis of the intermediate filament cytoskeleton in human hepatocellular carcinomas and in hepatic nodules of mice. Lab Invest 46:584–596

Denk H, Weybora W, Ratschek M, Sohar R, Franke WW (1985) Distribution of vimentin, cytokeratins, and desmosomal plaque proteins in human nephroblastoma as revealed by specific antibodies: co-existence of cell groups of different degrees of epithelial differentiation. Differentiation 29:88–97

Denk H, Moll R, Weybora W, Lackinger E, Vennigerholz F, Beham A, Franke WW (1986) Cytokeratin and vimentin filaments, and desmosomal plaque proteins in testicular seminomas and non-seminomatous germ cell tumors as revealed by immunofluorescence microscopy. Virchows Arch (Pathol Anat) 410:295–307

Doran TI, Vidrich A, Sun T-T (1980) Intrinsic and extrinsic regulation of the differentiation of skin, corneal and esophageal epithelial cells. Cell 22:17–25

Droese M, Altmannsberger M, Dralle H (1984) Verteilung der Intermediärfilamente in Schilddrüsencarcinomen. Verh Dtsch Ges Pathol 68:498

Eckert RL, Green H (1984) Cloning of cDNAs specifying vitamin A-responsive human keratins. Proc Natl Acad Sci USA 81:4321–4325

Erlandson RA (1981) Diagnostic transmission electron microscopy of human tumors. Masson, New York, pp 1–186

Erlandson RA (1984) Diagnostic immunohistochemistry of human tumors. An interim evaluation. Am J Surg Pathol 8:615–624

Espinoza CG, Azar HA (1982) Immunochemical localization of keratin-type proteins in epithelial neoplasms. Correlation with electron microscopic findings. Am J Clin Pathol 78:500–507

Eto H, Hashimoto K, Kobayashi H, Matsumoto M, Kanzaki T, Mehregan AH, Weiss RA (1985) Monoclonal antikeratin antibody: Production, characterization, and immunohistochemical application. J Invest Dermatol 84:404–409

Fidler IJ, Gerstein DM, Hart IR (1978) The biology of cancer invasion and metastasis. Adv Cancer Res 28:149–250

Franke WW, Schmid E, Osborn M, Weber K (1978a) Different intermediate-sized filaments distinguished by immunofluorescence microscopy. Proc Natl Acad Sci USA 75:5034–5038

Franke WW, Weber K, Osborn M, Schmid E, Freudenstein C (1978b) Antibody to prekeratin. Decoration of tonofilament-like arrays in various cells of epithelial character. Exp Cell Res 116:429–445

Franke WW, Appelhans B, Schmid E, Freudenstein C, Osborn M, Weber K (1979a) Identification and characterization of epithelial cells in mammalian tissues by immunofluorescence microscopy using antibodies to prekeratin. Differentiation 15:7–25

Franke WW, Grund C, Schmid E (1979b) Intermediate-sized filaments present in Sertoli cells are of vimentin-type. Eur J Cell Biol 19:269–275

Franke WW, Schmid E, Osborn M, Weber K (1979c) Intermediate-sized filaments of human endothelial cells. J Cell Biol 81:570–580

Franke WW, Schmid E, Weber K, Osborn M (1979d) HeLa cells contain intermediate-sized filaments of the prekeratin type. Exp Cell Res 118:95–109

Franke WW, Schmid E, Winter S, Osborn M, Weber K (1979e) Widespread occurrence of intermediate-sized filaments of the vimentin type in cultured cells from diverse vertebrates. Exp Cell Res 123:25–46

Franke WW, Schmid E, Freudenstein C, Appelhans B, Osborn M, Weber K, Keenan TW (1980) Intermediate-sized filaments of the prekeratin type in myoepithelial cells. J Cell Biol 84:633–654

Franke WW, Denk H, Kalt R, Schmid E (1981a) Biochemical and immunological identification of cytokeratin proteins present in hepatocytes of mammalian liver tissue. Exp Cell Res 131:299–318

Franke WW, Mayer D, Schmid E, Denk H, Borenfreund E (1981b) Differences of expression of cytoskeletal proteins in cultured rat hepatocytes and hepatoma cells. Exp Cell Res 134:345–365

Franke WW, Schiller DL, Moll R, Winter S, Schmid E, Engelbrecht I, Denk H, Krepler R, Platzer B (1981c) Diversity of cytokeratins. Differentiation specific expression of cytokeratin polypeptides in epithelial cells and tissues. J Mol Biol 153:933–959

Franke WW, Winter S, Grund C, Schmid E, Schiller DL, Jarasch E-D (1981d) Isolation and characterization of desmosome-associated tonofilaments from rat intestinal brush border. J Cell Biol 90:116–127

Franke WW, Schmid E, Schiller DL, Winter S, Jarasch ED, Moll R, Denk H, Jackson BW, Illmensee K (1982) Differentiation-related patterns of expression of proteins of intermediate-sized filaments in tissues and cultured cells. Cold Spring Harbor Symp Quant Biol 46:431–453

Franke WW, Moll R, Achtstaetter T, Kuhn C (1986) Cell typing of epithelia and carcinomas of the female genital tract using cytoskeletal proteins as markers. Banbury Report 21: Viral etiology of cervical cancer. Cold Spring Harbor Laboratory, Cold Spring Harbor, pp 121–148

Franz JK, Gall L, Williams MA, Picheral B, Franke WW (1983) Intermediate-sized filaments in a germ cell: Expression of cytokeratins in oocytes and eggs of the frog *Xenopus*. Proc Natl Acad Sci USA 80:6254–6258

Fuchs E, Green H (1978) The expression of keratin genes in epidermis and cultured epidermal cells. Cell 15:887–897

Fuchs E, Green H (1979) Multiple keratins of cultured human epidermal cells are translated from different mRNA molecules. Cell 17:573–582

Fuchs E, Green H (1981) Regulation of terminal differentiation of cultured human keratinocytes by vitamin A. Cell 25:617–625

Fuchs EV, Coppock SM, Green H, Cleveland DW (1981) Two distinct classes of keratin genes and their evolutionary significance. Cell 27:75–84

Fuchs E, Grace MP, Kim KH, Marchuk D (1984) Differential expression of two classes of keratins in normal and malignant epithelial cells and their evolutionary conservation. In: The transformed phenotype, cancer cells, vol 1. Cold Spring Harbor Laboratory, Cold Spring Harbor, pp 161–167

Fuchs E, Hanukoglu I, Marchuk D, Grace MP, Kim KH (1985) The nature and significance of differential keratin gene expression. Ann NY Acad Sci 455:436–448

Gabbiani G, Kapanci Y, Barazzone O, Franke WW (1981) Immunochemical identification of intermediate-sized filaments in human neoplastic cells. Am J Pathol 104:206–216

Gatter KC, Abdulaziz Z, Beverley P, Corvalan JRF, Ford C, Lane EB, Mota M, Nash JRG, Pulford K, Stein H, Taylor-Papadimitriou J, Woodhouse C, Mason DY (1982) Use of monoclonal antibodies for the histopathologic diagnosis of human malignancy. J Clin Pathol 35:1253–1267

Gatter KC, Pulford KAF, Vanstapel MJ, Roach B, Mortimer P, Woolston RE, Taylor-Papadimitriou J, Lane EB, Mason DY (1984) An immunohistological study of benign and malignant skin tumors. I. Epithelial aspects. Histopathology 8:209–227

Gatter KC, Dunnill MS, Pulford KAF, Heryet A, Mason DY (1985) Human lung tumors: a correlation of antigenic profile with histological type. Histopathology 9:805–823

Ghadially FN (1980) Diagnostic electron microscopy of tumours. Butterworths, London, pp 51–67

Gigi O, Geiger B, Eshhar Z, Moll R, Schmid E, Winter S, Schiller DL, Franke WW (1982) Detection of a cytokeratin determinant common to diverse epithelial cells by a broadly cross-reacting monoclonal antibody. EMBO J 1:1429–1437

Gigi-Leitner O, Geiger B, Levy R, Czernobilsky B (1986) Cytokeratin expression in squamous metaplasia of the human uterine cervix. Differentiation 31:191–205

Gould VE, Moll R, Moll I, Lee I, Franke WW (1985) Neuroendocrine (Merkel) cells of the skin: hyperplasias, dysplasias, and neoplasms. Lab Invest 52:334–353

Gown AM, Vogel AM (1982) Monoclonal antibodies to intermediate filament proteins of human cells: Unique and cross-reacting antibodies. J Cell Biol 95:414–424

Gown AM, Vogel AM (1984) Monoclonal antibodies to human intermediate filament proteins: II. Distribution of filament proteins in normal human tissues. Am J Pathol 114:309–321

Gown AM, Vogel AM (1985) Monoclonal antibodies to human intermediate filament proteins: III. Analysis of tumors. Am J Clin Pathol 84:413–424

Gusterson B, Mitchell D, Warburton M, Sloane J (1982) Immunohistochemical localisation of keratin in human lung tumours. Virchows Arch (Pathol Anat) 394:269–277

Hanukoglu I, Fuchs E (1983) The cDNA sequence of a type-II cytoskeletal keratin reveals constant and variable structural domains among keratins. Cell 33:915–924

Hatzfeld M, Franke WW (1985) Pair formation and promiscuity of cytokeratins: Formation in vitro of heterotypic complexes and intermediate-sized filaments by homologous and heterologous recombinations of purified polypeptides. J Cell Biol 101:1826–1841

Heid HW, Werner E, Franke WW (1986) The complement of native α-keratin polypeptides of hair-forming cells. A subset of eight polypeptides that differ from epithelial cytokeratins. Differentiation 32:101–119

Herman CJ, Moesker O, Kant A, Huysmans A, Vooijs GP, Ramaekers FCS (1983) Is renal cell (Grawitz) tumor a carcinosarcoma? Virchows Arch (Pathol Anat) 44:73–83

Hoefler H, Denk H (1984) Immunocytochemical demonstration of cytokeratin in gastrointestinal carcinoids and their probable precursor cells. Virchows Arch (Pathol Anat) 403:235–240

Hoefler H, Denk H, Walter GF (1984a) Immunohistochemical demonstration of cytoskeleton-proteins in normal pituitary gland and in pituitary adenomas. Virchows Arch (Pathol Anat) 404:359–368

Hoefler H, Kerl H, Rauch H-J, Denk H (1984b) New immunocytochemical observations with diagnostic significance in cutaneous neuroendocrine carcinoma. Am J Dermatopathol 6:525–530

Hoefler H, Kerl H, Lackinger E, Helleis G, Denk H (1985) The intermediate filament cytoskeleton of cutaneous neuroendocrine carcinoma (Merkel cell tumor): Immunohistochemical and biochemical analyses. Virchows Arch (Pathol Anat) 406:339–350

Holden J, Churg A (1984) Immunohistochemical staining for keratin and carcinoembryonic antigen in the diagnosis of malignant mesothelioma. Am J Surg Pathol 8:277–279

Holthöfer H, Miettinen M, Paasivuo R, Lehto VP, Linder E, Alfthan O, Virtanen I (1983) Cellular origin and differentiation of renal cell carcinomas. Lab Invest 49:317–326

Holthöfer H, Miettinen A, Lehto VP, Lehtonen E, Virtanen I (1984) Expression of vimentin and cytokeratin types of intermediate filament proteins in developing and adult human kidneys. Lab Invest 50:552–559

Huitfeldt HS, Brandtzaeg P (1985) Various keratin antibodies produce immunohistochemical staining of human myocardium and myometrium. Histochemistry 83:381–389

Huszar M, Gigi-Leitner O, Moll R, Franke WW, Geiger B (1986) Monoclonal antibodies to various acidic (type I) cytokeratins of stratified epithelia: Selective markers for stratification and squamous cell carcinomas. Differentiation 31:141–153

Jasani B, Edwards RE, Thomas ND, Gibbs AR (1985) The use of vimentin antibodies in the diagnosis of malignant mesothelioma. Virchows Arch (Pathol Anat) 406:441–448

Kahn HJ, Yeger H, Baumal R, Thom H, Phillips JM (1983) Categorization of pediatric neoplasms by immunostaining with antiprekeratin and antivimentin antisera. Cancer 51:645–653

Kahn HJ, Huang SN, Hanna WM, Baumal R, Phillips JM (1984) Immunohistochemical localization of epidermal and Mallory body cytokeratins in undifferentiated epithelial tumors. Comparison with ultrastructural features. Am J Clin Pathol 81:184–191

Kariniemi A-L, Ramaekers F, Lehto V-P, Virtanen I (1985) Paget cells express cytokeratins typical of glandular epithelia. Br J Dermatol 112:179–183

Karsten U, Papsdorf G, Roloff G, Stolley P, Abel H, Walther I, Weiss H (1985) Monoclonal anti-cytokeratin antibody from a hybridoma clone generated by electrofusion. Eur J Cancer Clin Oncol 21:733–740

Kartenbeck J, Schwechheimer K, Moll R, Franke WW (1984) Attachment of vimentin filaments to desmosomal plaques in human meningiomal cells and arachnoidal tissue. J Cell Biol 98:1072–1081

Kim KH, Rheinwald JG, Fuchs EV (1983) Tissue specificity of epithelial keratins: Differential expression of mRNAs from two multigene families. Mol Cell Biol 3:495–502
Kirsch J, Oehr P, Winkler C (1983) Localization of Tissue Polypeptide Antigen in interphase HeLa cells by immunofluorescence microscopy. Tumor Diagnostik 4:222–224
Knapp AC, Franke WW, Heid H, Hatzfeld M, Jorcano JL, Moll R (1986) Cytokeratin no. 9, an epidermal type I keratin characteristic of a special program of keratinocyte differentiation displaying body site specificity. J Cell Biol 103:657–667
Knight J, Gusterson B, Jones RR, Landells W, Wilson P (1985) Monoclonal antibodies specific for subsets of epidermal keratins: Biochemical and immunocytochemical characterization – applications in pathology and cell culture. J Pathol 145:341–354
Krepler R, Denk H, Artlieb U, Moll R (1982) Immunocytochemistry of intermediate filament proteins present in pleomorphic adenomas of the human parotid gland: Characterization of different cell types in the same tumor. Differentiation 21:191–199
Lane EB (1982) Monoclonal antibodies provide specific intramolecular markers for the study of epithelial tonofilament organization. J Cell Biol 92:665–673
Lane EB, Bartek J, Purkis PE, Leigh IM (1985) Keratin antigens in differentiating skin. Ann NY Acad Sci 455:241–258
Lang AB, Odermatt BF, Ruettner JR (1986) Monoclonal antibodies to human cytokeratins: Application to various epithelial and mesothelial cells. Exp Cell Biol 54:61–72
LaRocca P, Rheinwald JG (1984) Coexpression of simple epithelial keratins and vimentin by human mesothelium and mesothelioma in vivo and in culture. Cancer Res 44:2991–2999
Lauder I, Holland D, Mason DY, Gowland G, Cunliffe WJ (1984) Identification of large cell undifferentiated tumours in lymph nodes using leukocyte common and keratin antibodies. Histopathol 8:259–272
Lazarides E (1982) Intermediate filaments: a chemically heterogeneous, developmentally regulated class of proteins. Annu Rev Biochem 51:219–250
Lehto V-P, Stenman S, Miettinen M, Dahl D, Virtanen I (1983) Expression of a neural type of intermediate filament as a distinguishing feature between oat cell carcinoma and other lung cancers. Am J Pathol 110:113–118
Lehto V-P, Miettinen M, Virtanen I (1985) A dual expression of cytokeratin and neurofilaments in bronchial carcinoid cells. Int J Cancer 35:421–425
Leube RE, Bosch FX, Romano V, Zimbelmann R, Hoefler H, Franke WW (1986) Cytokeratin expression in simple epithelia. III. Detection of mRNAs encoding human cytokeratins nos. 8 and 18 in normal and tumor cells by hybridization with cDNA sequences in vitro and in situ. Differentiation 33:69–85
Loening T, Staquet M-J, Thivolet J, Seifert G (1980) Keratin polypeptide distribution in normal and diseased epidermis and oral mucosa. Immunohistochemical study on unaltered epithelium and inflammatory pre-malignant and malignant lesions. Virchows Arch (Pathol Anat) 388:273–288
Loening T, Caselitz J, Otto HF (1981) The epithelial framework of the thymus in normal and pathological conditions. Virchows Arch (Pathol Anat) 392:7–20
Loening T, Kühler C, Caselitz J, Stegner H-E (1983) Keratin and Tissue Polypeptide Antigen profiles of the cervical mucosa. Int J Gynecol Pathol 2:105–112
Lüning B, Nilsson U (1983) Sequence homology between tissue polypeptide antigen (TPA) and intermediate filament (IF) proteins. Acta Chem Scand 37:731–733
Lüning B, Wiklund B, Redelius P, Björklund B (1980) Biochemical properties of tissue polypeptide antigen. Biochem Biophys Acta 624:90–101
Madri JA, Barwick KW (1982) An immunohistochemical study of nasopharyngeal neoplasms using keratin antibodies. Epithelial versus nonepithelial neoplasms. Am J Surg Pathol 6:143–149
Magin TM, Jorcano JL, Franke WW (1986) Cytokeratin expression in simple epithelia. II. cDNA cloning and sequence characteristics of bovine cytokeratin A (no. 8). Differentiation 30:254–264
Makin CA, Bobrow LG, Bodmer WF (1984) Monoclonal antibody to cytokeratin for use in routine histopathology. J Clin Pathol 37:975–983
McNutt MA, Bolen JW, Gown AM, Hammar SP, Vogel AM (1985) Coexpression of intermediate filaments in human epithelial neoplasms. Ultrastruct Pathol 9:31–43
Miettinen M, Damjanov I (1985) Coexpression of keratin and vimentin in epithelioid sarcoma. Am J Surg Pathol 9:460–463

Miettinen M, Lehto V-P, Virtanen I (1982a) Keratin in the epithelial-like cells of classic biphasic synovial sarcoma. Virchows Arch (Cell Pathol) 40:157–162

Miettinen M, Lehto V-P, Virtanen I (1982b) Nasopharyngeal lymphoepithelioma. Histological diagnosis as aided by immunohistochemical demonstration of keratin. Virchows Arch (Cell Pathol) 40:163–169

Miettinen M, Lehto V-P, Dahl D, Virtanen I (1983a) Differential diagnosis of chordoma, chondroid, and ependymal tumors as aided by anti-intermediate filament antibodies. Am J Pathol 112:160–169

Miettinen M, Lehto V-P, Virtanen I (1983b) Monophasic synovial sarcoma of spindle-cell type: Epithelial differentiation as revealed by ultrastructural features, content of prekeratin and binding of peanut agglutinin. Virchows Arch (Cell Pathol) 44:187–199

Miettinen M, Franssila K, Lehto V-P, Paasivuo R, Virtanen I (1984a) Expression of intermediate filaments in thyroid gland and thyroid tumors. Lab Invest 50:262–270

Miettinen M, Lehto V-P, Virtanen I (1984b) Antibodies to intermediate filament proteins in the diagnosis and classification of human tumors. Ultrastruct Pathol 7:83–107

Miettinen M, Clark R, Lehto V-P, Virtanen I, Damjanov I (1985a) Intermediate-filament proteins in parathyroid glands and parathyroid adenomas. Arch Pathol Lab Med 109:986–989

Miettinen M, Lehto V-P, Dahl D, Virtanen I (1985b) Varying expression of cytokeratin and neurofilaments in neuroendocrine tumors of human gastrointestinal tract. Lab Invest 52:429–436

Miettinen M, Lehto V-P, Virtanen I (1985c) Immunofluorescence microscopic evaluation of the intermediate filament expression of the adrenal cortex and medulla and their tumors. Am J Pathol 118:360–366

Miettinen M, Virtanen I, Talerman A (1985d) Intermediate filament proteins in human testis and testicular germ-cell tumors. Am J Pathol 120:402–410

Miettinen M, Clark R, Virtanen I (1986) Intermediate filament proteins in choroid plexus, ependyma and their tumors. Am J Pathol 123:231–240

Moll I, Moll R (1985) Cells of extramammary Paget's disease express cytokeratins different from those of epidermal cells. J Invest Dermatol 84:3–8

Moll R, Franke WW (1985) Cytoskeletal differences between human neuroendocrine tumors: A cytoskeletal protein of molecular weight 46,000 distinguishes cutaneous from pulmonary neuroendocrine neoplasms. Differentiation 30:165–175

Moll R, Franke WW (1986) Cytochemical cell typing of metastatic tumors according to their cytoskeletal proteins. In: Lapis K, Liotta LA, Rabson AS (eds) Biochemistry and molecular genetics of cancer metastasis. Nijhoff, The Hague, pp 101–114

Moll R, Franke WW, Schiller DL, Geiger B, Krepler R (1982) The catalog of human cytokeratins: patterns of expression in normal epithelia, tumors and cultured cells. Cell 31:11–24

Moll R, Krepler R, Franke WW (1983a) Complex cytokeratin polypeptide patterns observed in certain human carcinomas. Differentiation 23:256–269

Moll R, Levy R, Czernobilsky B, Hohlweg-Majert P, Dallenbach-Hellweg G, Franke WW (1983b) Cytokeratins of normal epithelia and some neoplasms of the female genital tract. Lab Invest 49:599–610

Moll R, Moll I, Franke WW (1984) Identification of Merkel cells in human skin by specific cytokeratin antibodies: changes of cell density and distribution in fetal and adult plantar epidermis. Differentiation 28:136–154

Moll R, Cowin P, Kapprell H-P, Franke WW (1986a) Desmosomal proteins: New markers for identification and classification of tumors. Lab Invest 54:4–25

Moll R, Osborn M, Hartschuh W, Moll I, Mahrle G, Weber K (1986b) Variability of expression and arrangement of cytokeratin and neurofilaments in cutaneous neuroendocrine carcinomas: Immunocytochemical and biochemical analysis of 12 cases. Ultrastruct Pathol 10:473–495

Moll R, Lee I, Gould VE, Berndt R, Roessner A, Franke WW (1987) Immunocytochemical analysis of Ewing's tumors: Patterns of expression of intermediate filaments and desmosomal proteins indicate cell type heterogeneity and pluripotential differentiation. Am J Pathol 127:288–304

Moss F, Bobrow LG, Sheppard MN, Griffiths M, Rowe D, Beverly PCL, Addis B, Souhami RL (1986) Expression of epithelial and neural antigens in small cell and non small cell lung carcinoma. J Pathol 149:103–111

Nagle RB, McDaniel KM, Clark VA, Payne CM (1983) The use of antikeratin antibodies in the diagnosis of human neoplasms. Am J Clin Pathol 79:458–466

Nagle RB, Licas DO, McDaniel KM, Clark VA, Schmalzel BS (1985a) Paget's cells. New evidence linking mammary and extramammary Paget cells to a common cell phenotype. Am J Clin Pathol 83:431–438

Nagle RB, Moll R, Weidauer H, Nemetschek H, Franke WW (1985b) Different patterns of cytokeratin expression in the normal epithelia of the upper respiratory tract. Differentiation 30:130–140

Nagle RB, Böcker W, Davis JR, Heid HW, Kaufmann M, Lucas DO, Jarasch E-D (1986) Characterization of breast carcinomas by two monoclonal antibodies distinguishing myoepithelial from luminal epithelial cells. J Histochem Cytochem 34:869–881

Nathrath WBJ, Heidenkummer P, Björklund V, Björklund B (1985) Distribution of tissue polypeptide antigen (TPA) in normal human tissues: Immunohistochemical study on unfixed, methanol-, ethanol-, and formalin-fixed tissues. J Histochem Cytochem 33:99–109

Nelson WG, Sun T-T (1983) The 50- and 58-kdalton keratin classes as molecular markers for stratified squamous epithelia: cell culture studies. J Cell Biol 97:244–251

Nelson WG, Battifora H, Santana H, Sun T-T (1984) Specific keratins as molecular markers for neoplasms with a stratified epithelial origin. Cancer Res 44:1600–1603

Nicolson GL (1984) Cell surface molecules and tumor metastasis. Exp Cell Res 150:3–22

Osborn M, Weber K (1983) Biology of disease. Tumor diagnosis by intermediate filament typing: a novel tool for surgical pathology. Lab Invest 48:372–394

Osborn M, van Lessen G, Weber K, Klöppel G, Altmannsberger M (1986) Differential diagnosis of gastrointestinal carcinomas by using monoclonal antibodies specific for individual keratin polypeptides. Lab Invest 55:497–504

Overbeck J von, Stähli C, Gudat F, Carmann H, Lautenschlager C, Dürmüller U, Takacs B, Miggiano V, Staehelin T, Heitz PU (1985) Immunohistochemical characterization of an anti-epithelial monoclonal antibody (mAB lu-5). Virchows Arch (Pathol Anat) 407:1–12

Pinkus GS, Kurtin PJ (1985) Epithelial membrane antigen – a diagnostic discriminant in surgical pathology. Hum Pathol 16:929–940

Pinkus GS, O'Connor EM, Etheridge CL, Corson JM (1985) Optimal immunoreactivity of keratin proteins in formalin-fixed, paraffin-embedded tissue requires preliminary trypsinization. J Histochem Cytochem 33:465–473

Pinkus GS, Etheridge CL, O'Connor EM (1986) Are keratin proteins a better tumor marker than epithelial membrane antigen? A comparative immunohistochemical study of various paraffin-embedded neoplasms using monoclonal and polyclonal antibodies. Am J Clin Pathol 85:269–277

Pitz S, Moll R, Störkel S, Thoenes W (1987) Expression of intermediate filament proteins in subtypes of renal cell carcinomas and in renal oncocytomas: Distinction in two classes of renal cell tumors. Lab Invest 56:642–653

Pulford KAF, Gatter KC, Mason DY (1985) The characterization of two monoclonal anti-keratin antibodies and their use in the study of epithelial disorders. Histopathology 9:825–840

Quinlan RA, Schiller DL, Hatzfeld M, Achtstätter T, Moll R, Jorcano JL, Magin TM, Franke WW (1985) Patterns of expression and organization of cytokeratin intermediate filaments. Ann NY Acad Sci 455:282–306

Racadot J, Olivier L, Porcile E, De Grye C, Klotz HP (1964) Adénome hypophysaire de type "mixte" avec symptomatologie acromégalique. II. Étude au microscope optique et au microscope électronique. Ann Endocrinol (Paris) 25:503–507

Raju T, Adelman LS, Dahl D, Bignami A (1983) Localization of keratin in the notochord and in notochord derived tumors – Immunohistological study of rat embryo and human chordoma. Int J Dev Neurosci 1:375–382

Ramaekers FCS, Osborn M, Schmid E, Weber K, Bloemendal H, Franke WW (1980) Identification of the cytoskeletal proteins in lens-forming cells, a special epithelioid cell type. Exp Cell Res 127:309–327

Ramaekers FCS, Huysmans A, Moesker O, Kant A, Jap PHK, Herman C, Vooijs P (1983a) Monoclonal antibody to keratin filaments, specific for glandular epithelia and their tumors. Use in surgical pathology. Lab Invest 49:353–361

Ramaekers FCS, Puts JJG, Moesker O, Kant A, Jap PHK, Vooijs P (1983b) Demonstration of keratin in human adenocarcinomas. Am J Pathol 111:213–223

Ramaekers FCS, Puts JJG, Moesker O, Kant A, Huysmans A, Haag D, Jap PHK, Herman CJ, Vooijs GP (1983c) Antibodies to intermediate filament proteins in the immunohistochemical identification of human tumours: An overview. Histochem J 15:691–713

Ramaekers F, Feitz W, Moesker O, Schaart G, Herman C, Debruyne F, Vooijs P (1985a) Antibodies to cytokeratin and vimentin in testicular tumour diagnosis. Virchows Arch (Pathol Anat) 408:127–142

Ramaekers FCS, Moesker O, Huysmans A, Schaart G, Westerhof G, Wagenaar SJ SC, Herman CJ, Vooijs GP (1985b) Intermediate filament proteins in the study of tumor heterogeneity: An in-depth study of tumors of the urinary and respiratory tracts. Ann NY Acad Sci 455:614–634

Redelius P, Lüning B, Björklund B (1980) Chemical studies of tissue polypeptide antigen (TPA). II. Partial amino acid sequences of cyanogen bromide fragments of TPA subunit B_1. Acta Chem Scand (B) 34:265–273

Rosai J, Pinkus GS (1982) Immunohistochemical demonstration of epithelial differentiation in adamantinoma of the tibia. Am J Surg Pathol 6:427–434

Saba SR, Espinoza CG, Richman AV, Azar HA (1983) Carcinomas of the lung: An ultrastructural and immunocytochemical study. Am J Clin Pathol 80:6–13

Said JW, Nash G, Lee M (1982) Immunoperoxidase localization of keratin proteins, carcinoembryonic antigen, and Factor VIII in adenomatoid tumors: Evidence for a mesothelial derivation. Hum Pathol 13:1106–1108

Said JW, Nash G, Banks-Schlegel S, Sasson AF, Murakami S, Shintaku IP (1983) Keratin in human lung tumors. Pattern of localization of different molecular weight proteins. Am J Pathol 113:27–32

Schermer A, Galvin S, Sun T-T (1986) Differentiation-related expression of a major 64K corneal keratin in vivo and in culture suggests limbal location of corneal epithelial stem cells. J Cell Biol 103:49–62

Schiller DL, Franke WW, Geiger B (1982) A subfamily of relatively large and basic cytokeratin polypeptides as defined by peptide mapping is represented by one or several polypeptides in epithelial cells. EMBO J 1:761–769

Schlegel R, Banks-Schlegel S, McLeod JA, Pinkus GS (1980) Immunoperoxidase localization of keratin in human neoplasms. Am J Pathol 101:41–50

Schröder S, Dockhorn-Dworniczak B, Kastendieck H, Böcker W, Franke WW (1986) Intermediate-filament expression in thyroid gland carcinomas. Virchows Arch (Pathol Anat) 409:751–766

Seifert G, Caselitz J (1983) Tumor markers in parotid gland carcinomas: Immunohistochemical investigations. Cancer Detect Prevent 6:119–130

Sibley RK, Dahl D (1985) Primary neuroendocrine (Merkel cell?) carcinoma of the skin: II. An immunocytochemical study of 21 cases. Am J Surg Pathol 9:109–116

Sieinski W, Dorsett B, Ioachim HL (1981) Identification of prekeratin by immunofluorescence staining in the differential diagnosis of tumors. Hum Pathol 12:452–458

Spagnolo DV, Michie SA, Crabtree GS, Warnke RA, Rouse RV (1985) Monoclonal antikeratin (AE1) reactivity in routinely processed tissue from 166 human neoplasms. Am J Clin Pathol 84:697–704

Steinert PM, Steven AC, Roop DR (1985) The molecular biology of intermediate filaments. Cell 42:411–419

Sun T-T, Green H (1978) Keratin filaments of cultured human epidermal cells: formation of intermolecular disulfide bonds during terminal differentiation. J Biol Chem 253:2053–2060

Sun T-T, Shih C, Green H (1979) Keratin cytoskeletons in epithelial cells of internal organs. Proc Natl Acad Sci USA 76:2813–2817

Sun T-T, Eichner R, Schermer A, Cooper D, Nelson WG, Weiss RA (1984) Classification, expression and possible mechanisms of evolution of mammalian epithelial keratins: A unifying model. In: The transformed phenotype, cancer cells, vol 1. Cold Spring Harbor Laboratory, Cold Spring Harbor, pp 169–176

Tölle, H-G, Weber K, Osborn M (1985) Microinjection of monoclonal antibodies specific for one intermediate filament protein in cells containing multiple keratins allow insight into the composition of particular 10 nm filaments. Eur J Cell Biol 38:234–244

Thoenes W, Stoerkel S, Rumpelt HJ (1985) Human chromophobe cell renal carcinoma. Virchows Arch (Cell Pathol) 48:207–217

Thomas P, Said JW, Nash G, Banks-Schlegel S (1984) Profiles of keratin proteins in basal and squamous cell carcinomas of the skin: an immunohistochemical study. Lab Invest 50:36–41

Tseng SCG, Jarvinen MJ, Nelson WG, Huang J-W, Woodcock-Mitchell J, Sun T-T (1982) Correlation of specific keratins with different types of epithelial differentiation: monoclonal antibody studies. Cell 30:361–372

Upton MP, Hirohashi S, Tome Y, Miyazawa N, Suemasu K, Shimosato Y (1986) Expression of vimentin in surgically resected adenocarcinomas and large cell carcinomas of lung. Am J Surg Pathol 10:560–567

Van Muijen GNP, Ruiter DJ, Ponec M, Huiskens-van der Mey C, Warnaar SO (1984a) Monoclonal antibodies with different specificities against cytokeratins: An immunohistochemical study of normal tissues and tumors. Am J Pathol 114:9–17

Van Muijen GNP, Ruiter DJ, Van Leeuwen C (1984b) Cytokeratin and neurofilament in lung carcinomas. Am J Pathol 116:363–369

Van Muijen GNP, Ruiter DJ, Franke WW, Achtstaetter T, Haasnoot WHB, Ponec M, Warnaar SO (1986) Cell type heterogeneity of cytokeratin expression in complex epithelia and carcinomas as demonstrated by monoclonal antibodies specific for cytokeratins nos. 4 and 13. Exp Cell Res 162:97–113

Viac J, Reano A, Thivolet J (1982) Cytokeratins in human basal and squamous cell carcinomas: biochemical, immunohistological findings and comparisons with normal epithelia. J Cutan Pathol 9:377–390

Viac J, Reano A, Brochier J, Staquet M-J, Thivolet J (1983) Reactivity pattern of a monoclonal antikeratin antibody (KL 1). J Invest Dermatol 81:351–354

Virtanen I, Lehto V-P, Lehtonen E, Vartio T, Stenman S, Kurki P, Wagner O, Small JV, Dahl D, Badley RA (1981) Expression of intermediate filaments in cultured cells. J Cell Sci 50:45–63

Virtanen I, Miettinen M, Lehto V-P, Kariniemi A-L, Paasivuo R (1985) Diagnostic application of monoclonal antibodies to intermediate filaments. Ann NY Acad Sci 455:635–648

Vogel AM, Gown AM, Caughlan J, Haas JE, Beckwith JB (1984) Rhabdoid tumors of the kidney contain mesenchymal specific and epithelial specific intermediate filament proteins. Lab Invest 50:232–238

Waldherr R, Schwechheimer K (1985) Co-expression of cytokeratin and vimentin intermediate-sized filaments in renal cell carcinomas. Comparative study of the intermediate-sized filament distribution in renal cell carcinomas and normal human kidney. Virchows Arch (Pathol Anat) 408:15–27

Walts AE, Said JW, Shintaku IP, Sassoon AF, Banks-Schlegel S (1983) Keratins of different molecular weight in exfoliated mesothelial and adenocarcinoma cells – an aid to cell identification. Am J Clin Pathol 81:442–446

Warhol MJ, Rice RH, Pinkus GS, Robboy SJ (1984) Evaluation of squamous epithelium in adenoacanthoma and adenosquamous carcinoma of the endometrium: Immunoperoxidase analysis of involucrin and keratin localization. Int J Gynecol Pathol 3:82–91

Weber K, Geisler N (1984) Intermediate filaments from wool α-keratins to neurofilaments: A structural overview. In: The transformed phenotype, cancer cells, vol 1. Cold Spring Harbor Laboratory, Cold Spring Harbor, pp 153–159

Weber K, Osborn M, Moll R, Wiklund B, Lüning B (1984) Tissue polypeptide antigen (TPA) is related to the non-epidermal keratins 8, 18 and 19 typical of simple and non-squamous epithelia: Re-evaluation of a human tumor marker. EMBO J 3:2707–2714

Wernert N, Seitz G, Dhom G (1986) Verhalten verschiedener Cytokeratine und von Vimentin in der normalen Prostata von der Foetalzeit bis zum Erwachsenenalter und im Prostatakarzinom. Verh Dtsch Ges Pathol 70:613

Winter S, Jarasch E-D, Schmid E, Franke WW, Denk H (1980) Differences in polypeptide composition of cytokeratin filaments, including tonofilaments, from different epithelial tissues and cells. Eur J Cell Biol 22:371

Wu Y-J, Parker LM, Binder NE, Beckett MA, Sinard JH, Griffiths CT, Rheinwald JG (1982) The mesothelial keratins: A new family of cytoskeletal proteins identified in cultured mesothelial cells and nonkeratinizing epithelia. Cell 31:693–703

Zimmer KP, Caselitz J, Seifert G (1985) Subcellular localization of tissue polypeptide antigen and cytokeratins in epithelial cells (salivary and mammary glands). Virchows Arch (Cell Pathol) 49:161–173

Epithelial Tumor Markers: Oncofetal Antigens (Carcinoembryonic Antigen, Alpha Fetoprotein) and Epithelial Membrane Antigen

G. KLÖPPEL and J. CASELITZ

1 Introduction

Oncofetal antigens are substances that are produced in high quantities during fetal development as well as during evolution of neoplasms in adult life. The best known oncofetal antigens are CEA and AFP. Their discovery marks the beginning of the still continuing search for tumor antigens which may serve as markers for tumor diagnosis and surveillance. Compared with CEA and

AFP, the epithelial membrane antigen (EMA) is of minor significance, but has gained general recognition as an immunocytochemical marker for epithelial tissues. This article places its emphasis on reviewing the usefulness of CEA, AFP and EMA as tissue markers for malignancies.

2 Carcinoembryonic Antigen (CEA)

2.1 General Properties

CEA is distinguished by 3 properties: (1) It shows an unusually high degree of molecular size heterogeneity due to a variable carbohydrate content; (2) it belongs to a family of closely related glycoproteins displaying a wide tissue distribution; and (3) CEA as well as the CEA-related substances reveal an extraordinarily strong immunogenicity. This makes CEA difficult to define solely on the basis of one analytical method. As long as the genes for CEA have not been cloned and sequenced, a combination of methods, particularly including immunological techniques using polyclonal as well as monoclonal antibodies, are necessary to define the CEA molecule and the structure of the other members of the complex CEA family (for detailed reviews see Burtin and Gold 1978; von Kleist 1983; Rogers 1983; Shively and Beatty 1985).

2.2 Chemistry

Perchloric acid extraction of liver metastases of colon adenocarcinomas is the initial procedure most commonly used to obtain CEA (Hammarström 1985). Further purification steps include ion exchange chromatography, gelfiltration, Con A-Sepharose affinity chromatography and immunosorbent separation (Shiveley and Todd 1980). The purity then has to be controlled by SDS-PAGE and immunodiffusion resulting in an about 90% precipitation of the ^{125}I-labelled CEA preparation with specific polyclonal and some monoclonal anti-CEA antibodies. CEA prepared by this technique proved to be a large glycoprotein with beta-electrophoretic mobility and a molecular weight of approximately 180,000 daltons consisting of a single polypeptide chain of 829 residues, and about 40 N-acetylglucosamine-asparagine linked oligosaccharide chains (Chandrasekaran et al. 1983). Although only the N-terminal sequence of the intact molecule and of several tryptic fragments have been determined, it seems that the composition of the peptide chain is remarkably constant. The carbohydrate moiety, in constrast, which comprises about 50–60% of molecule shows a considerable microheterogeneity. These carbohydrate chains are predominantly of the tetra-antennary type bearing many terminal sialic acid residues and some fucose residues. In addition, some CEA molecules may express blood group specificity in their carbohydrate side chains, especially for the blood group precursor (H) antigen (Rogers 1983).

2.3 CEA-related Antigens

So far at least six well established (Table 1) and several putative CEA-related substances have been identified (Matsuoka et al. 1982; Shively and Beatty 1985; Neumaier et al. 1985). These antigens differ from CEA with respect to molecular weight and/or immunoreactivity. The first of the CEA related antigens was reported by von Kleist et al. (1972) and Mach and Pusztazseri (1972) and named nonspecific cross-reacting antigen (NCA) and normal glycoprotein (NGP), respectively. Subsequently a number of other substances with different molecular weight, different carbohydrate content and variations in the N-terminal sequence of the amino acids were described, which now constitute the CEA family.

Table 1. CEA and related antigens[a]

Molecular weight (KD)	Terms and synonyms	Carbohydrate content (%)
175–180	CEA (carcinoembryonic antigen)	50–60
150–170	NCA 160 (nonspecific cross-reacting antigen)	45–50
	NCA 2	
	NFA 2 (normal faecal antigen)	
	MA (meconium antigen)	
128	CRA-128 (colorectal antigen)	
	CEA-low	
100	Meconium-antigen 100	
90–110	NCA 95	20–30
	TEX (tumor-extracted antigen)	
80–85	BGP 1 (biliary glycoprotein)	–40
55–65	NCA 55	–30
	NCA 1	
	NGP (normal glycoprotein)	
	CEX	
	CCEA 2	
20–30	NFA 1	10–20

[a] from Rogers (1983); Buchegger et al. (1984); Shively and Beatty (1985); Neumaier et al. (1985)

At present, the exact molecular relationship of NCA to CEA is unclear, and it is far beyond the scope of this review to discuss the issue in detail (for reviews see Wagener and Breuer 1982; Rogers 1983; Shively and Beatty 1985). However, it should be mentioned here that on the basis of the most recent works two models for the structure of CEA and NCA have been proposed (Grunert et al. 1985). In the first model it is suggested that all NCA and CEA molecules share an unique polypeptide chain, but differ widely in the degree of their glycolysation. The second model suggests the existence of closely related repeat units in the polypeptide chain and explains the molecular heterogeneity of CEA and NCA by the different frequency of the occurrence of the repeats in the respective molecules. Evidence in support of the second model has been recently provided by gene cloning studies which demonstrated the existence of two very closely related repeats in CEA (Zimmermann et al. 1987).

2.4 Immunology

CEA and the other members of the CEA family are highly immunogenic in all species commonly used for immunization. The majority of the polyclonal antisera raised against CEA react with antigenic determinants both on CEA and CEA related substances. Similar reactivity patterns have become known for many monoclonal antibodies which are currently available. This demonstrates that the CEA related substances have several epitopes in common with CEA and that specificity of a monoclonal antibody for CEA can only be determined by exhaustively investigating all of the potentially cross-reacting antigens (Primus et al. 1983; Wagener et al. 1983a; Grunert et al. 1985; Muraro et al. 1985).

Hedin et al. (1982) generated eight monoclonal antibodies recognizing six determinants on native CEA, two of the epitopes probably occurring twice in CEA. Two antibodies also reacted with NCA. Recently, the same authors extended the number of epitopes to nine, six of which were also found on NCA 160, NCA 95 and/or NCA 55 (Hammarström 1985). Kuroki et al. (1984) reported 8 different epitopes, while Muraro et al. (1985) found at least 5 epitopes. Using a set of 18 monoclonal antibodies Haggerty et al. (1986) defined 9 epitopes when only the reaction with the intact molecule was considered. Twelve epitopes were found when data on binding to NCA, denaturated CEA and CEA fragments were taken into account as well. Seven binding sites appeared to be CEA specific. Since not all of the monoclonal antibodies recognized epitopes in native CEA, it was suggested that some epitopes may also be carbohydrate in nature.

Human anti-CEA autoantibodies were detected in the sera of some cancer patients. These antibodies were also capable of reacting with the CEA molecule in tissue sections (Endo et al. 1985).

2.5 Immunocytochemical Spectrum

The fact that CEA belongs to a family of closely related glycoproteins which have several epitopes in common but show a different tissue distribution necessitates the use of specific and well defined antibodies when CEA is localized by immunostaining at the cellular level. Thus it has been shown that with a commercial, unabsorbed polyclonal antiserum against CEA, which contained antibodies to NCA, 93% of breast carcinomas, 85% of mastopathic lesions and 66% of fibroadenomas stained positively, while, after absorption with normal tissues, the same antiserum labelled only 42% of the carcinomas and none of the mastopathic lesions or fibroadenomas (Nap et al. 1984). Another point which has to be considered when studying the immunocytochemical distribution of CEA is the influence exerted by fixation procedures on the antigenicity of CEA. Tsutsumi et al. (1984) demonstrated that noncancerous formalin fixed paraffin-embedded tissues generally had a lessened stainability with (polyclonal and monoclonal) antisera against CEA and NCA when compared with the reactivity pattern on frozen sections. On sections from formalin-fixed, paraffin-embedded pancreatic duct type carcinomas monoclonal antibodies specific for CEA gave a less intense staining reaction than on frozen sections (Tsutsumi et al. 1984; Bätge et al. 1985). No significant differences between the rate of CEA positivity in frozen and paraffin sections, however, were noted in gastric carcinomas (Heidl et al. 1986).

2.5.1 Fetal Tissues

Among the fetal tissues the gastrointestinal tract is the major source of CEA (Gold and Freedman 1965; Wagener et al. 1983b). Here it can be detected consistently from the 16th week of gestation onwards, with the greatest intensity in the colon (Kauf and Borchard 1985). The fetal pancreas, in contrast to the gut, reveals non immunostaining for CEA (Tsutsumi et al. 1984).

2.5.2 Normal Adult Tissues

Recent studies based on sensitive staining techniques and using polyclonal antisera absorbed with perchloric extracts of lung (or spleen) and normal colonic mucosa, or monoclonal antibodies specific for CEA demonstrated CEA in the superficial part of normal colonic epithelium both in patients with and without carcinomas (Wagener et al. 1978; Primus et al. 1983; Tsutsumi et al. 1984; Yachi et al. 1984) (Fig. 1). The failure of other authors to demonstrate CEA in normal colonic mucosa (Isaacson and LeVann 1976; Goldenberg et al. 1978; Wiley et al. 1981) may be explained by different specificities and sensitivi-

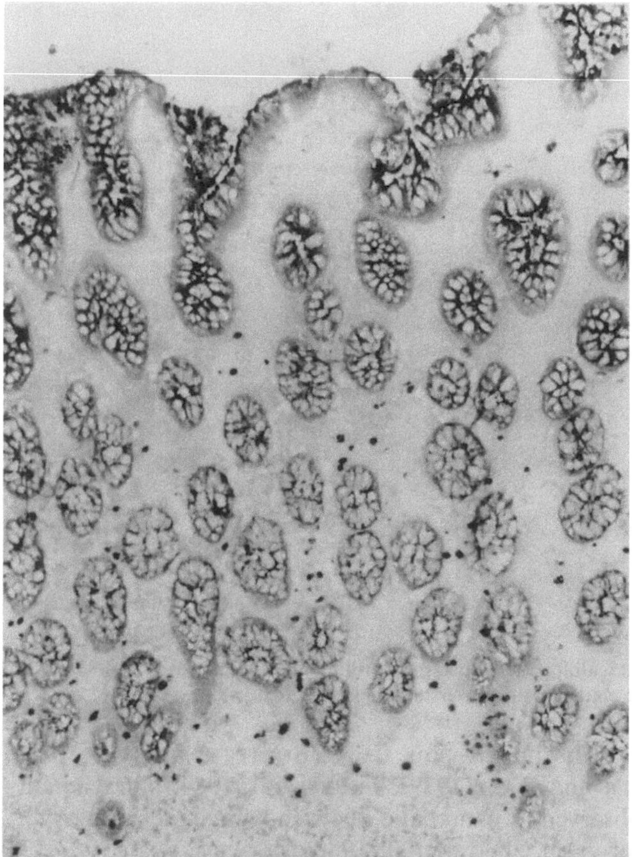

Fig. 1. Frozen section from normal colonic mucosa stained with monoclonal antibody C1P83 specific for CEA. Positive labelling of epithelial cell surfaces. ×125

ties of the various CEA antisera and immunocytochemical methods used. Conflicting results were also reported for the stomach and small intestine in adults. While BURTIN (1985) found no CEA at both sites, TSUTSUMI et al. (1984) applying a monoclonal antibody specific to CEA, observed CEA positivity along the luminal rims of the antral foveolar glands, and ISAACSON and JUDD (1978) using a polyclonal CEA antiserum with minimal anti-NCA activity reported on CEA positivity of the small intestine on the surface of the villi, along the lining of the crypt lumens and within goblet cells. The tissues that are normally free of CEA include those of lung, liver, pancreas, kidney, spleen, prostate and endocrine glands (Fig. 2).

2.5.3 Adult Tissues with Nonneoplastic Changes

Some epithelial tissues which are usually negative for CEA may become positive when undergoing metaplastic or hypoplastic changes. Thus gastric intestinal

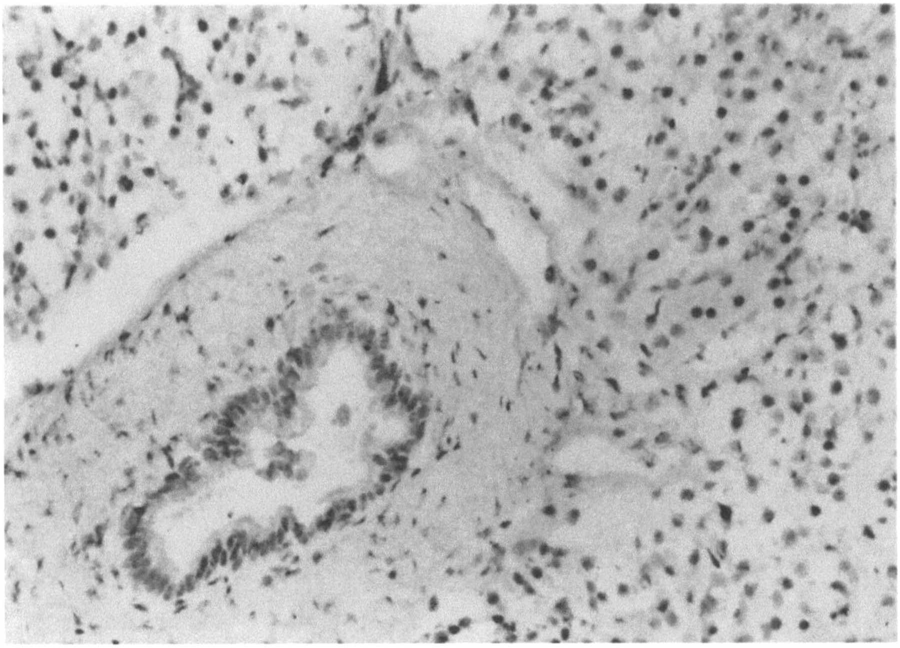

Fig. 2. Frozen section from normal adult pancreas, stained with monoclonal antibody C1P83 specific for CEA. No staining of the duct epithelium (counterstaining with hematoxylin). ×250

metaplasia was described to be positive for CEA (Burtin et al. 1973, 1977; Denk et al. 1973; Wurster and Rapp 1979; Yachi et al. 1984). Occasionally CEA positivity was also observed in pancreatic ducts showing mucinous hypertrophy and papillary hyperplasia (Bätge et al. 1986; Klöppel et al. 1987). In these lesions CEA positivity appeared to be discrete and was confined to the apical border of the cells (Fig. 3). Allum et al. (1986) observed a similar staining pattern in six of ten specimens of chronic pancreatitis. No CEA immunoreactivity could be identified in the hyperplastic prostate (Purnell et al. 1984).

2.5.4 Benign Neoplastic Lesions

Originally it was thought that adenomas of the colon were CEA negative. Using frozen sections and more sensitive immunocytochemical methods it has been shown that all epithelial lesions of the colon, irrespective of benign or malignant, stain for CEA (Burtin et al. 1972; Stenger et al. 1979). In the pancreas benign lesions such as serous cystadenoma or solid-cystic (papillary-cystic) tumor are consistently CEA negative, while in mucinous cystic tumors with their known malignant potential, those cells showing the greatest degree of atypia may stain for CEA (Bätge et al. 1986). Fibroadenomas of the breast and other benign lesions were found to be virtually or totally CEA negative (Walker 1980; von Kleist et al. 1982; Mansour et al. 1983; Nap et al. 1984). In pleomorphic adenomas of salivary glands, 11% were positively stained with a monoclonal antibody to CEA (Sumitomo et al. 1987). In adnexal tumors of the skin, CEA was only detected in sweat gland neoplasms (Penneys et al. 1982).

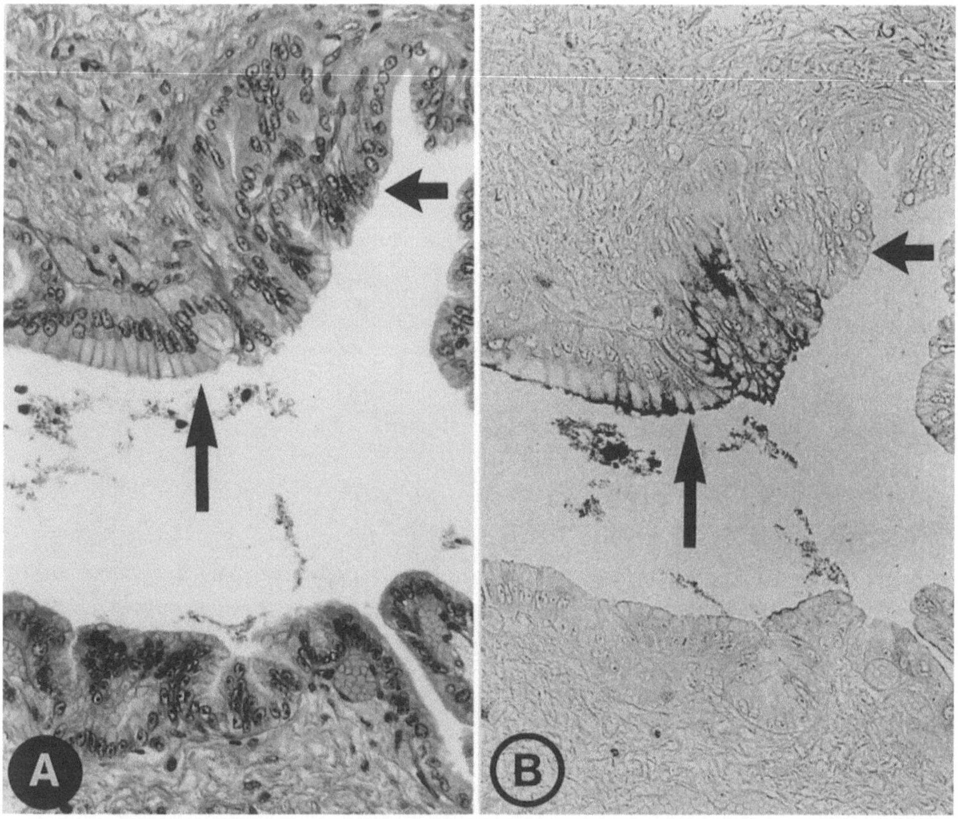

Fig. 3. Consecutive formalin-fixed, paraffin-embedded sections from normal pancreas, stained with Hematoxylin and Eosin (**A**) and monoclonal antibody BMA130c recognizing CEA (**B**). Apical labelling of papillary hyperplasia of duct epithelium (*long arrows*). Normal duct cells remain unstained (*short arrows*). ×250

2.5.5 Malignant Neoplastic Lesions

Common features for tumors that produce CEA are that they are of epithelial origin and belong predominantly to the variety of mucus producing adenocarcinomas. This association has become particularly evident when well absorbed polyclonal antisera or monoclonal antibodies specific to CEA were used. Consequently the vast majority of the adenocarcinomas arising in the gastrointestinal and biliary tract, the pancreas, lung, breast, salivary glands, ovary and endometrium belong to the CEA positive category. A noteworthy exception from this rule is the well known CEA production observed in the C-cell carcinoma of the thyroid (DeLellis et al. 1978) and in squamous and some small cell carcinomas of the lung (Wachner et al. 1984). In general, CEA expression is stronger in well than in poorly differentiated tumors. In well differentiated gland forming adenocarcinomas with mucus production most of the CEA is localized to the

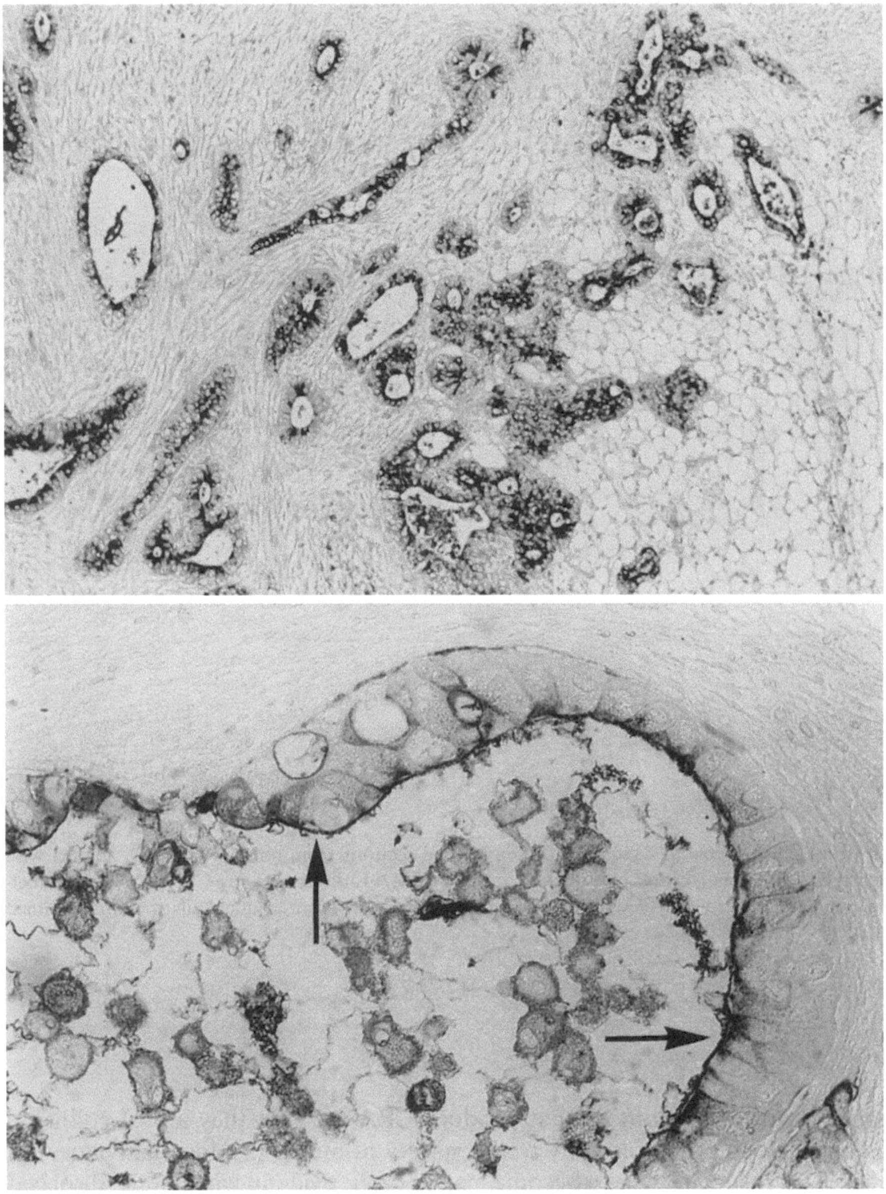

Fig. 4. Formalin-fixed, paraffin embedded sections from well differentiated ductal adenocarcinoma of pancreas, stained with monoclonal antibody BMA 130b recognizing CEA and NCA 95. Mainly apical labelling of neoplastic epithelial cells (*arrows*). ×125 and 250

apical cell pole, the luminal border of the epithelium and the luminal content (DENK et al. 1973; STENGER et al. 1979; O'BRIEN et al. 1981; NIELSEN and TEGLBJAERG 1982; BÄTGE et al. 1986) (Fig. 4). Less well differentiated and poorly differentiated adenocarcinomas including signet ring cell carcinomas, where some or most of the mucus is retained within the cell, predominantly show

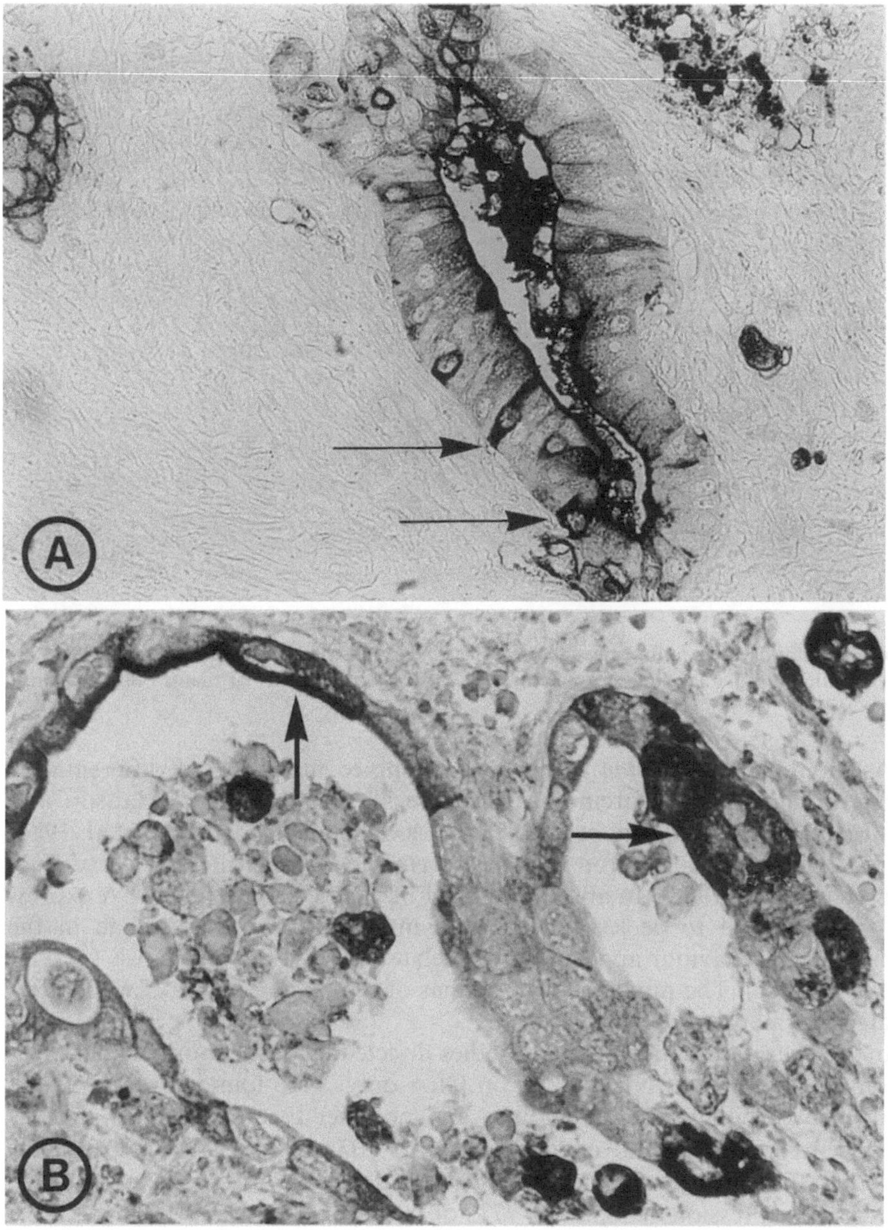

Fig. 5. Formalin-fixed, paraffin-embedded sections from two pancreatic ductal carcinomas of moderate (**A**) and poor differentiation (**B**), stained with BMA 130 b (CEA and NCA 95). Apical and strongly intracytoplasmic labelling (*arrows*). **A** × 250, **B** 640

intracytoplasmic immunostaining for CEA (Burtin et al. 1973; Denk et al. 1972 and 1973; Nielsen and Teglbjaerg 1982; Heidl et al. 1986) (Fig. 5). CEA is usually absent from anaplastic carcinomas unless there are foci of glandular differentiation where CEA may be positive.

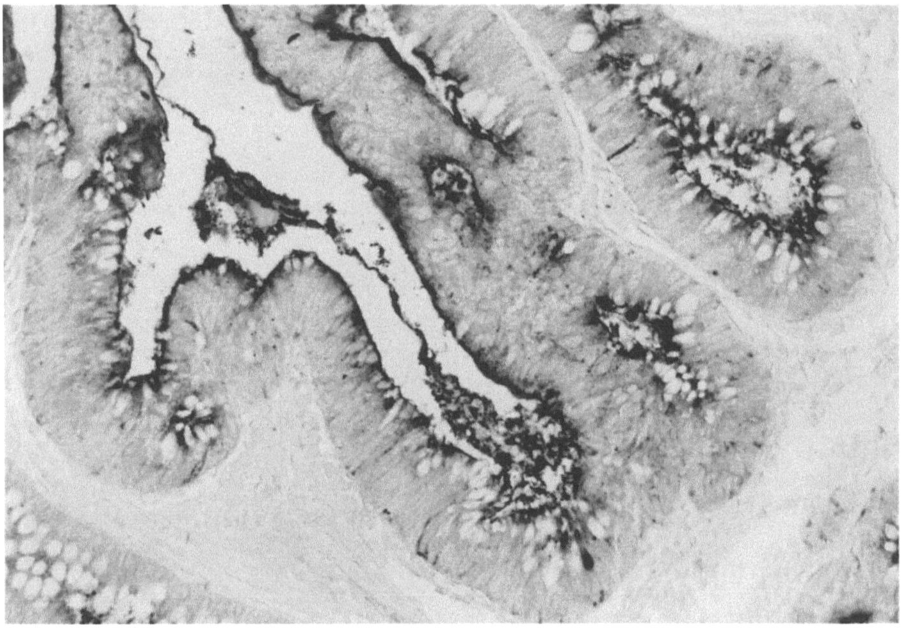

Fig. 6. Formalin-fixed, paraffin-embedded section from colon carcinoma, stained with BMA130c (CEA specific). Positive labelling is apparent at the apical brush border of tumor cells and within lumina. ×250

As the majority of *colon carcinomas* belong to the group of differentiated mucus producing adenocarcinomas they are all CEA positive (ISAACSON and LEVANN 1976; GOLDENBERG et al. 1978; MURARO et al. 1985; JOTHY et al. 1986) (Fig. 6). The same is true for *stomach carcinomas* of the intestinal or diffuse type (HEIDL et al. 1986). In about one third of gastric carcinomas CEA expression was reported to be less intense in lymph node metastases than in the primary. This behaviour appeared to correlate with a worse prognosis (WITTEKIND et al. 1986). The rare adenocarcinomas of the *small intestine*, if well differentiated, are all positive for CEA.

In the *pancreas*, monoclonal antibodies directed against epitopes either present on CEA as well as NCA 95 or on CEA only, were found to discriminate clearly between duct type carcinomas and nonduct type carcinomas such as acinar cell carcinomas and endocrine tumors (BÄTGE et al. 1986). Formalin-fixed, paraffin-embedded material of pancreatic ductal adenocarcinomas disclosed CEA expression in about 60–77% of the cases (BÄTGE et al. 1986; ALLUM et al. 1986). Using frozen sections of duct type adenocarcinomas all tumors reacted positively (KLÖPPEL, unpublished observation).

Earlier examinations on CEA expression in *hepatocellular carcinomas* reported positivity rates of 30% (THUNG et al. 1979; NAKANUMA et al. 1981; IMOTO et al. 1985). Recently however it was found that CEA was consistently absent from hepatocellular carcinomas if these tumors were stained with a monospecific CEA antiserum (KOELMA et al. 1986). This result has been confirmed in an own series of hepatocellular carcinomas which were tested using monoclonal

Table 2. Tumor markers in liver neoplasms[a] (Positivity given in percentage)

Neoplasm	Marker					
	AFP	CEA	α-HCG	AAT	KER	VIM
HCC (*n*=63)	24	0	2	6	95	11
CCC (*n*=18)	6	72	0	11	100	17
FNH (*n*=7)	0	0	0	0	100	0
HCA (*n*=4)	0	0	0	0	100	0

HCC, Hepatocellular carcinoma; *CCC*, cholangiocellular carcinoma; *FNH*, focal nodular hyperplasia; *HCA*, hepatocellular adenoma; *KER*, Keratin (MAb KL-1); *VIM*, Vimentin (MAb V9); *CEA*, Mab specific for CEA (BMA 130 c)

[a] Klöppel, Brumm, Schultz, Morohoshi (unpublished observations)

antibodies specific for CEA. In this study, none of 63 hepatocellular carcinomas and 13/18 *cholangiocellular carcinomas* expressed CEA (Table 2).

In *salivary gland tumors*, CEA is mainly expressed in carcinomas with glandular differentiation, but may also be observed in squamous cell carcinomas (Caselitz et al. 1981a, b; Seifert and Caselitz 1983). The latter carcinomas, however, appeared to be either weakly positive or negative for CEA when tested with NCA absorbed polyclonal antiserum or a monoclonal antibody to CEA (Tsukitani et al. 1985; Sumitomo et al. 1987). CEA is positive in *sweat-gland carcinomas* but not in adnexal neoplasms derived from other structures of the skin (sebaceous glands, pilar structures) (Penneys et al. 1982).

In *breast carcinomas*, the rate of CEA positivity ranges between 40–90% (Shousha and Lyssiotis 1978; von Kleist et al. 1982; Wahren et al. 1978; Walker 1980; Nap et al. 1984; Böcker et al. 1985). Using an absorbed polyclonal CEA antiserum Böcker et al. (1985) showed that 51% of the breast carcinomas contained more than 5 percent CEA-positive cells. In addition, CEA positivity was found to be correlated with the histologic tumor type, invasive ductal carcinomas being positive more frequently than invasive lobular, tubular and cribriform carcinomas. Shousha et al. (1979) suggested a relation between CEA positivity and prognosis. However, such a correlation could not be confirmed by other authors (Nap et al. 1984; van der Linden et al. 1985). CEA immunoreactivity could also not be correlated with oestrogen receptor status (Böcker et al. 1985; van der Linden et al. 1985) and other established prognostic parameters such as lymph node status, tumor type, histologic grade, mitotic activity and others (van der Linden et al. 1985).

Virtually all *adenocarcinomas of the lung* express CEA (Wachner et al. 1984; Kimura et al. 1985). A high rate of positivity, however, was also found in the other types of lung cancer, especially in squamous cell carcinomas. Here the reactivity appeared to be restricted to the epidermal cell nests. Pascal et al. (1977) therefore discussed a cross-reactivity of CEA with keratin. This assumption was not confirmed by Wachner et al. (1984) since preincubation of the sections with a polyclonal anti-keratin serum did not abolish the staining intensity for CEA. CEA staining was weak or absent from *mesotheliomas* (Corson and Pinkus 1982).

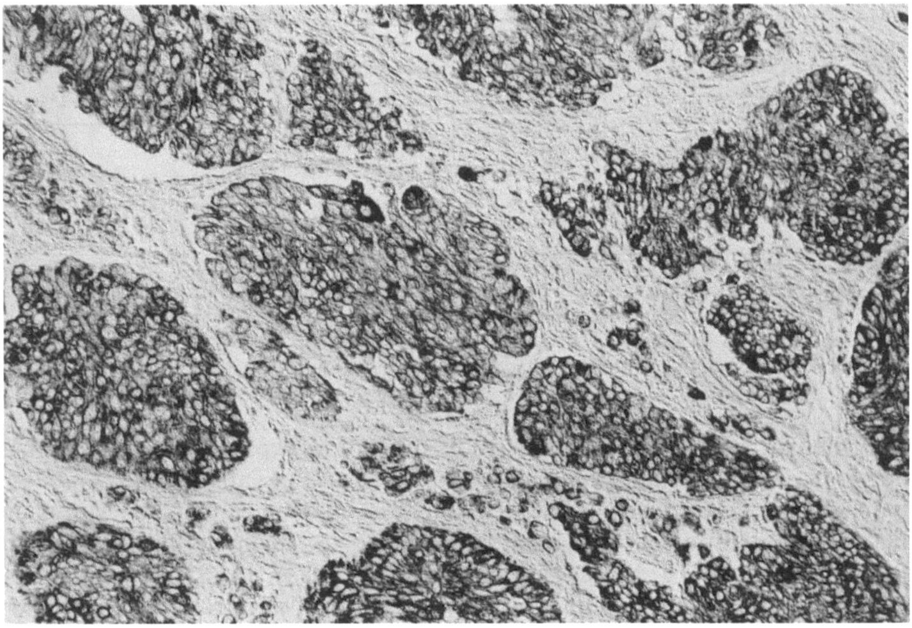

Fig. 7. Formalin-fixed, paraffin-embedded section from medullary thyroid carcinoma, stained with BMA 130 c (CEA specific). Intense intracytoplasmic labelling of all tumor cells. × 160

Ovarian carcinomas, with the exception of the serous, germ cell and anaplastic tumors, and *adenocarcinomas of the uterus* often express CEA (Fenoglio et al. 1981; Charpin et al. 1982). This is also seen in testicular teratomas containing glandular elements (Wittekind 1985).

CEA was not found in *renal cell carcinomas* and *melanomas* when tested with a monoclonal antibody to CEA (Bätge et al. 1986). In *prostate carcinomas*, only one out of 38 carcinomas stained with a polyclonal CEA antiserum absorbed for NCA (Purnell et al. 1984). Other authors also using an absorbed polyclonal CEA antiserum reported positive CEA values in up to 65% of prostate carcinomas (Vogel and Helpap 1986). The same authors also obtained high CEA scores (about 70%) in *bladder carcinomas* (Vogel et al. 1984).

Neuroendocrine tumors, except the medullary thyroid carcinomas, proved to be CEA negative (Bätge et al. 1986) but may contain some NCA (Bishopric and Ordonez 1986). *Medullary carcinomas of the thyroid* express CEA in about 80% of the cases (Schröder et al. 1986; DeLellis et al. 1978). The cellular distribution of CEA is very similar to that of calcitonin (Fig. 7). C-cell hyperplasia, the precursor lesion of the hereditary C-cell carcinoma (Wolfe et al. 1980), was reported to remain unstained with CEA antisera as do normal C-cells (DeLellis et al. 1978). Recent studies, however, revealed CEA positivity in those C-cell hyperplasias which are already accompanied by infiltrating C-cell carcinomas (Schröder and Klöppel 1987). C-Cell hyperplasias in the absence of an overt tumor proved to be CEA negative (Schröder and Klöppel 1987). *Nonmedullary thyroid carcinomas* are negative for CEA but may contain NCA (Schröder and Klöppel 1987).

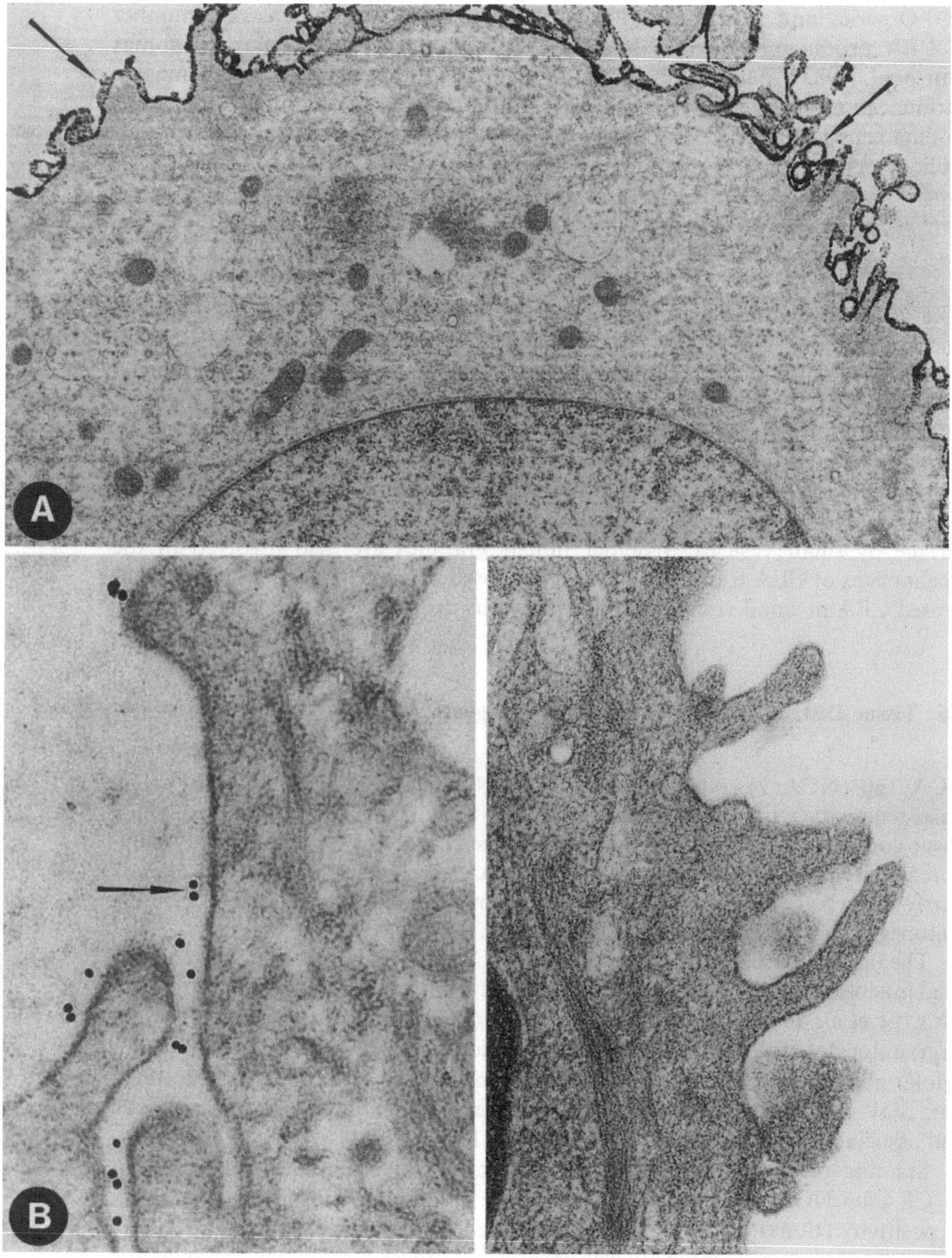

Fig. 8. Electron micrographs of cultured cells of pancreatic ductal adenocarcinoma immunostained with monoclonal antibody C1N3 recognizing epitopes on CEA and NCA. **A** Pre-embedding immunoperoxydase technique: Intense labelling of the surface of an ASPC cell (*arrows*), ×6290. **B** Pre-embedding immunogold technique: Reaction products (*arrow*) on the surface of a QGP1 cell (*left*). Control (*right*). ×17000

Apart from the well known cell lines such as LoVo, QGP-1, WIDR, the COLO series and others (SHIVELY and BEATTY 1985), an increasing number of CEA producing cell lines have been established from colonic carcinomas (SHI et al. 1983), pancreatic ductal carcinomas and other gut carcinomas in various laboratories during recent years. Immunostaining of cultured adenocarcinoma cells locates CEA primarily on the cell membrane and by electron immunocytochemistry CEA can be labelled on the surface of these cells (HERBERMAN et al. 1975; LOHSE, KLÖPPEL, KALTHOFF, SCHMIEGEL, unpublished observations) (Fig. 8).

2.5.6 Ultrastructural Localization

By ultrastructural immunocytochemistry CEA was localized primarily to the apical brush border and the glycocalyx (WOLF et al. 1984). Thus CEA might be anchored into the cell membrane. In addition, it was found in some mucus granules, in small vesicles and within the Golgi apparatus (HUITRIC et al. 1976; AHNEN et al. 1982; HAYNES et al. 1985). These findings suggest that CEA is packaged within the Golgi apparatus and transported from there to the cell surface by small vesicles. Whether mucus granules are also involved in the mechanisms of CEA transportation was doubted in a recent study which demonstrated CEA in small vesicles but not in mucus droplets (SCHULZ et al. 1986).

2.6 Tissue Distribution of CEA-Related Antigens

NCA 160 (NCA 2), NCA 95, NCA 55 (NCA) and BGP 1 show a strong cross-reactivity with CEA. Because of their wide tissue distribution and high tissue concentrations these CEA related antigens account for most of the immunoreactivity observed in normal or neoplastic tissues when unabsorbed or only partly absorbed polyclonal CEA antisera, or monoclonal antibodies detecting epitopes on both CEA and NCA are applied.

The normal tissue source of NCA 160 (NCA 2) is fetal and adult gastrointestinal mucosa (BURTIN et al. 1977). It is absent from monocytes and granulocytes (BURTIN et al. 1973). In contrast, NCA 95 (Fig. 9) and NCA 55 are present in granulocytes (BUCHEGGER et al. 1984). In addition, NCA 55 is found in bronchiolar and alveolar epithelium of normal lung as well as in pancreatic epithelium. BGP I is localized to the surfaces of hepatocytes lining biliary canaliculi and associated with the bile itself (SVENBERG 1976).

Staining of gastric and colonic carcinomas with antibodies specific either to CEA or NCA 160 (NCA 2) resulted in a similar incidence and intensity of positivity (BURTIN et al. 1977). A much higher incidence of immunoreactivity and a stronger staining intensity was observed in breast carcinomas or pancreatic carcinomas when labelled with antibodies recognizing CEA and CEA related substances, notably NCA 55 and NCA 95 (NAP et al. 1984; BÄTGE et al. 1986) (Fig. 10).

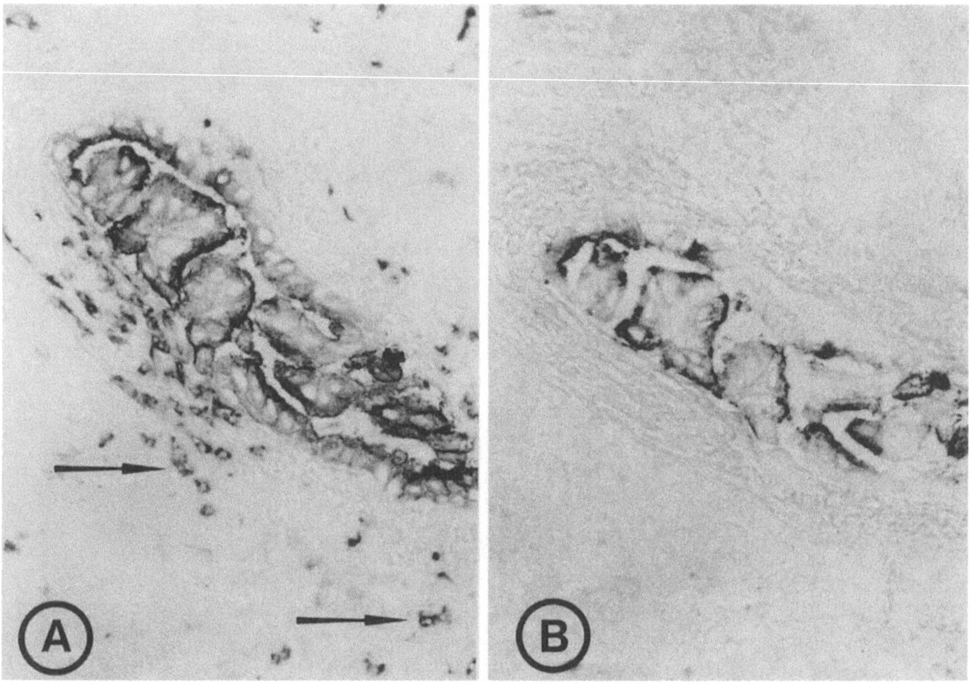

Fig. 9. Frozen section from pancreas with chronic pancreatitis, stained with BMA130b recognizing epitopes on CEA and NCA 95 (**A**), and monoclonal antibody C1P83 specific for CEA (**B**). BMA 130b stains papillary duct hyperplasia as well as monocytes (*arrows*), while C1P83 labels papillary hyperplasia only. ×250

2.7 Circulating CEA Levels and CEA Production

In the absence of imflammatory diseases involving epithelial tissues, elevated serum CEA levels usually indicate carcinomas originating from the gastrointestinal tract, the pancreas, the biliary system, the breast or the lung (NEVILLE and COOPER 1976). Complete removal of such a tumor correlates well with a decline of the elevated serum CEA level to the normal range. Correspondingly, tumor recurrence is usually marked by rising CEA values (STAAB et al. 1985; FLETCHER 1986). There is also a good correlation of serum CEA levels with regression or progression of tumor growth during chemotherapy (AHLEMANN et al. 1980; STAAB et al. 1980). In colorectal carcinomas the circulating CEA concentrations were found to be related to stage of disease (for review see FLETCHER 1986) and rising titres of CEA appeared most frequently in association with hepatic metastases (MACKAY et al. 1974). These observations suggest that the serum CEA levels reflect the general extent of the disease. However, the fact that serum CEA levels show great variability from one patient to another during the same stage of the disease (KALSER et al. 1978) indicates that the relationship between CEA level and tumor extent is rather complex. This assumption is further underlined by studies conducted in nude mice bearing trans-

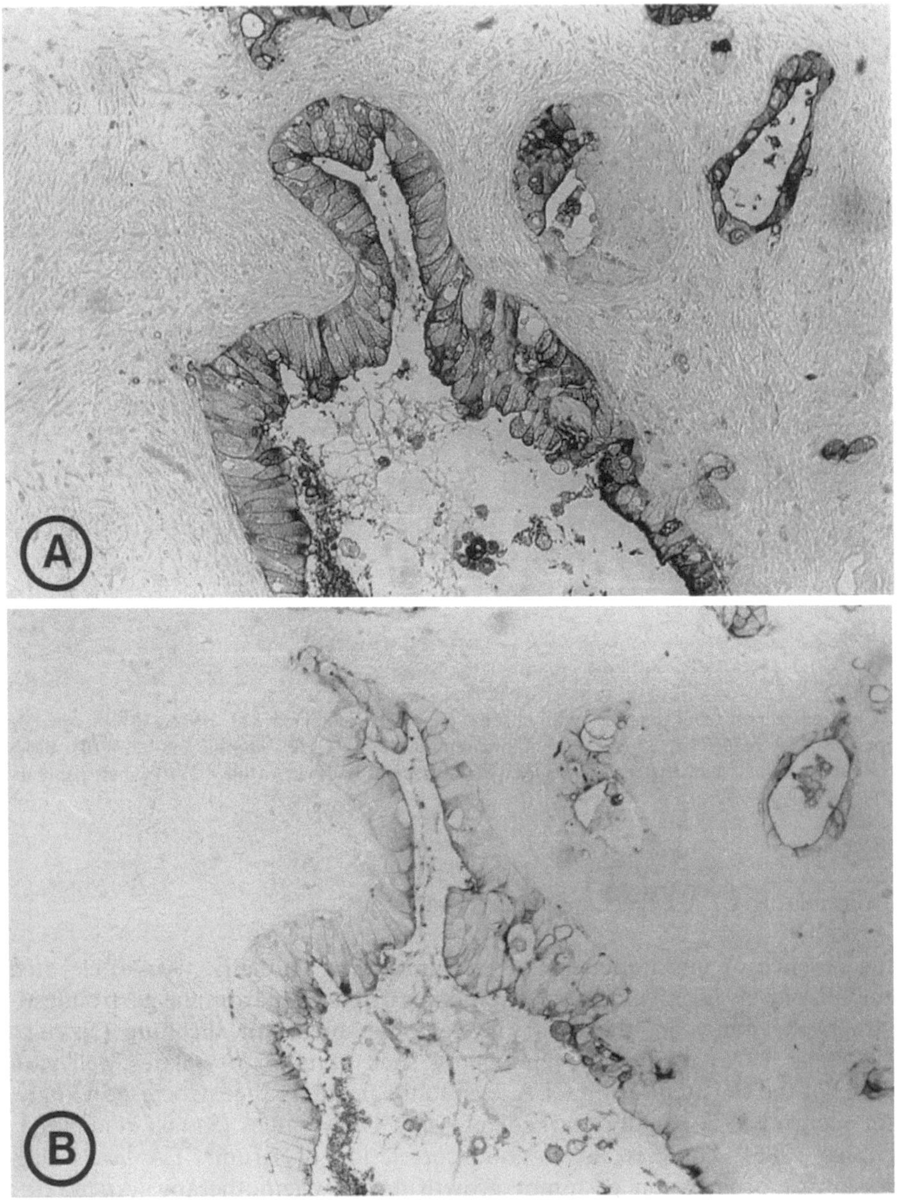

Fig. 10. Consecutive formalin-fixed, paraffin embedded sections from a ductal adenocarcinoma of the pancreas, stained with BMA130b (CEA and NCA 95) (**A**) and monoclonal antibody C1P83 (CEA) (**B**). Note the difference in staining intensity. ×250

planted colorectal carcinomas. While, in general, the elevation of the CEA level was found to be proportional to the increase in tumor size and tumor CEA content (STAAB and ANDERER 1982; FIEBIG and VON KLEIST 1983), considerable variability was noted between individual values, making a strict correlation often impossible (STRAGAND et al. 1980; LEWES and KEEP 1981). It is therefore most probable that serum CEA levels are influenced by a number of different factors. Basically these factors can be divided in two categories: One is related to the variable CEA production and secretion by the tumor, and the other to alterations in the CEA clearance and exretion by the liver (ZAMCHEK and MARTIN 1981). Factors belonging to the first group are (1) the degree of tumor differentiation (though some investigators reported a lack of correlation between tumor grade and CEA level (BIVINS et al. 1975; KALSER et al. 1978), the apparent inability of some carcinomas to release CEA into the extracellular space (VON KLEIST et al. 1982), the preferable secretion of CEA in glandular lumina without getting access to the circulation (BÖCKER et al. 1985), and the degree of necrosis and vascularisation within a tumor (WAGENER et al. 1981 b). The most important variables included in the second category are liver diseases such as acute and chronic hepatitis as well as obstructive jaundice resulting in an impaired capability of the liver to adequately clear the circulation of CEA and excrete it (ZAMCHECK and MARTIN 1981).

2.8 Function

The exact role of CEA during fetal development as well as the regulation of its pronounced expression and synthesis upon malignant transformation of normal cells are still obscure. It is of interest, however, that CEA induces the release of a suppressor factor from normal human lymphocytes (MEDOFF et al. 1984). Alpha-interferon treatment of human breast and colon tumor cells resulted in the enhanced expression of CEA on the cell surface (GREINER et al. 1984). The expression of CEA was found to be inversely related to the detection of histocompatibility antigens (HIRAI et al. 1980), and this heterogeneous expression of HLA-DR antigen and CEA in colonic carcinomas appeared to be associated with DNA-profile variations (ROGNUM et al. 1983). In a recent study it was observed that CEA stimulated the growth of thymic epithelial cells cultured in vitro (SAVINO et al. 1985).

2.9 Future Aspects

One of the most interesting issues with regard to CEA is whether there are CEA molecules which might be specifically associated with certain tumors. This question may be solved by molecular cloning studies which should render it possible to analyze the expression of these genes in normal and malignant tissues. In addition, gene transfer studies may aid in elucidating the biological function of CEA. Finally, such studies should also have implications for the development of new and more specific antibodies to CEA species which may further improve radioimmunodetection of tumors and possibly also tumor therapy.

3 Alpha Fetoprotein (AFP)

3.1 General Properties

AFP, like CEA, is an oncofetal glycoprotein synthesized in significant quantities in the liver, the yolk sac and the gastrointestinal tract of the fetus of many species (GITLIN and BOESMAN 1967). This protein which was first observed by BERGSTRAND and CZAR (1956) in the human fetus, appears as a normal constituent in fetal serum and amniotic fluid, reaching peak concentrations in the blood at about the 12th to 15th week of gestation. After birth AFP almost completely disappears from the blood. Its reappearance is usually associated with the development of hepatocellular carcinomas or yolk sac tumors but may also be seen as a result of hepatocellular damage and regeneration (SELL et al. 1974; KUHLMANN 1981; TUCZEK et al. 1981; BLOOMER et al. 1985a, b).

As a tumor associated antigen it was first described in mice bearing a transplantable hepatoma (ABELEV et al. 1963). Subsequently it was also found to be elevated in sera of patients with hepatocellular carcinoma and testicular germ cell tumors (ABELEV 1974; NEVILLE and COOPER 1976). During pregnancy raised AFP levels in amniotic fluid were observed to be associated with defects of the neural tube and are considered useful for antenatal detection of anencephally and spina bifida (BUAMAH et al. 1981).

3.2 Chemistry

AFP represents a group of closely-related molecular variants of an acidic glycoprotein with a molecular weight of approximately 70000 and an α1-globulin electrophoretic mobility (SMITH and KELLEHER 1980). It consists of a single polypeptide chain showing sequence homology with albumin (RUOSHLATI and TERRY 1976) and contains about 3–5% carbohydrate. The AFP of yolk sac tumors differs from that of hepatomas in that it has an additional carbohydrate residue at the epitope which in the hepatoma associated AFP bind Concanavalin A (TSUCHIDA et al. 1984; BUAMAH et al. 1984). The different content of neuraminoacid distinguishes embryonal AFP from hepatoma-AFP (PURVES et al. 1971). Recently a fucosylated variant of AFP was found to be useful for differentiating between benign liver diseases and hepatocellular carcinoma (AOYAGI et al. 1985). For the generation of antibodies, AFP isolated and purified from pools of amniotic fluid is used (KUHLMANN 1975; BELLET et al. 1984).

3.3 Immunocytochemical Spectrum

3.3.1 *Fetal Tissues*

The fetal liver expresses AFP in single or grouped hepatocytes, preferably accumulating around liver sinus (ENGELHARDT et al. 1971; KUHLMANN 1975; NAYAK et al. 1974; PURTILO and YUNIS 1971; NAYAK and MITAL 1977). The serum concentration was found to be directly proportional to the number of hepatocytes staining for AFP. Ultrastructurally, AFP was localized to the endosplasmic reticulum but not the Golgi region. This suggests that secretion of AFP may be achieved without the involvement of the Golgi apparatus (PEYROL et al. 1977).

3.3.2 *Adult Tissue*

In the normal adult hepatocyte AFP remains undetectable by immunocytochemical means because of the low quantity of AFP production. In phases of liver cell regeneration as observed after hepatitis, in liver cirrhosis or experimentally after partial hepatectomy or CCl_4 intoxication, AFP production is slightly enhanced (KUHLMANN 1981; NEVILLE and COOPER 1976; HARADA et al. 1980; RONCALLI et al. 1986). This increase in CEA levels is correlated with growth and cell multiplication of hepatocytes (SELL et al. 1974). Immunocytochemically, AFP staining hepatocytes occur shortly after damage to the liver cells and disappear after restoration of the tissue (KUHLMANN 1981). In chronic liver disease such as alcoholic hepatitis and secondary biliary cirrhosis, AFP positive liver cells were occasionally observed (KOELMA et al. 1986).

3.3.3 *Neoplastic Tissues*

Although 80–90% of the patients with *hepatocellular carcinomas* have elevated serum AFP levels, immunocytochemical demonstration of AFP is only achieved in 20 to maximally 50% of these tumors (CHU et al. 1974; THUNG et al. 1979; ESPINOZA et al. 1984; IMOTO et al. 1985) (Table 2). AFP expression in the tumors is very patchy and more likely to be found in poorly differentiated than in well differentiated hepatomas (ABELEV 1974; KOELMA et al. 1986; KLOEPPEL, BRUMM, SCHULTZE and MOROHOSHI, unpublished observations) (Fig. 11). Pronase digestion does not enhance the score of AFP positive hepatocellular carcinomas (KOELMA et al. 1986). In mice bearing xenotransplanted hepatocellular carcinomas serum AFP levels reflected the degree of tumor growth in the host (HIROHASHI et al. 1979). Experimentally, AFP positive tumors frequently develop in rats treated with different carcinogens (KUHLMANN 1981).

Hepatoblastoma, the liver neoplasm of young children, is nearly always AFP positive (SCHMIDT et al. 1985), while *fibrolamellar hepatocellular carcinoma,* a distinct variant of hepatocellular carcinoma occuring in adolescents and young adults, is frequently AFP negative (CABALLERO et al. 1985; SCHMIDT et al. 1985). *Cholangiocellular carcinomas* invariably behave AFP negative and this is also true for *liver cell adenoma* (Table 2). *Focal nodular hyperplasia* (FNH) was found to be AFP negative in our series (Table 2), while an AFP positive case was observed by KOELMA et al. (1986).

Approximately two-thirds of *testicular teratomas* are associated with elevated serum AFP levels (NEVILLE and COOPER 1976). The increased serum AFP levels correlate with the presence of *endodermal sinus (yolk sac) tumor* elements within these neoplasms. This is due to the fact that AFP is produced and localized in the cells of visceral endoderm lining the endodermal sinuses and in the various-sized intra- and extracytoplasmic PAS-positive hyaline globules in the tumor (ENGELHARDT et al. 1983; ITOH et al. 1974; NORGAARD-PEDERSON et al.

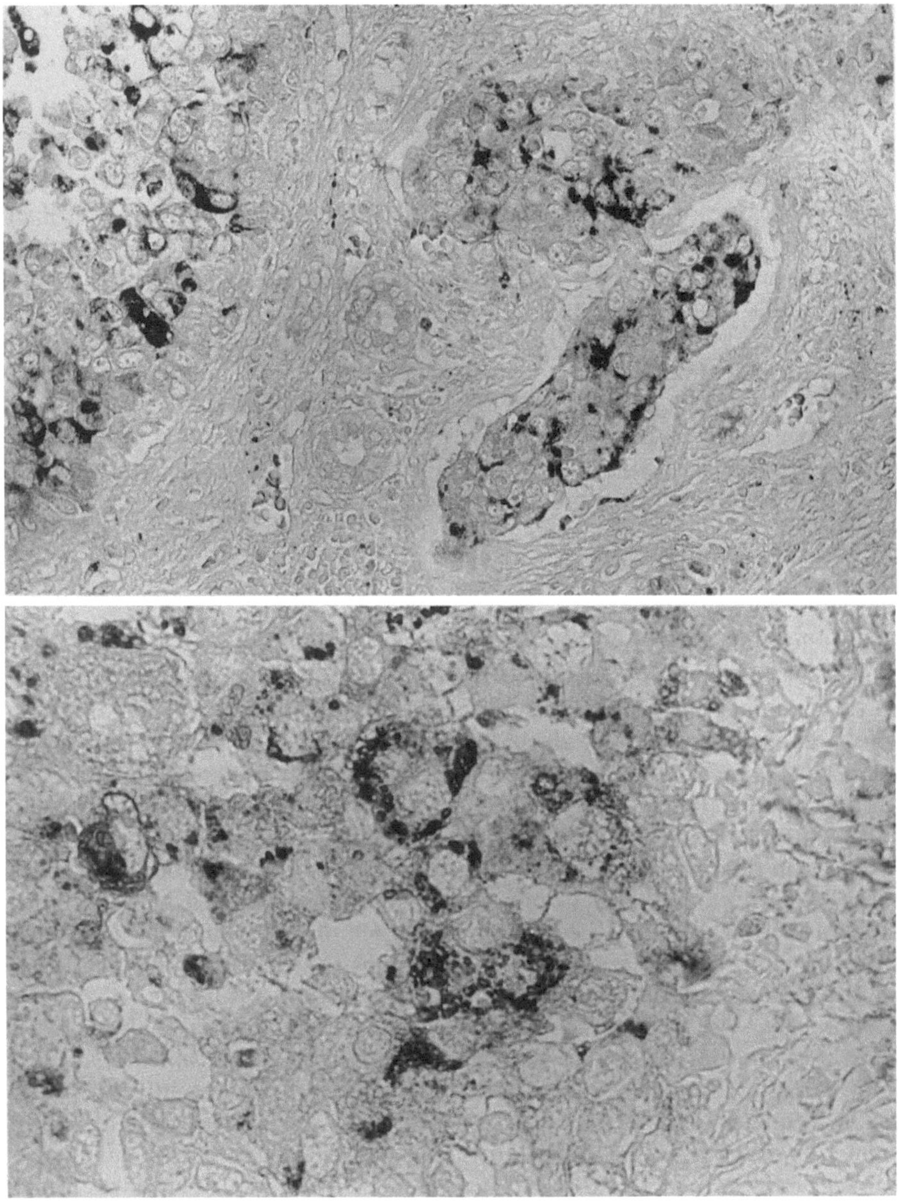

Fig. 11. Formalin-fixed, paraffin-embedded sections from poorly differentiated hepatocellular carcinoma, stained with polyclonal AFP antiserum. Granular intracytoplasmic AFP positivity. ×120 and 640

1974; TALERMAN 1980; BOSMAN et al. 1980; HORN et al. 1980; WITTEKIND 1985). Germ cell tumors located in the ovary and extragonadal sites containing homologous structures also produce AFP in high quantities. Pure *embryonal carcinomas* are associated either with normal or slightly elevated AFP levels and may show some AFP-positive cells (WAGENER et al. 1981a). AFP production was also reported in some rare germ cell tumors with borderline histology between seminoma and embryonal carcinoma (HOFSTÄDTER 1986; WALT et al. 1986). Pure *seminomas, teratomas* or *choriocarcinomas* are AFP negative (JACOBSEN et al. 1981).

Neoplasms not arising from liver or germ cells, e.g. *adenocarcinomas of the stomach, pancreas, colon and lung* occasionally produce AFP but rarely in large amounts (KODAMA et al. 1981).

3.4 Function

The exact role of AFP during fetal development is still obscure. As it shows a particular affinity for oestrogen and binds to bilirubin and fatty acids, it may serve as a carrier for these substances (for review see SMITH and KELLEHER 1980). In addition, it has been considered to be involved in immune regulation mechanisms because of its ability to suppress certain immune reactions in vitro (SHEPPARD et al. 1977; GERSHWIN 1980).

4 Epithelial Membrane Antigen (EMA)

4.1 General Properties

EMA has proved to be a global marker of epithelial cells of normal and neoplastic origin. It was isolated from human milk (membranes of human fat globules; HEYDERMAN et al. 1979) and represents a high molecular weight glycoprotein (molecular weight 265000 to 400000 Dalton, ORMEROD et al. 1983). In analogy to CEA, EMA may belong to a family of proteins with numerous epitopes in common. EMA can be detected by the use of polyclonal antibodies (HEYDERMAN et al. 1979; SLOANE and ORMEROD 1981) and by a number of monoclonal antibodies (HMFG 2, LICR LOM/M, E29; EDWARDS and BROOKS 1984; MOLL 1986).

4.2 Immunocytochemical Spectrum

4.2.1 Normal Tissues

In normal breast tissue, EMA labels the luminal cells, while the myoepithelial cells remain negative. The staining for EMA is concentrated at the apical cell membrane. The EMA staining pattern of other exocrine glands such as pancreas, sweat glands and salivary glands (Fig. 12) is analogous to that of the mammary gland. No staining is observed in the cells of the proximal kidney tubules and in hepatocytes. In the transitional epithelium of the urogenital tract, EMA is present in the superficial layer. Squamous epithelium is often negative for

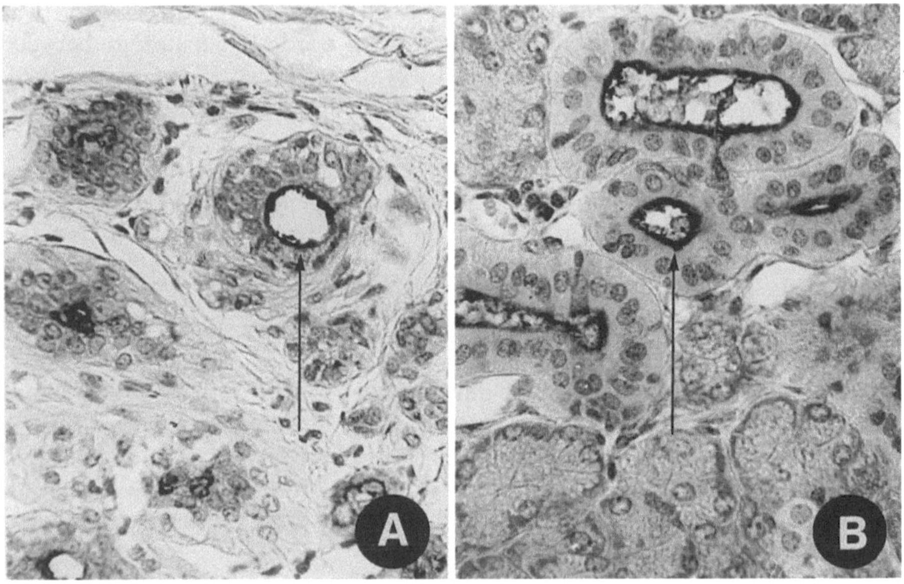

Fig. 12. Formalin-fixed, paraffin-embedded sections from mammary gland (**A**) and parotid gland (**B**), stained with polyclonal EMA antiserum. Labelling of apical border of duct cells (*arrows*). ×300

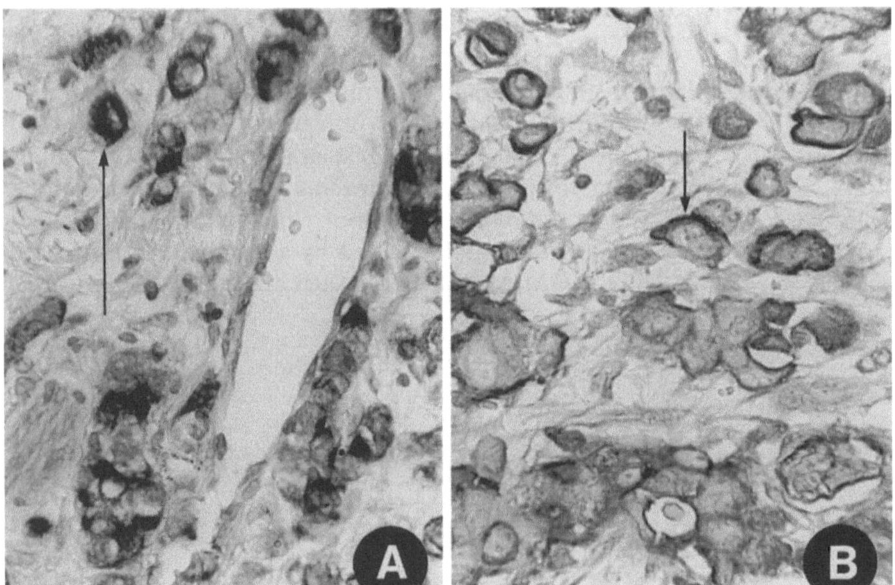

Fig. 13. Formalin-fixed, paraffin-embedded sections from signet ring cell carcinoma of stomach (**A**) and poorly differentiated carcinoma of ovary (**B**), stained with polyclonal EMA antiserum. Intracytoplasmic (*long arrow*) and apical (*short arrow*) labelling of tumor cells. ×300

EMA (Sloane and Ormerod 1981), and mesenchymal tissues lack EMA. Nonepithelial cells containing EMA are plasma cells and transformed lymphocytes (Delsol et al. 1984).

4.2.2 Neoplastic Tissues

As a rule, all tumors with glandular differentiation are EMA positive, while a few nonglandular tumors may be negative (Sloane and Ormerod 1981; Ormerod et al. 1982; To et al. 1982; Edwards and Brooks 1984; Pinkus et al. 1986). The staining pattern of EMA is comparable to that of total keratin, but its variability is greater. EMA has been found in adenocarcinomas of the breast, pancreas, lung, kidney, prostate and stomach, but also in squamous cell carcinomas of the esophagus, the cervix and the skin. It is found in benign and malignant salivary gland tumors (Gusterson et al. 1982; Caselitz 1987). Undifferentiated carcinomas, e.g. bronchial carcinomas of the large and small cell type, are often positive so that EMA turns out to be a valuable marker in identifying the epithelial nature of these tumors (Fig. 13). EMA is also positive in mesotheliomas. Tumors nonreactive with EMA include hepatocellular carcinomas, neuroendocrine tumors, embryonal carcinomas and sarcomas (with few exceptions). Among the malignant lymphomas, the T-cell lymphomas of the large cell variety, and plasmacytomas may stain for EMA (Delsol et al. 1984; Norton and Isaacson 1986).

4.2.3 Correlation with Keratin Markers

EMA is most intensively expressed in those tissues and tumors where keratin 8, 18 and/or 19 are present. Since antibodies against EMA react on sections of formalin-fixed, paraffin-embedded tissue, they are a helpful adjunct in establishing the diagnosis of a carcinoma. Although the distributions of keratin and EMA are overlaping, there are still some differences. The keratins belong to the group of intermediate filaments which are located in the cytoplasm, sometimes even in a fibrillar pattern (especially in cytological specimens). Thus antikeratin antibodies with a broad reactivity generally stain the cytoplasm of all epithelial cells. As EMA is a membrane glycoprotein, antibodies against EMA stain the cell membrane, in particular the apical border of normal glands. Only in tumors, it is occasionally detected in the cytoplasm as well (Fig. 13B).

4.2.4 Function

The functional role of EMA remains to be elucidated. As it seems to be a special membrane component of most glandular tissues, it might be involved in functional mechanisms like secretion or transport of secretory material (Heyderman et al. 1979).

References

Abelev GI (1974) Alpha-fetoprotein as a marker of embryo-specific differentiations in normal and tumor tissues. Transplant Rev 20:2–37

Abelev GI, Perova SD, Kharmkoyan NI, Postnikova ZA, Irlin IS (1963) Production of embryonal alpha-globulin by transplantable mouse hepatomas. Transplantation (Baltimore) 1:174–180

Ahlemann LM, Staab HJ, Koch HL, Anderer FA (1980) CEA measurements as an aid to management of patients with lung cancer treated by radiotherapy. Oncodevelop Biol Med 1:143–150

Ahnen DJ, Nakane PK, Brown WR (1982) Ultrastructural localization of carcinoembryonic antigen in normal intestine and colon cancer. Cancer 49:2077–2090

Allum WH, Stokes HJ, MacDonald F, Fielding JWL (1986) Demonstration of carcinoembryonic antigen (CEA) expression in normal, chronically inflamed, and malignant pancreatic tissue by immunohistochemistry. J Clin Pathol 39:610–614

Aoyagi Y, Isemura M, Suzuki Y, Sekine C, Soga K, Ozaki T, Ichida F 61985) Fucosylated alpha-fetoprotein as marker of early hepatocellular carcinoma. Lancet II:1353–1354

Bätge B, Bosslet K, Sedlacek HH, Kern HF, Klöppel G (1986) Monoclonal antibodies against CEA-related components discriminate between pancreatic duct type carcinomas and nonneoplastic duct lesions as well as nonduct type neoplasias. Virchows Arch (Pathol Anat) 408:361–374

Bellet DH, Wands JR, Isselbacher KJ, Bohoun C (1984) Serum alpha-fetoprotein levels in human disease: perspective from a highly specific monoclonal radioimmunoassay. Proc Natl Acad Sci 81:3869–3873

Bergstrand CG, Czar B (1956) Demonstration of a new protein fraction in serum from the human fetus. Scand J Clin Lab Invest 8:174

Bisphopric GA, Ordónez NG (1986) Carcinoembryonic antigen in primary carcinoid tumors of the lung. Cancer 58:1316–1320

Bivins BA, Meeker WR, Griffen WO (1975) Carcinoembryonic antigen (CEA) levels and tumor histology in colon cancer. J Surg Res 18:257–261

Bloomer JR, Waldmann TA, McIntire KR, Klatskin G (1975a) Relationship of serum alpha-fetoprotein to the severity and duration of illness in patients with viral hepatitis. Gastroenterology 68:342–350

Bloomer JR, Waldmann TA, McIntire KR, Klatskin G (1975b) Alphafetoprotein in nonneoplastic hepatic disorders. J Am Med Assoc 233:38–41

Böcker W, Schweikhart G, Pollow K, Kreienberg R, Klauberg A, Schröder S, Mitze M, Bahnsen J, Stegner HE (1985) Immunohistochemical demonstration of carcinoembryonic antigen (CEA) in 120 mammary carcinomas and its correlation with tumor type, grading, staging, plasma-CEA, and biochemical receptor status. Path Res Pract 180:490–497

Bosman FT, Giard RWM, Nieuwenhuuzen Kruseman AC, Knijnenberg G, Spaander P (1980) Human chorionic gonadotropin and alpha-fetoprotein in testicular germ cell tumors: a restrospective immunohistochemical study. Histopathology 4:673–684

Buamah PK, Taylor P, Milford WA (1981) Concanavalin A binding of alpha-fetoprotein in amniotic fluid as an aid in the diagnosis of neural tube defects. Clin Chem 27:1658–1660

Buamah PK, Gibb I, Bates G, Ward AM (1984) Serum alpha fetoprotein heterogeneity as a means of differentiating between primary hepatocellular carcinoma and hepatic secondaries. Clin Chim Acta 139:313–316

Buchegger F, Schreyer M, Carrel S, Mach J-P (1984) Monoclonal antibodies identify a CEA cross reacting antigen of 95 kD (NCA 95) distinct in antigenicity and tissue distribution from the previously described NCA of 55 kD. Int J Cancer 33:643–649

Burtin P (1985) Immunochemical basis of immunohistochemical reactions. Application to CEA, antigens cross-reacting to CEA, and CA 19-9 antigen. In: Holmgren J (ed) Tumor marker antigens. Studentliteratur, Lund, pp 50–57

Burtin P, Gold P (1978) Carcinoembryonic antigen. Scand J Immunol 8 (Suppl 8):27–38

Burtin P, Martin E, Sabine MC, von Kleist S (1972) Immunological study of polyps of the colon. J Natl Cancer Inst 48:25–32

Burtin P, Chavanel G, Hirsch-Marie H (1973) Characterization of a second normal antigen that cross-reacts with CEA. J Immunol 111:1926–1928

Burtin P, Sabine MC, Chavanel G (1977) A comparative study of the localization of CEA and NCA 2 in cancerous and normal gastrointestinal tissues. Int J Cancer 19:634–641

Caballero T, Aneiros J, Lopez-Caballero J, Gomez-Morales M, Nogales F (1985) Fibrolamellar hepatocellular carcinoma. An immunohistochemical and ultrastructural study. Histopathology 9:445–456

Caselitz J (1987) Das pleomorphe Adenom der Speicheldrüsen. Histogenese, zelluläre Differenzierung, Tumormarker. Fischer, Stuttgart New York

Caselitz J, Seifert G, Jaup T (1981 a) Presence of carcinoembryonic antigen (CEA) in the normal and inflamed human parotid gland. An immunohistochemical study of 31 cases. J Cancer Res Clin Oncol 100:205–211

Caselitz J, Jaup T, Seifert G (1981 b) Immunohistochemical detection of carcinoembryonic antigen (CEA) in parotid gland carcinomas. Analysis of 52 cases. Virchows Arch (Pathol Anat) 394:49–60

Chandrasekaran EV, Davita M, Nixon DW, Goldfarb M, Mendicino J (1983) Isolation and structures of the oligosaccharide units of carcinoembryonic antigen. J Biol Chem 258:7213–7222

Charpin C, Bhan AK, Zurawski VR Jr, Scully RE (1982) Carcinoembryonic antigen (CEA) and carbohydrate determinant 19-9 (CA 19-9) localization in 121 primary and metastatic ovarian tumors: an immunohistochemical study with the use of monoclonal antibodies. Int J Gynecol Pathol 1:231–245

Chu ML, Wen-Shih-Jen L, Yoshida TO, Chu SH, Lin TY (1974) Demonstration of alpha-foetoglobulin in hepatoma tissue by fluorescent antibody technique. Cancer 34:268

Corson JM, Pinkus GS (1982) Mesothelioma: profile of keratin proteins and carcinoembryonic antigen. An immunoperoxidase study of 20 cases and comparison with pulmonary adenocarcinomas. Am J Pathol 108:80–87

DeLellis RA, Rule AH, Spiler I, Nathanson L, Tashijan AT, Wolfe HJ (1978) Calcitonin and carcinoembryonic antigen as tumor markers in medullary thyroid carcinoma. Am J Clin Pathol 70:587–594

Delsol G, Gatter KC, Stein H, Erber WN, Pulford KAF, Zinne K, Mason DY (1984) Human lymphoid cells express epithelial membrane antigen. Implication for diagnosis of human neoplasms. Lancet II:1124–1128

Denk H, Tappeiner G, Eckerstorfer R, Holzner JH (1972) Carcinoembryonic antigen (CEA) in gastrointestinal and extragastrointestinal tumors and its relationsship to tumor-cell differentiation. Int J Cancer 10:262–272

Denk H, Tappeiner G, Davidovits A, Holzner JH (1973) The carcinoembryonic antigen (CEA) in carcinomata of the stomach. Virchows Arch (Pathol Anat) 360:339–347

Edwards PAW, Brooks IM (1984) Antigenic subsets of human breast epithelial cells distinguished by monoclonal antibodies. J Histochem Cytochem 32:531–537

Endo T, Imai K, Yachi A, Nakajima K, Watanabe T (1985) Immunohistologic distribution of CEA defined by human anti-CEA antibody in the serum of a colonic cancer patient. Cancer 56:2205–2211

Engelhardt NV, Goussev AI, Shipova LJA, Abelev GI (1971) Immunofluorescent study of alpha-fetoprotein (AFP) in liver and liver tumors. I. Technique of AFP localization in tissue sections. Int J Cancer 7:198–206

Engelhardt NV, Poltoranina VS, Yazova AK (1973) Localization of alphafetoprotein in transplantable murine teratocarcinoma. Int J Cancer 11:448–459

Espinoza CG, Pillarisetti SG, Azar HA (1984) Immunohistochemistry of hepatocellular carcinoma associated with cirrhosis. Ann Clin Lab Sci 14:467–473

Fenoglio CM, Crum CP, Pascal RR, Richart RM 61981) Carcinoembryonic antigen in gynecologic patients. II. Immunohistological expression. Diagn Gynecol Obstet 3:291–299

Fiebig HH, von Kleist S (1983) Carcinoembryonic antigen in human colorectal cancers growing subcutaneously in nude mice. J Cancer Res Clin Oncol 105:238–242

Fletcher RH (1986) Carcinoembryonic antigen. Ann Int Med 104:66–73

Gershwin ME (1980) The immunobiology of alpha-fetoprotein. Cancer Immunol Immunother 8:1–2

Gitlin D, Boesman M (1967) Sites of serum alpha-fetoprotein synthesis in the human and in the rat. J Clin Invest 46:1010–1016

Gold P, Freedman SO (1965) Specific antigen similarity between malignant adult and normal fetal tissues of the human digestive system. J Clin Invest 44:1051–1052

Goldenberg DM, Sharkey RM, Primus FJ (1978) Immunocytochemical detection of carcinoembryonic antigen (CEA) in conventional histopathologic specimens. Cancer 42:1546–1553

Greiner JW, Hand PH, Noguchi P, Fisher PB, Pestka S, Schlom J (1984) Enhanced expression of surface tumor-associated antigens on human breast and colon tumor cells after recombinant human leukocyte alpha-interferon treatment. Cancer Res 44:3208–3214

Grunert F, AbuHarfeil N, Schwarz K, von Kleist S (1985) Two CEA and three NCA species, although distinguishable by monoclonal antibodies, have nearly identical peptide patterns. Int J Cancer 36:357–362
Gusterson BA, Raleigh RB, Ormerod MG (1982) Distribution of epithelial membrane antigen in benign and malignant lesions of the salivary glands. Virchows Arch (Pathol Anat) 397:227–233
Haggarty A, Legler C, Krantz MJ, Fuks A (1986) Epitopes of carcinoembryonic antigen defined by monoclonal antibodies prepared from mice immunized with purified carcinoembryonic antigen or HCT-8R cells. Cancer Res 46:300–309
Hammarström S (1985) Chemistry and immunology of CEA, CA 19-9 and CA-50. In: Holmgren J (ed) Tomor marker antigens. Studentliteratur, Lund, pp 32–49
Harada T, Shigeta K, Noda K, Fukomoto Y, Nishimura H, Mizuta M, Takemoto T (1980) Clinical implications of alpha-fetoprotein in liver cirrhosis: five-year follow-up study. Hepato-Gastroenterol 27:169–175
Haynes WDG, Shertock KL, Skinner JM, Whitehead R (1985) The ultrastructural immunohistochemistry of oncofoetal antigens in large bowel carcinomas. Virchows Arch (Pathol Anat) 405:263–275
Hedin A, Hammarström S, Larsson A (1982) Specificities and binding properties of eight monoclonal antibodies against carcinoembryonic antigen. Mol Immunol 19:1641–1648
Heidl G, Davaris P, Düchting S, Jagoda MS, Bierhoff E, Krieg V, Grundmann E (1986) Zur Nachweisbarkeit des karzinoembryonalen Antigens (CEA) an Gefrier- und Paraffinschnitten von Magenkarzinomen. Verh Dtsch Ges Pathol 70:225–230
Herberman RB, Aoki T, Cannon G, Liu M, Sturm M (1975) Brief communication: location by immunoelectron microscopy of carcinoembryonic antigen on cultured adenocarcinoma cells. J Natl Cancer Inst 55:797
Heyderman E, Steele K, Ormerod MG (1979) A new antigen on the epithelial membrane; its immunoperoxidase localization in normal and neoplastic tissues. J Clin Pathol 32:35–39
Hirai T, Yamamoto H, Hamaoka T (1980) Relationship between carcinoembryonic antigen and major histocompartibility antigens. J Immun (Baltimore) 124:2765–2771
Hirohashi S, Shimosato Y, Kameya T, Koide T, Mukojima T, Taguchi Y, Kageyama K (1979) Production of alpha-fetoprotein and normal serum proteins by xenotransplanted human hepatomas in relation to their growth and morphology. Cancer Res 39:1819–1828
Hofstädter F (1986) Tumoren des männlichen Urogenitalsystems. Verh Dtsch Ges Pathol 70:172–183
Horn F, Hedinger CHR, Grob PJ (1980) Alpha-Fetoprotein und Hodentumoren. Schweiz Med Wochenschr 110:642–647
Huitric E, Laumonier R, Burtin P, von Kleist S, Chavanel G (1976) An optical and ultrastructural study of the localization of carcinoembryonic antigen (CEA) in normal and cancerous human rectocolonic mucosa. Lab Invest 76:97–107
Imoto M, Nishimura D, Fukuda Y, Sugiyama K, Kumada T, Nakano S (1985) Immunohistochemical detection of alpha-fetoprotein, carcinoembryonic antigen, and ferritin in formalin-paraffin sections from hepatocellular carcinoma. Am J Gastroenterol 80:902–906
Isaacson P, LeVann HP (1976) The demonstration of carcinoembryonic antigen in colonic polyps using immunoperoxidase technique. Cancer 38:1348–1356
Isaacson P, Judd MA (1978) Immunohistochemistry of carcinoembryonic antigen in the small intestine. Cancer 42:1554–1559
Itoh T, Shirai T, Naka A, Matsumoto S (1974) Yolk sac tumor and alpha-fetoprotein: clinicopathological study of four cases. Gann 65:215–226
Jacobsen GK, Jacobsen M, Clausen P (1981) Distribution of tumor-associated antigens in the various histologic components of germ cell tumors of the testis. Am J Surg Pathol 5:257–266
Jothy S, Brazinsky SA, Chin-A-Loy M, Haggarty A, Krantz MJ, Cheung, Fuks A (1986) Charakterization of monoclonal antibodies to carcinoembryonic antigen with increased tumor specificity. Lab Invest 54:108–117
Kalser MH, Barkin JS, Redlhammer D, Heal A (1978) Circulating carcinoembryonic antigen in pancreatic cancer. Cancer 42:1468–1471
Kauf FG, Borchard F (1985) Über das Auftreten gastrointestinaler Tumormarker und magenspezifischer Enzyme im gesamten Gastrointtestinaltrakt während der Fetalperiode. Verh Dtsch Ges Pathol 69:445
Kimura N, Ghandur-Mnaymen L (1985) An immunohistochemical study of keratin, carcinoembryonic antigen, human chorionic gonadotropin and alpha-fetoprotein in lung cancer. Tohoku J Exp Med 145:23–38

Kleist S von (1983) Das karzinoembryonale Antigen (CEA). Biologische Grundlagen und klinische Anwendung. Schattauer, Stuttgart New York

Kleist S von, Chavanel G, Burtin P (1972) Identification of an antigen from normal human tissue that cross-reacts with the carcinoembryonic antigen. Proc Natl Acad Sci USA 69:2492–2494

Kleist S von, Wittekind C, Sandritter W, Gropp H (1982) CEA positivity in sera and breast tumor tissues obtained from the same patients. Path Res Pract 173:390–401

Klöppel G, Lohse T, Bosslet K, Rückert K (1987) Ductal adenocarcinoma of the head of the pancreas: incidence of tumor involvement beyond the Whipple resection line. Histological and immunocytochemical analysis of 37 total pancreatectomy specimens. Pancreas 2:170–175

Kodama T, Kameya T, Hirota T, Shimosato Y, Ohkura H, Mukojima T, Kitaoka H (1981) Cancer 48:1647–1655

Koelma IA, Nap M, Huitema S, Krom RAF, Houthoff HJ (1986) Hepatocellular carcinoma, adenoma, and focal nodular hyperplasia. Comparative histopathologic study with immunohistochemical parameters. Arch Pathol Lab Med 110:1035–1040

Kuhlmann WD (1975) Purification of mouse alpha 1-fetoprotein and preparation of specific peroxidase conjugates for its cellular localization. Histochemistry 44:155–167

Kuhlmann WD (1981) Alpha-fetoprotein: cellular origin of a biological marker in rat liver under various experimental conditions. Virchows Arch (Pathol Anat) 393:9–26

Kuroki M, Kuroki M, Koga Y, Matsuoka Y (1984) Monoclonal antibodies to carcinoembryonic antigen: a systemic analysis of antibody specificities by using related normal antigens and evidence for allotypic determinants on carcinoembryonic antigen. J Immunol 133:2090–2097

Lewis JCM, Keep PA (1981) Relationship of serum CEA levels to tumor size and CEA-content in nude mice bearing colonic-tumor xenografts. Br J Cancer 44:381–387

Linden JC van der, Baak JPA, Lindeman J, Smeulders AWM, Meyer CJLM (1985) Carcinoembryonic antigen expression and peanut agglutinin binding in primary breast cancer and lymph node metastases; lack of correlation with clinical, histopathological, biochemical and morphometric features. Histopathology 9:1051–1059

Mach JP, Pusztaszeri G (1972) Carcinoembryonic antigen (CEA): demonstration of partial identity between CEA and a normal glycoprotein. Immunocytochemistry 9:1031–1033

MacKay AM, Patel S, Carter S, Stevens H, Laurence DJR, Cooper EH, Nevile AM (1974) Role of serial CEA assays in detection of recurrent and metastatic colorectal carcinomas. Br Med J 4:382–385

Mansour EG, Hastert M, Hopark CH, Koehler KA, Petrelli M (1983) Tissue and plasma carcinoembryonic antigen in early breast cancer. Cancer 51:1243–1247

Matsuoka Y, Kuroki M, Koga Y, Kuriyama H, Mori T, Kosaki G (1982) Immunochemical differences among carcinoembryonic antigen in tumor tissues and related antigens in meconium and adult feces. Cancer Res 42:2012

Medoff JR, Jegasothy BV, Roche JK (1984) CEA-induced release of a suppressor factor from normal human lymphocytes in vitro. Cancer Res 44:5822–5827

Muraro RD, Wunderlich A, Thor J, Lundy P, Noguchi R, Cunningham R, Schlomm J (1985) Definition by monoclonal antibodies of a repertoire of epitopes on carcinoembryonic antigen differentially expressed in human colon carcinomas versus normal adult tissues. Cancer Res 45:5769–5780

Moll R (1986) Epitheliale Tumormarker. Verh Dtsch Ges Pathol 70:28–50

Nakanuma Y, Tanaka Y, Ohta G (1981) Immunohistochemical examination of primary liver cancer. Jpn J Gastroenterol 78:1619–1626

Nap M, Keuning H, Burtin P, Oosterhuis JW, Fleuren G (1984) CEA and NCA in benign and malignant breast tumors. Am J Clin Pathol 82:526–534

Nayak NC, Mital I (1977) The dynamics of α-fetoprotein and albumin synthesis in human and rat liver during normal ontogeny. Am J Pathol 86:359–371

Nayak NC, Das PK, Bhuyan UN, Mittal A (1974) Localization of alphafetoprotein in human and rat fetal livers: An immunohistochemical method using horseradish peroxidase. J Histochem Cytochem 22:414–418

Neumaier M, Fenger U, Wagener C (1985) Monoclonal antibodies for carcinoembryonic antigen (CEA) as a model system: identification of two novel CEA-related antigens in meconium and colorectal carcinoma tissue by western blots and differential immunoaffinity chromatography. J Immunol 135:3604–3609

Neville AM, Cooper EH (1976) Biochemical monitoring of cancer: a review. Ann Clin Biochem 13:283–305

Nielsen K, Teglbjaerg PS (1982) Carcinoembryonic antigen (CEA) in gastric adenocarcinomas. Morphological patterns and their relationship to a histogenetic classification. Acta Pathol Microbiol Immunol Scand Sect A 90:393–396

Norgaard-Pederson B, Dabelsteen E, Edeling CJ (1974) Localization of human alpha-fetoprotein synthesis in hepatoblastoma cells by immunofluorescence and immunoperoxidase methods. Acta Pathol Microbiol Scand 82:169–174

Norton AJ, Isaacson PG (1986) An immunocytochemical study of T-cell lymphomas using monoclonal and polyclonal antibodies effective in routinely fixed was embedded tissues. Histopathology 10:1243–1260

O'Brien MJ, Zamcheck N, Burke B, Kirkham SE, Saravis CA, Gottlieb LS (1981) Immunocytochemical localization of carcinoembryonic antigen in benign and malignant colorectal tissues. Am J Clin Pathol 75:283–290

Ormerod MG, Bussolati G, Sloane JP, Steele K, Gugliotta P (1982) Similarities of antisera to casein and epithelial membrane antigen. Virchows Arch (Pathol Anat) 397:327–333

Ormerod MG, Steele K, Westgood JH, Mazzani MN (1983) Epithelial membrane antigen: Partial purification assay and properties. Br J Cancer 48:533–541

Pascal RR, Mesa-Tejada R, Bennett SJ, Garcea A, Fenoglio CM (1977) Carcinoembryonic antigen: immunohistologic identification in invasive and intra-epithelial carcinomas of the lung. Arch Pathol Lab Med 101:568–571

Penneys NS, Nadji M, Ziegels-Weissman J, Ketabchi M, Morales AR (1982) Carcinoembryonic antigen in sweat-gland carcinomas. Cancer 50:1608–1611

Peyrol S, Grimaud JA, Pirson Y, Chayvialle JA, Touillon C, Lambert R (1977) Ultrastructural immunoenzymatic study of alpha-fetoprotein-producing cells in the human fetal liver. Histochem Cytochem 25:432–438

Pinkus GS, Etheridge CL, O'Connor EM (1986) Are keratin proteins a better tumor marker than epithelial membrane antigen? A comparative immunohistochemical study of various paraffin-embedded neoplasms using monoclonal and polyclonal antibodies. Am J Clin Pathol 85:269–273

Primus FJ, Kuhns WJ, Goldenberg DM (1983) Immunological heterogeneity of carcinoembryonic antigen: immunohistochemical detection of carcinoembryonic antigen determinants in colonic tumors with monoclonal antibodies. Cancer Res 43:693–701

Purnell DM, Heatfield BM, Trump BF (1984) Immunocytochemical evaluation of human prostatic carcinomas for carcinoembryonic antigen, nonspecific cross-reacting antigen, β-chorionic gonadotrophin, and prostate-specific antigen. Cancer Res 44:285–292

Purtilo DT, Yunis EJ (1971) Alpha-fetoprotein: Its immunofluorescent localization in human fetal liver and hepatoma. Lab Invest 25:291–294

Purves IR, van der Merwe E, Bersden I (1971) More sensitive test for alpha-fetoprotein. Lancet II:464–468

Rogers GT (1983) Carcinoembryonic antigens and related glycoproteins. Molecular aspects and specificity. Biochim Biophys Acta 695:227–249

Rognum TO, Brandtzaeg P, Thorud E (1983) Is heterogenous expression of HLA-DR antigens and CEA along with DNA-profile variations evidence of phenotypic instability and clonal proliferation in large bowel carcinomas. Br J Cancer 48:543–551

Roncalli M, Borzio M, de Biagi G, Ferrari AR, Macchi R, Tombesi VM, Servida E (1986) Liver cell dysplasia in cirrhosis. A serologic and immunohistochemical study. Cancer 57:1515–1521

Ruoshlati E, Terry WD (1976) Alpha-fetoprotein and serum albumin show sequence homology. Nature 260:804–805

Savino W, Durand D, Dardenne M (1985) Immunohistochemical evidence for the expression of the carcinoembryonic antigen by human thymic epithelial cells in vitro and in neoplastic conditions. Am J Pathol 121:418–425

Schmidt D, Harms D, Lang W (1985) Primary malignant hepatic tumors in childhood. Virchows Arch (Pathol Anat) 407:387–405

Schröder S, Klöppel G, Heitz PU, Kastendieck H, Böcker W (1986) Das medulläre Schilddrüsencarcinom: Morphologische und funktionelle Heterogenität, Prognosekriterien. Verh Dtsch Ges Pathol 70:588

Schröder S, Klöppel G (1987) Carcinoembryonic antigen and nonspecific cross-reacting antigen in thyroid cancer. Am J Surg Pathol 11:100–108

Schulz A, Jeckel HM, Berghäuser KH, Jundt G (1986) Bildung und Abgabe von CEA durch Coloncarcinom-Zellen. Ultrastrukturell-immuncytochemische Untersuchungen mit der Protein-A-Gold-Methode. Verh Dtsch Ges Pathol 70:269–273

Seifert G, Caselitz J (1983) Tumor markers in parotid gland carcinomas: Immunohistochemical investigations. Cancer Detection and Prevention 6:119–130

Sell S, Nochols M, Becker FF, Leffer HL (1974) Hepatocyte proliferation and α1-fetoprotein in pregnant neonatal, and partially hepatectomized rats. Cancer Res 34:865–871

Sheppard HW, Sell S, Trefts P, Bahu R (1977) Effects of alpha-fetoprotein on murine immune response. I. Studies on mice. J Immunol 119:91

Shi ZR, Tsao D, Kim YS (1983) Subcellular distribution, synthesis and release of CEA in cultured human clonadenocarcinoma cell lines. Cancer Res 43:4045–4049

Shiveley JE, Todd CW (1980) Carcinoembryonic antigen. A: chemistry and biology. In: Sell S (ed) Cancer markers. Diagnostic and developmental significance. Humana Press, New Jersey, pp 295–314

Shively JE, Beatty JD (1985) CEA-related antigens: molecular biology and clinical significance. In: Critical reviews in oncology and hematology, vol 2, Issue 4. Chemical Rubber Company, pp 355–399

Shousha S, Lyssiotis T (1978) Correlation of carcinoembryonic antigen in tissue sections with spread of mammary carcinoma. Histopathology 2:433–447

Shousha S, Lyssiotis T, Godefrey VM, Scheuer PJ (1979) Carcinoembryonic antigen in breast cancer tissue: a useful prognostic indicator. Br Med J 1:777–779

Sloane JP, Ormerod MG (1981) Distribution of epithelial membrane antigen in normal and neoplastic tissues and its values in diagnostic tumor pathology. Cancer 45:297–305

Smith CJP, Kelleher PC (1980) Alpha-fetoprotein molecular heterogeneity. Physiologic correlations with normal growth, carcinogenesis, and tumor growth. Biochim Biophys Acta 605:1–3

Staab HJ, Anderer FA (1982) Growth of human colonic adenocarcinoma and development of serum CEA in athymic mice: Strict correlation of tumor size and mass with serum CEA concentration during logarithmic growth. Br J Cancer 46:841–847

Staab HJ, Ahlemann LM, Koch HL, Anderer FA (1980) Serial CEA determinations in the management of patients with breast cancer. Oncodevelop Biol Med 1:151–160

Staab HJ, Brümmendorf T, Anderer FA, Hornung A, Ahlemann A, Kieninger LM (1985) Differentiation of disease recurrence in various primary malignancies using CEA slope analysis. Tumor Biol 6:157–171

Stenger RJ, Chabon AB, Primus FJ, Wolff WI (1979) Immunoperoxidase localization of carcinoembryonic antigen in benign and malignant colorectal tumors and related mucosa. Mount Sinai J Med 46:185–189

Stragand JJ, Yang LY, Drewinko B (1980) Serum CEA levels in a human colonic adenocarcinoma (LoVo) xenograft system. Cancer Lett 10:45–48

Sumitomo S, Kumasa S, Mitani H, Mori M (1987) Comparison of CEA distribution in lesions and tumors of salivary glands as determined with monoclonal and polyclonal antibodies. Virchows Arch B 53:133–139

Svenberg T (1976) Carcinoembryonic antigen-like substances of human bile. Isolation and partial characterization. Int J Cancer 17:358–361

Talerman A (1980) Endodermal sinus (yolk sac) tumor elements in testicular germ-cell tumors in adults: comparison of prospective and retrospective studies. Cancer 46:1213–1217

Thung SN, Gerber MA, Sarno E, Popper H (1979) Distribution of five antigens in hepatocellular carcinoma. Lab Invest 41:101–105

To A, Dearnaley DP, Ormerod MG, Canti G, Coleman DV (1982) epithelial membrane antigen. Its use in cytodiagnosis of malignancy in serous effusions. Am J Pathol 78:214–219

Tsuchida Y, Yamashita K, Kobata A, Nishi S, Endo Y, Saito S (1984) Structure of the sugar chain of AFP purified from a human yolk sac tumor and its reactivity with concanavalin A. Tumor Biol 5:33–40

Tsukitani K, Kobayasi K, Murase N, Sumitomo S, Mitani H, Mori M (1985) Characterization of cells in salivary gland lesions by immunohistochemical identification of carcinoembryonic antigens. Oral Surg 59:595–599

Tsutsumi Y, Nagura H, Watanabe K (1984) Immunohistochemical observations of carcinoembryonic antigen (CEA) and CEA-related substances in normal and neoplastic pancreas. Am J Clin Pathol 82:535–542

Tuczek HV, Fritz P, Wagner T, Braun U, Grau A, Wegner G (1981) Synthesis of alpha-fetoprotein

(AFP) and cell proliferation in regenerating livers of NMRI mice after partial hepatectomy. An immunohistochemical and autoradiographic study with 3H-thymidine. Virchows Arch (Cell Pathol) 38:229–237

Vogel J, Helpap B (1986) Vergleichende histologisch-zytologische, autoradiographische und immunhistochemische Untersuchungen an Prostatakarzinomen. Verh Dtsch Ges Pathol 70:332–338

Vogel J, Helpap B, Oehr P (1984) Immunhistochemisches Verhalten von CEA und TPA sowie Proliferationskinetik in Harnblasenkarzinomen und begleitenden Epithelveränderungen. Verh Dtsch Ges Pathol 68:546

Wachner R, Wittekind C, von Kleist S (1984) Localization of CEA, β-HCG, SP1, and keratin in the tissue of lung carcinomas. An immunohistochemical study. Virchows Arch (Pathol Anat) 402:415–423

Wagener C, Breuer H (1982) Biochemical and immunochemical aspects of carcinoembryonic antigen (CEA) and CEA-related antigens: current status and further perspectives. J Clin Chem Clin Biochem 20:705–712

Wagener C, Csaszar H, Totovic V, Breuer H (1978) A highly sensitive method for the demonstration of carcinoembryonic antigen in normal and neoplastic colonic tissue. Histochemistry 58:1–11

Wagener C, Menzel B, Breuer H, Weissbach L, Tschubel K, Henkel K, Gedigk P (1981 a) Immunohistochemical localisation of alpha-fetoprotein (AFP) in germ cell tumours: evidence for AFP production by tissues different from endodermal sinus tumour. Oncology 38:236–239

Wagener C, Müller-Wallraf R, Nisson S, Groner J, Breuer H (1981 b) Localization and concentration of carcinoembryonic antigen (CEA) in gastrointestinal tumors: correlation with CEA levels in plasma. J Natl Cancer Inst 67:539–547

Wagener C, Yang YHJ, Grawford FG, Shively JE (1983 a) Monoclonal antibodies for carcinoembryonic antigen and related antigens as a model system: a systematic approach for the determination of epitope specificities of monoclonal antibodies. J Immunol 130:2308–2315

Wagener C, Hain F, Födisch HJ, Breuer H (1983 b) Localization of CEA in embryonic and fetal human tissues. Histochemistry 78:1–9

Wahren B, Lindbrink E, Wallgren A, Eneroth P, Zajizek J (1978) Carcinoembryonic antigen and other tumor markers in tissue and serum or plasma of patients with primary mammary carcinoma. Cancer 42:1870–1878

Walker RA (1980) Demonstration of carcinoembryonic antigen in human breast carcinomas by the immunoperoxidase technique. J Clin Pathol 33:356–360

Walt H, Arrenbrecht S, Delozier-Blanchet D, Keller PJ, Nauer R, Hedinger C (1986) A human testicular germ cell tumor with borderline histology between seminoma and embryonal carcinoma secreting betahuman chorionic gonadotropin and alphafetoprotein only as a xenograft. Cancer 58:139–146

Wiley EL, Mendelsohn G, Eggleston JC (1981) Distribution of carcinoembryonic antigens and blood group substances in adenomcarcinoma of the colon. Lab Invest 44:507–513

Wittekind C (1985) Immunhistologische Untersuchungen bei Hodentumoren. Lab Med 9:10–15

Wittekind C, Wachner R, Kirchner R, von Kleist S (1986) Zur prognostischen Relevanz des Markerverhaltens in Magencarcinomen und deren Lymphknotenmetastasen. Verh Dtsch Ges Pathol 70:221–224

Wolf B, Thompson JA, von Kleist S (1984) Differences in the ultrastructural localization of CEA in human breast and colon tumor cells. Tumor Biol 5:233–242

Wolfe HJ, DeLellis RA, Jackson CE, Greenawald KA, Block MA, Tashjian AH (1980) Immunocytochemical distinction of hereditary from sporadic medullary thyroid carcinoma. Lab Invest 42:161–162

Wurster K, Rapp W (1979) Histological and immunohistological studies on gastric mucosa. I. The presence of CEA in dysplastic surface epithelium. Path Res Pract 164:270–281

Yachi A, Imai K, Fujita H, Moriya Y, Tanda M, Endo T, Tsujisaki M, Kawaharada M (1984) Immunohistochemical distribution of the antigenic determinants detected by monoclonal antibodies to carcinoembryonic antigen. J Immunol 132:2998–3004

Zamcheck N, Martin EW (1981) Factors controlling the circulating CEA levels in pancreatic cancer: some clinical correlations. Cancer 47:1620–1627

Zimmermann W, Ortlieb B, Friedrich R, von Kleist S (1987) Isolation and characterization of cDNA clones coding for the human carcinoembryonic antigen reveal a highly preserved repeating structure. Proc Natl Acad Sci (to be published)

Epithelial Tumor Markers: Special Markers of Glandular Differentiation

T.O. ROGNUM, P.S. THRANE, F.R. KORSRUD and P. BRANDTZAEG

1 Introduction

There are principally two types of epithelial tumor markers: (1) Normal cell products with known functional properties which may appear in increased or reduced amounts during malignant development; and (2) So-called oncodevelopmental antigens. The latter markers are mainly expressed in fetal life but may appear in increased amounts in neoplastic lesions. Examples of such antigens are alpha fetoprotein (AFP) and carcinoembryonic antigen (CEA). In the present chapter, however, only the former category of tumor markers will be included and the discussion will be limited to salivary glands, breast and large bowel mucosa.

Immunohistochemical mapping of epithelial markers of differentiation has varied significance in different organs. Whereas contribution to increased understanding of tumor histogenesis is an important aspect in studies of salivary glands, the use of such markers in breast tumors and carcinomas or pre-cancerous lesions of the large bowel may be of diagnostic and prognostic value.

2 Normal Glandular Products and Their Epithelial Distribution

2.1 Secretory Component and Secretory Immunoglobulins

Secretory component (SC) is part of an epithelial transmembrane glycoprotein that binds J chain-containing dimeric IgA and pentameric IgM. It is thereby a key factor in the external transport of secretory IgA (SIgA) and SIgM (for review, see BRANDTZAEG 1985). Because of this receptor function of SC, its distribution in secretory epithelia will normally be mimicked by that of IgA (and to a lesser extent IgM), with the exception of the selective appearance of free SC in the Golgi zones (BRANDTZAEG 1974c). Some physicochemical properties of SC and SIgA are given in Table 1.

Recent studies have conclusively shown that SC is localized to both ductal and serous acinar elements in salivary glands (BRANDTZAEG 1977; KORSRUD and BRANDTZAEG 1982; FANTASIA and LALLY 1984). It is likewise widely distributed in mammary glands, and epithelial uptake and transport of IgA during lactation is well documented (BRANDTZAEG 1983). The columnar cells of intestinal crypts are similarly active in SC-mediated transport of IgA and to a lesser extent IgM (BRANDTZAEG 1974a; BRANDTZAEG and BAKLIEN 1977).

Table 1. Physicochemical properties of epithelial cell markers discussed in this chapter

Name	Type of protein	Molecular mass (daltons)	References
Secretory component (SC)	Glycoprotein	83000	BRANDTZAEG 1974b
Secretory IgA	Complex of dimeric IgA and SC	390000	TOMASI et al. 1965 NEWCOMB et al. 1968
Lysozyme (Muramidase)	Enzyme (EC 3.2.1.17)	14500	FLEMMING 1922 PETIT and JOLLES 1963
Amylase	Enzyme (EC 3.2.1.1)	50000	JACOBSEN et al. 1966 MUNZEL 1976
Lactoferrin	Iron-binding glycoprotein	80000	MASSON et al. 1966 BULLEN et al. 1972
MHC class I	Glycoprotein	45000	STROMINGER et al. 1980
MHC class II	Glycoprotein	33000+28000	STROMINGER et al. 1980

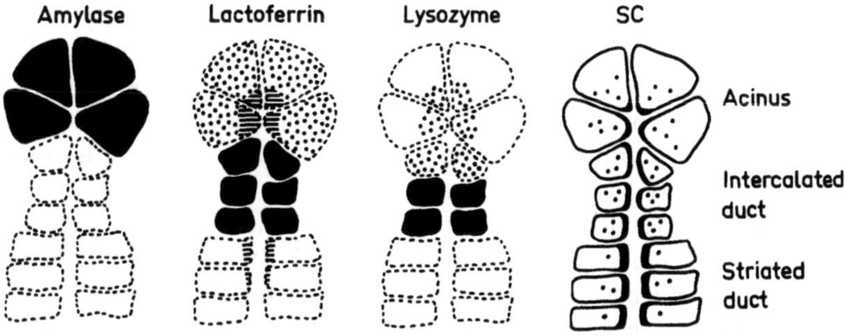

Fig. 1. Schematic distributional representation of various glandular markers in serous epithelial elements of salivary glands. The drawing is based on the average staining patterns: *dotted* (graded density), *shaded*, and *black areas* indicate, in that order, increasing occurrence and intensity. SC is particularly found apically and associated with the basolateral borders of the epithelial cells; the distribution of IgA (not shown) is similar except that SC appears selectively adjacent to the nuclei (not indicated in drawing). Adapted from KORSRUD and BRANDTZAEG (1982)

SC expression is often modulated during inflammation, and γ-interferon produced by activated T cells may be one of the stimulatory signals (SOLLID et al. 1987).

2.2 Lysozyme and Lactoferrin

Lysozyme (Ly), lactoferrin (Lf) and amylase (Am) will be discussed as tumor markers only in relation to salivary gland neoplasia. Ly is a bacteriolytic enzyme, Lf is an iron-binding glycoprotein (Table 1). Both have a broad spectrum of antibacterial properties (IACONO et al. 1980; ARNOLD et al. 1977) and are produced not only by phagocytes (MCCLELLAND and FURTH 1975) but also by certain secretory epithelia (LAI et al. 1976). In salivary glands Ly and Lf have by most authors been localized to intercalated ducts and to a much smaller extent to serous acini (REITAMO et al. 1977, 1980; KORSRUD and BRANDTZAEG 1982) (Fig. 1).

2.3 Amylase

Am is an enzyme (Table 1) produced by serous acini in salivary glands (KRAUS and MESTECKY 1971; KORSRUD and BRANDTZAEG 1982; CASELITZ et al. 1983; SEIFERT and CASELITZ 1983) (Fig. 1). Even in fetal glands it is mainly confined to primitive acini (THRANE et al. 1987a).

2.4 MHC Class I and II Molecules

Class I and II products (Table 1) of the major histocompatibility complex (MHC) play an important role in immune regulation; class I (HLA-A, -B,

-C) molecules serve as restriction elements for T cell-mediated cytotoxity (ZINKERNAGEL and DOHERTY 1979), whereas class II (HLA-DR, -DP, -DQ) molecules are required for the presentation of foreign antigen to helper T (Th) cells (BENACERRAF 1981). Loss of class I molecules may have an important impact on the malignancy and metastatic potential of a tumor (WALLICH et al. 1985; MOMBURG et al. 1986).

Class I determinants are normally expressed in a heterogeneous manner by epithelial elements in salivary glands (THRANE and BRANDTZAEG 1987a), mammary glands (FLEMING et al. 1981) and the gut (GHOSH et al. 1986). Although class II molecules are generally confined to B lymphocytes, activated T lymphocytes, various dendritic cells and macrophages, certain epithelial and endothelial cells may likewise express such determinants. In normal salivary glands DR expression has been noted in segments of intercalated ducts (THRANE and BRANDTZAEG 1987a), and scattered duct-lining cells are likewise positive in mammary glands (BHAN and DESMARAIS 1983; WHITWELL et al. 1984). In the normal gut, epithelial DR is virtually restricted to the entrocytes of the small intestinal villi (SCOTT et al. 1980; SELBY et al. 1981; GHOSH et al. 1986). In inflammatory conditions, however, increased expression of both class I and II molecules are observed in secretory epithelial cells of all these organs (SCOTT et al. 1981; SELBY et al. 1983; GHOSH et al. 1986; POULSEN et al. 1986; THRANE and BRANDTZAEG 1987a; ROGNUM et al. 1987a).

3 Salivary Gland Neoplasia

3.1 Secretory Component and Secretory Immunoglobulins

SC-positive epithelial elements have been found in pleomorphic adenoma (KORSRUD and BRANDTZAEG 1984a; FANTASIA and LALLY 1984; SAITO et al. 1984), Warthin's tumor or adenolymphoma (KORSRUD and BRANDTZAEG 1984b), and a few adenocystic and mucoepidermoid carcinomas (SAITO et al. 1984); IgA generally appears in a similar but less extensive distribution (Table 2). Large variability in co-expression of SC and the other secretory markers (Ly, Lf and Am) occurs in pleomorphic adenomas (KORSRUD and BRANDTZAEG 1984a), probably reflecting the complexity of this tumor in terms of differentiation. The most consistent epithelial positivity in Warthin's tumor was seen for SC and IgA, whereas the three other secretory markers were either absent or faintly expressed (KORSRUD and BRANDTZAEG 1984b). The staining pattern thus resembled that of normal striated ducts and supported the notion that this neoplasia is initiated in striated duct anlages (THOMPSON and BRYANT 1950). The number of malignant salivary gland tumors studied till now is too low to reveal any confident relation between SC expression and differentiation (Table 2).

3.2 Lysozyme and Lactoferrin

Epithelial Ly is expressed by relatively few pleomorphic adenomas (KORSRUD and BRANDTZAEG 1984a; SAITO et al. 1984; SEHESTED et al. 1985; CASELITZ et al.

1981) and Warthin's tumors (KORSRUD and BRANDTZAEG 1984b; SEHESTED et al. 1985) and it does seldom occur in malignant salivary gland tumors (CASELITZ et al. 1981; SAITO et al. 1984; SEHESTED et al. 1985) (Table 2). Positivity for Ly may thus corroborate exclusion of malignancy. Lf is more extensively distributed in both benign and malignant salivary gland tumors, although squamous cell carcinomas and anaplastic carcinomas are negative (Table 2). Since Ly and Lf may be considered as "terminal duct markers" (Fig. 1), the relatively consistent expression of Lf in pleomorphic adenomas supports a histogenetic relationship of this tumor to intercalated ducts (ERLANDSON et al. 1984) or, rather, to a duct-associated basal stem cell which may differentiate into both ductal elements and myoepithelial cells (REGEZI and BATSAKIS 1977; CASELITZ et al. 1985).

3.3 Amylase

Am is expressed almost exclusively by acinic cell carcinomas (CASELITZ et al. 1983; WARNER et al. 1985) and is present in more than half of the tumor specimens (Table 2), although in reduced amounts compared with normal parotid glands (FRANZEN et al. 1979). Its expression may depend on the degree of acinar cell differentiation (MORLEY et al. 1983). Nevertheless, Am seems to be a useful marker for acinic cell tumors.

3.4 MHC Class I and II Molecules

The neoplastic cells of pleomorphic adenoma (THRANE et al. 1987b) and Warthin's tumor (NATALI et al. 1981; THRANE and BRANDTZAEG 1987b) are virtually negative for MHC class II (HLA-DR) determinants (Table 2). These findings support the proposed histogenetic relationship between pleomorphic adenoma and myoepithelial cells (or their immediate precursors) and between Warthin's tumor and striated duct cells, respectively – i.e., cell types that under normal conditions show no DR expression (Table 2). However, pleomorphic adenomas are often rich in DR-positive dendritic histiocytic cells of apparently reactive nature (THRANE et al. 1987c).

Preliminary observations in our laboratory have indicated that epithelial elements in both pleomorphic adenoma and Warthin's tumor express MHC class I determinants. Studies of malignant salivary gland tumors are needed to see if loss of such determinants is related to malignant development.

3.5 Cell Lines and Transplantation Studies

Expression of various epithelial markers has been examined in cell lines isolated from salivary gland tumors (HAYASHI et al. 1985). Only neoplastic duct cells were found to express Lf and SC (SATO et al. 1984). Transplantation of such tumors to athymic mice gives further support for the notion that pleomorphic adenoma is derived from a multipotential duct stem cell able to differentiate

Table 2. Presence (+) or absence (−) of secretory component (*SC*), epithelial IgA, lysozyme, lactoferrin, amylase and MHC class I and II molecules in salivary gland epithelium and corresponding tumors

	Marker proteins							References
	SC	IgA	Lysozyme	Lactoferrin	Amylase	MHC class I	MHC class II	
Normal salivary glands								
Serous acini	+	+	−(+)[a]	−(+)	+			Korsrud and Brandtzaeg 1982
						+	−	Thrane and Brandtzaeg 1987a
Mucous acini	−	−	−	−	−			Korsrud and Brandtzaeg 1982
						+/−	−	Thrane and Brandtzaeg 1987a
Intercalated ducts	+	+	+	+	−			Korsrud and Brandtzaeg 1982
						+	+/−	Thrane and Brandtzaeg 1987a
Striated ducts	+	+	−	−(+)	−			Korsrud and Brandtzaeg 1982
						+	−	Thrane and Brandtzaeg 1987a
Salivary gland tumors								
Benign tumors								
Pleomorphic adenoma (59%)[b]					−(0/5)			Hodes et al. 1981
					−(0/10)			Morley et al. 1983
	+(11/15)[c]	+(9/15)	+(2/15)	+(11/15)	−/0/15)			Korsrud and Brandtzaeg 1984a
	+(16/16)							Fantasia and Lally 1984
	+(8/11)	+(8/11)	+(5/11)					Saito et al. 1984
			+(5/11)	+(6/11)				Sehested et al. 1985
	+(6/7)		+(5/13)	+(10/13)				Toto and Hsu 1985
Warthin's tumor (adenolymphoma) (6%)						+(10/10)	+(1/10)	Thrane et al. 1987c
							−(0/3)	Natali et al. 1981
	+(27/27)	+(27/27)			−(0/3)			Hodes et al. 1981
								Hsu et al. 1981
					−(0/7)			Morley et al. 1983
	+(6/6)	+(6/6)	+(2/6)	+(6/6)	−(0/6)			Korsrud and Brandtzaeg 1984b)
		+(22/22)						Caselitz et al. 1984
			−(0/3)	+(2/3)				Sehested et al. 1985
						+(7/7)	−(0/7)	Thrane and Brandtzaeg 1987b

Malignant tumors						
Malignant pleomorphic adenoma (7%)			−(0/21)	+(7/21)		Caselitz et al. 1981
			−(0/8)	+(5/8)		Caselitz et al. 1982
			+(1/1)	+(1/1)		Sehested et al. 1985
Adenocarcinoma (11%)					−(0/3)	Caselitz et al. 1983
– Adenoid cystic carcinoma (4%)				+(16/16)		Caselitz et al. 1981, 1982
					−(0/5)	Caselitz et al. 1983
	+(7/7)	+(7/7)	−(0/7)			Saito et al. 1984
	+(24/24)	+(24/24)	+(24/24)	+(24/24)	−(0/24)	Caselitz et al. 1986
– Cystadenocarcinoma (2.5%)			−(0/4)	+(2/4)		Caselitz et al. 1981
			−(0/5)	+(3/5)		Caselitz et al. 1982
– Salivary duct carcinoma (1%)			−(0/3)	+(2/3)		Caselitz et al. 1981
					−(0/3)	Caselitz et al. 1983
					−(0/3)	Hodes et al. 1981
– Acinic cell carcinoma (2.5%)					+(2/2)	Caselitz et al. 1983
					−(0/3)	Morley et al. 1983
	+(13/15)		+(4/15)	+(13/15)	+(11/15)	Warner et al. 1985
Mucoepidermoid carcinoma (11.5%)					−(0/1)	Hodes et al. 1981
				+(3/7)		Caselitz et al. 1981, 1982
					−(0/2)	Caselitz et al. 1983
	+(3/7)	+(3/7)	−(0/7)			Saito et al. 1984
Anaplastic carcinoma (2%)			−(0/3)	−(0/3)		Caselitz et al. 1981
Squamous cell carcinoma (4.5%)			−(0/8)	−(0/8)		Caselitz et al. 1981
					−(0/2)	Caselitz et al. 1983

[a] Infrequent finding given in parenthesis
[b] Occurrence frequency (adapted from Shafer, Hine and Levy: A textbook of oral pathology, 3rd edn. Saunders, Philadelphia, USA)
[c] No. of positive cases in relation to total no. tested

into both duct and myoepithelial cells (CASELITZ et al. 1985). A similar reserve duct cell has recently been proposed to be the origin of adenocystic carcinoma (CASELITZ et al. 1986).

3.6 General Remarks

Application of immunohistochemical markers has provided interesting information related to histogenesis in salivary gland neoplasia. However, further studies are needed because of the particular complexity of these tumors, especially when it comes to understanding of the nature of pleomorphic adenomas. It remains to be shown whether tumor markers will be of any value for classification and prognostic prediction in salivary gland neoplasia.

4 Breast Neoplasia

Several epithelial products have been studied in breast tumors but, with the exception of receptors for of oestrogen and progesteron (MCGUIRE et al. 1975; THORESEN et al. 1982), most of these markers have been of no value for prognostic and therapeutic considerations. In this chapter the discussion will be limited to the significance of SC, SIgA and MHC class I and II molecules.

4.1 Secretory Component and Secretory Immunoglobulins

RICHMAN (1976) claimed that epithelial IgA was similarly expressed by normal and malignant glandular epithelium, but most authors have reported that normal breast tissue and benign tumors contain more SC and epithelial IgA than carcinomas (HARRIS et al. 1975; LEE and BUEHRING 1981). We found about three quarters of the carcinomas to be positive for SC with a heterogeneous expression pattern (KVALE et al. 1987a), which accorded with the report of HARRIS and SOUTH (1981). Epithelial IgA positivity correlated well with SC expression but not with the histopathological grade of the breast carcinomas (KVALE et al. 1987a). Total serum SC, present in circulating SIgA and SIgM, showed no relation to tumor SC, indicating that the spillover to blood was generally low. A major increase in circulating SC reflected, instead, the presence of liver metastasis (KVALE et al. 1987a). The same was true for colonic carcinomas (KVALE et al. 1987b).

4.2 MHC Class I and II Molecules

Several laboratories have studied the distribution of MHC class I and II determinants in breast neoplasia (FLEMING et al. 1981; BHAN and DESMARAIS 1983; ROWE and BEVERLY 1984; WHITWELL et al. 1984; LWIN et al. 1985). In contrast to normal glandular epithelium and benign tumors, carcinomas show no consistent positivity (ROWE and BEVERLY 1984). In 28 breast carcinomas studied in our laboratory, 16 were class I-positive and 9 DR-positive (ROGNUM et al., unpublished observations). There seems to be no relation between the number and distribution of T cells and the expression of these determinants in the malignant breast epithelium (BHAN and DESMARAIS 1983; WHITWELL et al. 1984). Further studies are needed to confirm the hypothesis of SANDERSON and BEVERLY (1983), that impaired expression of class I and II molecules may be due to immunoselection of malignant cell clones.

4.3 Identification of Metastasis by Glandular Markers

Metastases from poorly differentiated or anaplastic adenocarcinomas are sometimes difficult to distinguish from malignant lymphomas by conventional histopathology. Preserved marker expression compatible with secretory epithelium may thus be of crucial value for the identification of secondary breast carcinomas in lymph nodes (BRANDTZAEG and ROGNUM 1983). It has been shown for colonic carcinomas that the original pattern of glandular markers is often preserved even after several years of tumor progression (ROGNUM et al. 1985).

5 Large Bowel Neoplasia

5.1 Carcinoma

Several attempts have been made to define tumor characteristics of prognostic value in patients with large bowel carcinoma. There is a positive relation between histological grade and survival time (RANKIN and BRODERS 1928; REMMELE and HEINE 1981), whereas changes of tumor mucus seem to be without prognostic value (SYRJANEN and HJELT 1978). Staging is considered as the most reliable way to assess prognosis (DUKES and BUSSEY 1958), but also production of SC (ARENDS et al. 1984), DNA ploidy pattern, (WOLLEY et al. 1982; ROGNUM et al. 1982a, 1987b), and epithelial expression of class I and II determinants (DAAR et al. 1982; ROGNUM et al. 1983; CSIBA et al. 1984; MOMBURG et al. 1986) are claimed to afford prognostic information.

5.1.1 Secretory Component and Secretory Immunoglobulins

SC and epithelial IgA are well correlated in the same carcinomas (ROGNUM et al. 1980a). Concordant expression of these markers indicates preservation of a relatively high degree of functional differentiation (Figs. 2, 3); which is in harmony with histopathological observations (POGER et al. 1976; WEISZ-CARRINGTON et al. 1976; ROGNUM et al. 1980a, 1982b; ISAACSON 1982). Interestingly, ARENDS et al. (1984) reported that patients with homogeneously SC-positive carcinomas had a better prognosis than those with SC-negative ones.

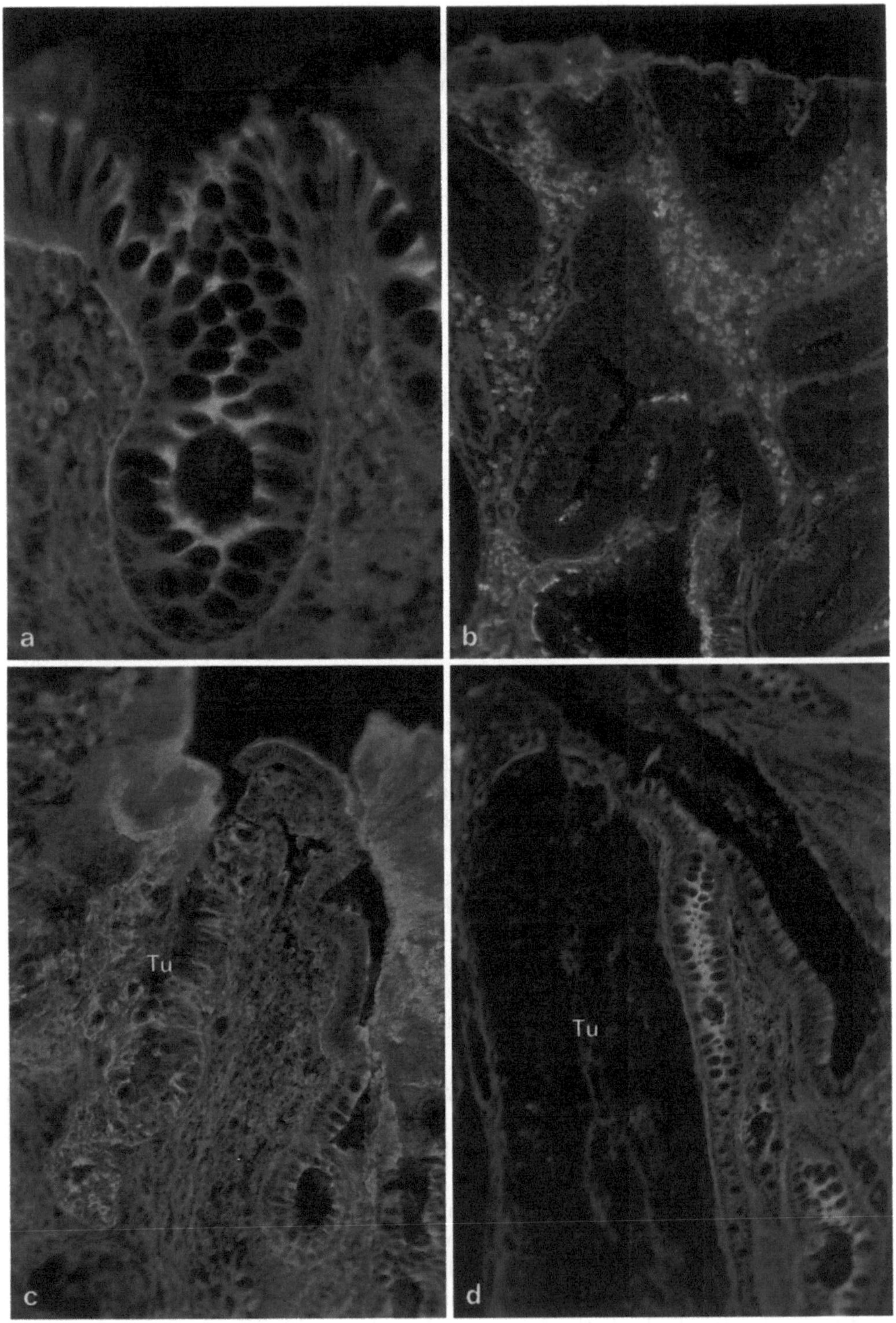
a
b
Tu
c
Tu
d

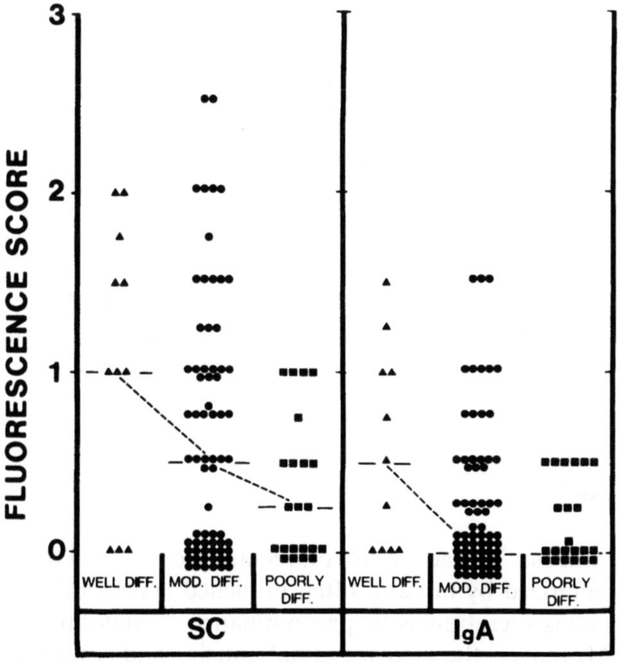

Fig. 3. Immunofluorescence scores given for epithelial staining of poorly (■), moderately (●), or well-differentiated (▲) carcinomas. Medians are connected by dashed lines. Well-differentiated carcinomas tended to express more SC and IgA than moderately differentiated ($p=0.08$ and $p=0.09$, respectively), and moderately and poorly differentiated carcinomas taken together were scored significantly lower than well-differentiated ones (SC, $p<0.05$; IgA, $p=0.06$). Adapted from ROGNUM et al. (1982b)

5.1.2 MHC Class I and II Molecules

Epithelial MHC class II determinants may appear in the large bowel in both inflammatory (SELBY et al. 1983) and neoplastic (ROGNUM et al. 1983; GHOSH et al. 1986) conditions. No simple relationship to tumor differentiation, clinicopathological stage or biological behaviour has as yet emerged from studies of these markers in large bowel carcinoma (DAAR et al. 1982; DAAR and FABRE 1983; CSIBA et al. 1984; UMPLEBY et al. 1985; MOMBURG et al. 1986; GHOSH et al. 1986).

Nevertheless, for HLA-DR we found a remarkable disparity according to histological tumor grade: well-differentiated carcinomas showed great intra-

◁ **Fig. 2.** Immunohistochemical staining for SC (*red*) and IgA (*green*). **a** Normal colon mucosa; yellow indicates presence of both SC and IgA and signifies normal epithelial transport. **b** Large bowel adenoma with severe dysplasia; the epithelium lacks both SC and IgA although there are numerous IgA-producing immunocytes in the stroma. **c** Well-differentiated large bowel carcinoma; yellow in tumor epithelium (*Tu*) signifies preserved SC-mediated transport of IgA. **d** Poorly differentiated large bowel carcinoma; tumor epithelium (*Tu*) lacks both SC and IgA. **a** ×260, **b, c, d** ×105 tumor staining variability whereas poorly differentiated and highly aggressive carcinomas were without exception homogeneously stained – either positively or negatively (ROGNUM et al. 1983). Differences in the composition of the carcinoma material and in the recording of staining heterogeneity might explain why previous studies failed to reveal these associations with DR expression. Our observations are in agreement with the concept of clonal evolution of tumor cell populations (NOWELL 1976, 1986). According to this hypothesis tumor progression is due to an acquired genetic lability permitting stepwise selection of variant sublines of neoplastic cells. One clone may possess selective advantages and finally become predominant. Another argument for this theory is that in cases with advanced carcinoma the staining patterns for SC, epithelial IgA, and HLA-DR were found to be remarkably similar in primary and secondary tumors from the same patient (ROGNUM et al. 1985).

5.2 The "Transitional Mucosa"

The so-called "transitional mucosa" adjacent to colorectal carcinoma (FILIPE and BRANFOOT 1976) is characterized by tall and often branched crypts, dilated goblet cells and increased mucus secretion with predominance of sialomucins (DAWSON and FILIPE 1976). Moreover, the increase in sialomucins and concomitant decrease in sulfomucins is directly relarted to extent of invasion by the adjacent carcinomas (FILIPE and BRANFOOT 1974). These features may either reflect a pre-neoplastic phase (FILIPE and BRANFOOT 1974) or merely represent secondary changes due to toxic influences from the adjacent carcinoma. Furthermore, CEA seems to increase in the transitional mucosa with decreasing degree of tumor differentiation (ROGNUM et al. 1982b).

5.2.1 Secretory Component and Secretory Immunoglobulins

Decreased expression of SC and epithelial IgA was seen in the transitional mucosa both with decreasing degree of tumor differentiation and with increasing clinico-pathological stage (ROGNUM et al. 1982b). Expression of SC and IgA in this zone also seemed to be related to the DNA ploidy pattern of the carcinoma as less epithelial staining was observed adjacent to aneuploid than near diploid tumors (ROGNUM et al. 1982a).

5.3 Pre-cancerous Lesions in Ulcerative Colitis

Patients with long-standing total ulcerative colitis (UC) are at increased risk of developing large bowel carcinomas (LENNARD-JONES et al. 1983). This risk is related to development of epithelial dysplasia (MORSON and PANG 1967; RIDDEL et al. 1983). All patients with long-standing UC, therefore, should be subjected to regular biopsy sampling from the various segments of the large bowel to detect and select those who would benefit from prophylactic colectomy.

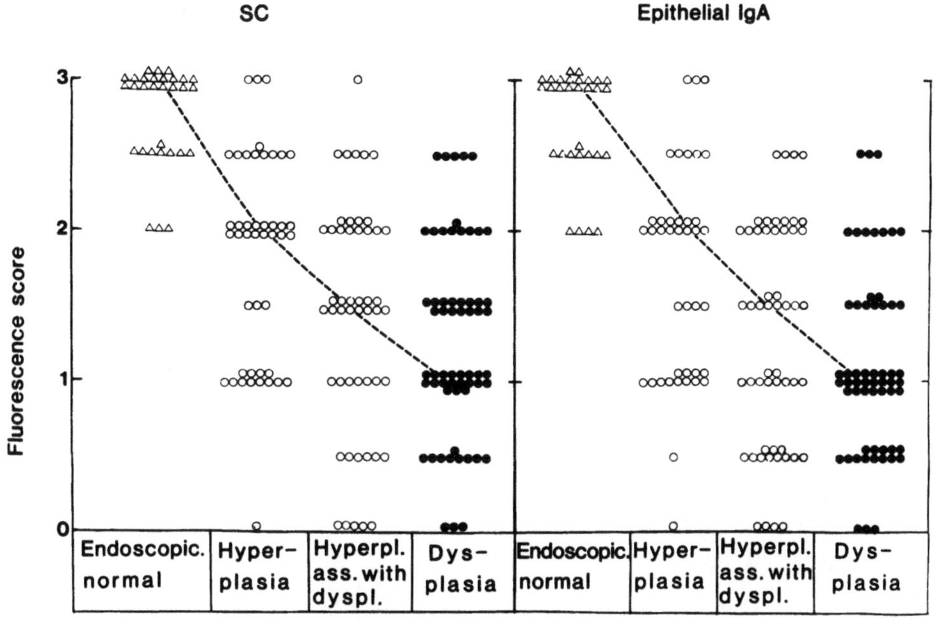

Fig. 4. Relation between immunofluorescence scores for SC and epithelial IgA in various large-bowel lesions of patients with long-standing ulcerative colitis (epithelial hyperplasia, hyperplasia associated with dysplasia in other areas, or dysplasia) compared with endoscopically normal biopsy specimens. Dashed lines connect median scores for each category. SC and IgA were scored significantly higher for normal mucosa than for the three lesions and also higher for hyperplasia than for dysplasia ($p < 0.01$). Adapted from Rognum et al. (1982c)

Inflammation and crypt destruction result in regenerative epithelial alterations which may imitate dysplasia (Riddel et al. 1983), and many attempts have been made to find markers of a truly dysplastic development. Isaacson (1976) claimed that epithelial expression of CEA was indicative of pre-cancer but we could not confirm this observation (Rognum et al. 1982c). Binding of peanut lectin likewise failed to afford a precise diagnosis (Jass et al. 1986).

5.3.1 Secretory Component and Secretory Immunoglobulins

Dysplastic epithelium was reported by Isaacson (1982) to be consistently negative for SC. Also Rognum et al. (1982c) found significantly reduced expression of SC and epithelial IgA compared with reactive hyperplasia, but large individual variations were observed in both types of lesions (Fig. 4) and precluded application of these markers in diagnostic work.

5.3.2 MHC Class I and II Molecules

Evaluation of heterogeneous expression of SC and class II determinants (Figs. 5 and 6) appears more promising in a diagnostic context (Rognum et al. 1987a).

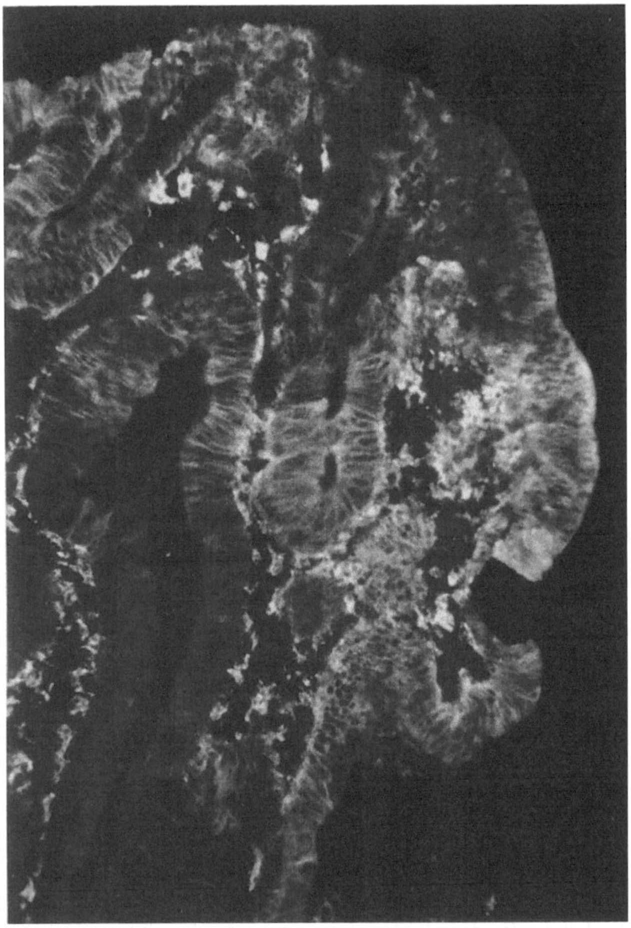

Fig. 5. Heterogeneous expression of epithelial HLA-DR determinants in ulcerative colitis with severe dysplasia. ×150

In high grade dysplasia abrupt transitions between intensely positive and completely negative epithelial cells were often seen. The similarities between dysplastic lesions and early large bowel carcinomas (ROGNUM et al. 1983) indicated that such heterogeneous marker expression is characteristic of a neoplastic development in the colon epithelium. The clinical significance of this observation in distinguishing regenerative from dysplastic epithelium will have to be evaluated in a larger follow-up study.

5.4 Adenomas

The theories of an "adenoma-carcinoma sequence" (MORSON 1974) or a "dysplasia-carcinoma sequence" (MORSON and KONISHI 1980) postulate that most

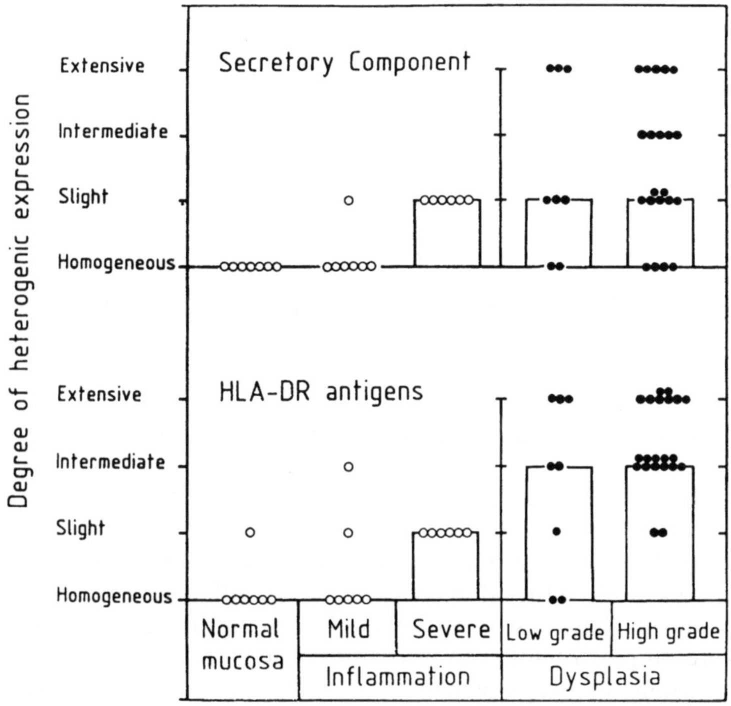

Fig. 6. Scatter diagram of degree of heterogeneous expression of HLA-DR and secretory component in normal colonic mucosa and inflammatory or dysplastic lesions of various severity. From ROGNUM et al. (1987a)

large-bowel carcinomas develop through increasing degrees of epithelial dysplasia. This notion has been supported by recent advances in the field of molecular biology. Increased expression of the *ras* oncogene in the course of neoplastic transformation is of particular interest (SPANDIDOS and KERR 1984). Studies of DNA ploidy patterns (GOH and JASS 1986), SC and epithelial IgA (ISAACSON 1982; ROGNUM et al. 1982d) may support the stepwise concept as cancer-like alterations are more pronounced in high grade dysplasia. Nevertheless, the decreased expression of SC and epithelial IgA, both with increased degree of dysplasia in adenomas (Fig. 2b) and with decreased degree of differentiation in adenocarcinomas (Figs. 2d, 3), indicates that invasiveness requires a second and as yet unknown fundamental change in the epithelial cell.

5.5 Metaplastic Polyps

The metaplastic polyp is regarded as a non-neoplastic disorder of differentiation (KAYE et al. 1973). Nevertheless, metaplastic polyps and colorectal carcinomas show several common features such as mucin alterations (BOLAND et al. 1982), increased CEA expression (ROGNUM et al. 1982d; JASS et al. 1984), and decreased SC and epithelial IgA (ROGNUM et al. 1982d; JASS and FAULDY 1985).

This does not necessarily imply that a metaplastic polyp is precancerous but rather that it develops as a result of environmental factors which may be implicated in carcinogenesis.

6 Conclusions

Markers of glandular differentiation are useful in several ways:
- they may contribute to better understanding of the histogenesis of certain tumors
- they may corroborate histopathological and clinical evaluation of tumors by enhancing the diagnostic and prognostic information
- they may be of significant value in the search for origin of anaplastic secondary tumors
- they may be an adjunct in distinguishing pre-cancerous epithelial alterations (dysplasia) from reactive hyperplasia

It is important to stress that several of these markers may likewise show either increased or decreased epithelial expression during inflammatory conditions. Such changes may imitate those found in tumors and precancerous lesions. It is crucial, therefore, to pay attention to the distribution pattern of the actual markers (e.g., homogenous vs. heterogeneous) and to the histopathological features of the lesions.

Methodological considerations have been beyond the scope of this chapter although tissue preparation and immunological staining technology are important variables in immunohistochemical work. The markers discussed above may thus be denatured or masked by routinely used formalin fixation and paraffin embedding. Some of these problems have been discussed in detail elsewhere (ROGNUM et al. 1980b; BRANDTZAEG 1981; BRANDTZAEG and ROGNUM 1984a, 1984b). With regard to the choice between immunoenzyme or immunofluorescence staining, the former method affords the possibility of evaluating morphological features in the same section. However, the observation of coexpression of two markers in a single cell is usually unreliable with paired immunoenzyme staining (VALNES et al. 1983; VALNES and BRANDTZAEG 1984). Paired immunofluorescence is superior for such studies (BRANDTZAEG 1981; BRANDTZAEG and ROGNUM 1983).

References

Arends JW, Wiggers T, Thijs CT, Verstijnen C, Swaen GJV, Bosman FT (1984) The value of secretory component (SC) immunoreactivity in diagnosis and prognosis of colorectal carcinomas. Am J Clin Pathol 82:267–274

Arnold RR, Cole MF, McGee JR (1977) A bactericidal effect for human lactoferrin. Science 197:263–265

Benacerraf B (1981) Role of MHC gene products in immune regulation. Science 212:1229–1238

Bhan AK, DesMarais CL (1983) Immunhistologic characterization of major histocompatibility antigens and inflammatory cellular infiltrate in human breast cancer. JNCI 71:507–516
Boland CR, Montgomery CK, Kim YS (1982) A cancer-associated mucin alteration in benign colonic polyps. Gastroenterology 82:664–672
Brandtzaeg P (1974a) Mucosal and glandular distribution of immunoglobulin components. Immunohistochemistry with a cold ethanol-fixation technique. Immunology 26:1101–1114
Brandtzaeg P (1974b) Human secretory component. II Physicochemical characterization of free secretory component purified from colostrum. Scand J Immunol 3:707–716
Brandtzaeg P (1974c) Mucosal and glandular distribution of immunoglobulin components. Differential localization of free and bound SC in secretory epithelial cells. J Immunol 112:1553–1559
Brandtzaeg P (1977) Immunohistochemical studies on various aspects of glandular immunoglobulin transport in man. Histochemical J 9:553–572
Brandtzaeg P (1981) Prolonged incubation time in immunohistochemistry: effects on fluorescence staining of immunoglobulins and epithelial components in ethanol- and formalin-fixed paraffin-embedded tissues. J Histochem Cytochem 29:1302–1315
Brandtzaeg P (1983) The secretory immune system of lactating human mammary glands compared with other exocrine organs. Ann NY Acad Sci 409:353–383
Brandtzaeg P (1985) Role of J Chain and secretory component in receptor-mediated glandular and hepatic transport of immunoglobulins in man. Scand J Immunol 22:111–146
Brandtzaeg P, Baklien K (1977) Intestinal secretion of IgA and IgM: A hypothetical model. In: Immunology of the gut. Ciba Found Symp 46:106–113
Brandtzaeg P, Rognum TO (1983) Evaluation of tissue preparation methods and paired immunofluorescence staining for immunocytochemistry of lymphomas. Histochem J 15:655–689
Brandtzaeg P, Rognum TO (1984a) Evaluation of nine different fixatives. 1. Preservation of immunoglobulin isotypes, J chain, and secretory component in human tissues. Pathol Res Pract 179:250–266
Brandtzaeg P, Rognum TO (1984b) Evaluation of nine different fixatives. 2. Preservation of IgG, IgA and secretory component in artificial immunohistochemical test substrate. Histochemistry 81:313–319
Bullen JJ, Rogers HJ, Leigh L (1972) Iron-binding proteins in milk and resistance to Escherichia coli infection in infants. Br Med J 1:69–75
Caselitz J, Jaup T, Seifert G (1981) Lactoferrin and lysozyme in carcinomas of the parotid gland. A comparative immunocytochemical study with the occurence in normal and inflamed tissue. Virchows Arch (Pathol Anat) 394:61–73
Caselitz J, Seifert G, Jaup T (1982) Tumour antigens in neoplasms of the human parotid gland. J Oral Pathol 11:374–386
Caselitz J, Seifert G, Grenner G, Schmidtberger R (1983) Amylase as an additional marker of salivary gland neoplasms. Pathol Res Pract 176:276–283
Caselitz J, Salfelder A, Seifert G (1984) Adenolymphoma: a immunohistochemical study with monoclonal antibodies against lymphocyte antigens. J Oral Pathol 13:438–447
Caselitz J, Lunau U, Hamper K, Schmiegelow P (1985) The pleomorphic adenoma of salivary glands transplanted on athymic mice. A lightmicroscopical and immunohistochemical investigation. Virchows Arch (Pathol Anat) 408:191–209
Caselitz J, Schulze I, Seifert G (1986) Adenoid cystic carcinoma of the salivary glands: an immunohistochemical study. J Oral Pathol 15:308–318
Csiba A, Whitwell HL, Moore M (1984) Distribution of histocompatibility and leucocyte differentiation antigens in normal human colon and in benign and malignant colonic neoplasms. Br J Cancer 50:699–709
Daar AS, Fabre JW (1983) The membrane antigens of human colorectal cancer cells: Demonstration with monoclonal antibodies of heterogeneity within and between tumors and of anomalous expression of HLA-DR. Eur J Cancer Clin Oncol 19:209–222
Daar AS, Fuggle SV, Ting A, Fabre JW (1982) Anomalous expression of HLA-DR antigens on human colorectal cancer cells. J Immunol 129:447–449
Dawson PA, Filipe MI (1976) An ultrastructural and histochemical study of the mucous membrane adjacent to and remote from carcinoma of the colon. Cancer 37:2388–2398
Dukes CE, Bussey HJR (1958) The spread of rectal cancer and its effect on prognosis. Br J Cancer 12:309–320

Erlandson RA, Cardon-Cardo C, Higgins PJ (1984) Histogenesis of benign pleomorphic adenoma (mixed tumor) of the major salivary glands. An ultrastructural and immunohistochemical study. Am J Surg Pathol 8:803–820

Fantasia JE, Lally ET (1984) Localization of free secretory component in pleomorphic adenomas of minor salivary gland origin. Cancer 53:1786–1789

Filipe MI, Branfoot AC (1974) Abnormal patterns of mucus secretion in apparently normal mucosa of large intestine with carcinoma. Cancer 34:282–290

Filipe MI, Branfoot AC (1976) Mucin histochemistry of the colon. Curr Top Pathol 63:143–178

Flemming A (1922) On a remarkable bacteriolytic element found in tissues and secretions. Proc R Soc B93:306–317

Fleming KA, McMichael A, Morton JA, Woods J, McGee JOD (1981) Distribution of HLA class I antigens in normal human tissue and in mammary cancer. J Clin Pathol 34:779–784

Franzen L, Carlssø B, Ångstrøm T (1979) Cytology and cytochemistry of acinic cell carcinoma. Acta Otolaryngol Suppl 360:174–177

Ghosh AK, Moore M, Street AJ, Howat JMT, Schofield PF (1986) Expression of HLA-D sub-region products on human colorectal carcinoma. Int J Cancer 38:459–464

Gion M, Mione R, Dittadi R, Fasan S, Pallini A, Bruscagnin G (1986) Carcinoembryonic antigen, ferritin, and tissue polypeptide antigen in serum and tissue. Relationship with the receptor content in breast carcinoma. Cancer 57:917–922

Goh HS, Jass JR (1986) DNA content and the adenoma-carcinoma sequence in the colorectum. J Clin Pathol 39:387–392

Harris JP, South MA (1981) Secretory component. A glandular epithelial cell marker. Am J Pathol 105:47–53

Harris JP, Caleb MH, South MA (1975) Secretory component in human mammary carcinoma. Cancer Res 35:1861–1864

Hayashi Y, Yanagawa T, Yoshida H, Yura Y, Nitta T, Sato M (1985) Induction of other differentiation stages in neoplastic epithelial duct and myoepithelial cells from the human salivary gland grown in athymic nude mice. Cancer 55:2575–2583

Hodes JE, Hull MT, Karn RC, Merritt AD, Hodes ME (1981) Immunohistochemical demonstration of the loss of immunoreactive amylase from neoplastic human salivary gland. Experientia 37:184–187

Hsu SM, Hsu PL, Nayak RN (1981) Warthin's tumour: an immunohistochemical study of its lymphoid stroma. Hum Pathol 12:251–257

Iacono VJ, MacKay BJ, DiRienzo S, Pollock JJ (1980) Selective antibacterial properties of Lysozyme for oral microorganisms. Infect Imm 29:623–632

Isaacson P (1976) Tissue demonstration of carcinoembryonic antigen (CEA) in ulcerative colitis. Gut 17:561–567

Isaacson P (1982) Immunoperoxidase study of the secretory immunoglobulin system in colonic neoplasia. J Clin Pathol 35:14–25

Jacobsen N, Lyche Melvaer K, Hensten-Pettersen A (1976) Some properties of salivary amylase: A survey of the literature and some observations. J Dent Res 51:381–388

Jass JR (1986) Lymphocyte infiltration and survival in rectal cancer. J Clin Pathol 39:585–589

Jass JR, Fauldy J (1985) Immunohistochemical demonstration of IgA and secretory component in relation to epithelial cell differentiation in normal colorectal mucosa and metaplastic polyp: A semiquantitative study. Histochem J 17:373–380

Jass JR, Filipe MI, Abbas S, Falcon CAJ, Wilson Y, Lovell D (1984) A morphological and histochemical study of metaplastic polyps of the colorectum. Cancer 53:510–515

Kaye GI, Fenoglio CM, Pascal RR, Lane N (1973) Comparative electron microscopic features of normal, hyperplastic, and adenomatous human colonic epithelium. Gastroenterology 64:926–945

Korsrud FR, Brandtzaeg P (1982) Characterization of epithelial elements in human major salivary glands by functional markers: Localization of amylase, lactoferrin, lysozyme, secretory component and secretory immunoglobulins by paired immunofluorescence staining. J Histochem Cytochem 30:657–666

Korsrud FR, Brandtzaeg P (1984a) Immunofluorescence study of secretory epithelial markers in pleomorphic adenomas. Virchows Arch (Pathol Anat) 403:291–300

Korsrud FR, Brandtzaeg P (1984b) Immunohistochemical characterization of cellular immunoglobulins and epithelial marker antigens in Warthin's tumour. Hum Pathol 15:361–367

Kraus FW, Mestecky J (1971) Immunohistochemical localization of amylase, lysozyme and immunoglobulins in the human parotid gland. Arch Oral Biol 16:781–789

Kvale D, Rognum TO, Thorud E, Fosså SD, Roe JS (1987a) Circulating secretory component (SC) in breast neoplasms. Secretory IgA and secretory IgM in serum related to CEA and to histological grade, DNA ploidy, and expression of SC, IgA, and CEA. Am J Clin Pathol (to be published)

Kvale D, Rognum TO, Brandtzaeg P (1987b) Elevated levels of secretory immunoglobulins A and M in serum of patients with large bowel carcinoma indicate liver metastasis. Cancer 59:203–207

Lai A, Fat RFM, McClelland DBL, van Furth R (1976) In vitro synthesis of immunoglobulins, secretory component, complement and lysozyme by human gastrointestinal tissues. I Normal tissues. Clin Exp Immunol 23:9–19

Lee PP, Buehring GC (1981) Local immunoglobulin A in normal and neoplastic breast tissues. J Natl Cancer Inst 67:1231–1239

Lennard-Jones JE, Morson BC, Ritchie JK, Williams CB (1983) Cancer surveillance in ulcerative colitis. Lancet II:149–152

Lwin KY, Zuccarini O, Sloane JP, Beverly PCL (1985) An immunohistological study of leukocyte localization in benign and malignant breast tissue. Int J Cancer 36:433–438

Masson PL, Heremans JF, Dive C (1966) An iron-binding protein common to many external secretions. Clin Chim Acta 14:735–739

McClelland DBL, van Furth R (1975) In vitro synthesis of lysozyme by human and mouse tissues and leucocyts. Immunology 28:1099–1114

McGuire WL, Pearson OH, Segaloff A (1975) Predicting hormone responsiveness in human breast cancer. In: WL McGuire, PP Carbone, EP Vollmer (eds) Estrogen receptors in human breast cancer. Raven Press, New York, pp 17–30

Momburg F, Degener T, Bacchus E, Moldenhauer G, Hämmerling GJ, Möller P (1986) Loss of HLA-A, B, C and de novo expression of HLA-D in colorectal cancer. Int J Cancer 37:179–184

Morley DJ, Hodes JE, Calland J, Hodes ME (1983) Immunohistochemical demonstration of ribonuclease and amylase in normal and neoplastic parotid glands. Hum Pathol 14:969–973

Morson BC (1974) Evolution of cancer of the colon and rectum. Cancer 34:845–849

Morson BC, Konishi F (1980) Dysplasia in the colorectum. Clin Gastroenterol (Suppl I) 9:331–343

Morson BC, Pang LSC (1967) Rectal biopsy as an aid to cancer control in ulcerative colitis. Gut 8:423–434

Muenzel M (1976) Die Biochemie der menschlichen Speicheldrüsensekrete. Arch Oto-Rhino-Laryngol 213:209–285

Natali PG, De Martino C, Quaranta V, Bigotti A, Bellegrino MA, Ferrone S (1981) Changes in Ia-like antigen expression on malignant human cells. Immunogenetics 12:409–413

Newcomb RW, Normansell D, Stanworth DR (1968) A structural study of human exocrine IgA globulin. J Immunol 101:905–914

Nowell PC (1976) The clonal evolution of tumour cell populations. Acquired genetic lability permits stepwise selection of variant sublines and underlies tumour progression. Science 194:23–28

Nowell PC (1986) Mechanisms of tumor progression. Cancer Res 46:2203–2207

Petit JF, Jolles P (1963) Purification and analysis of human saliva lyzozyme. Nature Lond 200:168–169

Poger ME, Hirsh BR, Lamm ME (1976) Synthesis of secretory component by colonic neoplasms. Am J Pathol 82:327–338

Poulsen LO, Elling P, Brandt Sørensen F, Høedt-Rasmussen K (1986). HLA-DR expression and disease activity in ulcerative colitis. Scand J Gastroenterol 21:364–368

Rankin FW, Broders AC (1928) Factors influencing prognosis in carcinoma of the rectum. Surg Gynecol Obstet 46:660–667

Regezi JA, Batsakis JG (1977) Histogenesis of salivary gland neoplasms. Otolaryngol Clin North Am 10:297–307

Reitamo S, Klockars M, Raeste AM (1977) Immunohistochemical identification of lysozyme in the minor salivary glands of man. Archs Oral Biol 22:515–519

Reitamo S, Konttinen YT, Segerberg-Konttinen (1980) Distribution of lactoferrin in human salivary glands. Histochemistry 66:285–291

Remmele W, Heine M (1981) Pathologisch-anatomische Begutachtung Kolorectaler Karzinome unter prognostischen Gesichtspunkten. Med Welt 32:771–777

Richman AB (1976) Immunofluorescence studies of benign and malignant human mammary tissue. J Natl Cancer Inst 57:263–267

Riddell RH, Goldman H, Ranshoff DF et al. (1983) Dysplasia i inflammatory bowel disease: standarized classification with provisional clinical applications. Hum Pathol 14:931–968

Rognum TO, Brandtzaeg P, Ørjasaeter H, Elgjo K, Hognestad J (1980a) Immunohistochemical study of secretory component, secretory IgA and carcinoembryonic antigen in large bowel carcinomas. Pathol Res Pract 170:126–145

Rognum TO, Brandtzaeg P, Ørjasaeter H, Fausa O (1980b) Immunohistochemistry of epithelial cell markers in normal and pathological colon mucosa. Comparison of results based on routine formalin- and cold ethanol-fixation methods. Histochemistry 67:7–21

Rognum TO, Thorud E, Elgjo E, Brandtzaeg P, Ørjasaeter H, Nygaard K (1982a) Large-bowel carcinomas with different ploidy, related to secretory component, IgA, and CEA in epithelium and plasma. Br J Cancer 45:921–934

Rognum TO, Elgjo K, Brandtzaeg P, Ørjasaeter H, Bergan A (1982b) Plasma carcinoembryonic antigen concentrations and immunhistochemical patterns of epithelial marker antigens in patients with large bowel carcinoma. J Clin Pathol 35:922–933

Rognum TO, Elgjo K, Fausa O, Brandtzaeg P (1982c) Immunohistochemical evaluation of carcinoembryonic antigen, secretory component, and epithelial IgA in ulcerative colitis with dysplasia. Gut 23:123–133

Rognum TO, Fausa O, Brandtzaeg P (1982d) Immunohistochemical evaluation of carcinoembryonic antigen, secretory component, and epithelial IgA in tubular and villous large bowel adenomas with different grades of dysplasia. Scand J Gastroenterol 17:341–348

Rognum TO, Brandtzaeg P, Thorud E (1983) Is heterogeneous expression of HLA-DR antigens and CEA along with DNA-profile variations evidence of phenotypic instability and clonal proliferation in human large bowel carcinomas? Br J Cancer 48:453–551

Rognum TO, Thorud E, Brandtzaeg P (1985) Preservation of cytometric DNA Distribution and epithelial marker expression after tumour progression of human large bowel carcinomas. Cancer 56:1658–1666

Rognum TO, Brandtzaeg P, Elgjo K, Fause O (1987a) Heterogeneous expression of class II (HLA-DR) determinants and secretory component related to dysplasia in ulcerative colitis. Br J Cancer 56

Rognum TO, Thorud E, Lund E (1987b) Survival of large bowel carcinoma patients with different DNA ploidy. Br J Cancer (to be published)

Rowe DJ, Beverly PCL (1984) Characterisation of breast cancer infiltrates using monoclonal antibodies to human leucocyte antigens. Br J Cancer 49:149–159

Saito I, Teratini K, Inoue M, Saito A, Funatsu K, Moro I (1984) Immunohistochemical characterization of functional markers in human minor salivary gland tumors. J Oral Pathol 13:525–534

Sanderson AR, Beverly PCL (1983) Interferon, β-2-microglobulin and immunoselection in the pathway to malignancy. Immunology Today 4:211–213

Sato M, Hayashi Y, Yoshida H, Yanagawa T, Yura Y, Nitta T (1984) Search for specific markers of neoplastic epithelial duct and myoepithelial cell lines established from human salivary gland and characterization of their growth *in vitro*. Cancer 54:2959–2967

Scott H, Solheim BG, Brandtzaeg P, Thorsby E (1980) HLA-DR-like antigens in the epithelium of the human small intestine. Scand J Immunol 12:77–82

Scott H, Brandtzaeg P, Solheim BG, Thorsby E (1981) Relation between HLA-DR-like antigen and secretory component (SC) in jejunal epithelium of patients with coeliac disease or dermatitis herpatiformis. Clin Exp Immunol 44:233–238

Sehested M, Barfoed C, Krogdahl A, Bretlau P (1985) Immunohistochemical investigation of lysozyme, lactoferrin, α-1-antitrypsin, α-1-antichymotrypsin and ferritin in parotid gland tumours. J Oral Pathol 14:459–465

Seifert G, Caselitz J (1983) Tumour markers in parotid gland carcinomas: Immunohistochemical investigations. Cancer Detect Prev 6:119–130

Selby WS, Janossy G, Goldstein G, Jewell DP (1981) T lymphocyte subsets in human intestinal mucosa: the distribution and relationship to MHC-derived antigens. Clin Exp Immunol 44:453–458

Selby WS, Janossy G, Mason DY, Jewel DP (1983) Expression of HLA-DR antigen by colonic epithelium in inflammatory bowel disease. Clin Exp Immunol 53:614–618

Sollid LM, Kvale D, Brandtzaeg P, Markussen G, Thorsby E (1987) Interferon-γ enhances expression of secretory component, the epithelial receptor for polymeric immunoglobulins. J Immunol 138:4303–4306

Spandidos DA, Kerr IB (1984) Elevated expression of the human ras oncogene family in premalignant and malignant tumours of the colorectum. Br J Cancer 49:681–688

Strominger JL, Orr HT, Robb RJ, Pasham P, Ploegh HL, Mann OL, Bilofsky H, Saroft HA, Wytt, Kabat EA (1980) An evaluation of the significance of amino acid sequence homologies in human histocompatibility antigens (HLA-A and HLA-B) with immunoglobulins and other proteins, using relatively short sequences. Scand J Immunol 11:537–592

Syrjanen KJ, Hjelt LH (1978) Tumour-host relationship in colorectal carcinoma. Dis Colon Rect 21:29–36

Thompson AS, Bryant HC (1950) Histogenesis of papillary cystadenoma lymphomatosum (Warthin's tumor) of the parotid salivary gland. Am J Pathol 26:807–829

Thoresen S, Thorsen T, Tangen M, Hartveit F (1982) Oestrogen and progesterone receptor content and the distribution of histological grade in breast cancer. Breast Cancer Res Treatment 2:251–255

Thrane PS, Brandtzaeg P (1987a) Distribution of MHC class I and II molecules in fetal, postnatal and normal and inflamed adult major and minor salivary glands. Submitted

Thrane PS, Brandtzaeg P (1987b) Presumably activated (HLA-DR positive) T cells occur adjacent to HLA-DR negative duct epithelium in Warthin's tumor of the salivary gland. Abstract. Scand J Immunol 24:480

Thrane PS, Rognum TO, Huitfeldt HS, Brandzaeg P (1987a) The distribution of HLA-DR determinants, secretory component (SC), amylase, lysozyme, IgA, IgM, and IgD in fetal and postnatal salivary glands. Adv Exp Med Biol (to be published)

Thrane PS, Sollid LM, Huitfeldt HS, Brandtzaeg P (1987b) Paired immunofluorescence marker study of various cell populations in pleomorphic adenomas. Submitted

Thrane PS, Sollid LM, Brandtzaeg P (1987c) HLA-DR-positive dendritic histiocytic cells in pleomorphic adenomas. Immunohistochemical characterization and positive selection of DR-positive cells by immunomagnetic beads. Submitted

Tomasi TB, Tan EM, Solomon A, Prendergast RA (1965) Characteristics of an immune system common to certain external secretions. J Exp Med 121:101–124

Toto P, Hsu DJ (1985) Product definition of pleomorphic adenoma of minor salivary glands. J Oral Pathol 14:818–832

Umpleby HC, Heinemann D, Symes MO, Williamson RCN (1985) Expression of histocompatibility antigens and characterization of mononuclear cell infiltrates in normal and neoplastic colorectal tissues of humans. J Natl Cancer Inst 74:1161–1168

Valnes K, Brandtzaeg P (1984) Paired indirect immunoenzyme staining with primary antibodies from the same species. Application of horseradish peroxidase and alkaline phosphatase as sequential labels. Histochem J 16:477–487

Valnes K, Brandtzaeg P, Hanssen LE, Stave R, Larsen S, Londong W (1983) Quantitation of immunoglobulin- and peptide hormone-producing cells in gastrointestinal mucosa. Comparison of direct immunofluorescence and unlabelled antibody peroxidase-antiperoxidase method. Histochem J 15:1011–1020

Wallich R, Bulbuc N, Hammerling GJ, Katzav S, Segal S, Feldman M (1985) Abrogation of metastatic properties of tumour cells by *de novo* expression of H-2K antigens following H-2 gene transfection. Nature 315:301–305

Warner TFCS, Seo IS, Azen EA, Hafez GR, Zarling TA (1985) Immunocytochemistry of acinic cell carcinomas and mixed tumours of salivary glands. Cancer 56:2221–2227

Weisz-Carrington P, Poger ME, Lamm ME (1976) Secretory immunoglobulins in colonic neoplasms. Am J Pathol 85:303–316

Whitwell HL, Huges HPA, Moore M, Ahmed A (1984) Expression of major histocompatibility antigens and leucocyte infiltration in benign and malignant human breast disease. Br J Cancer 49:161–172

Wolley RG, Schreiber K, Koss LG, Kara M, Sherman A (1982) DNA distribution in human colon carcinomas and its relationship to clinical behaviour. J Natl Cancer Inst 69:15–22

Zinkernagel RM, Doherty PC (1979) MHC cytotoxic T cells: studies on the biological role of polymorphic major transplantation antigens determining T cell restriction-specificity, function and responsiveness. Adv Immunol 27:51–177

Mesenchymal Tumor Markers: Intermediate Filaments

M. ALTMANNSBERGER and M. OSBORN

1 Introduction

By far the majority of mesenchymal tumors can be classified by their typical morphological patterns in routine stains. In a minority of such tumors, and in particular those with an undifferentiated morphology, a differential diagnosis may be difficult of one relies only on the light microscopical appearance. In such situations immunohistochemical techniques using polyclonal or monoclonal antibodies can provide additional information; here we emphasize the use of antibodies specific for the different intermediate filament proteins in human tumor diagnosis and concentrate in particular on the use of antibodies to desmin and vimentin in the differential diagnosis of mesenchymal tumors.

Intermediate filaments (IF), also referred to as 10 nm filaments or 100 Å filaments, are one of the three fibrous elements of the cytoskeleton. They are present in nearly all cells of vertebrates (FRANKE et al. 1978; BENNETT et al. 1978; HOLTZER et al. 1982; OSBORN et al. 1982a; WEBER and GEISLER 1984) and seem to occur at least in some of the invertebrates. In electron micrographs they can be distinguished from the other two filament systems by their diameter (7–11 nm) which is intermediate between that of microfilaments (6 nm) and microtubules (20–22 nm). Using immunological techniques all three systems can be distinguished by the use of appropriate antisera. In the electron microscope it is difficult to distinguish different intermediate filament types since all IFs show very similar morphologies. However immunohistological studies on normal tissues as well as biochemical and protein chemical data have demonstrated that intermediate filament proteins can be subdivided in five major categories: keratins, vimentin, desmin, GFAP and neurofilaments (see above references). The most important intermediate filaments markers for mesenchymal tumors are desmin and vimentin.

2 Definition

2.1 Desmin

Desmin (skeletin) is the intermediate filament protein typical of skeletal, visceral and some vascular smooth muscle cells. In 1977 SMALL and SOBIESZEK showed that a 55 kd band in an SDS gel from smooth muscle preparations disappeared after in situ proteolysis of muscle. The 55 kd protein extracted from muscle by acetic acid could also be reconstituted into filamentous arrays (for later experiments which yielded reconstituted 10 nm filaments with a morphology equivalent to those seen in vivo, see GEISLER and WEBER 1981b; IP et al. 1985). Since the contractility of the smooth muscle was not reduced by the removal of the 55 kd protein and in addition extraction of actin did not influence the general cell shape, this protein was assumed to have a cytoskeletal function and given the name "skeletin" (SMALL and SOBIESZEK 1977). The same protein was also shown to be present in the cytoskeleton of heart Purkinje cells and in normal myocardium (THORNELL et al. 1978, 1983). At about the same time LAZARIDES and HUBBARD (1976) demonstrated an analogous protein in chicken smooth muscle. Using antibodies specific for this protein they showed that it was present at the Z-discs of isolated skeletal myofibrils. In heart muscle cells in culture desmin was found not only in the Z-discs, but also in intercalated discs. Since this protein appeared to link myofibrils and thus resembled the desmosomes in epithelial tissues, it was named desmin (Greek: desmos = link, band). It soon become clear that desmin and skeletin were identical proteins. As the name desmin has gained more general acceptance it will be used here.

2.2 Vimentin

Vimentin filaments are the only IF type characteristically present in nonmuscular mesenchymal cells. Two groups originally described a major cellular protein different from actin and tubulin which was enriched in the detergent resistant cytoskeleton of cells grown in culture (OSBORN and WEBER 1977; BROWN et al. 1976). Rabbit autoantibodies which decorated fibres in certain permanent cell lines and which were different from microfilaments and microtubules were also described. Comparing the immunofluorescence pictures with electron microscopic data it became clear that the decorated structures were bundles of intermediate filaments of the keratin type (OSBORN et al. 1977). In addition several human hepatitis sera were shown also this time to contain high levels of antibodies recognizing vimentin intermediate filaments in human fibroblasts (OSBORN, unpublished data). These studies with autoantibodies soon led to the production of antibodies raised against the different intermediate filament proteins. Thus antibodies specific for the 55 Kd band from mouse 3T3 cells clearly decorated intermediate filaments in a variety of human and animal cells (FRANKE et al. 1978). This antibody also decorated perinuclear whorls or aggregates after treatment of the cells with colcemid, a characteristic feature associated with intermediate filaments in cells, known from previous electronmicroscopic studies (GOLDMAN and KNIPE 1972; HOLTZER et al. 1975). In contrast this antibody did not react on cells of epithelial, muscle or neuronal origin. To emphasize the difference between this protein and the proteins forming other intermediate filament types it was named vimentin (latin vimentum – wickerwork), a name which also reflects the appearance of intermediate filaments in fibroblasts (FRANKE et al. 1978).

3 Biochemical and Molecular Structure of Desmin and Vimentin Intermediate Filaments

For biochemical work sources rich in only one IF protein are optimal. Desmin has been usually purified either from chicken gizzard or from pig stomach, while for vimentin porcine eye lenses are the starting material of choice. The purification schemes usually involve gel filtration and DEAE cellulose chromatography performed in the presence of 6 to 8 M urea.

The first amino acid sequence work in porcine vimentin and on desmin from chicken and pig established several points important for the later development of the IF field (GEISLER and WEBER 1981 a). First vimentin and desmin were shown to be different gene products but related in sequence. Second, from the linear sequence it was clear that part of the desmin and vimentin molecules contained a seven residue repeat pattern (a–g), where the residues in a and d positions are primarily hydrophobic. The presence of these repeating heptade sequences argues for a coiled coil formation of the helical domains. Third, as already argued prior to 1981 from immunological data, tissue specificity overrode species divergence (for review see WEBER and GEISLER 1984; GEISLER and WEBER 1986).

Because the partial sequences suggested that IF molecules might contain a central α-helical rod domain conserved in sequence principle and in length, rather than in actual sequence, it was necessary to obtain a complete sequence for one IF protein. This was first done for desmin in 1982 by GEISLER and WEBER. The desmin sequence, taken together with the then available partial sequences of other IF proteins showed that IF proteins are characterized by the prototype structure shown in Fig. 1. In this structure the canonic principle is the rod domain – the central α-helical rod region some 310 residues in length. This rod domain has been documented in all of the 20–30 IF sequences currently known from protein and from DNA sequencing work (see especially HANUKOGLU and FUCHS 1982, 1983; STEINERT et al. 1983). In contrast to the rod region, the terminal domains – the head and the tail – can be hypervariable both in sequence and length. Thus those features common to all IF proteins e.g. the 21 nm repeat seen in electron microscopy for metal shadowed *in vitro* reconstituted IFs, the 7–11 nm diameter seen for filaments both *in vivo* and after *in vitro* assembly, as well as the apparently similar morphologies in the electron microscope can be attributed to the rod domain. Consequently differences in function or in stability of different IF proteins are more likely to reside in the end domains.

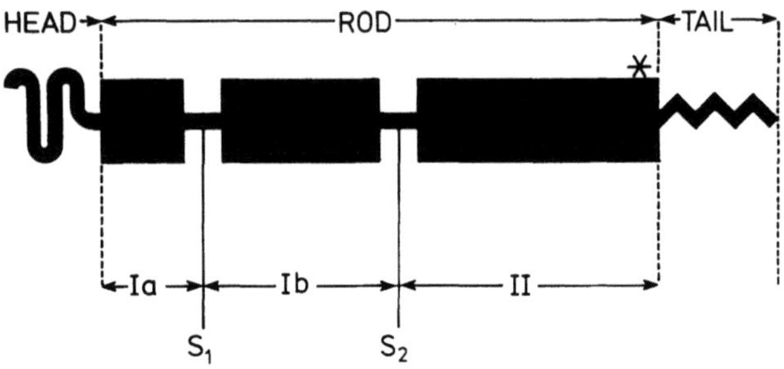

Fig. 1. A schematic representation of the domain structure of IF-proteins. The α-helical rod domain common to all IF proteins is interrupted by the nonhelical spacers S_1 and S_2 situated between helices Ia and Ib, and Ib and II respectively. The rod domain is flanked at the aminoterminal side by the nonhelical headpiece (head) and at the carboxyterminal side by the nonhelical tailpiece (tail). The asterisk marks the approximate position of the epitope of monoclonal antibody (IFA) that recognizes an epitope common to many proteins (PRUSS et al. 1981). (This figure is adapted from GEISLER und WEBER 1986)

Desmin and vimentin, together with glial fibrillary acidic protein (GFAP), form a subgroup of closely related non-epithelial, or type III, IF proteins. When sequence identities of the rod region of these IF polypeptides are compared identity within this group is around 63% whereas when the same sequences are compared either to the keratin type I or to the keratin type II sequences, the identity drops to around 30% (GEISLER and WEBER 1982; WEBER and GEISLER 1984; HANUKOGLU and FUCHS 1982, 1983). Although all three neurofilament proteins also share the same structural motif the identity between the rod regions of NF-L and vimentin is only 53%. A comparison of the DNA gene sequences shows that while the introns are very strongly conserved in number and position between keratin types I and II and the non-epithelial desmin, vimentin and GFAP group, NF-L (LEWIS and COWAN 1986) and NF-M show a totally different intron arrangement. Thus currently desmin, vimentin and GFAP are assigned to a IF group III separate from neurofilament proteins which are placed in group IV. Complete DNA sequences have been provided for hamster desmin and for hamster vimentin by QUAX et al. 1983, 1984, 1985. Hybridization studies suggest that vertebrates have only one gene for desmin and one for vimentin. (for review see BLOEMENDAL et al. 1985; for chicken m-RNA see NGAI et al. 1985; ZEHNER and PATERSON 1985).

The considerations important in arriving at a model of the 10 nm filament have been detailed elsewhere. Cross linking experiments, performed on a desmin derivative containing the whole rod region, showed that this was a tetramer and not a trimer (GEISLER and WEBER 1982). Thus the triple stranded model (STEINERT 1978) that had dominated IF structure for almost a decade could be laid to rest. The inference that the tetramer arose from double stranded coiled coils was soon confirmed by other methods. In the electron microscope the tetramers are indeed rod-shaped and have a length of around 50 nm (GEISLER et al. 1982). Decoration of these structures with Fab fragments from a desmin monoclonal antibody suggested that the two coiled coils in the tetramer are antiparallel since antibody molecules could be seen on both ends (GEISLER et al. 1985). Although details of the exact arrangements of the tetramer units, and of the importance of the tetramer arrangement for filament assembly are still under discussion, it is generally agreed that each 10 nm filament contains on average around 8 tetramers per diameter or 32 polypeptide chains/diameter (GEISLER et al. 1985; for review see IP et al. 1985; GEISLER and WEBER 1986).

The principles important in deciding the filament structure are assumed to be the same for all IF proteins with one important exception. Whereas the non-epithelial proteins in groups III and IV (desmin, vimentin, GFAP and the NF-L polypeptide) form homopolymers i.e. each protein can form a 10 nm filament by itself, the keratins are obligatory heteropolymers requiring one type I and one type II polypeptide for filament formation (for keratin nomenclature see HANUKOGLU and FUCHS 1982, 1983).

4 IF Arrangement in Vitro

Certain cell types can contain more than one IF polypeptides (for review see OSBORN et al. 1982a). In such cells it is interesting to know whether the different IF polypeptides are present in the same or in different 10 nm filaments. Several approaches have been used to try to answer this question. In some cell types it is already clear at the light microscopical level (OSBORN et al. 1980) that keratin and vimentin are present in different IFs a fact that has been confirmed by immunoelectron microscopy. However when the non-epithelial proteins occur together in the same cell e.g. desmin and vimentin in the rhabdomyosarcoma cell line RD (Fig. 2), or GFAP and vimentin in a glioma cell line, double labelling at the light microscopic level usually shows very similar profiles for the two IF proteins. Using the electron microscope it could be shown that individual 10 nm filaments in the glioma line contained both GFAP and vimentin (SHARP et al. 1982) and that individual 10 nm filaments in the RD cells contained both desmin and vimentin (TÖLLE et al. 1986). These experiments, suggesting that type III IFs can copolymerize into the same 10 nm filament are supported by chemical cross linking experiments (QUINLAN and FRANKE 1982, 1983) and also by experiments in which antibodies specific for a single intermediate filament type III protein have been microinjected. Thus for example if antibodies specific for either vimentin or desmin are microinjected into a cell line containing both IF proteins, perinuclear caps are formed in which the two different IF proteins colocalize, again suggesting that the two intermediate filament proteins were present in the same 10 nm filament (TÖLLE et al. 1986).

When cell types which have only vimentin filaments are injected with antibodies that recognize vimentin, the intermediate filaments are cross linked and within 2–4 hours retract into perinuclear caps. Cells with such caps seem perfectly capable of carrying out basic cellular functions such as cell division and movement over a substratum (GAWLITTA et al. 1981; KLYMKOWSKY 1981; LIN and FERAMISCO 1981). These experiments as well as the discovery of cell lines which lack IFs and yet divide without problem, argue strongly that IF function is more subtle than that of the other two filament systems present in the cell, i.e. microfilaments and microtubules. Perhaps, therefore, it may be more appropriate to search for a putative function of IFs in tissues rather than cells. Undoubtedly in certain situations IFs, through their links to the plasma membrane and the nucleus, contribute to tissue integrity, and the combined IF-membrane linked system may then serve to restrict the mobility of individual cells. Monoclonal antibodies specific for vimentin (OSBORN et al. 1984a; GOWN and VOGEL

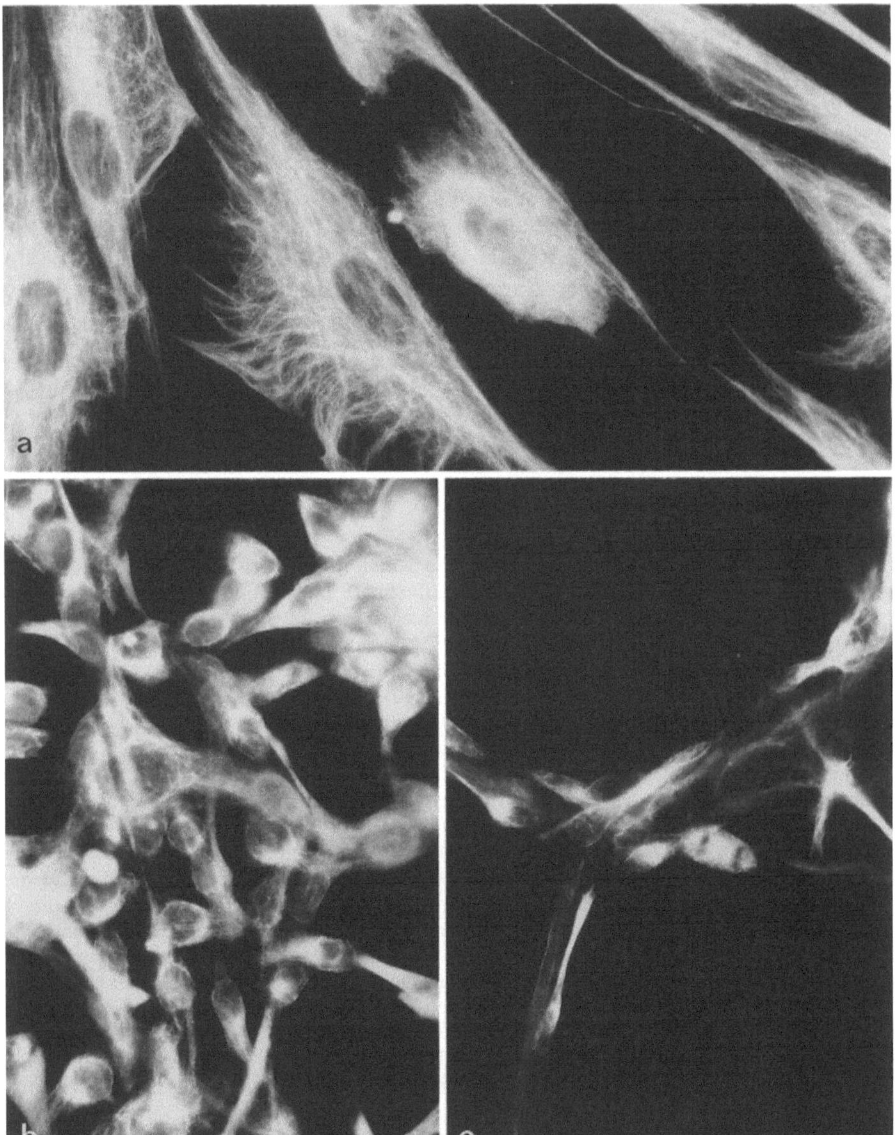

Fig. 2. Human cell lines labelled by antibodies to intermediate filaments. Fibroblast cell line (HS 27) stained by vimentin antibodies (**a**) and human rhabdomyosarcoma (RD) cell line incubated with antibodies against vimentin (**b**) and desmin (**c**). The fibroblast cell line expresses vimentin. In the RD cell line all cells are stained by the vimentin antibody (**b**), while the majority of cells additionally coexpress desmin (**c**) (**a–c**, FITC) (polyclonal antibodies against vimentin and desmin)

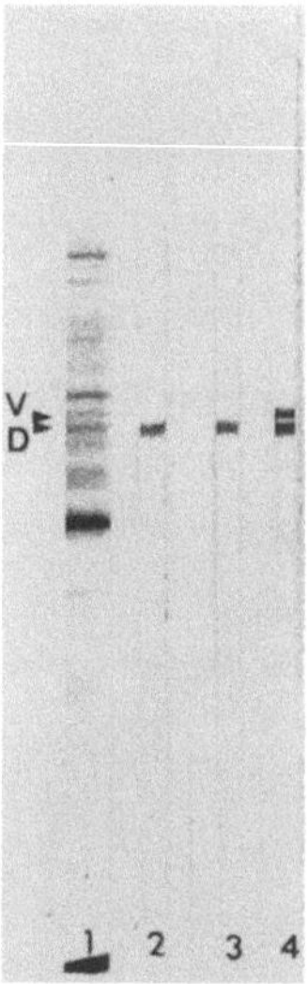

Fig. 3. Lanes 1–4 show an extract of the human rhabdomyosarcoma line RD stained with Coomassie Blue (lane 1) and immunoblots of the same RD extract after transfer to nitrocellulose and reaction with the desmin monoclonal antibodies DE-A-7 (lane 2). DE-C-3 (lane 3) and a mixture of the desmin monoclonal antibody DE-B-5 together with a vimentin monoclonal antibody (lane 4). V.D. indicate the positions of vimentin and desmin respectively (for further details see DEBUS et al. 1983a). Note that the RD cell line contains both vimentin and desmin as shown by immunblotting

1984; VIRTANEN et al. 1985) or specific for desmin have been described (DEBUS et al. 1983) (Fig. 3). Whether all monoclonal vimentin antibodies display the same specifity when tested on a large number of normal cells and tumors has not been definetely determined (see below).

5 Distribution of Desmin and Vimentin IFs in Normal Tissue

5.1 Vimentin

In normal tissue almost all mesenchymal cells express either vimentin or desmin or simultaneously both intermediate filament proteins. The majority of nonmuscular cell types are characterized by a positive reaction with antibodies against vimentin. This is true for the "cytic" as well as the "blastic" cell forms i.e. vimentin positive cell types include: fibrocytes, fibroblasts (Fig. 4), osteocytes and osteoblasts, chondrocytes and chondroblasts (FRANKE et al. 1978; OSBORN et al. 1984b). Expression of vimentin filaments seems to be independent of germ layer derivation and even cells coming from the neuroectoderm express only vimentin, e.g. Schwann cells (Fig. 5c) (OSBORN et al. 1982b, 1986a). The situation appears more complex in bone marrow. Using polyclonal vimentin antibodies DELLAGI et al. (1983) have shown that mature cells of erythropoesis and thrombopoesis express neither vimentin nor other intermediate filaments types. In addition in primary cultures of plasma cells 30% of the cells could not be labeled with a vimentin antibody.

Currently the question of whether all lymphatic cells express vimentin cannot be answered with certainty. Because of the close spatial relationship of hetero-

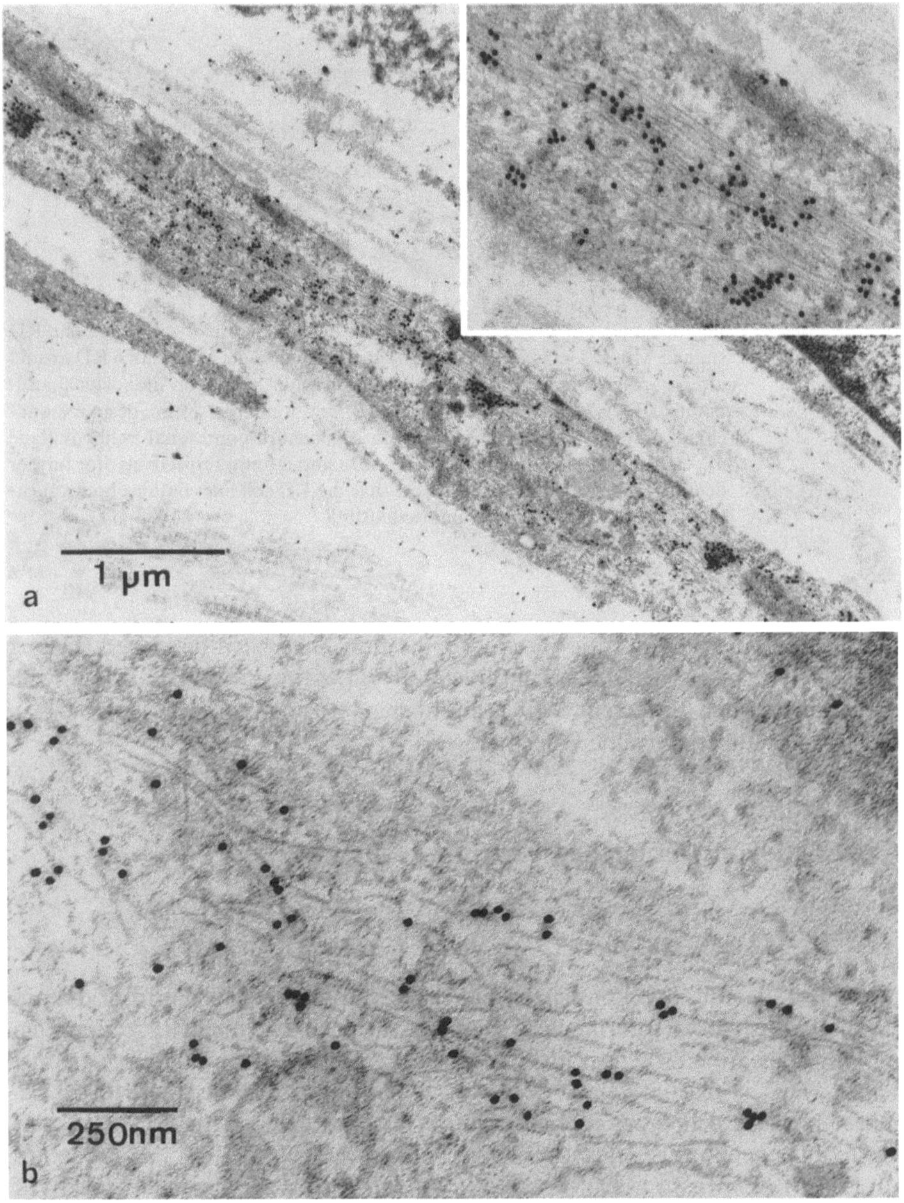

Fig. 4. Ultrastructural immunogold staining of vimentin. Tissue from an infantile fibromatosis was fixed in phosphate buffered saline containing 1% glutaraldehyde and 0,2% picric acid. It was then embedded at low temperature in Lowicryl K4M by standard procedures. Thin sections were incubated with the vimentin monoclonal antibody V 9 (diluted 1:100 in 1% ovalbumin in PBS), with rabbit antiserum to mouse IgG_S [Dako, diluted 1:100 (**a**) and 1:1000 (**b**) in 1% ovalbumin in PBS, and subsequently with protein-A-colloidal gold complex, particle size 15 nm (Janssen Life Science Products) diluted 1:25]. The sections were counterstained with uranyl acetate and lead citrate. **a** Fibroblast, ×21600, *inset* ×48900, **b** fibroblast, ×66600. (Courtesy R. ROMANOWSKI, Department of Pathology, University of Giessen)

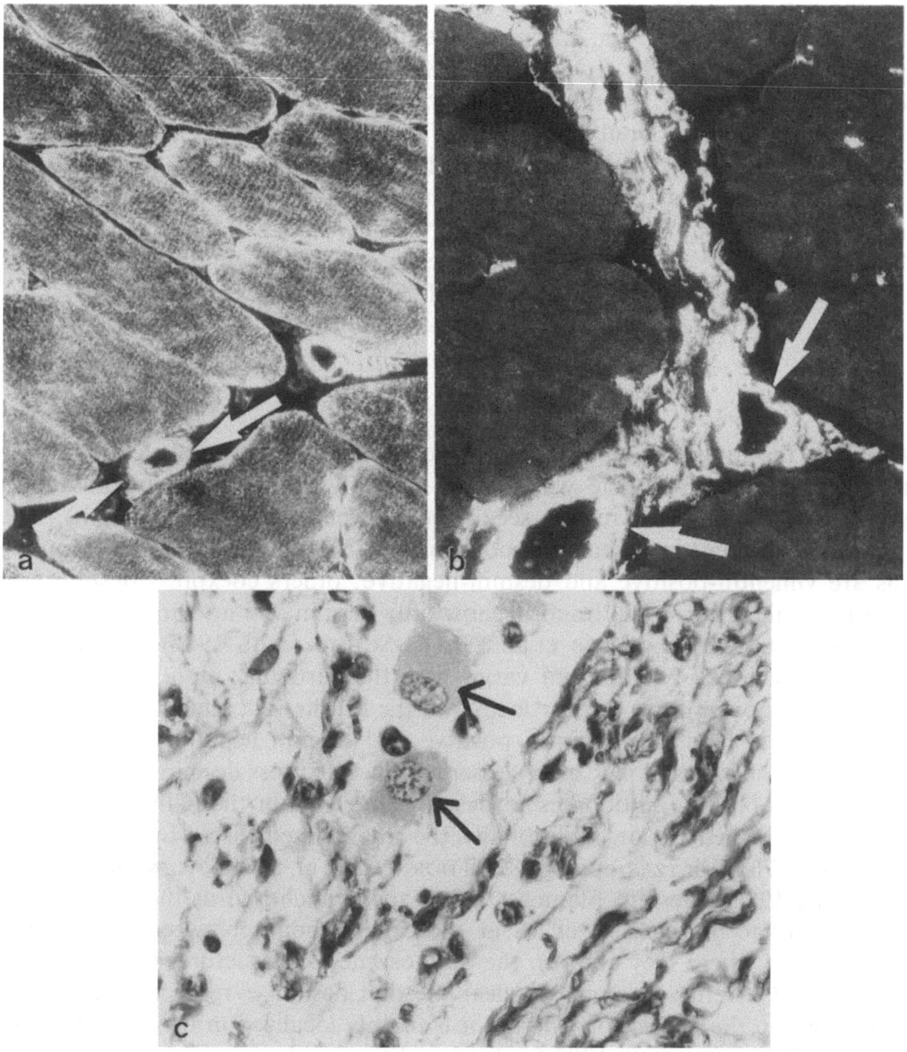

Fig. 5. Distribution of desmin and vimentin in skeletal muscle. Muscle fibres are desmin positive (**a**) and vimentin negative (**b**). In contrast fibrous tissues are not stained by the desmin antibody but are clearly labelled by the vimentin antibody (**b**). Note the coexpression of desmin and vimentin in vascular smooth muscle (**a**, **b**, *arrow*). **c** Schwann cells in a case of ganglioneuroblastoma also express vimentin; however the ganglion cells (*arrow*) are unstained (**a**, **b** FITC; **c** peroxidase) (polyclonal antibodies against desmin and vimentin)

genous cell populations in lymph nodes biochemical investigation using microdissection is not possible. Thus all studies are at the moment based on immunohistological findings using polyclonal or monoclonal antibodies against vimentin and it remains to be demonstrated whether the vimentin antibodies used in such studies have equivalent specifities. On the one hand, GIORNO (1985) has stated that nearly all lymphatic cells in the lymph node showed no staining with the vimentin antibody he used; on the other hand our preliminary results on lymph nodes using several different vimentin antibodies, show that the majority of lymphatic cells do have vimentin filaments although there is a subset of lymphocytes which are not stained by such antibodies. This problem can only be fully solved by double labelling of the cells with monoclonal antibodies against vimentin and with antibodies directed against appropriate surface markers to define the lymphatic cells. Vimentin is also the only intermediate filament type present in granulosa cells of the ovary (Czernobilsky et al. 1985; Moll et al. 1986), Sertoli cells of the testis, as well as in Langerhans cells (Löning et al. 1982) and in melanocytes of skin (Caselitz et al. 1983).

5.2 Desmin

Desmin is the intermediate filament protein characteristic of skeletal, cardiac and visceral smooth muscle where it is the only IF-type usually present in mature muscle cells (for review see Thornell et al. 1983) (Fig. 5a, b). The situation is different in vascular smooth muscle. Many vascular smooth muscle cells are vimentin-positive and desmin-negative; others coexpress desmin and vimentin while a very small fraction show only desmin and no vimentin (Frank and Warren 1981; Gabbiani et al. 1981a; Schmid et al. 1982). The relative number of each cell type present varies with the particular blood vessel examined. Thus for example in the aortic arch of the rat the vast majority of vascular smooth muscle cells are vimentin positive and desmin negative, while in the femoral and the external iliac arteries many cells coexpress desmin and vimentin (Osborn et al. 1981). Antibodies to desmin clearly stain the Z-line of skeletal muscle, and in heart stain additionally the intercalated discs (Lazarides and Hubbard 1976; Holtzer et al. 1982; Thornell et al. 1978; Altmannsberger et al. 1982, 1985). Desmin filaments seems to be anchored at the intercalated discs of cardiac muscle as well as at the cell boundary between cells of the Purkinje fibre where they abut on the desmosome-like structures (Kartenbeck et al. 1983). Currently it is not yet clear whether desmin is restricted in skeletal muscle only to the Z-discs or whether it is mainly localized in the interfibrillary space (see Tokuyasu et al. 1983 for further discussion).

6 Intermediate Filament Typing in Mesenchymal Tumors

6.1 Desmin-Positive Tumors (Tables 1, 2; Fig. 6)

A positive reaction with desmin antibodies is characteristic for rhabdomyosarcoma, rhabdomyoma, leiomyosarcoma and leiomyoma (for review see Altmannsberger et al. 1986b, c). We have investigated 31 cases of embryonal and alveolar rhabdomyosarcomas including the sarcoma botryoides subtype, and have shown that polyclonal or monoclonal antibodies against desmin positively identify tumor cells irrespective of their state of differentiation (Alt-

Table 1. Desmin-positive tumors

1. Rhabdomyosarcoma	(ALTMANNSBERGER et al. 1982, 1985; MIETTINEN et al. 1982c)
2. Leiomyosarcoma, leiomyoma	(EVANS et al. 1983; MIETTINEN et al. 1982b, c)

Table 2. Desmin positive tumors (own results, cryostat sections or ethanol-fixed and paraffin-embedded material)

Diagnosis	No. of cases	Keratin[b] KL1/ polycl.	Vimentin V 9/ polycl.	Desmin DE-B-5/ polycl.	Neuro-filaments[b] NR 4/ polycl.
Leiomyoma (stomach/small intestine)	7	–	(+)[a]	+	–
Leiomyosarcoma	6	–	(+)[a]	+	–
Embryonal rhabdomyosarcoma	26	–	(+)[a]	+	–
Alveolar rhabdomyosarcoma	6	–	(+)[a]	+	–
Sarcoma botryoides	2	–	(+)[a]	+	–

[a] In some tumor cells a coexpression of desmin and vimentin is seen. The percentage of cells in which such coexpression is observed varies with the tumor (see ALTMANNSBERGER et al. 1985)
[b] Not all tumors have been examined with all IF antibodies. However when they have been tested they proved negative
Polyclonal antibodies see OSBORN et al. (1982a)
Monoclonal antibodies V 9 OSBORN et al. (1984a); DE-B-5 DEBUS et al. (1983a); NR 4/G-A-5 DEBUS et al. (1983b); KL 1 VIAC et al. (1983)

MANNSBERGER et al. 1985). In alveolar rhabdomyosarcomas a coexpression of desmin and vimentin is consistently found in nearly all tumor cells (ALTMANNSBERGER et al. 1982, 1985; MIETTINEN et al. 1982c). Pleomorphic rhabdomyosarcomas are extremely rare and thus far no studies on intermediate filaments in this entitity have been published. MOLENAAR et al. (1985b) have reinvestigated tumors in which the first diagnosis done by routine light microscopic stains was pleomorphic rhabdomyosarcoma. All the tumors that were reinvestigated could not be labelled by the desmin antibody but clearly express vimentin. Since these tumors expressed 1-antitrypsin and 1-antichymotrypsin the authors reclassified these tumors as malignant fibrous histiocytomas, although it has to be emphasized that both markers can also be found in other mesenchymal tumors and even in carcinomas (ROHOLL et al. 1985a–c). The results on human rhabdomyosarcomas can be further supported by results in an animal model system. Rhabdomyosarcomas can be specifically induced by implantation of different heavy metals, including nickel or nickel-sulfide. When the heavy metal is injected into skeletal muscle rhabdomyosarcomas develop, but when it is implanted in bone or subcutaneous tissue fibrosarcomas are observed (see ALTMANNSBERGER et al. 1985). The histologic features of the tumors induced by injection of nickelsulfide were of three different types. In the rhabdomyoblastic

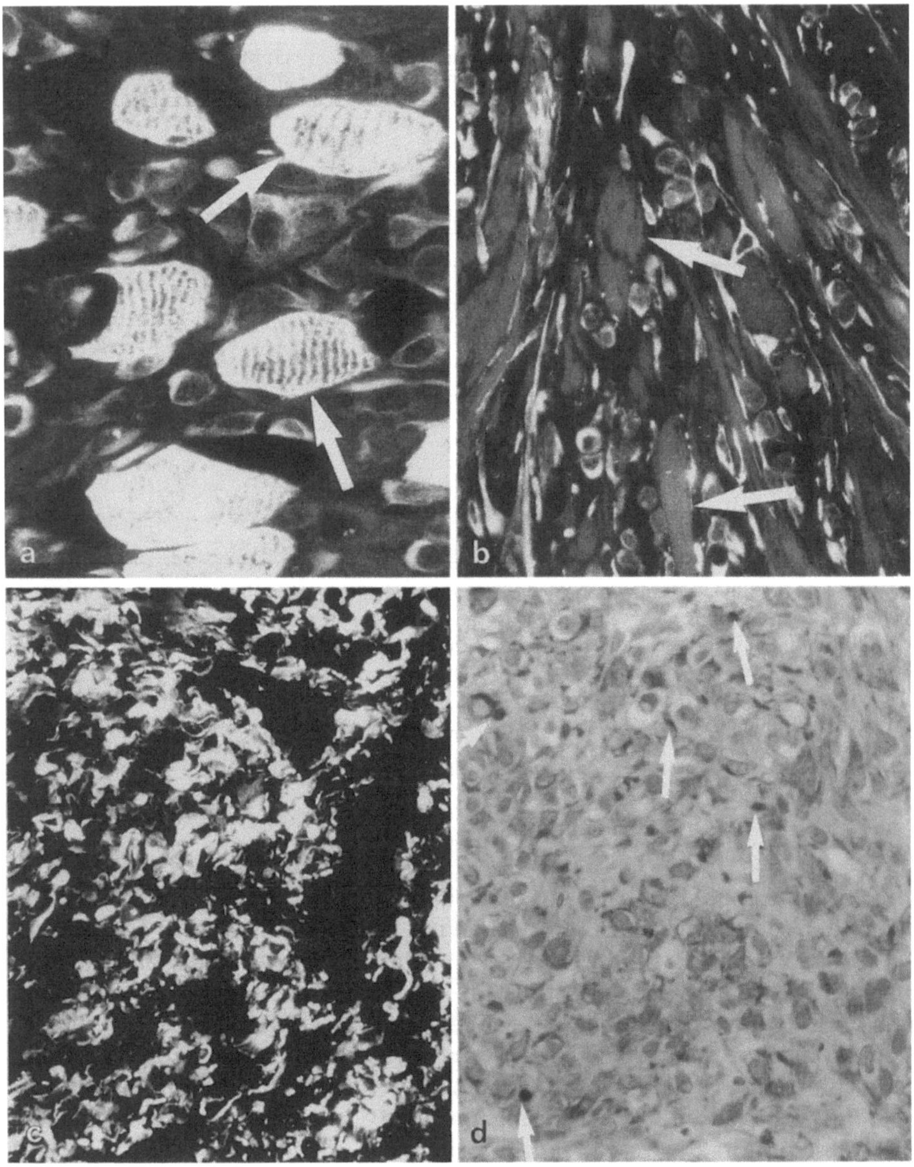

Fig. 6. Human (**a**, **b**) and rat rhabdomyosarcoma (**c**) as well as a case of epitheloid leiomyosarcoma stained by antibodies to desmin (**a**, **c**, **d**) and vimentin (**b**). The human embryonal rhabdomyosarcoma (**a**, **b**) infiltrates preexisting skeletal muscle (**a**, **b**, *arrow*), while the tumor cells coexpress desmin and vimentin. The preexisting muscle is desmin positive and vimentin negative. Rat rhabdomyosarcomas (spindle cell types) are clearly labeled by the desmin antibody. In a case of epithelioid leiomyosarcoma desmin antibody stains intracytoplasmatic globules (**d**, *arrow*) (**a–c**, FITC; **d**, peroxidase labeled second, antibodies; polyclonal desmin antibody)

type tumor cells were highly differentiated mimicking rhabdomyoblasts, and in a few scattered tumor cells cross striations could even be detected. The spindle cell type was characterized by elongated nuclei and a small rim of eosinophilic cytoplasm, whereas the round cell type, showed oval cells with eccentrically located nuclei. Both the spindle as well as the round cell type correspond to a lower degree of differentiation. A small percentage of tumors was built up of only a single histological type. In the majority of tumors the histologic type changed from area to area when these tumors were studied by immunohistology using antibodies against desmin and vimentin. We found using polyclonal desmin antibodies that all tumor cells were desmin-positive. Even in the less differentiated round and spindle cell type desmin expression could be documented. In contrast vimentin staining depended on the histologic type. It was consistently seen in the spindle and round cell type but was restricted to a few small cells in the rhabdomyoblastic type. Thus also in nickelsulfide-induced rat rhabdomyosarcomas undifferentiated tumors coexpressed desmin and vimentin, while in the more differentiated tumors only desmin was detected. These findings should be considered in relation to a similar pattern of IF expression in skeletal muscle embryogenesis where in early stages of development desmin and vimentin are coexpressed, but in mature muscle only desmin is observed.

Our results with human and rat rhabdomyosarcomas do not support the idea of a class of primitive rhabdomyosarcomas, which are vimentin positive and desmin negative. In all 31 cases of rhabdomyosarcomas which we have examined, a positive desmin staining was observed, regardless whether frozen sections, or ethanol fixed and paraffin embedded material (ALTMANNSBERGER et al. 1981, 1985) or smears were investigated. The advantages of desmin as a marker for rhabdomyosarcomas has been confirmed in studies from other laboratories (MIETTINEN et al. 1982c; MOLENAAR et al. 1985a) although some authors have not always seen desmin in undifferentiated tumor cells of rhabdomyosarcomas (GABBIANI et al. 1981b; DENK et al. 1983; KAHN et al. 1983). These negative reactions may be attributable to the fact that different desmin antibodies were used and, more importantly perhaps, to the different fixation methods used in the studies cited above. Thus MOLENAAR et al. (1985a) detected desmin in primitive undifferentiated tumor cells of one rhabdomyosarcoma when frozen sections were used, but another portion of the same tumor fixed in formaldehyde and embedded in paraffin was desmin-negative. Since in the same study in well differentiated rhabdomyosarcomas desmin could be detected also after formalin fixation, the different reactivities may reflect quantitative differences in the amount of desmin present.

When desmin is compared with other markers used in the differential diagnosis of muscle sarcomas, it is important to note that desmin expression is an early event in muscle differentiation, as shown by its presence in replicating cell lines of muscle origin, such as the hamster cell line BHK and the human RD rhabdomyosarcoma line (FRANK et al. 1981; OSBORN et al. 1982a). From studies of muscle myogenesis in culture it seems clear that desmin is expressed in the first generation of myoblasts and its synthesis probably precedes that of sarcomeric specific proteins such as the sarcomeric specific isoforms of actin and myosin (HOLTZER et al. 1982) as well as of titin, a high molecular weight constituent protein of sarcomeric muscle (WANG et al. 1979; HILL and WEBER 1986). Myoglobin expression seems to occur at an even later time point in sarcomeric muscle development. We suggest that the protein complement seen in different rhabdomyosarcomas can be used to assay how differentiated the tumor is. Undifferentiated tumors express desmin as well as in some cases muscle specific forms of actin, myosin and also the high molecular weight protein titin (OSBORN et al. 1986b). Moderately and well differentiated rhabdomyosarcomas are characterized by increasing levels of the muscle specific forms of actin and myosin and by myoglobin expression. In the last few years there are several reports in which antibodies against actin and myosin have been used in the diagnosis of rhabdomyosarcoma (BUSSOLATTI et al. 1980; DONNER et al. 1983; TSOKOS et al. 1983). Unfortunately by far the majority of these antibodies react not only with the sarcomeric isoforms of actin or myosin respectively, but also crossreact with other actin and myosin isoforms, which form a part of the microfilament system in all cells of the organism. This crossreactivity is not suprising, since for example the isoforms of actin in vertebrates have nearly identical sequences (VANDEKERCKHOVE and WEBER 1979). Using such antibodies only a small percentage of rhabdomyosarcomas can be labeled but other mesenchymal tumors and even some epithelial neoplasms also show a strong positive reaction (BUSSOLATTI et al. 1980; DONNER et al. 1983). Polyclonal antibodies to sarcomeric myosin and actin cross absorbed on non-sarcomeric actin or myosin are probably skeletal muscle specific (DE JONG et al. 1984, 1985), but again these antibodies only label mature forms of rhabdomyosarcoma, because the tumor cells in undifferentiated cases have not started to produce the sarcomeric specific proteins. However a recent report by TSUKADA et al. (1987) encourages further trials with actin specific antibodies. Although both myoglobin and titin are specific for sarcomeric muscle they share the disadvantage that they are expressed relatively late in muscle development and consequently are only detected in mature forms of rhabdomyosarcoma (KINDBLOM et al. 1982; MUKAI et al. 1979; OSBORN et al. 1986b).

Thus desmin currently appears to be the best available marker for diagnosis of rhabdomyosarcomas especially for undifferentiated neoplasms.

Desmin antibodies do not allow the separation of rhabdomyosarcomas from smooth muscle tumors, since leiomyomas and leiomyosarcomas in different locations also express desmin (MIETTINEN et al. 1982b; BONAZZI DEL POGETTO et al. 1983; EVANS et al. 1983). The distinction of rhabdomyosarcoma from leiomyosarcoma generally can be done on the grounds of morphology and from the clinical history. Preliminary data suggest that antibodies to titin may also allow a distinction, since expression appears characteristic of skeletal muscle and its tumors, but not of of smooth muscle and its tumors. Spindle cell tumors in the gastric wall which are classified as smooth muscle tumors when routine stains are used are of particular interest. Surprisingly these tumors are desmin-negative and vimentin-positive and in addition show positive staining with the S-100 protein. Thus they should better be classified as Schwann cell tumors (MAZUR and CLARK 1983).

Table 3. Vimentin-positive tumors

1. All nonmuscular soft tissue tumors (ALTMANNSBERGER et al. 1986b, c) including granular cell tumor and glomus tumor
   But: synovial sarcoma and epithelioid sarcoma coexpress K and V
2. Bone tumors (LÖNING et al. 1985)
   But: adamantinoma and chordoma K^+
3. Lymphomas and leukemias in some cases heterogenous staining (GABBIANI et al. 1981b)
4. Malignant melanoma (CASELITZ et al. 1983)

K = Keratin; *V* = Vimentin

Table 4. Vimentin Positive Tumors (own results, cryostat sections or ethanol-fixed and paraffin-embedded material)

Diagnosis	No. of cases	Vimentin V 9/ polycl.	Desmin DE-B-5/ polycl.	Neuro-filaments[b] NR 4/ polycl.	Keratin[b] KL 1/ polycl.
Non-Hodgkin lymphoma (LB, CB, IB, CC-CB, LP-immunocytoma)	19	(+)[a]	–	–	–
Hodgkin lymphoma	6	+	–	–	–
Malignant histiocytosis	2	+	–	–	–
Ewing's sarcoma	10	+	–	–	–
Osteosarcoma	2	+	–	–	–
Chondrosarcoma	2	+	–	–	–
Soft tissue sarcoma (without muscle sarcomas, Synovial sarcomas, Epithelioid sarcoma)	25	+	–	–	–

[a] In some cases not all lymphatic cells are stained
[b] Not all tumors have been examined with all IF-antibodies. However when they have been tested they proved negative
[c] For polyclonal antibodies see OSBORN et al. (1982a)
Monoclonal antibodies V9 OSBORN et al. (1984a); DE-B-5 DEBUS et al. (1983a); NR 4/G-A-5 DEBUS et al. (1983b); KL 1 VIAC et al. (1983)

6.2 Vimentin-Positive Mesenchymal Tumors (Tables 3, 4; Fig. 7)

These tumors are listed in Tables 3 and 4.

The overwhelming majority of soft tissue tumors and of bone tumors (LÖNING et al. 1985; ALTMANNSBERGER et al. 1986b) are vimentin-positive corresponding to their proposed origin from mesenchymal cells. Additionally in tumors such as glomus tumor and granular cell tumor (MIETTINEN et al. 1983c, 1984; SLOOTWEG et al. 1983), in which for the former a derivation from muscle and for the latter a neural origin was assumed by ultrastructural studies, vimentin is the only intermediate filament protein present. Exceptions expressing addi-

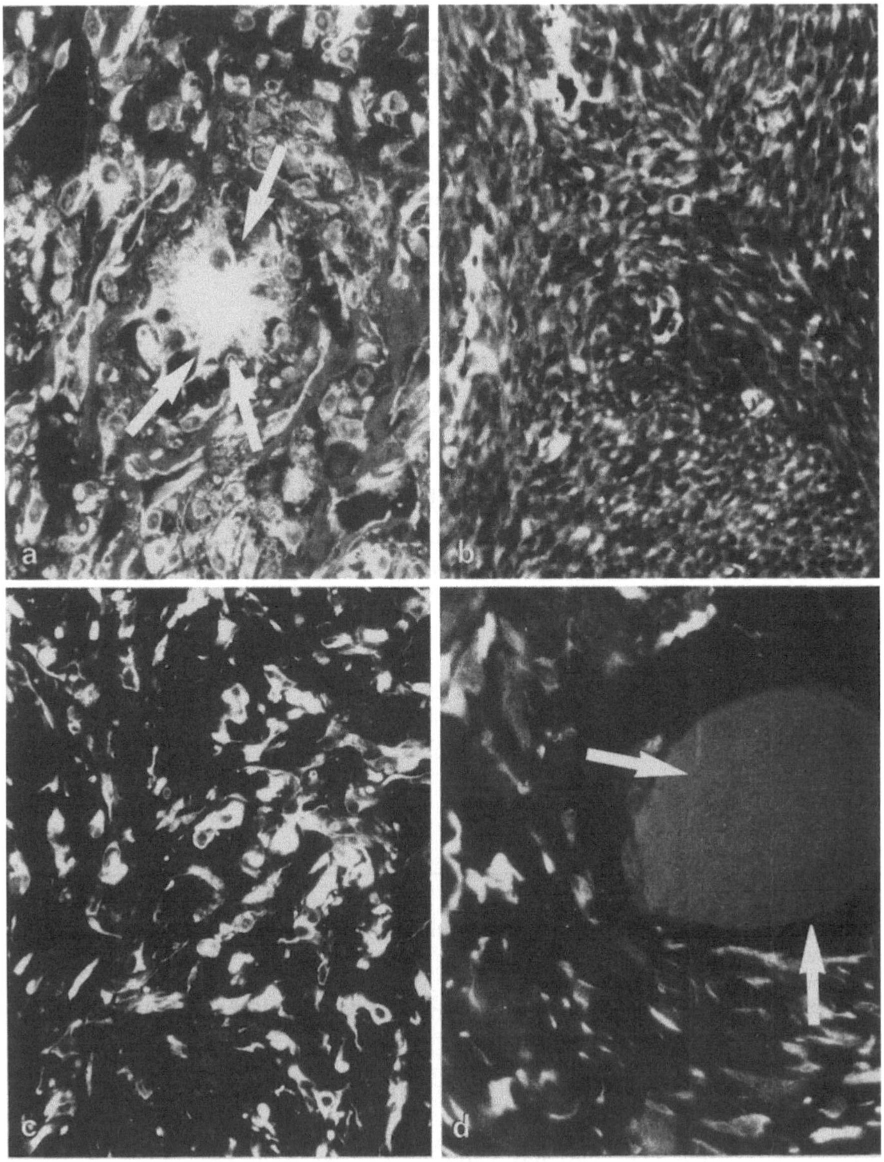

Fig. 7. Liposarcoma (**a**), fibrosarcoma (**b**), malignant fibrous histiocytoma (**c**) and fibromatosis (**d**) stained by a polyclonal vimentin antibody. Note the positive staining of lipoblasts in liposarcoma (**a**, *arrow*). Muscle fibres in fibromatosis (**d**, *arrow*) are negative. (Polyclonal vimentin antibody: **a–d** FITC)

tionally keratins include synovial sarcomas (MIETTINEN et al. 1982a; MIETTINEN and VIRTANEN 1984) and epitheloid sarcomas (CHASE et al. 1984; CHASE and ENZINGER 1985) in the group of soft tissue tumors, and chordomas (MIETTINEN et al. 1983b) as well as adamantinomas (ROSAI and PINKUS 1982) in the group of bone tumors.

Most Hodgkin and Non-Hodgkin lymphomas are vimentin-positive, but in some cases the staining is heterogenous and not all tumor cells are stained. It cannot be excluded that there is a group of Non-Hodgkin lymphomas lacking intermediate filaments, because in normal lymph nodes there are lymphocyte subsets which appear unstained by one or the other monoclonal vimentin antibodies. Our own data thus far do not support the findings of GIORNO (1985), and GIORNO and SCIOTTO (1985), who state that the majority of lymphomas express neither vimentin nor other intermediate filament polypeptides. It seems more likely that the negative reaction on these tumors was caused by the limited reactivity pattern of the vimentin antibody used by these authors, since this antibody did not react with any lymphatic cell in normal lymph nodes.

Further differential diagnosis of the entities in the group of vimentin-positive tumors requires the use of additional markers. Suitable markers which can yield further diagnostic information include endothelial markers (ALLES and BOSSLET 1986) S-100 (e.g. WEISS et al. 1983) histiocytic markers (ROHOLL et al. 1985a–c) or leucocyte common antigen (for review see ALTMANNSBERGER et al. 1986b, c).

6.3 Keratin-Positive Mesenchymal Tumors (Tables 5, 6; Fig. 8)

The third main group is the keratin-positive mesenchymal tumors. The existence of this group was a surprising finding, because originally keratin-positivity ap-

Table 5. Keratin-positive mesenchymal tumors

1. Synovial sarcoma K+V+ (MIETTINEN and VIRTANEN 1984)
   Monophasic synovial sarcoma most cells V+K−, some cells coexpress V and K (MIETTINEN et al. 1983a)
2. Epithelioid sarcoma K+V+ (CHASE et al. 1984)
3. Chordoma K+V+ (MIETTINEN et al. 1983b)

K = Keratin; *V* = Vimentin

Table 6. Keratin-positive mesenchymal tumors (own results, cryostat sections or ethanol-fixed and paraffin-embedded material)

Diagnosis	No. of cases	Keratin KL 1/ polycl.	Vimentin V 9/ polycl.	Desmin DE-B-5/ polycl.	Neuro-filaments NR5/NF-MIX
Biphasic synovial sarcoma	2	+ [a]	+ [b]	−	−
Monophasic synovial sarcoma	5	(+)3/5	+5/5	−	−
Epithelioid sarcoma	1	+	+	−	−

[a] Epithelial part + [b] Mesenchymal part +
+ 95% of tumor cells positive; (+) 5% tumor cells positive
Polyclonal antibodies see OSBORN et al. (1982a)
Monoclonal antibodies V9 OSBORN et al. (1984a); DE-B-5 DEBUS et al. (1983a); NR4/G-A-5 DEBUS et al. (1983b); KL 1 VIAC et al. (1983)

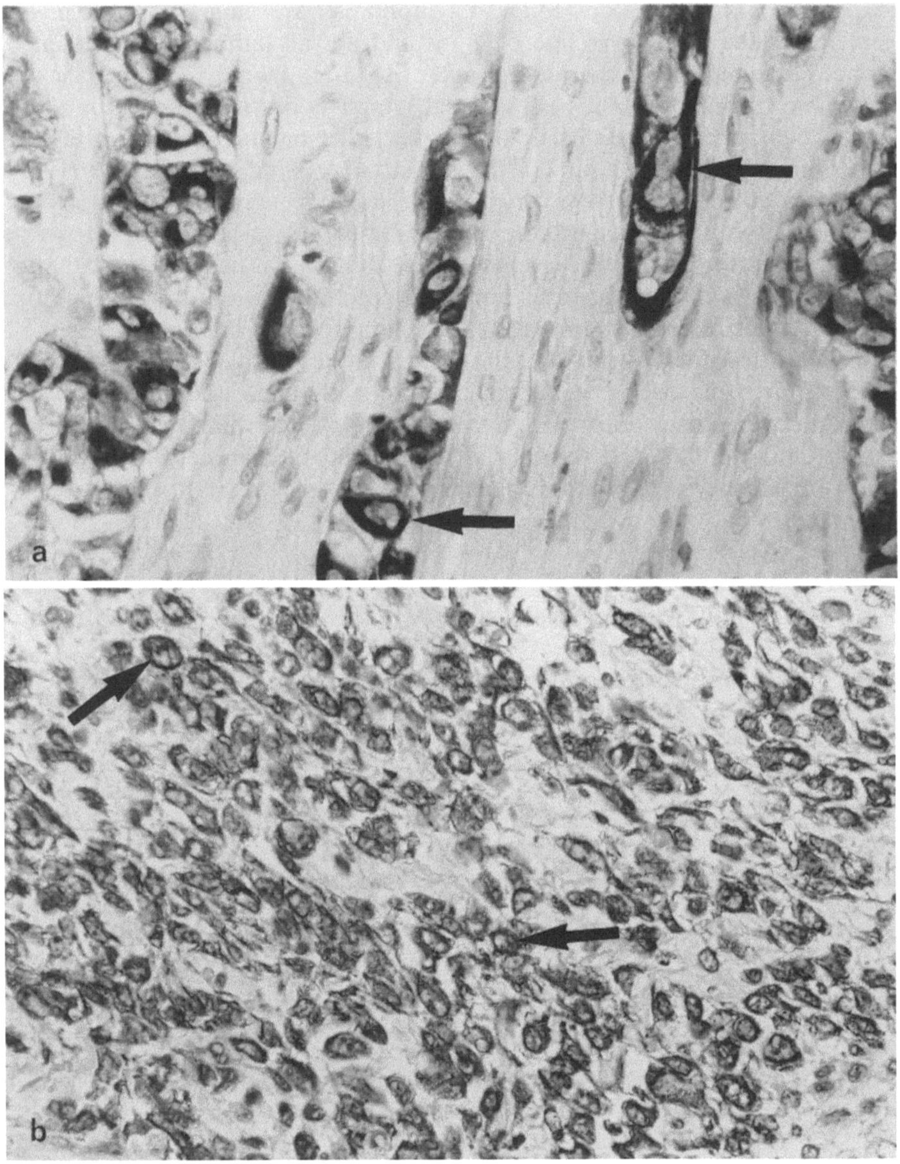

Fig. 8. Epithelioid sarcoma (**a**) and biphasic synovial sarcoma (**b**) stained by the monoclonal keratin antibody *KL1*. Tumor cells are positive (*arrow*), mesenchymal cells are negative (**a**, **b** peroxidase labeled second antibodies)

peared restricted to carcinomas (ALTMANNSBERGER et al. 1982, 1986a, 1987; OSBORN and WEBER 1983; MOLL et al. 1982, 1986). However keratin was soon detected in synovial sarcomas (MIETTINEN et al. 1982a; MIETTINEN and VIRTANEN 1984; SALISBURY and ISAACSON 1985) in epitheloid sarcomas (CHASE et al. 1984; CHASE and ENZINGER 1985), in adamantinoma of tibia (ROSAI and PINKUS

1982) and in chordomas (MIETTINEN et al. 1983b). Most of these tumors coexpress keratin and vimentin. In monophasic synovial sarcomas the majority of tumor cells were vimentin-positive with a few scattered tumor cells expressing keratin and vimentin (MIETTINEN et al. 1983a).

The transformation of mesenchymal cells to epithelial cells is perhaps best studied in nephroblastomas. In a first step cells of the undifferentiated blastema only synthesize vimentin. In a second step keratin and vimentin are coexpressed and finally when tubules are formed vimentin expression is lost and keratin is the only intermediate filament system found (ALTMANNSBERGER et al. 1984).

6.4 Alveolar Soft Part Sarcoma

Alveolar soft part sarcoma is a tumor of unknown histogenesis. The hope that intermediate filament typing would have solved the problem, has not yet been fulfilled. Controversal results have been published by two different groups. We have investigated two cases, one of which was negative when tested with different intermediate filament antibodies (OSBORN et al. 1982b) whereas the second case showed a focally positive reaction with the vimentin antibody (JUNDT et al. 1984). This is in contrast to two other reports, which demonstrated coexpression of desmin and vimentin and suggested an origin from muscle cells for alveolar soft part sarcoma (DENK et al. 1983; MÜLLER and STUTTE 1984).

7 Conclusions

There is abundant evidence that intermediate filaments can be used as cell type specific markers both for normal tissue and for tumors. The results of intermediate filament typing in mesenchymal tumors and its diagnostic relevance is shown above. This system can also be used to solve other problems in surgical pathology (for review see OSBORN and WEBER 1983). One of the most useful applications of intermediate filament typing is to differentiate the round cell tumors of children, thus rhabdomyosarcomas are desmin positive, the majority of malignant lymphomas as well as Ewing sarcomas contain only vimentin and neuroblastomas show positivity for neurofilaments (OSBORN et al. 1986a).

Acknowledgments. We would like to thank Dr. KLAUS WEBER for discussion. S. ISENBERG, P. HAHN-KOHLBERGER and S. LÜBKE for expert technical assistance and M. OBERSCHELP for help with the manuscript.

References

Alles JU, Bosslet K (1986) Immunohistochemical and immunochemical characterization of a new endothelial cell-specific antigen. J Histochem Cytochem 34:209–214

Altmannsberger A, Osborn M, Schauer A, Weber K (1981) Antibodies to different intermediate proteins are cell type specific markers on fixed and paraffin embedded human tissues. Lab Invest 45:427–434

Altmannsberger M, Osborn M, Treuner J, Hölscher A, Weber K, Schauer A (1982) Diagnosis of human childhood rhabdomyosarcoma by antibodies to desmin, the structural protein of muscle specific intermediate filaments. Virchows Arch (Cell Pathol) 39:203–215

Altmannsberger M, Osborn M, Schäfer HJ, Schauer A, Weber K (1984) Distinction of nephroblastomas from other childhood tumors using antibodies to intermediate filaments. Virchows Arch (Cell Pathol) 45:113–124

Altmannsberger M, Weber K, Droste R, Osborn M (1985) Desmin is a specific marker for rhabdomyosarcomas of human and rat origin. Am J Pathol 118:85–95

Altmannsberger M, Dirk T, Droese M, Weber M, Osborn M (1986a) Keratin polypeptide distribution in benign and malignant breast tumors – subdivision of ductal carcinomas using monoclonal antibodies. Virchows Arch (Cell Pathol) 51:265–275

Altmannsberger M, Dirk M, Osborn M, Weber K (1986b) Immunohistochemistry of cytoskeletal filaments in the diagnosis of soft tissue tumors. Sem Diagn Pathol 3:306–316

Altmannsberger M, Alles JU, Fitz H, Jundt G, Osborn M (1986c) Mesenchymale Tumormarker. Verh Dtsch Ges Pathol 70

Altmannsberger M, Droese M, Dralle H, Osborn M, Weber K (1987) Intermediate filament in thyroid carcinomas. Diagn Cytopathol (to be published)

Bennett GS, Fellini SA, Croop JM, Otto JJ, Bryan J, Holtzer H (1978) Differences among 100 A filament subunit from different cell types. Proc Natl Acad Sci USA 76:4364–4368

Bloemendal H, Quax W, Quax-Jeuken Y, van den Heuvel R, Egberts WV, van den Broek L (1985) Vimentin and desmin cDNA clones: Structural aspects of corresponding proteins and genes. Ann NY Acad Sci 455:95–105

Bonnazi del Poggetto C, Virtanen I, Lehto VP, Wahlström T, Saksela E (1983) Expression of intermediate filaments in ovarian and uterine tumors. Int J Gynecol Pathol 1:359–366

Brown S, Levinson W, Spudich JA (1976) Cytoskeletal elements of chick embryo fibroblasts revealed by detergent extraction. J Supramol Struct 5:110–130

Bussolati G, Alfani V, Weber K, Osborn M (1980) Immunocytochemical detection of actin on fixed and embedded tissues: Its potential use in routine pathology. J Histochem Cytochem 28:169–173

Caselitz J, Jänner M, Breitbart E, Weber K, Osborn M (1983) Malignant melonomas contain only the vimentin type of intermediate filaments: Implications for histogenesis and diagnosis. Virchows Arch (Pathol Anat) 400:43–51

Chase DR, Enzinger FM (1985) Epitheloid sarcoma. Am J Surg Pathol 9:241–263

Chase DR, Weiss SW, Enzinger FM, Langloss IM (1984) Keratin in epithelioid sarcoma. Am J Surg Pathol 8:435–441

Czernobilsky B, Moll R, Levy R, Franke WW (1985) Coexpression of cytokeratin and vimentin filaments in mesothelial, granulosa and rete ovarii cells of the human ovaries. Eur J Cell Biol 37:175–190

Debus E, Weber K, Osborn M (1983a) Monoclonal antibodies to desmin, the muscle-specific intermediate filament protein. EMBO J 12:2305–2312

Debus E, Weber K, Osborn M (1983b) Monoclonal antibodies specific for glial fibrillary acidic (GFA) protein and for each of the neurofilament triplet polypeptides. 25:193–203

De Jong ASH, van Vark M, Albus-Letter CH, van Raamsdonk W, Voute PA (1984) Myosin and myoglobin as tumor markers in the diagnosis of rhabdomyosarcoma. Am J Surg Pathol 8:521–528

De Jong ASA, van Kessel-van Vark M, van Raamsdonk CHE, Voute PA (1985) Skeletal muscle actin as tumor marker in the diagnosis of rhabdomyosarcoma in childhood. Am J Surg Pathol 9:467–474

Dellagi K, Vainchenker W, Vinci W, Paulin D, Broute JC (1983) Alteration of vimentin intermediate filament expression during differentiation of human hemopoietic cells. EMBO J 2:1509–1514

Denk H, Krepler R, Artlieb U, Gabbiani G, Rungger-Brandle E, Leocini P, Franke WW (1983) Proteins of intermediate filaments: an immunohistochemical and biochemical approach to the classification of soft tissue tumors. Am J Pathol 110:193–208
Donner L, de Lanerolle P, Cost J (1983) Immunoreactivity of paraffin-embedded normal tissue and mesenchymal tumors for smooth muscle myosin. Am J Clin Pathol 80:677–681
Evans DJ, Lambert IA, Jacob M (1983) Intermediate filaments in smooth muscle tumors. J Clin Pathol 36:57–61
Frank ED, Warren L (1981) Aortic smooth muscle cells contain vimentin instead of desmin. Proc Natl Acad Sci USA 78:3020–3024
Franke WW, Schmid E, Osborn M, Weber K (1978) Different intermediate-size filaments distinguished by immunofluorescence microscopy. Proc Natl Acad Sci USA 75:5034–5038
Franke WW, Schmid E, Schiller DL, Winter S, Jarasch ED, Moll R, Denk H, Jackson B, Illmensee K (1982) Differentiation-related patterns of expression of proteins of intermediate-sized filaments in tissues and cultured cells. Cold Spring Harbor Symp Quant Biol 46:431–453
Gabbiani G, Schmid E, Winter S, Chaponnier C, de Chastonay, Vandekerckhove K, Weber K, Franke WW (1981 a) Vascular smooth muscle cells differ from other smooth muscle cells: Predominance of vimentin filaments and a specific α-type actin. Proc Natl Acad Sci USA 78:298–302
Gabbiani G, Kapanci Y, Barrazone P, Franke WW (1981 b) Immunohistochemical identification of intermediate-size filaments in human neoplastic cells. A diagnostic aid for the surgical pathologist. Am J Pathol 104:206–216
Gawlitta W, Osborn M, Weber K (1981) Coiling of intermediate filaments induced by microinjection of a vimentin-specific antibody does not interfere with locomotion and mitosis. Eur J Cell Biol 26:83–90
Geisler N, Weber K (1981 a) Comparison of the proteins of two immunologically distinct intermediate-sized filaments by amino acid sequence analysis: desmin and vimentin. Proc Natl Acad Sci USA 78:4120–4123
Geisler N, Weber K (1981 b) Selfassembly in vitro of the 68,000 molecular weight component of the mammalian neurofilament triplet proteins into intermediate-sized filaments. J Mol Biol 151:565–571
Geisler N, Weber K (1982) The amino acid sequence of chicken muscle desmin provides a common structural model for intermediate filament proteins including the wool α-keratins. EMBO J 1:1649–1656
Geisler N, Weber K (1986) Structural aspect of intermediate filaments cell and molecular biology of the cytoskeleton. In: Shay JW (ed) Plenum Press, New York London, pp 41–68
Geisler N, Kaufmann E, Weber K (1982) Protein chemical characterization of three structurally distinct domains along the protofilament unit of desmin 10 nm filaments. Cell 30:277–286
Geisler N, Kaufmann E, Weber K (1985) Antiparallel orientation of the two doublestranded coiled-coils in the tetrameric protofilament unit of intermediate filaments. J Mol Biol 182:173–177
Giorno R (1985) Immunohistochemical analysis of the distribution of vimentin in human peripheral lymphoid tissues. Anat Rec 211:43–47
Giorno R, Sciotto CG (1985) Use of monoclonal antibodies for analyzing the distribution of the intermediate filaments protein vimentin in human Non-Hodgkin lymphomas. Am J Pathol 120:351–355
Goldman RD, Knipe DM (1972) Functions of cytoplasmic fibers in non-muscle cell motility. Cold Spring Harbor Symp Quant Biol 37:523–534
Gown AM, Vogel AM (1984) Monoclonal antibodies to human intermediate filaments proteins. II. Distribution of filament proteins in normal human tissue. Am J Pathol 114:309–321
Hanukoglu I, Fuchs E (1982) The cDNA sequences of a human epidermal keratin: divergence of sequence but conservation of structure among intermediate filament proteins. Cell 31:243–252
Hanukoglu I, Fuchs E (1983) The cDNA sequence of a type II cytoskeletal keratin reveals constant and variable structural domains among keratins. Cell 33:915–924
Hill C, Weber K (1986) Monoclonal antibodies distinguish titins from heart and skeletal muscle. J Cell Biol 102:1099–1108
Holtzer H, Croop J, Diestmann S, Ishikawa H, Somlyo AP (1975) Effects of cytochalasin B and colcemid on myogenic cultures. Proc Natl Acad Sci USA 72:513–517
Holtzer H, Bennet GS, Tapscott SJ, Croop JM, Toyama (1982) Intermediate-size filaments: Changes in synthesis and distribution in cells of the myogenic and neurogenic lineage. Cold Spring Harbor Symp Quant Biol 46:317–329

Ip W, Heuser JE, Pang YYS, Hartzer MK, Robson RM (1985) Subunit structure of desmin and vimentin protofilaments and how they assemble into intermediate filaments. Ann NY Acad Sci 455:185–199

Jundt G, Schulz A, Altmannsberger M (1984) Alevolar soft part sarcoma. Immunocytochemical and ultrastructural studies. J Cancer Res Clin Oncol 107:83

Kahn HJ, Yeger H, Kassim O, Jorgensen AO, McLennan MC, Baumal R, Smith CR, Philipp MJ (1983) Immunohistochemical and electron microscopic assessment of childhood rhabdomyosarcoma. Cancer 51:1897–1903

Kartenbeck J, Franke WW, Moser JG, Stoffels U (1983) Specific attachment of desmin filaments to demosomal plaques in cardiac myocytes. EMBO J 2:735–742

Kindblom LG, Seidel T, Karlsson K (1982) Immunohistochemical localization of myoglobin in human muscle tissue and embryonal and alveolar rhabdomosarcoma. Acta Pathol Microbiol Immunol Scand A 90:167–174

Klymkowsky MW (1981) Intermediate filaments in 3T3 cells collapse after intracellular injection of monoclonal anti-intermediate filament antibody. Nature 291:249–251

Lazarides E, Hubbard BD (1976) Immunological characterization of the subunit of the 100 A filaments from muscle cells. Proc Natl Acad Sci USA 73:4344–4348

Lewis SA, Cowan NJ (1986) Anomalous placement of introns in a member of the intermediate filament multigene family: an evolutionary conundrum. Molec Cell Biol 6:1529–1534

Lin JJC, Feramisco (1981) Disruption of the in vivo distribution of the intermediate filaments fibroblasts through the microinjection of a specific monoclonal antibody. Cell 24:185–193

Löning T, Caselitz J, Seifert G, Weber K, Osborn M (1982) Identification of Langerhans cells: Simultaneous use of sera to intermediate filaments, T6 and HLA-DR antigens on oral mucosa, human epidermis and their tumors. Virchows Arch (Pathol Anat) 398:119–128

Löning T, Liebsch J, Delling G (1985) Osteosarcomas and Ewing's sarcomas. Virchows Arch (Pathol Anat) 407:323–336

Mazur MT, Clark HB (1983) Gastric stroma tumors. Reappraisal of histogenesis. Am J Surg Pathol 7:507–520

Miettinen M, Virtanen I (1984) Synovial sarcoma – A misnomer. Am J Pathol 117:18–25

Miettinen M, Lehto VP, Virtanen I (1982a) Keratin positivity in the epithelial-like cells of classical biphasic synovial sarcoma. Virchows Arch (Cell Pathol) 40:157–161

Miettinen M, Lehto VP, Badley RA, Virtanen I (1982b) Expression of intermediate filaments in soft tissue sarcomas. Int J Cancer 30:541–546

Miettinen M, Lehto VP, Badley RA, Virtanen I (1982c) Alveolar rhabdomyosarcoma. Demonstration of the muscle type of intermediate filament protein. Desmin as a diagnostic aid. Am J Pathol 108:246–251

Miettinen M, Lehto VP, Virtanen I (1983a) Monophasic synovial sarcoma of spindle-cell type. Virchows Arch (Cell Pathol) 44:187–199

Miettinen M, Lehto VP, Dahl D, Virtanen I (1983b) Differential diagnosis of chordoma, chondroid, and ependymal tumors as aided by anti-intermediate filament antibodies. Am J Pathol 112:160–169

Miettinen M, Lehto VP, Virtanen I (1983c) Glomus tumor cells: Evaluation of smooth muscle and endothelial cell properties. Virchows Arch (Cell Pathol) 43:139–149

Miettinen M, Lehtonen E, Lehtola H, Ekblom P, Lehto VP, Virtanen I (1984) Histogenesis of granular cell tumor: An immunohistochemical and ultrastructural study. J Pathol 142:221–229

Molenaar WM, Oosterhuis JW, Oosterhuis AM, Ramaekers FCS (1985a) Mesenchymal and muscle-specific intermediate filaments (vimentin and desmin) in relation to differentiation in childhood rhabdomyosarcomas. Hum Pathol 16:838–843

Molenaar WM, Oosterhuis AM, Ramakers FCS (1985b) The rarity of rhabdomyosarcomas in the adult. Path Res Pract 180:400–404

Moll R, Franke WW (1986) Cytochemical cell typing of metastatic tumors according to their cytoskeletal proteins. In: Lapsis K, Liotta LA, Rabson AS (eds) Biochemistry and molecular genetics of cancer metastasis. Nijhoff, Boston, pp 101–114

Moll R, Franke WW, Schiller DL, Geiger B, Krepler R (1982) The catalogue of human cytokeratins. Patterns of expression in normal epithelia, tumors and cultures cells. Cell 31:11–24

Moll R, Cowin P, Kapprell HP, Franke WW (1986) Biology of disease: Desmosomal proteins; New markers for identification and classification of tumors. Lab Invest 54:4–25

Müller H, Stutte HJ (1984) Untersuchungen zur Histogenese des alveolären Weichteilsarkoms. Verh Dtsch Ges Pathol 68:515

Mukai K, Rosai J, Hallaway BE (1979) Localization of myoglobin in normal and neoplastic human skeletal muscle cells using an immunoperoxidase method. Am J Surg Pathol 3:373–376

Ngai J, Captanak YG, Lazarides (1985) Expression of the genes coding for the intermediate filament proteins vimentin and desmin. Ann NY Acad Sci 455:144–157

Osborn M, Weber K (1977) The detergent-resistant cytoskeleton of tissue culture cells includes the nucleus and the microfilament bundles. Exp Cell Res 106:339–349

Osborn M, Weber K (1983) Tumor diagnosis by intermediate filament typing. A novel tool for surgical pathology. Lab Invest 48:372–394

Osborn M, Franke WW, Weber K (1977) Visualization of a system of filaments 7–10 nm thick in cultured cells of an epithelioid line (PtK_2) by immunofluorescence microscopy. Proc Natl Acad Sci USA 74:2480–2494

Osborn M, Franke WW, Weber K (1980) Direct demonstration of the presence of two immunologically distinct intermediate-sized filament systems in the same cell by double immunofluorescence microscopy. Exp Cell Res 125:37–46

Osborn M, Caselitz J, Weber K (1981) Heterogeneity of intermediate filaments expression in vascular smooth muscle: A gradient in desmin positive cells from the rat aortic arch to the level of arteria iliaca communis. Differentiation 20:196–202

Osborn M, Geisler N, Shaw G, Sharp GA, Weber K (1982a) Intermediate filaments. Cold Spring Harbor Symp Quant Biol 49:423–429

Osborn M, Altmannsberger M, Shaw G, Schauer A, Weber K (1982b) Various sympathetic derived human tumors differ in neurofilament expression. Virchows Arch (Pathol Anat) 40:141–156

Osborn M, Debus E, Weber K (1984a) Monoclonal antibodies specific for vimentin. Eur J Cell Biol 34:137–143

Osborn M, Altmannsberger M, Debus E, Weber K (1984b) Conventional and monoclonal antibodies to intermediate filament proteins in human tumor diagnosis. In: Cancer cells, vol 1, The transformed phenotype. Cold Spring Harbor Laboratory, Cold Spring Harbor, pp 191–200

Osborn M, Dirk T, Käser M, Weber K, Altmannsberger M (1986a) Immunohistochemical localization of neurofilaments and neuron-specific enolase in 29 cases of neuroblastoma. Am J Pathol 122:433–442

Osborn M, Hill C, Altmannsberger M, Weber K (1986b) Monoclonal antibodies to titin in conjunction with antibodies to desmin separate rhabdomyosarcomas from other tumor types. Lab Invest 55:101–108

Pruss RM, Mirsky R, Raff MC, Thorpe R, Dowding AJ, Anderton BH (1981) All classes of intermediate filaments share a common antigenic determinant defined by a monoclonal antibody. Cell 37:419–428

Quax W, Egbert WV, Hendriks W, Quax-Jeuken A, Bloemendal H (1983) The structure of the vimentin gene. Cell 35:215–223

Quax W, van den Heuvel R, Egbert WV, Quax-Jeuken A, Bloemendal H (1984) Intermediate filament cDNAs from BHK-21 cells: Demonstration of distinct genes for desmin and vimentin in all vertebrate classes. Proc Natl Acad Sci USA 81:5970–5974

Quax W, Van der Broeck L, Egbert WV, Ramaekers F, Blomendal H (1985) Characteriztion of the hamster desmin gene: expression and formation of desmin filaments in nonmuscle cells after gene transfer. Cell 43:327–338

Quinlan RA, Franke WW (1982) Heteropolymer filaments of vimentin and desmin in vascular smooth muscle tissue and cultured hamster kidney cells demonstrated by chemical crosslinking. Proc Natl Acad Sci USA 79:3452–3456

Quinlan RA, Franke WW (1983) Molecular interactions in intermediate-sized filaments revealed by chemical crosslinking. Eur J Biochem 132:477–484

Roholl PJM, Kleyne J, Pipjers HW, van Unnik JAM (1985a) Comparative immunohistochemical investigation of markers for malignant histiocytes. Hum Pathol 16:763–771

Roholl PJM, Keyne J, Elbers H, van der Vegt MCD, Albus-Lutter CH, van Unnik JAM (1985b) Characterization of tumor cells in malignant fibrous histiocytomas and other soft tissue tumors in comparison with malignant histiocytes. J Pathol 147:87–95

Roholl PJM, Kleyne J, van Unnik JAM (1985c) Characterization of tumor cells in malignant fibrous histiocytomas and other soft tissue tumors in comparison with malignant histiocytes. II. Immunoperoxidase study on cryostat sections. Am J Pathol 121:269–274

Rosai J, Pinkus GS (1982) Immunohistochemical demonstration of epithelial differentiation in adamantinoma of the tibia. Am J Pathol 6:427–434

Salisbury JR, Issacson PG (1985) Synovial sarcoma. J Pathol 147:49–57

Schmid E, Osborn M, Rungger-Brändle E, Gabbiani G, Weber K, Franke WW (1982) Distribution of vimentin filaments in smooth muscle tissue of mammalian and avian aorta. Exp Cell Res 137:329–340

Sharp G, Osborn M, Weber K (1982) Occurrence of two different intermediate filament proteins in the same filament in situ within a human glioma cell line. Exp Cell Res 141:385–395

Slootweg P, de Wilde P, Vooijs, Ramaekers F (1983) Oral granular cell lesion. Virchows Arch (Pathol Anat) 402:35–45

Small JV, Sobieszek A (1977) Studies on the function and composition of the 10 nm (100 A; filaments of vertebrate smooth muscle. J Cell Sci 23:243–268

Steinert PM (1978) Structure of the three-chain unit of the bovine epidermal keratin filament. J Mol Biol 123:49–70

Steinert PM, Rice RH, Roop DR, Trus BL, Steven AC (1983) Complete amino acid sequence of a mouse epidermal keratin subunit and implication for the structure of intermediate filaments. Nature 302:794–800

Thornell LE, Eriksson A, Stigbarnd T, Sjöstrom M (1978) Structural proteins in cow Purkinje and ordinary ventricular fibres – a marked difference. J Mol Cell Cardiol 10:605–616

Thornell LE, Erikson A, Edström L (1983) Intermediate filaments in human myopathies in cell and muscle motility, vol 4. Dowben RM, Shay JW (eds) Plenum Press, New York London

Tölle HG, Weber K, Osborn M (1986) Microinjection of monoclonal antibodies to vimentin, desmin, and GFAP in cells which contain more than one IF-type. Exp Cell Res 162:462–474

Tokuyasu KT, Dutton AH, Singer SJ (1983) Immunelectronmicroscopic studies of desmin (skeletin) localization and intermediate filament organization in chicken cardiac muscle. J Cell Biol 96:1736–1742

Tsokos M, Howard R, Costa J (1983) Immunohistochemical study of alveolar and embryonal rhabdomyosarcoma. Lab Invest 48:148–155

Tsukada T, McNutt MA, Ross R, Gown AM (1987) HHF, a muscle actin-specific monoclonal antibody. II: Reactivity in normal, reactive, and neoplastic human tissues. Am J Pathol (to be published)

Vandekerckhove J, Weber K (1979) The complete amino acid sequence of actin from bovine aorta, bovine heart, bovine skeletal muscle. Differentiation 14:123–133

Viac J, Reano A, Brochier J, Staquet MJ, Thivolet J (1983) Reactivity pattern of a monoclonal antikeratin antibody (Kl 1). J Invest Dermatol 81:351–354

Virtanen I, Miettinen M, Lehto VP, Kariniemi AL, Paasivo R (1985) Diagnostic application of monoclonal antibodies to intermediate filaments. Ann NY Acad Sci 455:635–648

Wang K, McClure J, Tu A (1979) Titin major myofibrillar components of striated muscle. Proc Natl Acad Sci USA 76:3698–3702

Weber K, Geisler N (1984) Intermediate filaments – from wool – α-keratins to neurofilaments: structural overview. In: Cancer cells, vol 1, The transformed phenotype. Cold Spring Harbor Laboratory, Cold Spring Harbor, pp 153–159

Weiss SM, Langloss JM, Enzinger FM (1983) Value of S-100 protein in the diagnosis of soft tissue tumors with particular reference to benign and malignant Schwann cell tumors. Lab Invest 49:200–308

Zehner ZE, Paterson BM (1985) The chicken vimentin gene: aspects of organization and transcription during myogenesis. Ann NY Acad Sci 455:79–94

Mesenchymal Tumor Markers: Special Proteins and Enzymes *

H.F. OTTO, R. BERNDT, K. SCHWECHHEIMER and P. MÖLLER

1 Introduction

The differential diagnosis of sarcomas is difficult. Different sarcomas can exhibit almost identical histological features when examined with light microscopic techniques. The visualization of structural and functional proteins in the tumor cells, using immunohistochemical methods, is an important factor in establishing a precise diagnosis. However, it is imperative to be familiar with the reaction range of the antibodies of the different tumors. The literature does not provide uniform findings with regard to the reaction range of "histiocytic" markers and thus far, there are no systematic investigations for ferritin and transferrin when applied to soft tissue tumors. This is why, in addition to the literature, we rely on our investigations carried out on 73 soft tissue sarcomas and 16 benign soft tissue lesions. Antisera have been used against the following proteins: desmin (Euro Diagnostics), myoglobin, lysozyme, α_1-antitrypsin, α_1-antichymotrypsin, ferritin, transferrin, factor VIII-related antigen, and S-100 protein (all from Dako). Myosin was identified on frozen sections of selected tumors with monoclonal antibodies.

* This study was supported by the *Deutsche Forschungsgemeinschaft* (MO 384/1–2) and the *Tumorzentrum* Heidelberg/Mannheim

2 Myosin

Myosin is a globular enzyme acting both as ATPase and a fibrous structural protein simultaneously. Myosin from smooth muscle and myosin from striated muscle have slightly different molecular properties which account for their different contractile regulatory mechanisms. A single monomeric myosin molecule always contains two identical heavy chains (MW 200 kD) and two pairs of light chains of two different types (MW ≈22 kD). Each heavy chain consists of a globular headpiece and a long rodlike α-helical part. Bound to each myosin head are the two molecules of light chains, one of each type. Movements of the head are important in generating the force of contraction in muscle. Myosin reversibly binds to actin. The regular array of actin and myosin filaments form the element of a myofibril (Darnell et al. 1986).

Myosin has been isolated from smooth muscle (Pinkus et al. 1986) and non-muscle cells (Weber and Groeschel-Steward 1974), although myosin filaments are not visible in the latter cells by using standard electron microscopic techniques (Longtine et al. 1985). Since myosins have a high variability in primary structure and enzymatic activity in different cell types (Syvory 1979), antibodies have been raised. They discriminate among myosins (Longtine et al. 1985; Saku et al. 1985). Anti-myosin antibodies have then been used in diagnostic pathology (Hiramoto et al. 1961; Unsicker et al. 1978; Macartney et al. 1979; Drenckhahn et al. 1980; Donner et al. 1983; Lemanski and Tu 1983; De Jong et al. 1984; Hayashi et al. 1984; Saku et al. 1985; Schiaffino et al. 1986; Scupham et al. 1986).

The heterogeneity of myosins and antibodies hampers the analysis of the data in literature concerning the diagnostic value of myosin in mesenchymal tumors. In the diagnosis of rhabdomyosarcoma, anti-myosins have been found superior to anti-myoglobins (see below); (De Jong et al. 1984) since they were found in both the well-differentiated and poorly-differentiated variants; the latter failed to show staining for myoglobin. Pleomorphic rhabdomyosarcomas were also positive for myosin, while malignant fibrous histiocytomas, liposarcomas, schwannomas, and leiomyosarcomas were negative. Others, however, found myoglobin in leiomyomatous tumors (Bures et al. 1981). Even epithelia reacted with some of the antisera (Drenckhahn et al. 1980; Tsokos et al. 1983). In recent studies (Scupham et al. 1986; Tsokos and Triche 1986) the superiority of myosin to myoglobin was questioned since myosin failed to be detectable in poorly differentiated rhabdomyosarcomas.

Apart from the nosological or diagnostic question of whether and to what extent myoglobin occurs in myogenic tumors, there is now a problem in specifying the antibodies described and used. Those monoclonal antibodies that are sufficiently characterized (Longtine et al. 1985; Pinkus et al. 1986) are not yet commercially available. Therefore, the diagnostic and differential diagnostic value of myosin is still unclear.

Our own experience with monoclonal antibodies to myosin is very preliminary. Four monoclonal antibodies, kindly provided by H.A. Katus and his co-workers (Dept. of Internal Medicine III (Cardiology) at Heidelberg Universi-

Fig. 1. a *Myosin* immunoreactivity in *normal skeletal muscle fibers* by the monoclonal antibody ▷ 1B2/1H4; ×160. **b** *Myosin* immunoreactivity in *normal skeletal muscle* is restricted to some fibers; anti-myosin 4B8/1A7; ×160. **c** Demonstration of *myosin* microfilaments in the *muscle layer of small intestine*; monoclonal anti-myosin antibody 5C9/3E2; ×160. **d** *Rhabdomyosarcoma: desmin* immunoreactivity is seen in some cells; ×250. **a–d** PAP method; AEC; haematoxylin counterstain

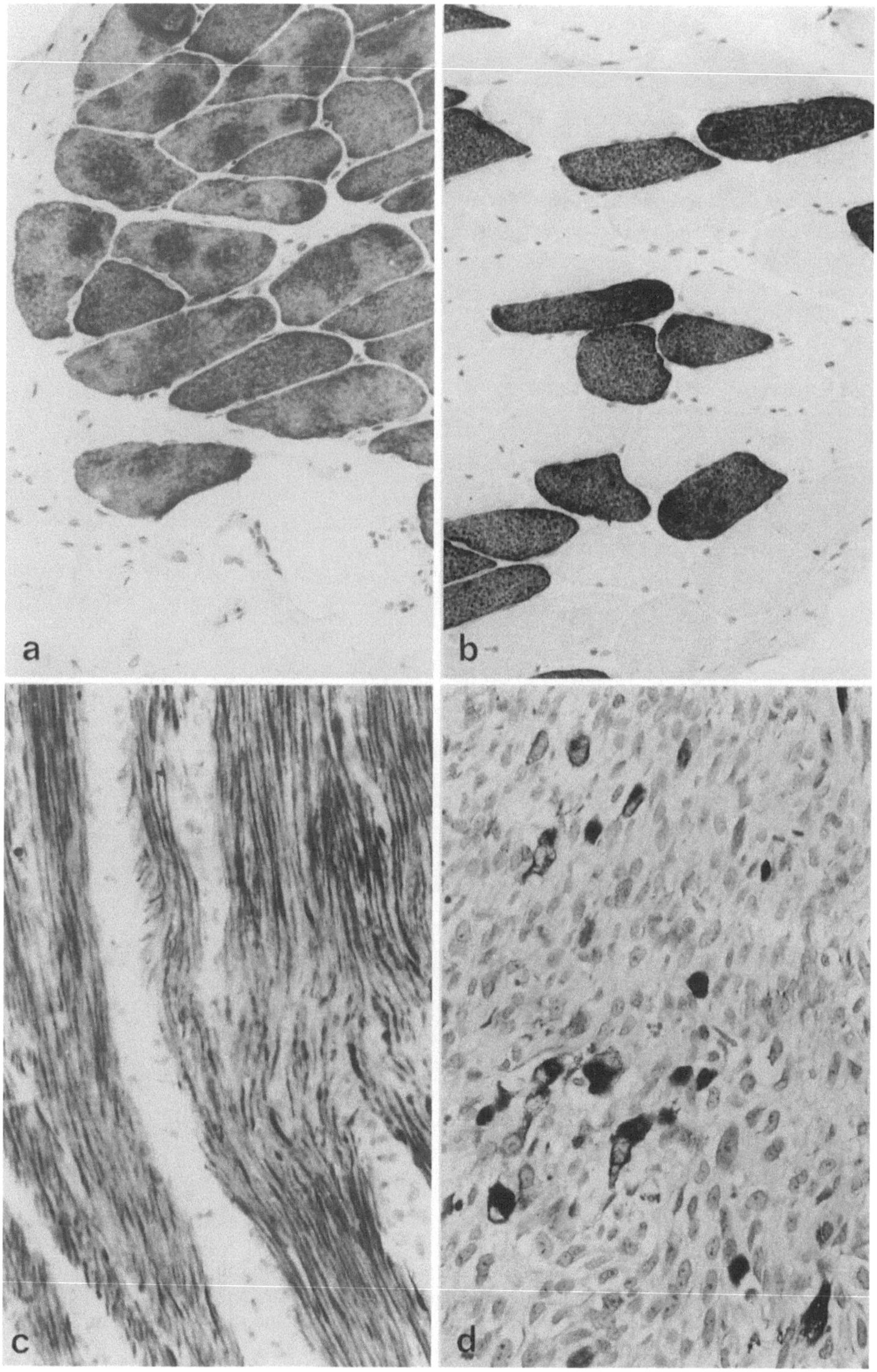
a
b
c
d

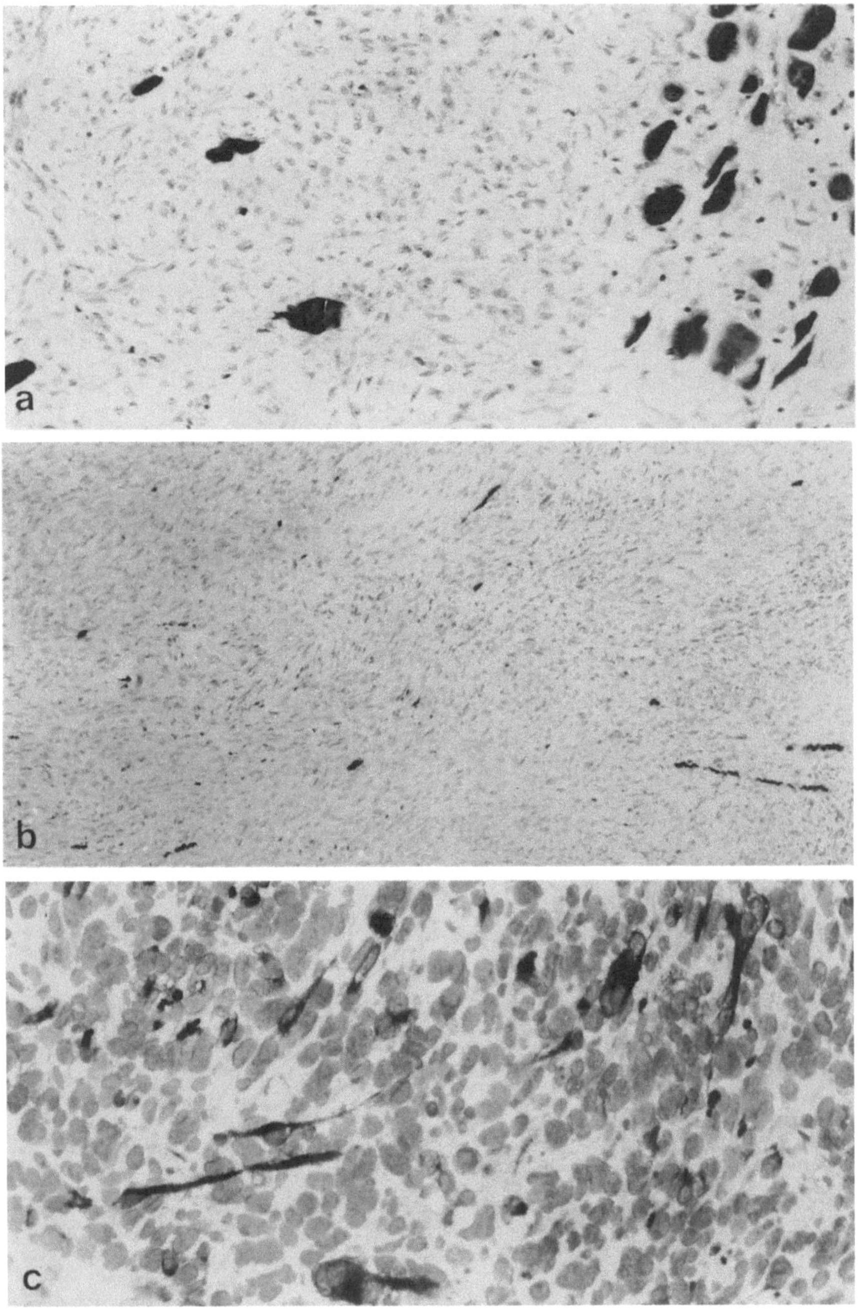

Fig. 2. a *Aggressive fibromatosis: myoglobin* reaction in the underlying muscle fibers; tumor cells negative; ×250. **b** Same tumor as **a**: note the *myoglobin* immunoreaction in *normal skeletal muscle fibers;* ×160. **c** *Rhabdomyosarcoma: myosin* immunoreaction in some spindle-shaped or strap-like tumor cells; ×250. **a–c** PAP method; AEC; haematoxylin counterstain

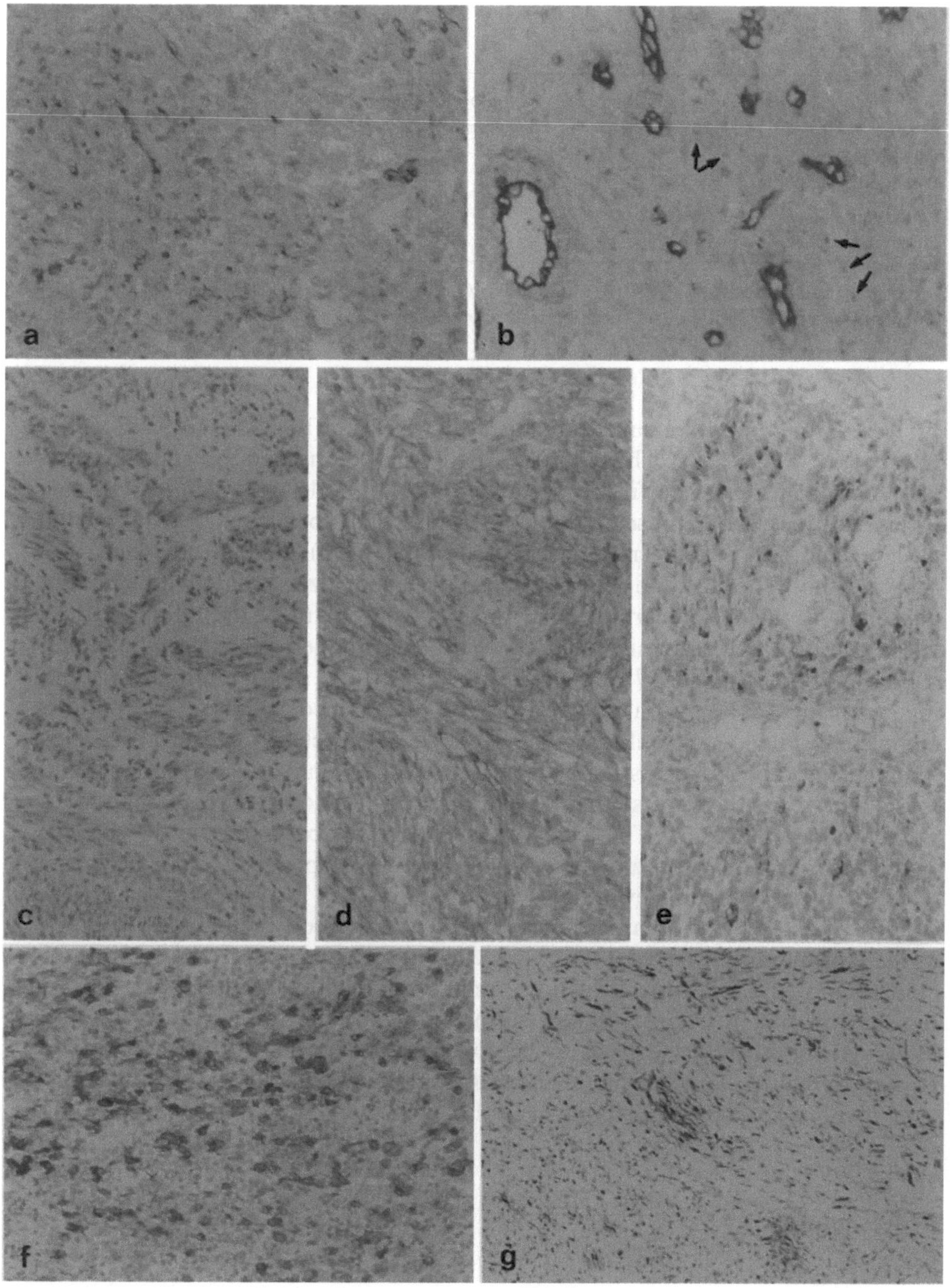

Fig. 3. a Immunoreaction of *factor VIII-associated antigen* in a *malignant haemangioendothelioma;* × 140. **b** *Kaposi sarcoma* (note the discontinuity of swollen endothelial cells): *FVIIIaAg* in the endothelium as well as in the cytoplasm of some tumor cells *(arrows)*; × 140. **c, d** *Desmin* immunoreaction in *leiomyosarcomas*; × 140. **e** *Myosin* immunoreaction in *embryonal rhabdomyosarcoma* is restricted to some cells; × 140. **f** *Protein S-100* immunoreactivity in *interdigitating reticulum cells* of a *dermatopathic lymphadenitis;* × 140. **g** *Neurofibroma* with numerous *protein S-100-positive Schwann cells;* × 90. **a–g** PAP method; AEC; haematoxylin counterstain

Table 1. Myogenic marker proteins

	Desmin	Myoglobin	Myosin
Rhabdomyoma Rhabdomyosarcoma	+	(+)	+/−[a]
Leiomyoma Leiomyosarcoma	+	−	+/−[a]

\+ positive in most cases; (+) positive in the minority of cases
[a] depends on myosin subtype and specifity of antibody

ty) were initially raised against human cardiac myosin (KATUS et al. 1982, 1984; KHAW et al. 1982, 1983). The tissue specificity was determined by using immunohistological methods on frozen sections. The antibody 5C9/3E2 is restricted in binding to smooth muscle of the gastrointestinal, bronchial and urogenital tract, excluding vascular smooth muscle cells (Fig. 1c). It does not react with myocardium. The antibody 1B2/1H4 always recognizes striated muscle cells and binds to atrial myocardium and to the myocardium of the right ventricle, whereas it does not recognize the muscle fibres of the left ventricle. The monoclonal antibodies 4B8/1A7 and 2D7/1C2 have a very similar binding pattern in normal organs: smooth muscle cells are not stained, while myocardium is completely stained, but only 50% of the skeletal muscle fibres show a specific binding (Fig. 1b). No cross-reactivity with non-myogenic cells was noted. With these four reagents we tested three embryonic rhabdomyosarcomas and one proliferating leiomyoma. The leiomyoma was stained by neither antibody. The rhabdomyosarcomas contained a varying number of positive tumor cells that were, however, scarce or even nearly absent (Figs. 2c, 3e). Best results for the positive staining were obtained by 1B2/1H4, the antibody that binds to 100% of the skeletal muscle fibres (Fig. 1a). A control series of liposarcomas, fibrosarcomas and malignant fibrous histiocytomas was completely unreactive, indicating that myosin, if ever expressed, seems to be specific for myogenic tumors. However, more data will have to be collected to support this view. For this reason, statements concerning the diagnostic value in comparison to other myocytic markers should be postponed (Table 1).

3 Myoglobin

Myoglobin as a momomeric oxygen-binding protein is evolutionarily related to haemoglobins (BLANCHETOT et al. 1983). It has a molecular weight of 17.8 kD and is a cytoplasmic constituent of normal skeletal and cardiac muscle fibres. Although all skeletal muscle fibres contain myoglobin, its concentration varies. The type I (red) fibres are rich in myoglobin and show a slow-sustained contraction, whereas type II (white) fibres exhibit a rapid short contraction and are poor in myoglobin (KAGEN 1973). Since most muscles contain a mixture of fibre types, immunohistology gives rise to a checkerboard pattern for myoglobin.

Since the first observation of myoglobin in a myogenic tumor (Mukai et al. 1979) it has since been proclaimed as a marker for tumors of skeletal muscle differentiation (Corson and Pinkus 1981; Brooks 1982), although the sera applied showed, at least in some hands, cross-reactivity with nonmuscular cells (Eusebi et al. 1984). Using antisera to myosin, Mukai et al. (1980a) succeeded in demonstrating rhabdomyoblastic cells in 40% of 25 uterine and ovarian müllerian tumors. Thus, the immunohistochemical method was superior to the conventional proof of cross-striations which are indicative of this special differentiation. Likewise, Giangaspero et al. (1981) found myoglobin in rhabdomyoblastic cells of Wilms' tumors. Dickinson et al. (1983) successfully applied anti-myoglobin to demonstrate rhabdomyoblastical differentiation in a medulloblastoma. Daimaru et al. (1984) used myoglobin antisera to stain the rhabdomyoblastically differentiated parts of "Triton" tumors.

The data on frequency of myoglobin containing rhabdomyosarcomas vary considerable in literature: Corson and Pinkus (1981): 13/17; Brooks (1982): 25/28; Kahn et al. (1983b): 16/53 of the embryonal vs. 8/12 of the alveolar type; Kagawa et al. (1983): 13/26; Jong et al. (1984): 11/23; Royds et al. (1985) 18/26; Eusebi et al. (1986): 10/15.

Our data, based on 73 soft tissue sarcomas and 16 benign lesions with a polyclonal antiserum (Dakopatts, Denmark), illustrate the high specificity but low sensitivity of myoglobin as a marker for malignant skeletal tumors. Only 2/8 rhabdomyosarcomas were specifically stained in a considerable but minor part of the neoplastic cell population, while normal skeletal muscle regularly served as intrinsic or extrinsic positive control. It is important not to regard stained cells as positive when they are apparently necrotic or are adjacent to areas of necrosis, since this can be observed with a multitude of antisera. On the other hand, specifically stained fibers or cellular buds must not be regarded as intrinsic parts of the tumor while they are actually degenerating or regenerating non-neoplastic muscle fibers (Fig. 2a). This is not always easy to assess. In order to exclude errors, the direction of stained fibers proved to be helpful (Fig. 2b). In case of unidirectional arrangement, specifically stained fibers or single cells ought to be regarded as remnants of normal muscle. Keeping these pitfalls in mind, we had no false positive results in our series.

It is still controversial whether myoglobin is a more sensitive marker for rhabdomyosarcoma than myosin or vice versa. Some authors (Jong et al. 1984; Eusebi et al. 1986) who have anti-fast myosin antibodies at their disposal, claim that this special type of myosin is superior to myoglobin in detecting rhabdomyoblasts. Royds et al. (1985) favour β-enolase and Kahn et al. (1983b) prefer creatine kinase to myoglobin. There is, nevertheless, agreement that desmin is the most reliable marker for myogenic tumors (see also chapter 7). In our series, 10/11 tumors of this category stained for desmin, and only 1 rhabdomyosarcoma was unreactive. However, our series contained 2 leiomyosarcomas and one myogenic sarcoma that could not further specified (Figs. 1d, 3c, d). This indicates that the desmin reaction cannot distinguish between leiomyogenic and rhabdomyogenic tumors (Gabbiani et al. 1981; Altmannsberger et al. 1982; Miettinen et al. 1982; Denk et al. 1983).

4 Lysozyme

Lysozyme (muramidase), first described by FLEMING (1922) as a bacteriolytic substance, is a carbohydrase acting on cell wall of susceptible bacteria through cleavage of $\beta(1-4)$linkages between N-actetyl muraminic acid and N-acetyl glucosamine (STROMINGER and TIPPER 1974). Among mesenchymal cells, lysozyme is found in monocytes (FLANAGAN and LIONETTI 1955), mature myeloid cells (BRIGGS et al. 1966) and histiocytes (OSSERMAN 1975). Since the classical immunohistochemical studies of KLOCKARS and REITAMO (1974) and MASON and TAYLOR (1975) on the tissue distribution of lysozyme, it is regarded as a reliable marker indicating the myelomonocytic origin of mononuclear cells in the lymphoid system and in inflammatory tissues.

There are reports of marked positivity for lysozyme in neoplastic histiocytes of malignant histiocytosis (MEISTER et al. 1980; MENDELSON et al. 1980; VILPO et al. 1980; BURGDORF et al. 1981 a; ROHOLL et al. 1985). Juvenile xanthofibroma was found to consist predominantly of lysozyme-positive histiocytic cells (SONODA et al. 1985; SEO et al. 1986). While BROOKS et al. (1984) detected lysozyme in the tumor cells of a malignant giant cell tumor of the bone, LING et al. (1986) found lysozyme only in the giant cell tumors of the tendon sheath among the 24 giant cell tumors they examined. The tumor cells of histiocytosis X, however, do not contain lysozyme (MOTOI et al. 1980; NAKANISHI et al. 1982). In proliferative and neoplastic fibrohistiocytic lesions, lysozyme expression turned out to be a rare phenomenon. NAKANISHI et al. (1982) detected lysozyme-positive cells in 13/100 dermatofibromas, 1/4 xanthogranulomas and 8/33 malignant fibrous histiocytomas; 13 cases of dermatofibrosarcoma protuberans were unreactive. FLETCHER et al. (1985) confirmed the lack of lysozyme in dermatofibrosarcoma protuberans. In the series of ROHOLL et al. (1985), again, only 2/77 malignant fibrous histiocytomas were classified as lysozyme-positive, in the series of KINDBLOM et al. (1982) only 1/31 cases. Lysozyme could thus be characterized as a marker reliable only for monocytes/macrophages. When raising the question of whether tissue histiocytes are of myeloid origin or not, WOOD et al. (1985) emphasized that the histiocytic cells of malignant fibrous histiocytoma do not express the antigenic or enzyme-histochemical features of the myelogenic macrophages and are thus derived from primitive mesenchymal cells as suggested by electron microscopy (FU et al. 1975; TAXY and BATTIFORA 1977; ALGUACIL-GARCIA et al. 1978). This handsome concept, however, is challenged by recent data on the expression of differentiation antigens detected by monoclonal antibodies (STRAUCHEN and DIMITRIU-BONA 1986).

Using a well-established antiserum (Dakopatts, Kopenhagen, Denmark), we observed in our series (Table 2) scattered lysozyme containing tumor cells in 2/5 malignant fibrous histiocytomas, and in considerable proportions of atypical cells in 2/5 sarcomas not further specified, 1 sarcoma of spindle-cell type, and 1 sarcoma of pleomorphic type (Fig. 4d). In this context, it is noteworthy that 1/3 cases of dermatofibrosarcoma protuberans were clearly lysozyme-positive. Nevertheless, all our myogenic, lipogenic and fibrogenic tumors were unreactive. It is important not to misinterpret intermingled phagocytic histiocytes, mastocytes and granulocytes as part of the neoplastic population. Furthermore, areas of necrosis of necrotic cells may be artifactually stained.

Table 2. Pattern of reactivity of 73 soft tissue sarcomas and 16 benign lesions with antisera to lysozyme, AAT, ACT, ferritin, and transferrin

Diagnose[a]	n	Lysozyme n:pos.[b]	AAT n:pos.	ACT n:pos.	Ferritin n:pos.	Transferrin n:pos.
Lipoma	1	0	0	0	0	0
Leiomyoma	1	0	0	0	0	0
Fibroma	1	0	0	0	0	0
Dermatofibroma/cut. histiocytoma	3	0	2	2	1	1
Fasciitis	2	0	0	0	1	0
Neurofibroma	1	0	0	1	1	0
Schwannoma	1	0	0	1	0	1
Liposarcoma	23	0	6	22	12	9
Leiomyosarcoma	2	0	0	1	0	0
Rhabdomyosarcoma	8	0	1	5	3	4
Myogenic sarcoma, n.f.sp.[b]	1	0	0	1	0	1
Fibrosarcoma	2	0	1	1	1	1
Dermatofibrosarcoma protuberans	3	1	0	2	1	1
Aggressive fibromatosis	3	0	0	2	2	3
Malignant fibrous histiocytoma	5	2*	4	5	4	4
Infantile hemangiopericytoma	1	0	0	1*	0	0
Hemangiopericytoma	2	0	1	2	1	0
Malignant hemangioendothelioma	1	0	1	1	1	1
Malignant synovialoma	2	0	0	2	0	1
Neurogenic sarcoma	2	1	0	1	0	0
Neuroblastoma	1	0	0	0	0	0
Epithelioid sarcoma	2	0	0	0	2	0
Spindle cell sarcoma, n.f.sp.	11	1	4	6	4	6
Pleomorphic sarcoma, n.f.sp.	4	1	2	3	1	2
Sarcoma, n.f.sp.	6	0	0	3	3	3
Total	89					

[a] Diagnosis made on conventional histological grounds
[b] n.f.sp.: not further specified; n: pos.: number of tumors characterized as "positive"; *: weak reactivity

To conclude, we regard lysozyme as a marker which, among mesenchymal cells, is generally restricted to the monocyte/marcophage/epithelioid cell system (MASON and TAYLOR 1975). It is very useful in detecting myeloid metaplasia of lymph nodes in the course of myelomonocytic leukaemia and chloromas in other sites (Fig. 5d). Lysozyme may become the most reliable conventional marker for "true" histiocytic lymphomas, since α_1-antitrypsin has recently lost its alledged cell type specificity among the cellular constituents of the immune system (STEIN et al. 1984). In initial immunodiagnosis of soft tissue tumors, lysozyme is a marker of second choice, because of its low sensitivity (Figs. 4d, 5a, b). In case of positive immunoreactivity, however, it may be an aid in discriminating between malignant fibrous histiocytoma and other tumors of comparable morphology (Table 2).

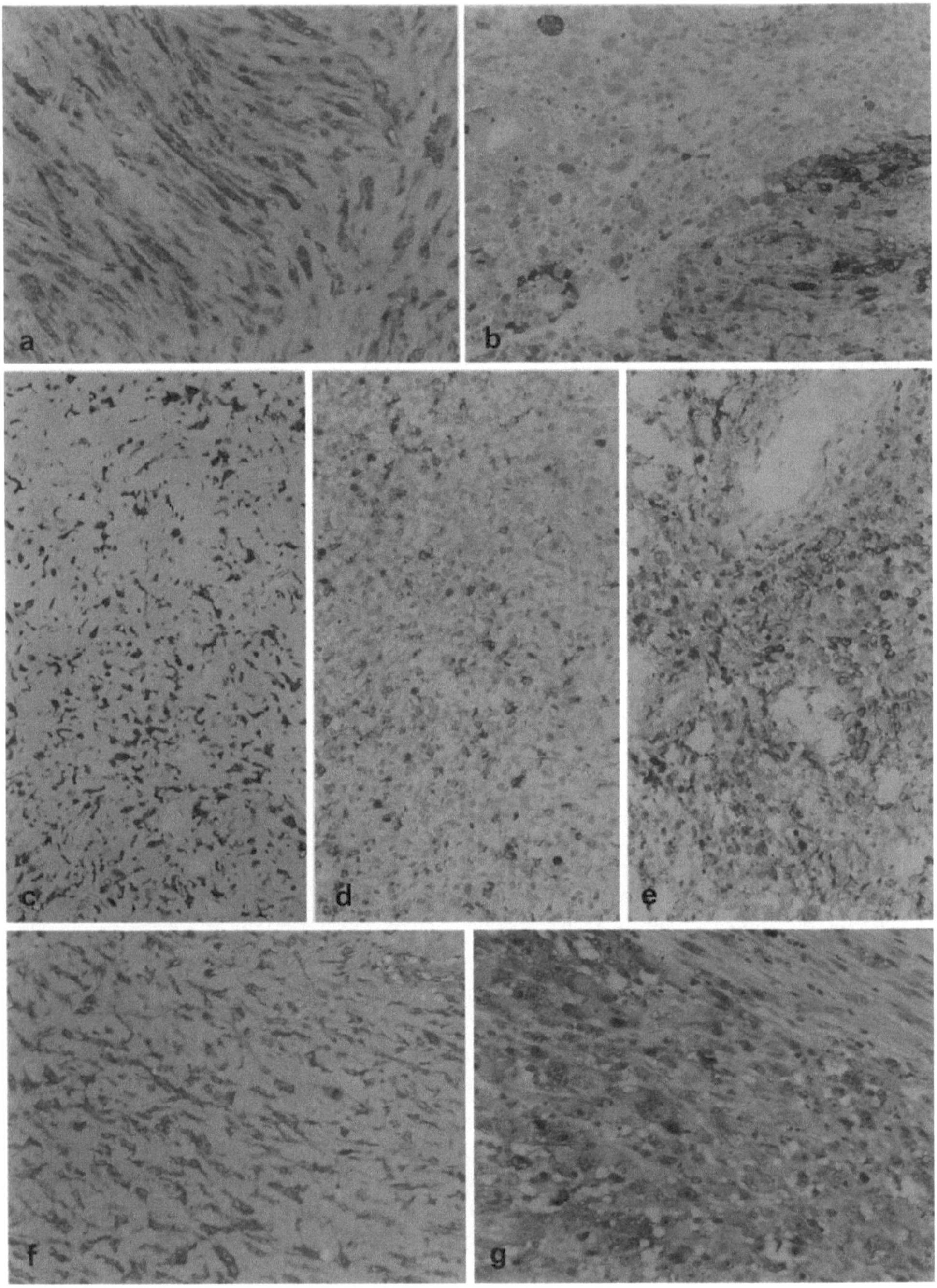

Fig. 4. **a** *Transferrin* reaction in a *rhabdomyosarcoma*; ×140. **b** *Ferritin* immunoreactivity in individual cells of a *pleomorphic liposarcoma*; ×140. **c** *Ferritin* immunoreaction in the majority of tumor cells of a *malignant fibrous histocytoma*; ×90. **d** *Lysozyme* reaction in randomly distributed cells of a *malignant fibrous histiocytoma*; ×160. **e** *ACT* reaction in most cells of a round cell *liposarcoma*; ×90. **f** *AAT* immunoreaction in slender processes of *fibrosarcoma* cells; ×140. **g** *AAT* in a *pleomorphic liposarcoma* is restricted to some cells; ×140. **a–g** PAP method; AEC; haematoxylin counterstain

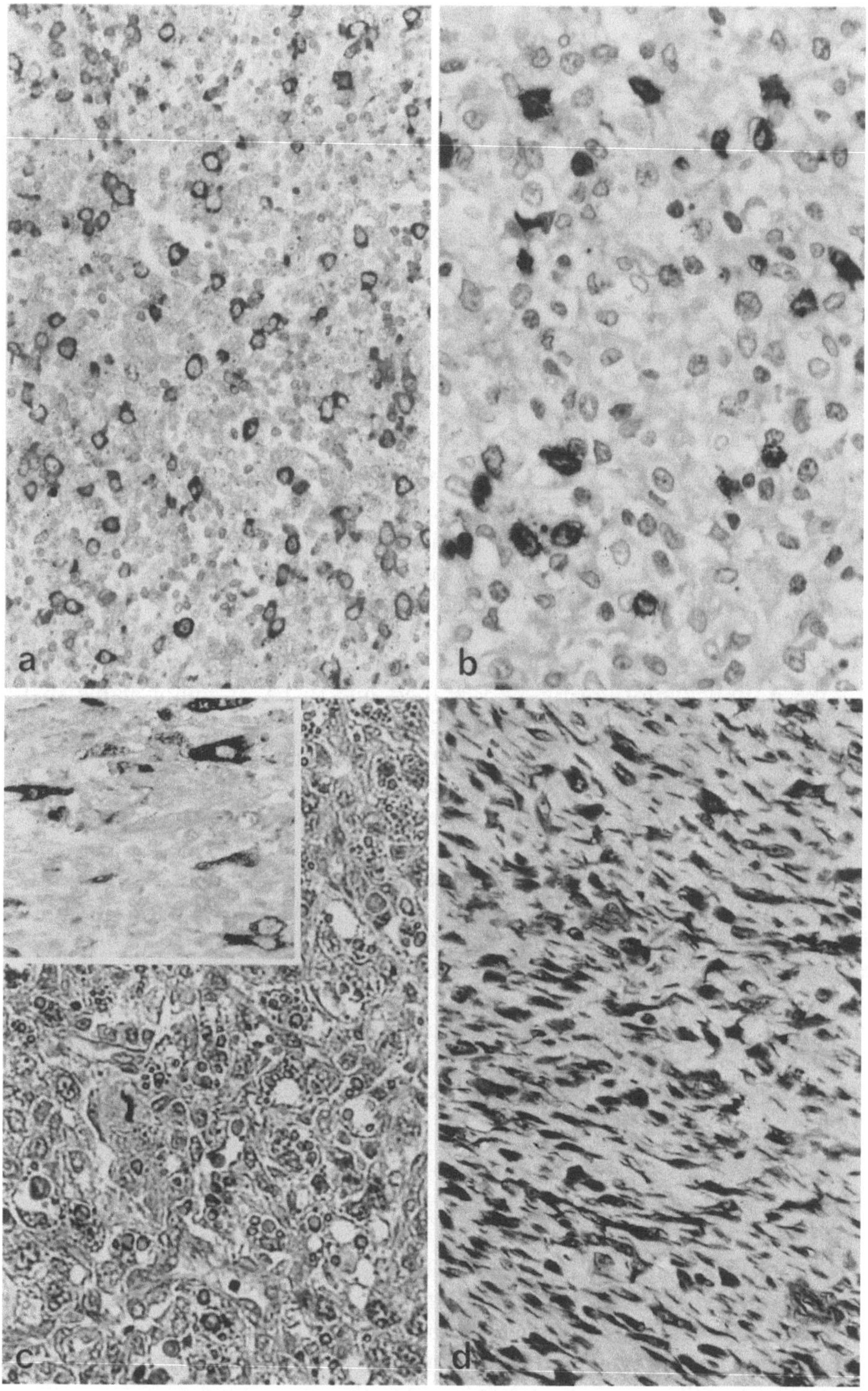

Fig. 5. a Tumor-forming subtype of *myelo-monocytic leukaemia: lysozyme* in scattered leukaemia cells; ×250. **b** *Malignant histiocytosis* of the intestine: *lysozyme* immunoreactivity in some tumor cells; ×400. **c** *Angiogenic sarcoma*, H & E stain, ×250; *inset:* Demonstration of *AAT* in individual tumor cells; ×250. **d** *Fibrosarcoma: ACT* immunoreaction in a large number of spindle-shaped tumor cells; ×250. **a–d** PAP method; AEC; haemetoxylin counterstain

5 Alpha-1-Antitrypsin

α_1-antitrypsin (AAT), a glycoprotein of 51 kD (CARRELL et al. 1982), is the major component of the α_1-electrophoretic band of the human plasma proteins and is capable of inhibiting a variety of proteolytic enzymes, such as trypsin, chymotrypsin, kallikrein and the neutral proteases of polymorphonuclear leukocytes (MORSE 1978). By this effect, AAT inhibits the inflammatory process, in the course of which these enzymes are released by various participating cell types (BREIT et al. 1982). Furthermore, AAT inhibits chemokinesis and chemotaxis of neutrophils and monocytes (BREIT et al. 1983). Reduced levels of AAT as a consequence of a genetic deficiency are associated with pulmonary emphysema and juvenile cirrhosis of the liver (for review see CARRELL et al. 1982). This protein is synthesized by liver cells (ASOFSKY and THORBECKE 1961; ALPER et al. 1980), monocytes (WILSON et al. 1980; VAN FURTH et al. 1983) and macrophages (GUPTA et al. 1979; ISAACSON et al. 1979, 1981), but not in normal peripheral lymphocytes (VAN FURTH et al. 1983). This is why AAT was regarded for several years as a marker – among mesenchymal cells – to be specific for monocytes and histiocytes.

Since the tumor cells contained AAT, a highly malignant lymphoma of the intestine was called "malignant histiocytosis" (ISAACSON and WRIGHT 1978; ISAACSON et al. 1981, 1982). AAT was then detected in Hodgkin and Sternberg-Reed cells (POPPEMA et al. 1978; PAYNE et al. 1982), a finding interpreted as a consequence of their histiocytic origin. At present, both concepts are challenged since STEIN et al. (1984) claimed to have detected AAT in malignant T cells.

Like α_1-antichymotrypsin, AAT turned out to be of minor value in the differential diagnosis of soft tissue tumors. ROHOLL et al. (1985) found both markers in the majority of cases of "true" malignant histiocytosis, whereas they were detectable only in the minority of cases of malignant fibrous histiocytoma. Within the latter group, cases of the storiform subtype were less frequently stained then the pleomorphic or giant cell subtypes. These data are in sharp contrast to the results of KINDBLOM et al. (1982) who found AAT in 18/22 pleomorphic fibrous histiocytomas, but in 0/5 spindle-cell types of this tumor. NATHRATH and REMBERGER (1986) discussed whether some rare cases of granular cell tumor were of histiocytic origin since they found AAT in 1/5 such cases. In our series (Table 2) we observed AAT in soft tissue tumors of different types including rhabdomyosarcoma, liposarcoma, hemangiopericytoma, endothelioma and even Schwannoma (Figs. 4f, g, 5c). At present we doubt whether this marker is likely to be a helpful tool for a specific differential diagnostic problem.

6 Alpha-1-Antichymotrypsin

α_1-antichymotrypsin (ACT) like AAT acts as a protease inhibitor on chymotrypsin, catepsin D, elastase, collagenase, urokinase, renin, kallikrein, Hageman-factor, thrombin and plasmin (TRAVIS et al. 1978; BEATTY et al. 1980). Among non-epithelial cells, ACT is presently regarded as a monocyte/macrophage marker (PAPADIMITRIOU et al. 1978). However, there are very few reports on this issue in the literature (BRAUNHUT et al. 1984; MEISTER et al. 1980). Cytoplasmic presence of ACT was used as an additional argument to support theories on the monocyte origin of Hofbauer cells in the placenta (BRAUNHUT et al. 1984) and of Hodgkin cells (PAPADIMITRIOU et al. 1978). ROHOLL et al. (1985) found co-expression of ACT and AAT in "true" malignant histiocytosis, and ISAACSON et al. (1982) in malignant histiocytosis of the intestine. Only a minority of the malignant fibrous histiocytomas examined by ROHOLL et al. (1985) contained ACT-positive tumor cells. In our series of soft tissue tumors ACT, unlike AAT, proved to be synthesized by 5/5 malignant fibrous histiocytomas, 2/3 dermatofibromas, in 2/3 cases of dermatofibrosarcoma protuberans, 5/8 rhabdomyosarcomas, 22/23 liposarcomas, in 2/2 haemangiopericytomas and others (Table 2; Figs. 4e, 5d). Therefore, ACT was of no help in discriminating among these. Our numbers were too small to allow statements concerning the negative tumors (epithelioid sarcomas, neuroblastoma, myoma, fibroma, lipoma). If these tumor types were consistently ACT-negative, ACT could claim some differential diagnostic relevance. This point needs further study.

7 Protein S-100

In 1965, MOORE was able to isolate a protein from rat brain which has been called "S-100" due to its solubility in 100% ammonium sulfate at neutral pH. The heterogeneous protein fraction consists of two subunits (α and β) which form heterodimers in solutions (ISOBE et al. 1978, 1981, 1983). From ox brain, three S-100 isoforms can be purified, i.e. S-100 a_0, S-100a and S-100b that are characterized by the same molecular weight (21 kD) and similar pH (about 4.3). They differ, however, in their subunit composition ($\alpha\alpha$, $\alpha\beta$ and $\beta\beta$, respectively; ISOBE et al. 1978, 1981, 1983). The physiological role of this dimerization is not clear. Amino acid sequence analysis of the α- and β-subunits (ISOBE et al. 1978, 1981) demonstrated that both are characterized by a sequence of 24 amino acids highly homologous to the calcium-binding domain (TUFTY and KRETSINGER 1975), suggesting that the S-100 protein is evolutionarily related to the family of calcium-binding proteins such as calmodulin. A crude S-100 fraction containing S-100a and S-100b has been shown to be involved in the Ca^{2+}-mediated control of the assembly and disassembly of brain microtubule proteins in vitro (BAUDIER et al. 1982; DONATO 1983; ENDO and HIDAKA 1983). In addition, S-100 (a+b) interferes with the nucleation and the elongation of microtubule proteins by interacting with tubulin (DONATO 1984a, b). Furthermore, S-100 (a+b) is a Zn^{2+}-binding protein. Recent reports by DONATO and co-workers (1985) demonstrate that single S-100 isoforms have similar, if not identical, effects in the presence of Ca^{2+}, i.e. the inhibition of assembly and promotion of disassembly of microtubule proteins. In the presence of Zn^{2+}, rat S-100 and ox S-100a and S-100b inhibit assembly, while S-100a_0 yields no effect.

There have been many efforts to demonstrate the cellular and tissue distribution of S-100 by immunocytochemical methods. Originally regarded as a brain-specific protein (BOCK 1978; ZOMZELY-NEURATH and WALKER 1980) and demonstrated in astrocytes, oligodendrocytes and ependymal cells (MATUS and MUGHAL 1975; YAMAGUCHI 1980; LUDWIN et al. 1981), protein S-100 was then observed in Schwann cells and supportive cells of the peripheral (NAKAJIMA et al. 1982; STEFANSSON et al. 1982b) as well as in the autonomic nervous system (COCCHIA and MICHETTI 1981; FERRI et al. 1982; KONDO et al. 1982). Since that time, an increasing number of mesenchymal and epithelial cell types are reported to show S-100 immunoreactivity such as interstitial cells of the pineal gland (MOLLER et al. 1978), stellate cells of the adenohypophysis (NAKAJIMA et al. 1980), melanocytes (NAKAJIMA et al. 1982), satellite cells of the adrenal medulla (COCCHIA and MICHETTI 1981), chondrocytes (STEFANSSON et al. 1982c), fat cells (MICHETTI et al. 1983), interdigitating reticulum cell in T-dependent areas of peripheral lymphoid tissue (Tabl. IIf) and in human thymic medulla (TAKAHASHI et al. 1981; NAKAJIMA et al. 1982; MECHTERSHEIMER et al. 1986), Langerhans cells of the epidermis (NAKAJIMA et al. 1982) and tubular epithelium of rat kidney (MOLIN et al. 1985) as well as salivary gland epithelium (MOLIN et al. 1985; NAKAZATO et al. 1985).

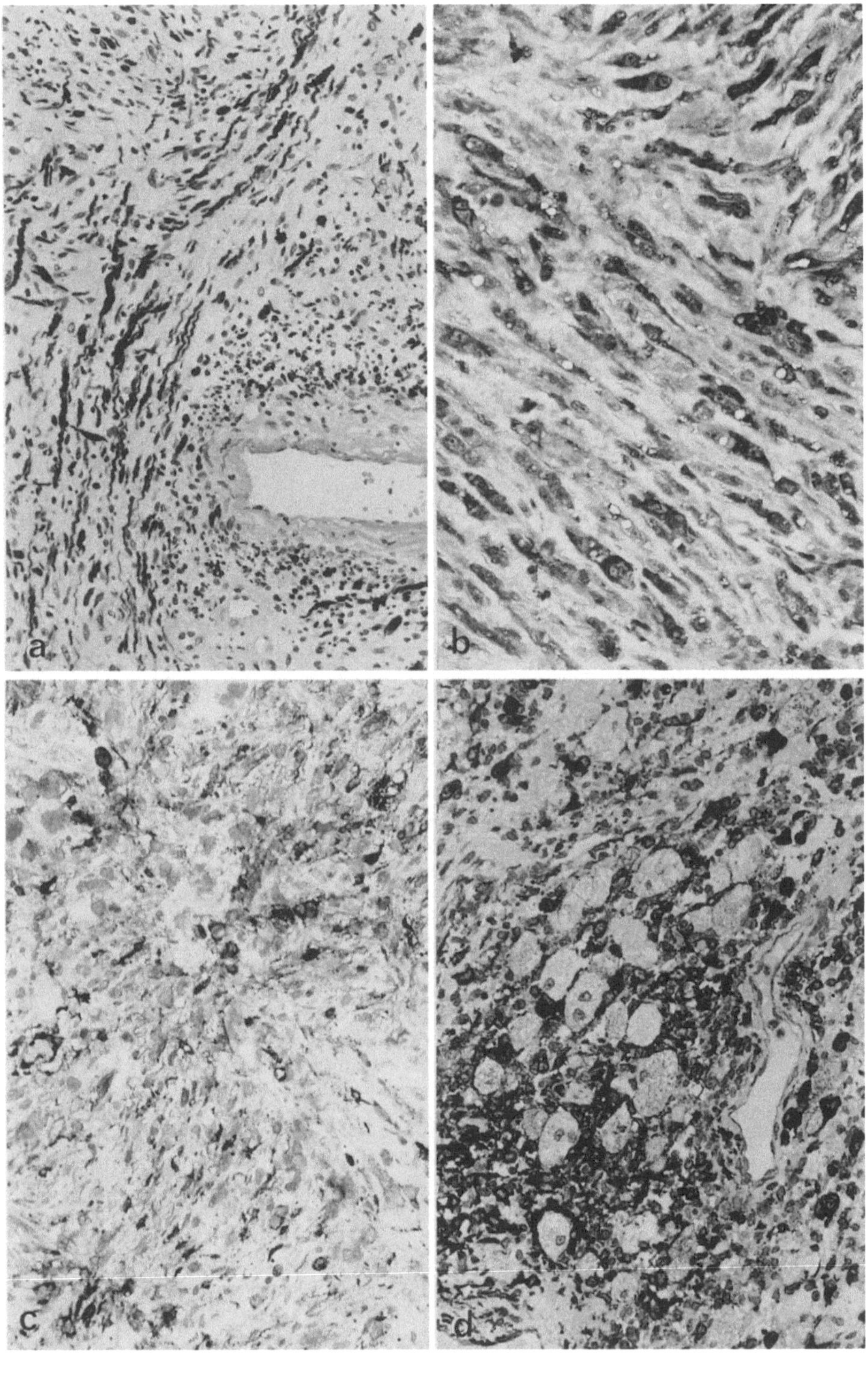
a
b
c
d

In correlation with its distribution in normal cells, protein S-100 immunoreactivity has been found in neuroectodermal and Schwann cell tumors, malignant melanomas, lipo- and chondrosarcomas, chordomas, histiocytosis X as well as salivary gland tumors (for reviews see: HAGLID et al. 1973; NAKAJIMA et al. 1982; STEFANSSON et al. 1982a; KAHN et al. 1983a; NAKAMURA et al. 1983; WEISS et al. 1983; TERENGHI et al. 1984; NAKAZATO et al. 1985; SCHMIDT et al. 1985; SHIMADA et al. 1985; KIMURA et al. 1986).

Within the spectrum of soft tissue tumors, protein S-100 immunoreactivity has been detected in benign tumors of the nerve sheath including neurinoma, schwannoma, neurofibroma (Figs. 3g, 6a), nerve sheath myxoma and traumatic neuroma (NAKAJIMA et al. 1982; STEFANSSON et al. 1982a; WEISS et al. 1983; KAHN et al. 1983a; ANGERVALL et al. 1984). This finding is highly important concerning the differential diagnosis of protein S-100 negative meningiomas and leiomyomas which can exhibit similar or nearly identical histiological features and a very similar topographical distribution (YAMAGUCHI 1980; NAKAJIMA et al. 1982; STEFANSSON et al. 1982a; KAHN et al. 1983a).

In malignant schwannomas and neurogenic sarcomas, however, protein S-100 shows a wide range of expression including neoplasms completely lacking this protein. In neurogenic sarcomas associated with von Recklinghausen's disease and malignant nerve sheath tumors originating from a nerve trunk, the great majority of tumors expresses protein S-100 when investigated with immunocytochemical methods (WEISS et al. 1983; HERRERA and PINTO DE MORAES 1984; NAKAJIMA et al. 1984; DAIMARU et al. 1985; MATSUNOU et al. 1985). In neurogenic sarcomas which were exclusively diagnosed on histological grounds the number of tumor cells containing S-100 protein was markedly reduced (DAIMARU et al. 1985; MATSUNOU et al. 1985). In such cases, the majority of S-100-positive cells corresponded to the wavy spindle cells in the loosely arranged areas. These cells were frequently interspersed among cells forming characteristic palisading, plexiform, storiform, or nodular patterns and tactile differentiation. Protein S-100 can be completely negative in the fibrosarcoma-like areas of neurogenic sarcomas. Since, on the other hand, non-neurogenic spindle cell soft tissue tumors such as fibrosarcomas, leiomyosarcomas, synovial sarcomas and dermatofibroma protuberans completely lack protein S-100, the demonstration of a small number of S-100-positive cells can be regarded in accordance with the clinical data and their morphology as indicative of a neurogenic sarcoma (NAKAJIMA et al. 1982; STEFANSSON et al. 1982a; WEISS et al. 1983). Although granular cell tumors show a characteristic morphology, the consistent immunoreactivity of protein S-100 can be a valuable contribution for the diagnosis. This finding is used to support the presumptive Schwann cell derivation of these tumors (NAKAJIMA et al. 1982; NAKAZATO et al. 1982; STEFANSSON et al. 1982a; ARMIN et al. 1983; MUKAI 1983).

◁ **Fig. 6. a** *Neurofibroma: protein S-100* reaction in the majority of spindle-shaped wavy Schwann cells; ×160. **b** *Rhabdomyosarcoma: transferrin* reaction in most, possibly all tumor cells; ×250. **c** *Unclassified sarcoma: ferritin* in some scattered tumor cells; ×160. **d** *Pleomorphic liposarcoma: ferritin* in most tumor cells; ×250. **a–d** PaP method; AEC; haematoxylin counterstain

Table 3. Pattern of reactivity of 73 soft tissue sarcomas and 16 benign lesions with antiserum to S-100 protein

Diagnosis	positive/ cases	Diagnosis	positive/ cases
Fasciitis	0/2	Leiomyoma	0/1
Aggressive fibromatosis	0/3	Leiomyosarcoma	0/2
Fibroma	0/1	Rhabdomyosarcoma	0/8
Dermatofibrosarcoma protub.	0/3	Myogenic sarcoma, n.f.sp.	0/1[a]
Fibrosarcoma	0/2	Malignant synovialoma	0/2
Dermatofibroma/cut. histiocyt.	0/3	Epithelioid sarcoma	0/2
Malignant fibrous histiocytoma	2/5[a]	Spindle cell sarcoma, n.f.sp.	4/11[a]
Neurofibroma	1/1	Pleomorphic sarcoma, n.f.sp.	0/4
Cellular schwannoma	0/1	Unclassified sarcoma	2/6[a]
Neurogenic sarcoma	0/2	Hemangiopericytoma	0/3
Neuroblastoma	0/1	Malignant hemangioendothelioma	0/1
Lipoma	1/1		
Liposarcoma	8/23[a]	Total	88

[a] few scattered cells
n.f.sp.: not further specified

In the literature, the expression of S-100 protein in lipomas and liposarcomas is controversial (Nakajima et al. 1982; Cocchia et al. 1983; Kahn et al. 1983a; Weiss et al. 1983; Hashimoto et al. 1984). In lipomas of various differentiation, a rather consistent S-100 immunoractivity can be demonstrated. In liposarcomas, protein S-100 is found in mature adipocytes, vacuolated cells and lipoblasts, predominantly. Our series, which includes 23 liposarcomas among 73 soft tissue sarcomas, showed scattered S-100 positive tumor cells in 8 sarcomas (Table 3). In accordance with the literature, myxoid and spindle-cell areas were negative, and, 2 of 5 malignant fibrous histiocytomas exhibited rare positive cells. These data suggest that protein S-100 is not a very reliable marker to identify less differentiated liposarcomas. In spite of the exceptions mentioned above, S-100 protein may be of some value in the differential diagnosis between the myxoid variant of liposarcoma and the myxoid type of malignant fibrous histiocytoma (Hashimoto et al. 1984).

Since its first detection in tissue cultures of malignant melanomas (Gaynor et al. 1980), protein S-100 has been regarded as a marker for malignant melanomas (Gaynor et al. 1981; Stefansson et al. 1982a; Springall et al. 1983; Weiss et al. 1983; Kindblom et al. 1984; Rode and Dhillon 1984). According to these reports, protein S-100 is a relatively consistent marker of malignant melanomas including both the epitheloid and spindle cell type. Protein S-100 can therefore be used as a reliable marker in differential diagnosis between malignant melanoma and carcinoma. In this context it is interesting that also the clear-cell sarcoma of aponeuroses and tendon sheaths, which is sometimes interpreted as a variant of malignant melanoma, expresses protein S-100 (Kindblom et al. 1983; Weiss et al. 1983).

For principal and diagnostic reasons it is important that the histiocytosis X cells are S-100-positive (Watanabe et al. 1981, 1983; Nakajima et al. 1982).

This finding allows the correlation of histiocytosis X cell to the T-zone histiocyte and the separation from lysozyme- and non-specific cross-reacting antigen-positive histiocytes of the monocyte-macrophage system (WATANABE et al. 1981). S-100 positive variants of histiocytosis X can be distinguished from juvenile xanthogranuloma.

Despite an increasing number of protein S-100 positive cell types and tumors, the immunocytochemical demonstration may be helpful in selected problems of tumor differential diagnosis, e.g. (1) the distinction between benign nerve sheath tumors and leiomyomas or meningiomas, (2) cellular schwannomas and neurogenic sarcomas, (3) neurogenic sarcomas and other spindle-cell sarcomas, (4) histiocytosis X and juvenile xanthogranuloma. Scattered S-100 positive cells in spindle cell sarcomas can be used to support the diagnosis of neurogenic sarcoma; negative immunoreaction, however, does not exclude this diagnosis. Some S-100-positive cells in liposarcomas are of limited value for the differential diagnosis, since in malignant fibrous histiocytomas rare S-100-positive cells can also be identified by immunocytochemical methods.

8 Ferritin

Ferritin, a protein containing more than 20% iron, was first isolated by LAUFBERGER (1937) from horse spleen. It is widely distributed in mammalian tissues. The ferritin molecule's inner core is about 55 Å in diameter and consists of ferric hydroxid micelles surrounded by an almost spherical shell of protein with a molecular weight of 462 kD. The overall diameter of the molecule is approximately 110 Å. Because of its electron dense iron core, ferritin has been used as a tracer in electron microscopy since the early sixties. Since abnormally high levels of ferritin are associated with early stages of many malignant neoplasias (for review see DRYSDALE 1979) and, since tissue specific iso-ferritins have been isolated (BULLOCK et al. 1980), the question has been raised whether ferritins might be exploited as tumor markers.

So far, there have been few attempts to use ferritin as tumor marker in immunohistochemistry. Ferritin was demonstrated in hepatocellular carcinoma (COHEN et al. 1984a; IMOTO et al. 1985) and in testicular seminoma (JACOBSEN and JACOBSEN 1983; COHEN et al. 1984b). SEHESTED et al. (1985) detected ferritin in both epithelial and mesenchymal components of pleomorphic adenomas, but not in normal parotid tissue. The only report on in situ demonstration of ferritin in soft tissue tumors we could find was that published by KINDBLOM et al. (1982). The authors investigated malignant fibrous histiocytomas and detected ferritin in 12/22 of the pleomorphic and 3/5 tumors of the spindle cell type.

In our series of soft tissue tumors, ferritin could be detected in a considerable number of cases of nearly every type of benign and malignant lesions (Table 2; Figs. 4b, c, 6c, d). Accordingly there seems to be no possibility of using ferritin as a diagnostic aid. This phenomenon, however, does not exclude the possibility of existing differences in tumor biology between the tumors that contain high amounts of ferritin and those that do not. This issue as well as the practical relevance of tissue specific iso-ferritins, against which at present there are no specific antibodies available, remain to be elucidated.

9 Transferrin

Transferrin is one of the β_1 serum globulins. It has a molecular weight of 88 kD and is synthesized in liver cells (VASSY et al. 1984). Each molecule carries two binding sites with probably different functions. Its main function is iron transfer. The iron-free, apoferritin, binds to Fe^{3+} ions that form ferrotransferrin. All growing cells express surface transferrin receptors that bind ferrotransferrin which is then subjected to receptor-mediated endocytosis (IACOPETTA et al. 1983). The iron atoms are afterwards transferred to cytoplasmic ferritin and within a few minutes, apoferritin is secreted from the cell (WILLINGHAM and PASTAN 1985). The transferrin receptor is a homo-dimeric transmembrane glycoprotein (MW 180 kD) (SCHNEIDER et al. 1984). One antibody with a binding affinity for it is OKT9. These phenomena make it evident that a cytoplasmic localization of transferrin does not reflect a synthesis of this protein but the degree of transmembranous iron transfer (PARMLEY et al. 1979). In neoplastic cells, an additional effect might account for a high cytoplasmic transferrin content, i.e. its accumulation as a consequence of a defective or functionally impaired transferrin receptor. An immunohistochemical proof of increased cytoplasmic transferrin content in a tumor cell is thus a finding that cannot be further interpreted and that might explain the absence of data on tissue distribution of transferrin in tumors.

As can be seen in Table 2, we observed transferrin in many different malignant soft tissue tumors (Figs. 4a, 6b) whereas the benign tumors did not contain detectable intracytoplasmic transferrin. Our data, however, do not suffice to determine whether cytoplasmic transferrin might be an indicator of malignancy, but it is quite obvious that transferrin is of little diagnostic value because it does not discriminate between diverse types of sarcomas.

10 Factor VIII Associated Antigen

Together with the clot-promoting factor and the von Willebrand-factor, the factor VIII associated antigen (FVIIIaAg) which is synthesized by endothelial cells (JAFFE 1977; GIDDINGS 1982), forms the antihaemophilic agent factor VIII. The antigen, a glycoprotein with a molecular weight of approximatively 1100 kD (MCCOMB et al. 1982), has been detected within these cells (HOYER et al. 1973; MUKAI et al. 1980b), and also in platelets, megakaryocytes and mast cells (PIOVELLA et al. 1978; KINDBLOM 1982). FVIIIaAg has been used as a marker for tumors of vascular origin (FEIGL et al. 1976; BURGDORF et al. 1981b; SEHESTED and HOU-JENSEN 1981), of fibrous haematopoetic tumors (MEISTER and GOESSNER 1983) and Kaposi's sarcoma (NADJI et al. 1980; GUARDA et al. 1981; MODLIN et al. 1983; FLOTTE et al. 1984; MILLARD and HERYET 1985; BECKSTEAD et al. 1985).

In the literature, FVIIIaAg is regarded as a reliable marker for highly differentiated angiogenic tumors, excluding lymphangioma (LITTLE et al. 1986; BURGDORF et al. 1981b). However, staining was most marked in endothelial cells of normal vessels and was much weaker in neoplastic endothelia of poorly differentiated angiosarcoma and in the proliferating spindle cells of Kaposi's sarcoma which irregularly exhibit a dot-like or granular cytoplasmic staining for FVIIIaAg (BECKSTEAD et al. 1985; MILLARD and HERYET 1985; LITTLE et al. 1986). Some authors (HOSAKA et al. 1985; LITTLE et al. 1986) prefer Ulex europaeus lectin I (UEA I) as a more sensitive marker.

In our series of 89 mesenchymal tumors, we could readily demonstrate the vasculature of each tumor, but found FVIIIaAg-containing tumor cells only three times: in a poorly differentiated angiosarcoma and in two additional cases of Kaposi's sarcoma (Fig. 3a, b). The latter contained spindle cells with a granular positivity or a dot-like staining pattern for this marker. However, in the cases of haemangiopericytoma, the tumor cells were FVIIIaAg-negative. We suggest that in a case of poorly differentiated sarcoma, the detection of clearly FVIIIaAg-positive tumor cells might suffice for a diagnosis of malignant endothelioma. It is, however, important not to regard collapsed atypical capillaries in the "stromal" compartment of the tumor as part of the neoplastic cell population. Since FVIIIaAg stains platelets as well, areas of intratumoral haemorrhage should be avoided in evaluation. Apart from the fact that abnormal cells often produce smaller amounts of a cytotypical product than does the normal counterpart, a weak or negative staining result may be caused not only by a decrease in antigenic density alone but also by a low affinity of the antiserum. We agree with LITTLE (1986) that UEA I is a far more sensitive marker but that it has the great disadvantage of reacting with diverse epithelial cell types (MÖLLER et al. 1984; WIRBEL et al. 1984). Additionally, UEAI recognizes lymphatic endothelium at least within areas of inflammation and sclerosis (MÖLLER and LENNERT 1984), whereas FVIIIaAg does not. Therefore, the combination of these two markers might facilitate the differential diagnosis between lymphangioma and haemangioma.

11 Conclusions

Routine light microscopy, together with special stains, is a prerequisite for the classification of soft tissue tumors. Additional immunohistological investigations can either substantiate the diagnosis or lead to a decision between two alternatives if the antigenic pattern of respective tumors differs. The negative result of a reaction does not necessarily exclude a diagnosis and a histogenetic diagnosis is confirmed by a positive result only if the antibody is specific. Desmin, myosin and myoglobin are antibodies with a good specificity for myogenetic tumors. Desmin, the most sensitive antibody, can recognize tumors with smooth muscle as well as skeletal muscle differentiation whereas myoglobin only recognizes tumors with skeletal muscle differentiation and only highly differentiated tumor cells. Myosin is also a reliable myogenic marker, but it is not yet commercially available. Protein S-100 is a valuable adjunct in the diagnosis of benign and malignant nerve sheath tumors, histiocytosis X-related lesions, and malignant melanomas. The factor VIII-associated antigen is highly specific for tumors originating from blood vessel endothelium, while it cannot always be detected in poorly differentiated tumors. Lysozyme, AAT, ACT, ferritin and transferrin, because of their low specificity, are only of limited value for differential diagnosis of soft tissue tumors.

References

Alguacil-Garcia A, Unni KK, Goellner JN (1978) Malignant fibrous histiocytoma: An ultrastructural study of six cases. Am J Clin Pathol 69:121–129

Alper CA, Raum D, Awdeh ZL, Petersen BH, Taylor PD, Starzl TE (1980) Studies of hepatic synthesis in vivo of plasma proteins, including orosomucoid, transferrin, α_1-antitrypsin, C 8, and factor B. Clin Immunol Immunopathol 16:84–89

Altmannsberger M, Osborn M, Treuner J, Hölscher A, Weber K, Schauer A (1982) Diagnosis of human childhood rhabdomyosarcoma by antibodies to desmin, the structural protein of muscle specific intermediate filaments. Virchows Arch (Cell Pathol) 39:203–215

Altmannsberger M, Weber K, Droste R, Osborn M (1985) Desmin is a specific marker for rhabdomyosarcomas of human and rat origin. Am J Pathol 118:85–95

Angervall L, Kindblom LG, Haglid K (1984) Dermal nerve sheath myxoma. A light and electron microscopic, histochemical and immunohistochemical study. Cancer 53:1752–1759

Armin A, Connelly E, Rowden G (1983) An immunoperoxidase investigation of S-100 protein in granular cell myoblastomas: evidence for Schwann cell derivation. Am J Clin Pathol 79:37–44

Asofsky R, Thorbecke GJ (1961) Sites of formation of immuneglobulins and of a component of C3.II. Production of immunoelectrophoretically identified serum proteins by human and monkey tissues in vitro. J Exp Med 114:471–487

Baudier J, Briving C, Deinum J, Haglid K, Sorskog L, Wallin M (1982) Effect of S-100 proteins and calmodulin on Ca^{++}-induced assembly and disassembly of brain microtubule proteins in vitro. FEBS Lett 147:165–167

Beatty K, Bieth J, Travis J (1980) Kinetics of association of serine proteases with native and oxidized α_1-proteinase inhibitor and α_1-antichymotrypsin. J Biol Chem 255:3931–3934

Beckstead JH, Wood GS, Fletcher V (1985) Evidence of the origin of Kaposi's sarcoma from lymphatic endothelium. Am J Pathol 119:294–300

Blanchetot A, Wilson V, Wood D, Jeffreys AJ (1983) The seal myoglobin gene: an unusually long globin gene. Nature 301:732–734

Bock E (1978) Nervous system specific proteins. J Neurochem 30:7–14

Braunhut SJ, Blanc WA, Ramanarayanan M, Marboe C, Mesa-Tejada R (1984) Immunocytochemical localization of lysozyme and alpha-1-antichymotrypsin in the human placenta: an attempt to characterize the Hofbauer cell. J Histochem Cytochem 32:1204–1210

Breit NS, Robinson JP, Luckhurst E, Clark P, Penny R (1982) Immunoregulation by α_1-antitrypsin. J Clin Lab Immunol 7:127–131

Breit NS, Robinson JP, Penny R (1983) The effect of α_1-antitrypsin on phagocyte function. J Clin Lab Immunol 10:147–149

Briggs RS, Perillie PE, Finch SC (1966) Lysozyme in bone marrow and peripheral blood cells. J Histochem Cytochem 14:167–170

Brooks JJ (1982) Immunohistochemistry of soft tissue tumors. Myoglobin as a tumor marker for rhabdomyosarcoma. Cancer 50:1757–1763

Brooks JP, Pascal RR (1984) Malignant giant cell tumor of bone: ultrastructural and immunohistologic evidence of histiocytic origin. Hum Pathol 15:1098–1100

Bullock S, Bomford A, Williams R (1980) A biochemical comparison of normal human liver and hepatocellular carcinoma ferritins. Biochem J 185:639–645

Bures JC, Barnes L, Mecer D (1981) A comparative study of smooth muscle tumors utilizing light and electron microscopy, immunocytochemical staining and enzymatic assay. Cancer 48:2420–2426

Burgdorf WHC, Duray P, Rosai J (1981a) Immunohistochemical identification of lysozyme in cutaneous lesions of alleged histocytic nature. Am J Clin Pathol 75:162–167

Burgdorf WHC, Mukai K, Rosai J (1981b) Immunohistochemical identification of factor VIII-related antigen in endothelial cells of cutaneous lesions of alleged vascular nature. Am J Clin Pathol 75:167–171

Carrell RW, Jeppson JO, Laurell CB, Brennan SO, Owen MC, Vaughan L, Ross Boswell D (1982) Structure and variation of human α_1-antitrypsin. Nature 2986:329–334

Cocchia D, Michetti F (1981) S-100 antigen in satellite cells of the adrenal medulla and the superior cervical ganglion of the rat. An immunochemical and immunocytochemical study. Cell Tissue Res 215:103–112

Cocchia D, Lauriola L, Stolfi VM, Tallini G, Michetti F (1983) S-100 antigen labels neoplastic cells in liposarcoma and cartilaginous tumors. Virchows Arch (Pathol Anat) 402:139–145

Cohen C, Berson SD, Shulman G, Budgeon LR (1984a) Immunohistochemical ferritin in hepatocellular carcinoma. Cancer 53:1931–1935

Cohen C, Shulman G, Budgeon LR (1984b) Immunohistochemical ferritin in testicular seminoma. Cancer 54:2190–2194

Corson JM, Pinkus GS (1981) Intracellular myoglobin – a specific marker for skeletal muscle differentiation in soft tissue sarcomas. Am J Pathol 103:384–389

Daimaru Y, Hashimoto H, Enjoji M (1984) Malignant "Triton" tumors: a clinicopathologic and immunohistochemical study of nine cases. Hum Pathol 15:768–778

Daimaru Y, Hashimoto H, Enjoji M (1985) Malignant peripheral nerve sheath tumors (malignant schwannomas). An immunohistochemical study of 29 cases. Am J Surg Pathol 9:434–444

Darnell J, Lodish H, Baltimore D (eds) (1986) Molecular cell biology. Scientific American Books, New York, pp 839–852

De Jong ASH, Van Raamsdonk W, Van Vark M, Voûte PA, Albus-Lutter CE (1984) Myosin and myoglobin as tumor markers in the diagnosis of rhabdomyosarcoma. Am J Surg Pathol 8:521–528

Denk H, Krepler R, Artlieb U, Gabbiani G, Rungger-Brändle E, Leoncini P, Franke WW (1983) Proteins of intermediate filaments. An immunohistochemical and biochemical approach to the classification of soft tissue tumors. Am J Pathol 110:193–208

Dickinson DW, Hart MN, Menezes A, Cancilla PA (1983) Medulloblastoma with glial and rhabdomyoblastomaic differentiation. A myoglobin and glial fibrillary acidic protein immunohistochemical and ultrastructural study. J Neuropathol Exp Neurol 42:639–647

Donato R (1983) Effect of S-100 protein on assembly of brain microtubule proteins in vitro. FEBS Lett 162:310–313

Donato R (1984a) Chlorpromazine inhibits the calcium-mediated effect of S-100 protein(s) on assembled brain microtubule proteins, but not those on microtubule protein assembly. Biochem Biophys Res Commun 122:983–990

Donato R (1984b) Mechanism of action of S-100 protein(s) on brain microtubule protein assembly. Biochem Biophys Res Commun 124:850–856

Donato R, Isobe T, Okuyama T (1985) S-100 proteins and microtubules: analysis of the effects of rat brain S-100b and ox brain S-100ao, S-100a and S-100b on microtubule assembly – disassembly. FEBS Lett 186:65–69

Donner L, de Lanerolle P, Costa J (1983) Immunoreactivity of paraffin-embedded normal tissue and mesenchymal tumors for smooth muscle myosin. Am J Clin Pathol 80:677–681

Drenckhahn D, Steffans R, Groeschel-Steward U (1980) Immunocytochemical localization of myosin in the brush border region of the intestinal epithelium. Cell Tissue Res 205:163–166

Drysdale J (1979) Ferritin in cancer: biochemical perspectives. In: Carcinoembryonic proteins: Chemistry, biology, clinical application, vol I, pp 557–588

Endo T, Hidaka H (1983) Effect of S-100 protein on microtubule assembly – disassembly. FEBS Lett 161:235–238

Eusebi V, Bondi A, Rosai J (1984) Immunohistochemical localization of myoglobin in non muscular cells. Am J Surg Pathol 8:51–55

Eusebi V, Ceccarelli C, Gorza L, Schiaffino S, Bussolati G (1986) Immunocytochemistry of rhabdamyosarcoma. The use of four different markers. Am J Surg Pathol 10:293–299

Feigl W, Denk H, Davidovits A, Holzner JH (1976) Blood group isoantigens in human benign and malignant tumors. Virchows Arch (Pathol Anat) 370:323–332

Ferri GL, Probert L, Cocchia D, Michetti F, Marangos PJ, Polak JM (1982) Evidence for the presence of S-100 protein in the glial component of the human enteric nervous system. Nature 297:409–410

Flanagan P, Lionetti F (1955) Lysozyme distribution in blood. Blood 10:497–501

Fleming A (1922) On a remarkable bacteriolytic element found in tissues and secretions. Proc R Soc Lond (Biol) 93:306–317

Fletcher CD, Evans BJ, MacArtney JC, Smith N, Wilson Jones E, McKee PH (1985) Dermatofibrosarcoma protuberans: A clinicopathological and immunohistochemical study with a review of the literature. Histopathol 9:921–938

Flotte TJ, Hatcher VA, Friedman-Kien AE (1984) Factor VIII-related antigen in Kaposi's sarcoma in young homosexual men. Arch Dermatol 120:180–182

Fu Y-S, Gabbiani G, Kaye GI, Lattes R (1975) Malignant soft tissue tumors of probable histiocytic origin (malignant fibrous histiocytomas): General considerations and electron microscopic and tissue culture studies. Cancer 35:176–198

Gabbiani G, Kapanci Y, Barazzone P, Franke WW (1981) Immunochemical identification of intermediate-sized filaments in human neoplastic cells: A diagnostic aid for the surgical pathologist. Am J Pathol 104:206–216

Gaynor R, Irie R, Morton D, Herschman HR (1980) S-100 protein is present in cultured human malignant melanomas. Nature 286:400–401

Gaynor R, Irie R, Morton D, Herschman HR, Jones P, Cochran A (1981) S-100 protein: a marker for human malignant melanomas. Lancet 18:869–871

Giangaspero F, Zanetti G, Mancini A, Rosito P (1981) Sarcomatous variant of Wilms' tumor. A light microscopical and immunohistochemical study of four cases. Tumori 67:367–373

Giddings JC (1982) The analysis of factor VIII. Med Lab Sci 39:339–343

Guarda LG, Silva EG, Ordoñez NG, Smith JL (1981) Factor VIII in Kaposi's sarcoma. Am J Clin Pathol 76:197–200

Gupta PK, Frost JK, Geddes S, Aracil B, Davidovski F (1979) Morphological identification of α_1-antitrypsin in pulmonary macrophages. Hum Pathol 10:345–347

Haglid K, Carlsson CA, Stavrou D (1973) An immunological study of human brain tumors concerning the brain specific proteins S-100 and 14.3.2. Acta Neuropathol (Berl) 24:187–196

Hashimoto H, Daimaru Y, Enjoji M (1984) S-100 protein distribution in liposarcoma. An immunoperoxidase study with special reference to the distinction of liposarcoma from myxoid malignant fibrous histiocytoma. Virchows Arch (Pathol Anat) 405:1–10

Hayashi Y, Aoki Y, Eto R, Tokuoka S (1984) Findings of myoepithelial cells in human breast cancer. Ultrastructural and immunohistochemical study by means of anti-myosin antibody. Acta Pathol Jpn 34:537–552

Herrera GA, Pinto de Moraes H (1984) Neurogenic sarcomas in patients with neurofibromatosis (von Recklinghausen's disease). Virchows Arch (Pathol Anat) 403:361–376

Hiramoto R, Jurandowski J, Bernecky J, Pressman D (1961) Immunochemical differentiation of rhadomyosarcomas. Cancer Res 21:383–389

Hosaka M, Murase N, Orito T, Mori M (1985) Immunohistochemical evaluation of factor VIII related antigen, filament proteins and lectin binding in haemangiomas. Virchows Arch (Pathol Anat) 407:237–247

Hoyer LV, de los Santos RP, Hoyer JR (1973) Antihemophilic factor antigen: localization in endothelial cells by immunofluorescent microscopy. J Clin Invest 52:2737–2744

Iacopetta BJ, Morgan EH, Yeoh GCT (1983) Receptor-mediated endocytosis of transferrin by developing erythroid cells from the fetal rat liver. J Histochem Cytochem 31:336–344

Imoto M, Nishimura D, Fukuda Y, Sugiyama K, Kumada T (1985) Immunohistochemical detection of alpha-fetoprotein, carcinoembryonic antigen, and ferritin in formalin-paraffin sections from hepatocellular carcinoma. Am J Gastroenterol 80:902–906

Isaacson P, Wright DH (1978) Malignant histiocytosis of the intestine. Its relationship to malabsorption and ulcerative jejunitis. Hum Pathol 9:661–677

Isaacson P, Jones DB, Judd MA (1979) $Alpha_1$-antitrypsin in human macrophages. Lancet II:964–965

Isaacson P, Jones DB, Millward-Sadler GH, Judd MA, Payne S (1981) $Alpha_1$-antitrypsin in human macrophages. J Clin Pathol 34:982–989

Isaacson PG, Jones DB, Sworn MJ, Wright DH (1982) Malignant histiocytosis of the intestine: report of three cases with immunological and cytochemical analysis. J Clin Pathol 35:510–516

Isobe T, Tsugita T, Okuyama T (1978) The amino acid sequence and the subunit structure of bovine brain S-100 protein (PAP I-b). J Neurochem 30:921–923

Isobe T, Ishioka N, Okuyama T (1981) Structural relation of two S-100 proteins in bovine brain: subunit composition of S-100a protein. Eur J Biochem 115:469–474

Isobe T, Ishioka N, Masuda T, Takahashi Y, Ganno S, Okuyama T (1983) A rapid separation of S-100 subunits by high performance liquid chromatography: the subunit composition of S-100 proteins. Biochem Int 6:419–426

Jacobsen GK, Jacobsen M (1983) Ferritin (FER) in testicular germ cell tumors. An immunohistochemical study. Acta Pathol Microbiol Immunol Scand (A) 91:177–181

Jaffe EA (1977) Endothelial cells and the biology of factor VIII. N Engl J Med 296:377–383

Jong AS, van Vark M, Albus-Lutter CE, van Raamsdonk W, Voute PA (1984) Myosin and myoglobin as tumor markers in the diagnosis of rhabdomyosarcoma. A comparative study. Am J Surg Pathol 8:521–528

Kagawa N, Sano T, Inaba H, Mori K, Hizawa K (1983) Immunohistochemistry of myoglobin in rhabdomyosarcomas. Acta Pathol Jpn 33:515–522

Kagen L (1973) Myoglobin. Biochemical, physiological and clinical aspects. Columbia University Press, New York

Kahn HJ, Marks A, Thom H, Baumal R (1983a) Role of antibody to S-100 protein in diagnostic pathology. Am J Clin Pathol 79:341–347

Kahn HJ, Yeger H, Kassin O, Jørgensen AM, McLennan D, Baumal R, Smith CR, Phillips MJ (1983b) Immunohistochemical and electron microscopic assessment of childhood rhabdomyosarcoma. Increased accuracy of diagnosis over routine histologic methods. Cancer 51:1897–1903

Katus HA, Hurrell JG, Matsueda GR, Ehrlich P, Zurawski VR, Khaw BA, Haber E (1982) Increased specificity in cardiac-myosin radioimmunoassay utilizing two monoclonal antibodies in double sandwich assay. Mol Immunol 19:451–455

Katus HA, Yasuada I, Gold HK, Leinbach RC, Stauss HW, Waksmonski C, Haber E, Khaw BA (1984) Diagnosis of acute myocardial infarction by detection of circulating cardiac myosin light chains. Am J Cardiol 54:964–970

Khaw BA, Strauss HW, Carvallo A, Locke E, Goel HK, Haber E (1982) Technetium 99m labelling of antibodies to cardiac myosin Fab and to human fibrinogen. J Nucl Med 23:1011–1019

Khaw BA, Strauss HW, Pohost GM, Fallon JT, Katus HA, Haber E (1983) Relation of immediate and delayed thallium-201 distribution to localization of iodine-125 antimyosin antibody in acute experimental myocardial infarction. Am J Cardiol 51:1428–1432

Kimura T, Budka H, Soler-Federspiel S (1986) An immunohistochemical comparison of the glia-associated proteins glial fibrillary acidic protein (GFAP) and S-100 protein (S-100 P) in human brain tumors. Clin Neuropathol 5:21–27

Kindblom L (1982) Factor VIII related antigen and mast cells. Acta Path Microbiol Immunol Scand 90A:437–441

Kindblom LG, Jacobsen GK, Jacobsen M (1982) Immunohistochemical investigations of tumors of supposed fibroblastic-histiocytic origin. Hum Pathol 13:834–840

Kindblom LG, Lodding P, Angervall L (1983) Clear-cell sarcoma of tendons and aponeuroses. An immunohistochemical and electron microscopic analysis indicating neural crest origin. Virchows Arch (Pathol Anat) 401:109–128

Kindblom LG, Lodding P, Rosengren L, Baudier J, Haglid K (1984) S-100 protein in melanocyte tumors. Acta Pathol Microbiol Immunol Scand 92A:219–230

Klockars M, Reitamo S (1974) Tissue distribution of lysozyme in man. J Histochem Cytochem 23:932–940

Koh SJ, Johnson WW (1980) Antimyosin and antirhabdomyoblast sera: Their use for the diagnosis of childhood rhabdomyosarcoma. Arch Pathol Lab Med 104:118–122

Kondo H, Iwanaga T, Nakajima T (1982) An immunocytochemical study of the localization of neuron-specific enolase and S-100 protein in the carotid body of rats. Cell Tissue Res 227:291–295

Laufberger V (1937) Sur la cristallisation de la ferritin. Bull Soc Chim Biol 19:1575–1583

Lemanski LF, Tu ZH (1983) Immunofluorescent studies for myosin, actin, tropomyosin and α-actinin in cultured cardiomyopathic hamster heart cells. Dev Biol 97:338–348

Ling L, Klein MJ, Sissons HA, Steiner GC (1986) Lysozyme and α_1-antitrypsin in giant-cell tumor of bone and in other lesions that contain giant cells. Arch Pathol Lab Med 110:713–718

Little D, Said JW, Siegel RJ, Fealy M, Fishbein MC (1986) Endothelial cell markers in vascular neoplasms: an immunohistochemical study comparing factor VIII-related antigen, blood group specific antigens, 6-keto-PGF1 alpha, and Ulex europaeus I lectin. J Pathol 149:89–95

Longtine JA, Pinkus GS, Fujiwara K, Corson JM (1985) Immunohistochemical localization of smooth muscle myosin in normal human tissue. J Histochem Cytochem 33:179–184

Ludwin SK, Kosek JC, Eng LF (1981) The topographical distribution of S-100 and GFA proteins in the adult rat brain: an immunohistochemical study using horseradish peroxidase-labelled antibodies. J Comp Neurol 165:197–208

Macartney JC, Trevithick MA, Kricka L, Curran RC (1979) Identification of myosin in human epithelial cancers with immunofluorescence. Lab Invest 41:437–445

Mason DY, Taylor CR (1975) The distribution of muramidase (lysozyme) in human tissues. J Clin Pathol 28:124–132
Matsunou H, Shimoda T, Kakimoto S, Yamashita H, Ishikawa E, Mukai M (1985) Histopathologic and immunohistochemical study of malignant tumors of peripheral nerve sheath (malignant schwannoma). Cancer 56:2269–2279
Matus A, Mughal S (1975) Immunohistochemical localization of S-100 protein in brain. Nature 258:764–768
McComb RD, Jones TR, Pizzo SV, Bigner DD (1982) Specificity and sensitivity of immunohistochemical detection of factor VIII/von Willebrand factor antigen in formalin-fixed paraffin-embedded tissue. J Histochem Cytochem 30:371–377
Mechtersheimer G, Brandt I, Möller P (1986) Differences in marker expression among branched histiocytic cells in the T-areas of the lymphoreticular system and among their epidermis- and mucosa-associated equivalents. Cell Tissue Res 244:471–478
Meister P, Goessner W (1983) Factor VIII staining of fibrous haematopoietic tumors. Hum Pathol 14:924
Meister P, Huhn D, Nathrath W (1980) Malignant histiocytosis. Immunohistological characterization on paraffin embedded tissue. Virchows Arch (Pathol Anat) 385:233–246
Mendelson G, Eggleston JC, Mann RB (1980) Relationship of lysozyme (muramidase) to histiocytic differentiation in malignant histiocytosis. An immunohistochemical study. Cancer 45:273–279
Michetti F, Dell' Anna E, Tiberio G, Cocchia D (1983) Immunochemical and immunocytochemical study of S-100 protein in rat adipocytes. Brain Res 262:352–356
Miettinen M, Lehto V-P, Badley RA, Virtanen I (1982) Alveolar rhabdomyosarcoma: demonstration of the muscle type of intermediate filament protein, desmin, as a diagnostic aid. Am J Pathol 108:246–251
Millard PR, Heryet AR (1985) An immunohistological study of factor VIII related antigen and Kaposi's sarcoma using polyclonal and monoclonal antibodies. J Pathol 146:31–38
Modlin RL, Hofman FM, Kempf RA, Taylor CR, Conant MA, Rea TH (1983) Kaposi's sarcoma in homosexual men: an immunohistochemical study. J Am Acad Dermatol 8:620–627
Molin SO, Rosengren L, Baudier J, Hamberger A, Haglid K (1985) S-100 alpha-like immunoreactivity in tubules of rat kidney. The clue to the function of a "brain-specific" protein. J Histochem Cytochem 33:367–310
Möller P, Lennert K (1984) On the angiostructure of lymph nodes in Hodgkin's disase. An immunohistochemical study using the lectin I of Ulex europaeus as endothelial marker. Virchows Arch (Pathol Anat) 403:257–270
Möller P, Wirbel R, Hofmann W, Schwechheimer K (1984) Lymphoepithelial carcinoma (Schmincke type) as a derivate of the tonsillar crypt epithelium. Virchows Arch (Pathol Anat) 405:85–93
Moller M, Ingrid A, Bock E (1978) Immunohistochemical demonstration of S-100 protein and GFA protein in interstitial cells of rat pineal gland. Brain Res 140:1–12
Moore BW (1965) A soluble protein characteristic of the nervous system. Biochem Biophys Res Comm 19:739–744
Morse JO (1978) Alpha$_1$-antitrypsin deficiency. N Engl J Med 299:1045–1048
Motoi M, Helbron D, Kaiserling E, Lennert K (1980) Eosinophilic granuloma of lymph nodes – a variant of histiocytosis X. Histopathol 4:585–606
Mukai M (1983) Immunohistochemical localization of S-100 protein and peripheral nerve myelin proteins (P2 protein, PO protein) in granular cell tumors. Am J Pathol 112:139–146
Mukai K, Rosai J, Hallaway BE (1979) Localization of myoglobin in normal and neoplastic human skeletal muscle cells using an immunoperoxidase method. Am J Surg Pathol 3:373–376
Mukai K, Varela-Duran J, Nochomovitz LE (1980a) The rhabdomyoblast in mixed muellerian tumors of the uterus and ovary. An immunohistochemical study of myoglobin in 25 cases. Am J Clin Pathol 74:101–104
Mukai K, Rosai J, Burgdorf WHC (1980b) Localization of factor VIII-related antigen in vascular endothelial cells using an immunoperoxidase method. Am J Surg Pathol 4:273–276
Nadji M, Morales AR, Ziegels J, Penneys NS (1980) Kaposi's sarcoma: Immunohistochemical evidence for an endothelial origin. J Invest Dermatopathol 74:254–255
Nakajima T, Yamaguchi H, Takahashi K (1980) S-100 protein in folliculostellate cells of the rat pituitary anterior lobe. Brain Res 191:523–531
Nakajima T, Watanabe S, Yuichi S, Kameye T, Hirota T, Shimosato Y (1982) An immunoperoxidase study of S-100 protein distributed in normal and neoplastic tissues. Am J Surg Pathol 6:715–727

Nakamura Y, Becker LE, Marks A (1983) Distribution of immunoreactive S-100 protein in pediatric brain tumors. J Neuropathol Exp Neurol 42:136–145

Nakanishi S, Shinomiya S, Sano T, Hizawa K (1982) Immunohistochemical observation of intracytoplasmic lysozyme in proliferative and neoplastic fibrohistiocytic lesions. Acta Pathol Jpn 32:949–959

Nakazato Y, Ishizeki J, Takahashi K, Yamaguchi H (1982) Immunohistochemical localization of S-100 protein in granular cell myoblastomas. Cancer 49:1624–1628

Nakazato Y, Ishida Y, Takahashi K, Suzuki K (1985) Immunohistochemical distribution of S-100 protein and glial fibrillary acidic protein in normal and neoplastic salivary glands. Virchows Arch (Pathol Anat) 405:299–310

Nathrath WB, Remberger K (1986) Immunohistochemical study of granular cell tumors. Demonstration of neurone specific enolase, S-100 protein, laminin and alpha-1-antichymotrypsin. Virchows Arch (Pathol Anat) 408:421–434

Osserman EF (1975) Lysozyme. N Engl J Med 292:424–425

Papadimitriou CS, Stein H, Lennert K (1978) The complexity of immunohistochemical staining pattern of Hodgkin and Sternberg-Reed cells – demonstration of immunoglobulin, albumin, $alpha_1$-antichymotrypsin and lysozyme. Int J Cancer 21:531–541

Parmley RT, Ostroy F, Gams RA, de Lucas L (1979) Ferrocyanide staining of transferrin and ferritin-conjugated antibody to transferrin. J Histochem Cytochem 27:681–685

Payne SV, Wright DH, Jones KJM, Judd MA (1982) Macrophage origin of Reed-Sternberg cells: an immunohistochemical study. J Clin Pathol 35:159–166

Pinkus GS, Warhol MJ, O'Connor EM, Etheridge CL, Fujiwara K (1986) Immunohistochemical localization of smooth muscle myosin in human spleen, lymph node, and other lymphoid tissues. Am J Pathol 123:440–453

Piovella F, Nalli G, Malamani GD, Majdino I, Frassoni F, Sitar GM, Ruggeri A, Dell Orbo C, Asceri E (1978) The ultrastructural localization of Factor VIII antigen in human platelets, megakaryocytes and endothelial cells utilizing a ferritin-labelled antibody. Br J Haematol 39:209–213

Poppema S, Elema JD, Halie MR (1978) The significance of intracytoplasmic proteins in Reed-Sternberg cells. Cancer 42:1793–1803

Rode J, Dhillon AP (1984) Neurone specific enolase and S-100 protein as possible prognostic indicators in melanoma. Histopathology 8:1041–1052

Roholl PJ, Kleyne J, Elbers H, Van der Vegt MC, Albus-Lutter C, Van Unnik JA (1985) Characterization of tumor cells in malignant fibrous histiocytomas and other soft tissue tumors in comparison with malignant histiocytomas. I. Immunohistochemical study on paraffin sections. J Pathol 147:87–95

Royds JA, Variend S, Timperley WR, Taylor CB (1985) Comparison of beta enolase and myoglobin as histological markers of rhabdomyosarcoma. J Clin Pathol 38:1258–1260

Saku T, Tsuda N, Anami M, Okabe H (1985) Smooth and skeletal muscle myosin in spindle cell tumors of soft tissue. An immunohistochemical study. Acta Pathol Jpn 35:125–136

Schiaffino S, Gorza L, Sartore S, Saggin L, Carli M (1986) Embryonic myosin heavy chain as a differentiation marker of developing human skeletal muscle and rhabdomyosarcoma. A monoclonal antibody study. Exp Cell Res 163:211–220

Schmidt D, Harms D, Burach St (1985) Malignant peripheral neuroectodermal tumours of childhood and adolescence. Virchows Arch (Pathol Anat) 404:351–365

Schneider C, Owen MJ, Banville D, Williams JG (1984) Primary structure of human transferrin receptor deduced from the mRNA sequence. Nature 311:675–678

Scupham R, Gilbert EF, Wilde J, Wiedrich TA (1986) Immunohistochemical studies of rhabdomyosarcoma. Arch Pathol Lab Med 110:818–821

Sehested M, Hou-Jensen K (1981) Factor VIII related antigen as an endothelial cell marker in benign and malignant diseases. Virchows Arch (Pathol Anat) 391:217–225

Sehested M, Barfoed C, Krogdahl A, Bretlau P (1985) Immunohistochemical investigation of lysozyme, lactoferrin, alpha1-antitrypsin, alpha1-antichymotrypsin and ferritin in parotid gland tumors. J Oral Pathol 14:459–465

Seo S, Min KW, Mirkin D (1986) Juvenile xanthogranuloma. Ultrastructural and immunocytochemical studies. Arch Pathol Lab Med 110:911–915

Shimada H, Aoyama C, Chiba T, Newton WA (1985) Prognostic subgroups for undifferentiated neuroblastoma: Immunohistochemical study with anti-S-100 protein antibody. Hum Pathol 16:471–476

Sonoda T, Hashimoto H, Enjoji M (1985) Juvenile xanthogranuloma: Clinicopathologic analysis and immunohistochemical study of 57 patients. Cancer 56:2280–2286

Springall DR, Gu J, Cocchia D, Michetti F, Levene A, Levene MM, Marangos PJ, Bloom SR, Polak JM (1983) The value of S-100 immunostaining as a diagnostic tool in human malignant melanomas. Virchows Arch (Pathol Anat) 400:331–343

Stefansson K, Wollmann R, Jerkovic M (1982a) S-100 protein in soft tissue tumors derived from Schwann cells and melanocytes. Am J Pathol 106:261–268

Stefansson K, Wollmann RL, Moore BW (1982b) Distribution of S-100 protein outside the central nervous system. Brain Res 234:309–317

Stefansson K, Wollmann RL, Moore BW, Arnason BGW (1982c) S-100 protein in human chondrocytes. Nature 295:63–64

Stein H, Lennert K, Feller AC, Mason DY (1984) Immunohistological analysis of human lymphoma: correlation of histological and immunological categories. Adv Cancer Res 42:67–73

Strauchen JA, Dimitriu-Bona A (1986) Malignant fibrous histiocytoma. Expression of monocyte/macrophage differentiation antigens detected with monoclonal antibodies. Am J Pathol 124:303–309

Strominger JL, Tipper DJ (1974) Structure of the bacterial walls: The lysozyme substrate. In: Osserman EF, Canfield RE, Beychok S (eds) Lysozyme. Academic Press, New York, pp 169–184

Syrovy I (1979) Polymorphism and specificity of myosin: A review. Int J Biochem 10:383–389

Takahashi K, Yamaguchi H, Ishizeki J, Nakajima T, Nakazato Y (1981) Immunohistochemical and immunoelectron microscopic localization of S-100 protein in the interdigitating reticulum cells of the human lymph node. Virchows Arch (Cell Pathol) 37:125–135

Taxy JB, Battifora H (1977) Malignant fibrous histiocytoma: An electron microscopic study. Cancer 40:254–267

Terenghi G, Polak JM, Ballestra J, Cocchia D, Michetti F, Dahl D, Marangos PJ, Garner A (1984) Immunocytochemistry of neuronal and glial markers in retinoblastomas. Virchows Arch (Pathol Anat) 405:61–73

Travis J, Garner D, Bowen J (1978) Human α_1-antichymotrypsin: purification and properties. Biochemistry 17:5647–5651

Tsokos M, Triche TJ (1986) Immunocytochemical and ultrastructural study of primitive "solid variant" rhabdomyosarcoma. Lab Invest 54:65A

Tsokos M, Howard R, Costa J (1983) Immunohistochemical study of alveolar and embryonal rhabdomyosarcoma. Lab Invest 48:148–155

Tufty RM, Kretsinger RH (1975) Troponin and parvalbumin calcium binding regions predicted in myosin light chain and T4 lysozyme. Science 187:167–169

Unsicker K, Drenckhahn D, Groeschel-Steward U (1978) Further immunofluorescence – microscopic evidence for myosin in various peripheral nerves. Cell Tissue Res 188:341–344

van Furth R, Kramps JA, Dieselhof-Den Dulk MMC (1983) Synthesis of α_1-antitrypsin by human monocytes. Clin Exp Immunol 51:551–557

Vassy J, Rissel M, Kraemer M, Foucrier J, Guillouzo A (1984) Ultrastructural indirect immunolocalization of transferrin in cultured hepatocytes permeabilized with saponin. J Histochem Cytochem 32:538–540

Vilpo JA, Klemi P, Lassila O, Fränki J, Salmi TT (1980) Cytological and functional characterization of three cases of malignant histiocytosis. Cancer 46:1795–1801

Watanabe S, Nakajima T, Shimosato Y, Sato Y, Ise T (1981) A case report of histiocytic medullary reticulosis defined as neoplasm of T-zone histiocytes. Jpn J Clin Oncol 11:411–418

Watanabe S, Nakajima T, Shimosato Y, Sato Y, Shimizu K (1983) Malignant histocytosis and Letterer-Siwe disease. Neoplasms of T-zone histiocyte with S100 protein. Cancer 51:1412–1424

Weber K, Groeschel-Steward U (1974) Antibody to myosin: The specific visiualization of myosin-containing filaments in non-muscle cells. Proc Natl Acad Sci USA 71:4561–4564

Weiss SW, Langloss JM, Enzinger GM (1983) Value of S-100 protein in the diagnosis of soft tissue tumors with particular reference to benign and malignant Schwann cell tumors. Lab Invest 49:299–308

Willingham MC, Pastan I (1985) Ultrastructural immunocytochemical localization of the transferrin receptor using a monoclonal antibody in human KB cells. J Histochem Cytochem 33:59–64

Wilson GB, Walker JH, Watkins JH, Wolgroch D (1980) Determination of subpopulations of leukocytes involved in the synthesis of α_1-antitrypsin in vitro. Proc Soc Exp Biol Med 164:105–114

Wirbel R, Möller P, Schwechheimer K (1984) Lectin-binding spectra in the hyperplastic human tonsil. Effect of formalin fixation and paraffin embedding on lectin affinity of tissue components. Histochem 81:551–560

Wood GS, Beckstead JH, Turner RR, Hendrickson MR, Kempson RL, Warnke RA (1985) Malignant fibrous histiocytoma cells do not express the antigenic or enzyme histiochemical features of cells of monocyte/macrophage lineage. Lab Invest 52:78A

Yamaguchi H (1980) Studies on the immunohistochemical localization of S-100 and glial fibrillary acidic protein in the rat nervous system and human brain tumors. Brain Nerve (Tokyo) 32:287–303

Zomzely-Neurath CE, Walker WA (1980) Nervous system-specific proteins: 14.3.2 protein, neuron-specific enolase, and S-100 protein. In: Bradshaw RA, Schneider DM (eds) Proteins of the nervous system, 2nd edn. Raven Press, New York, pp 30–52

Proteoglycans and the Intercellular Tumor Matrix

R.V. Iozzo

1 Introduction

The intercellular matrix is a highly organized, dynamic system that dictates the overall shape and structure of organs. The spatial and chemical signals which are present in the extracellular matrix modulate the phenotypic expression of epithelial and mesenchymal cells. In turn, these cells regulate the chemistry and structural organization of the extracellular matrix. This active interplay between cells and their products leads to the formation of a highly complex environment in which normal and neoplastic cells can thrive and proliferate. Proteoglycans are suitable candidates for mediating this constant exchange of information. For instance, they influence the diffusion of molecules across tissues, the migration of cells along defined pathways, the surface properties of cells, and the structure of biological filters. These effects are primarily mediated by the polyanionic nature of the proteoglycan, by their expanded configuration in tissues and body fluids, and by their ability to interact with a variety of important matrix macromolecules. They are indeed constituents of both the stromal matrix and the cell surface (Fig. 1), and their biosynthesis, secretion and metabolism are highly regulated by intrinsic and extrinsic signals. These molecules are composed of a protein backbone to which a number of glycosaminoglycan chains and oligosaccharides are covalently attached, much like the branches to a tree. The proteoglycan-rich intercellular matrix can thus be conceived as a microcosmic forest, the properties of which depend primarily on the number, size and structure of the various glycosaminoglycan chains and their complex intermolecular affiliations. This "forest" is in a state of continuous remodeling particularly during embryogenesis and cancer development, and is finely regulated by a combination of biosynthetic and catabolic events. Thus, it would be easier to imagine the extracellular matrix as exerting

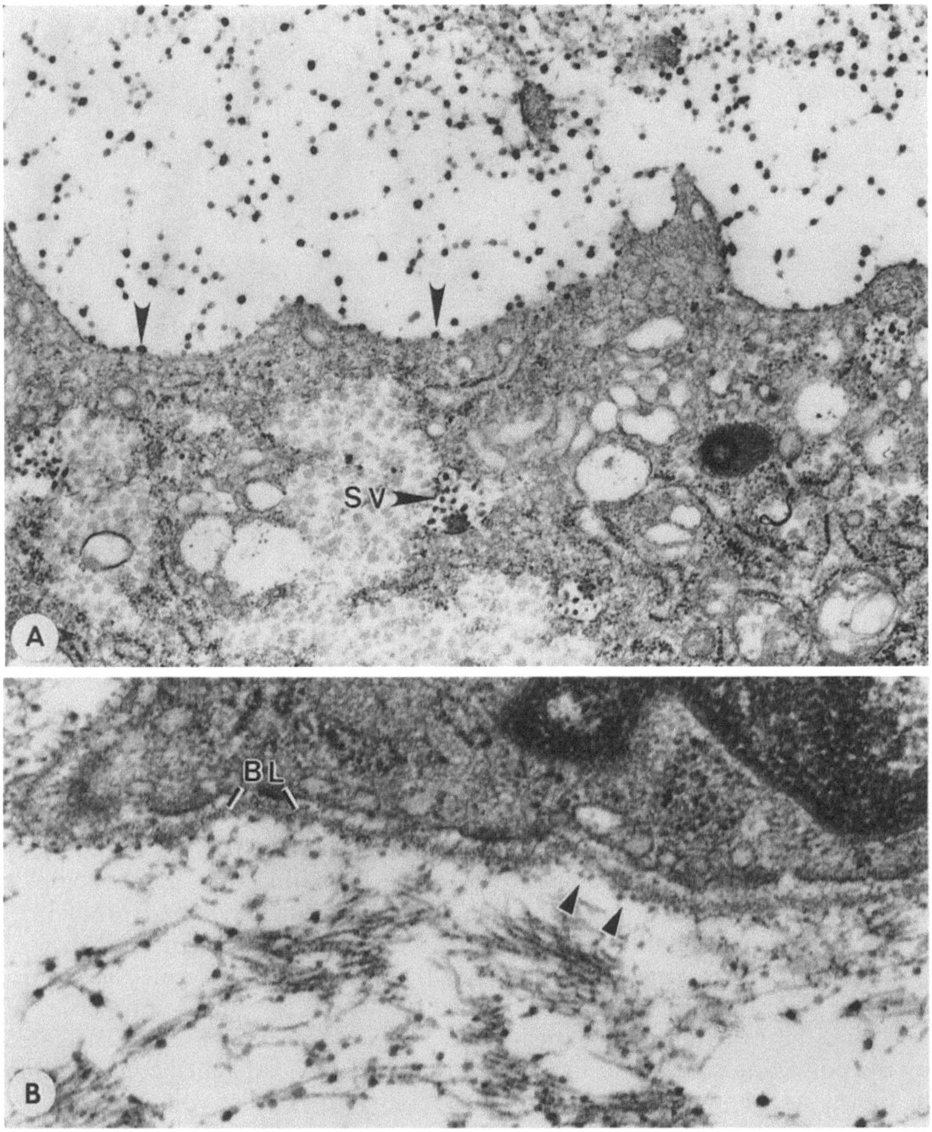

Fig. 1 A, B. Electron micrographs of human chondrosarcoma fixed and processed in the presence of ruthenium red, a cationic dye that retains proteoglycans. **A** shows portion of a chondrosarcoma cell containing numerous secretory vesicles (*SV*) filled with proteoglycans which appear as round, electron dense particles. Note the association of the proteoglycans with the cells surface (*arrowheads*) and the presence of the same particles dispersed throughout the extracellular matrix. **B** shows the association of proteoglycans (*arrowheads*) with the basal lamina (*BL*) of an endothelial cell and with fibrillar matrix constituents. **A** ×23000; **B** ×34000

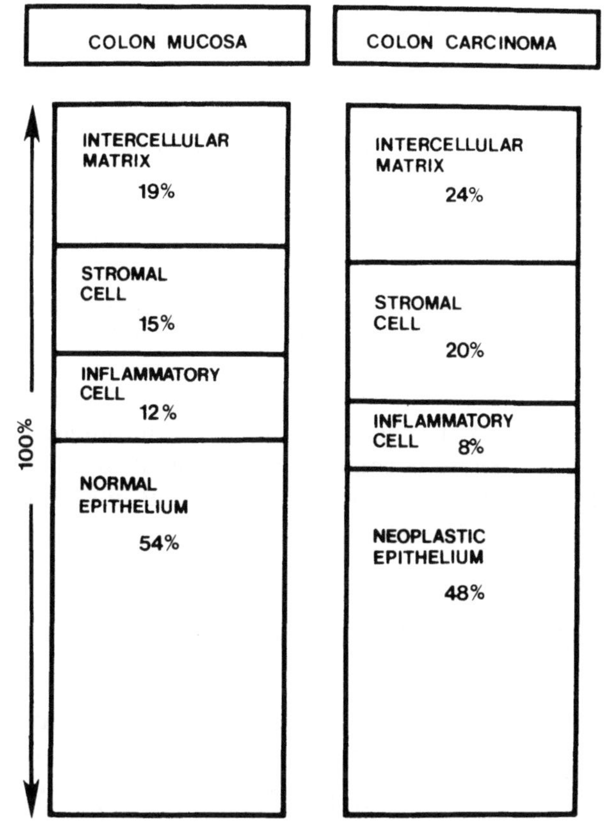

Fig. 2. Stereologic analysis of human colon mucosa and colon carcinoma. The data represent the summary of seven different cases (Iozzo et al. 1982) and show that about half of the volume in both the normal and neoplastic tissue is composed of stromal tissue and intercellular matrix rich in proteoglycans and collagen. This figure stresses the concept that in certain human malignancies, particularly those of epithelial origin, the associated connective tissue stroma occupies a significant proportion of the "tumor"

an active role, permissive or restrictive, on neoplastic growth and invasion, a far cry from the traditional notion of its being an amorphous ground substance.

This chapter will briefly review the alterations of proteoglycans and glycosaminoglycans in cancer, and it will describe some of the mechanisms through which neoplastic cells modulate their extracellular environment. For extended reviews on proteoglycan structure and function please see previous published work (Hascall and Hascall 1981; Heinegård and Paulsson 1984; Iozzo 1985a).

2 Proteoglycans and Neoplasia

If one analyzes quantitatively the various compartments of human colon carcinoma, one of the most common malignant neoplasms in the Western world, one finds that surprisingly only about half of the tissue is composed of neoplastic cells (Fig. 2). The other half is composed of an acellular matrix, rich in proteoglycans, and a cellular compartment including fibroblasts, endothelial and smooth muscle cells and a variety of immunocompetent cells. It follows that each component must contribute, in either a negative or positive way, to the growth and progression of the neoplastic cells. Two fundamental questions arise from the above considerations: a) is the tumor connective tissue a specific response of the host, and b) how is this response regulated? In answer to the first question, indirect evidence suggests that this response is specific, since it can be produced in vitro by factors released by the tumor cells (Iozzo 1985b) and it is found only in certain tumors, particularly those of epithelial origin. In response to the second question, current research (Liotta et al. 1983; Iozzo 1984a, 1985a; Wewer et al. 1986) indicates that neoplastic cells modulate their extracellular milieu by the following three, often interrelated, mechanisms: a) abnormal production of extracellular matrix by tumor cells, b) enhanced destruction of surrounding matrix, and c) neoplastic induction of matrix biosynthesis. These three mechanisms can be present in the same tumor, or in different locations of the various neoplasms where one could be predominant over the others. The significance of these basic processes and the implications in cancer growth and metastasis is the focus of the next sections.

3 Altered Proteoglycan and Glycosaminoglycan Levels in Tumors

It has been frequently observed that tumor cells grow and move within a hydrated interstitial space, analogous to the migrating cells during embryogenesis (Toole 1981). The ultrastructural appearance of the proteoglycans in these matrix spaces varies widely in size, arrangement and degree of packing (Fig. 3). Furthermore, these spaces are often different from normal tissues in that they contain abnormal amounts of specific proteoglycan or glycosaminoglycan (Table 1). For example, elevated amounts of hyaluronic acid in pleural effusion of mesotheliomas have been known for about fifty years (Meyer and Chaffee 1939), and detection of this glycosaminoglycan has been a useful marker in establishing the diagnosis of mesothelioma (Thompson et al. 1969). Furthermore, proteolytic extracts of human mesothelioma have been shown to contain elevated concentrations of hyaluronic acid (Waxler et al. 1979). However, the

Table 1. Reported proteoglycan changes in human tumors[a]

Mesothelioma	1. elevated hyaluronic acid in tissue and effusions 2. elevated chondroitin sulfate in tissue
Wilms' tumor	elevated hyaluronic acid in tissue, blood and urine
Chondrosarcoma	decrease in keratan sulfate content and chain length
Colon carcinoma	elevated chondroitin sulfate proteoglycan, with smaller chains
Hepatoma	abnormal levels of heparan sulfate with lower degree of sulfation
Plasmocytoma	abnormal circulating levels of heparan sulfate with anticoagulant activity

[a] Additional information about each individual abnormality and references are provided in the text

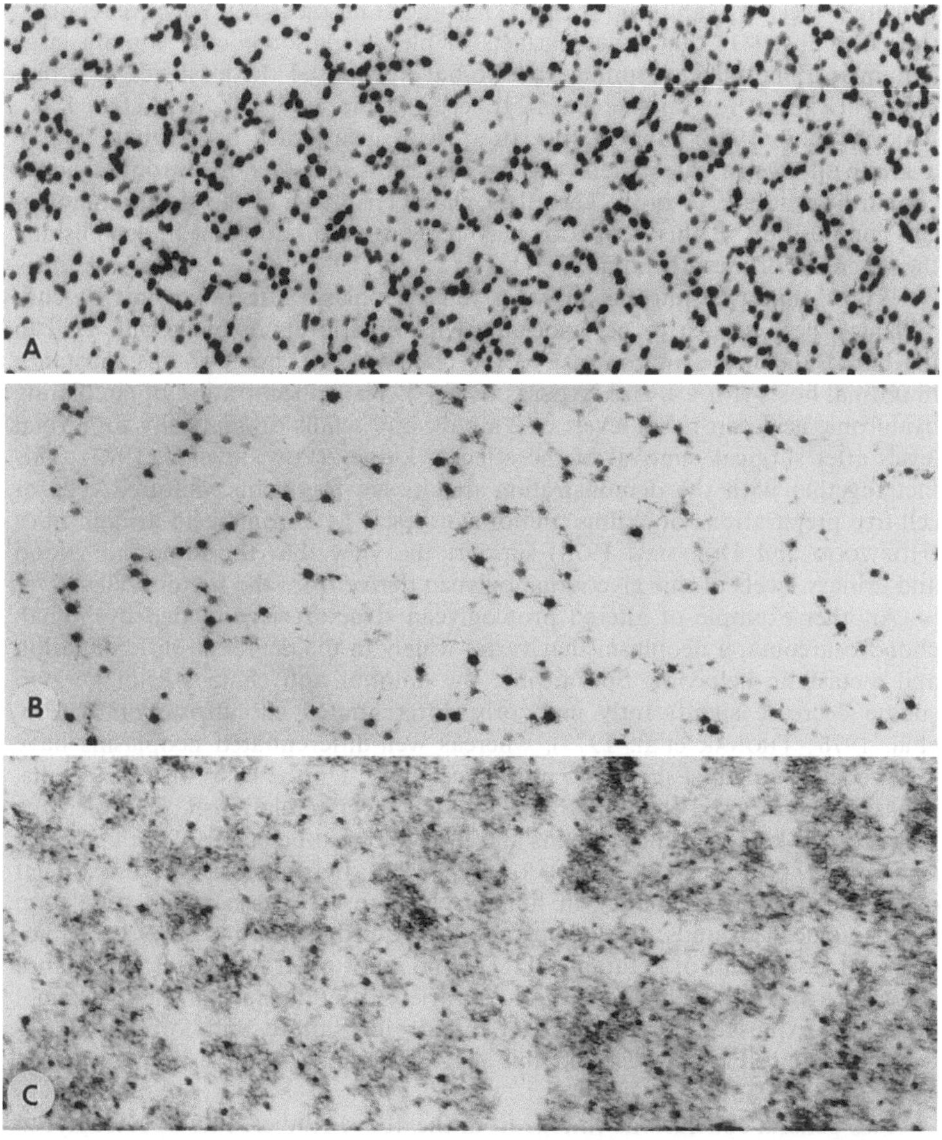

Fig. 3. Electron micrographs of ruthenium red-stained extracellular matrix from human chondrosarcoma (**A**), normal human colon (**B**) and human colon carcinoma (**C**). Notice the marked difference in the density and size of the proteoglycan granules among these three conditions. The three micrographs were taken at the same magnification of × 36000

specificity of hyaluronic acid as a marker for mesothelioma is questioned by the following contrasting reports: a) there are cases of mesothelioma that apparently lack the high levels of hyaluronic acid (Thompson et al. 1969); b) neoplastic tissues other than mesothelioma may express high levels of this glycosaminoglycan (Chiu et al. 1984); c) pleural effusions unrelated to mesothelioma can contain high levels of hyaluronic acid (Castor and Naylor 1977). We have reported (Iozzo et al. 1981) that tissue extracts of pleural mesothelioma contain large amounts of chondroitin sulfate and we have proposed the existence of a biochemical variant for this neoplasm. Indeed, recent studies (Kawai et al. 1985; Nakano et al. 1986) support the latter view since it has been found that in a group of twelve cases there was an elevation in either hyaluronic acid or chondroitin sulfate, or both. Therefore, characterization of the glycosaminoglycan composition in this neoplasm may represent a useful tool in establishing the diagnosis of mesothelioma.

Wilms' tumor (nephroblastoma), a renal neoplasm affecting primarily children, has been shown to be associated with high levels of hyaluronic acid in the blood and urine, in contrast to the trace-amount levels of this substance in normal hosts (Morse and Nussbaum 1967). The concentration of circulating hyaluronic acid can reach levels of 2 mg/dl, but it falls dramatically to normal levels after surgical removal of the affected kidney (Powars et al. 1972). This fact together with the demonstration that tissue fragments, cultured cells or cell-free preparations of Wilms' tumor synthesize hyaluronic acid at high rates (Hopwood and Dorfman 1978) support the view that the abnormal blood and urinary levels of this glycosaminoglycan derive from the tumors cells.

Another example of altered proteoglycan structure is provided by human chondrosarcoma, a neoplasm that varies widely in the degree of differentiation and metastatic behavior. Specifically, the amount and chain size of keratan sulfate decrease significantly in poorly differentiated chondrosarcomas (Pal et al. 1978; Thonar et al. 1979), whereas well-differentiated neoplasms have a proteoglycan profile similar to normal cartilage. On the basis of these results, it has been proposed that there might be an inverse relationship between the length of the keratan sulfate chains and the biological behavior of the chondrosarcoma (Pal et al. 1978; Thonar et al. 1979). Inasmuch as the assessment of the degree of malignancy in human chondrosarcoma is based purely on morphological grounds, the use of alternative prognostic indicators, i.e. biochemical assessment of keratan sulfate, could be of great biological importance (Rosenberg et al. 1979). In addition, a variety of epithelial and mesenchymal neoplasms express abnormal levels of chondroitin sulfate, and correlations between tumor cell properties and this glycosaminoglycan have been proposed (Iozzo 1984, 1985a).

Two general conclusions emerge from the above studies. First, proteoglycan changes do indeed occur in human malignancy and may be closely related to some biological properties of the tumor-matrix complex. Second, it appears that a common alteration in proteoglycan structure is not universally shared by all the different neoplasms; rather, the changes summarized above appear to be tissue specific.

4 Abnormal Synthesis of Proteoglycans by Neoplastic Cells

A vast body of literature on the abnormal biosynthesis of proteoglycans and glycosaminoglycans has been accumulating over the past years. Indeed, in vitro studies of either virally or spontaneously transformed cells have elucidated several aspects of the biosynthetic pathways of proteoglycans as well as provided important clues as to what might occur in an in vivo situation. Since a comprehensive review of the literature falls outside the scope of this chapter, only a few examples will be described below.

One of the most consistent qualitative abnormalities in proteoglycan structure which is directly related to the transformed phenotype is a reduced degree of sulfation in the heparan sulfate molecule. Over a decade ago, UNDERHILL and KELLER (1975) reported this change in 3T3 cells transformed with simian virus 40. Since then, a number of investigators (WINTERBOURNE and MORA 1981; DAVID and VAN DEN BERGHE 1983; ROBINSON et al. 1984) have reported similar findings in different cell systems. Specifically, the degree of sulfation of the heparan sulfate chains is markedly reduced upon transformation, and the reduced sulfation occurs in the 0-sulfate groups (KELLER et al. 1980). The link between the neoplastic transformation and these structural abnormalities is supported by the findings that cells selected for high tumorigenicity also produce heparan sulfate with lower degree of sulfation (WINTERBOURNE and MORA 1981). Interestingly, in vivo studies of human hepatoma (NAKAMURA and KOJIMA 1981) have shown the presence of undersulfated heparan sulfate, a finding also reported in ascites hepatoma (NAKAMURA et al. 1978; HURST et al. 1981) or hepatoma cells in vitro (ROBINSON et al. 1984). Finally, spontaneously transformed mouse mammary cells synthesize undersulfated basement membrane heparan sulfate (DAVID and VAN DEN BERGHE 1983). Two major questions arise from the above reports: a) what is the specific cause of this alteration, and (b) what are the possible functional implications of it? Although the specific cause of this molecular alteration is not clear, it is likely that transformation may generate some inhibitor or reduce the activity of sulfotransferases involved in the transfer of sulfate to the growing polymer or in the level of 3′-phosphoadenosine-5′-phosphosulfate (PAPS), the major sulfate donor in most of the cells (WINTERBOURNE and MORA 1981). Indeed, it has been demonstrated that the availability of PAPS plays a fundamental role in the sulfation of glycosaminoglycans (SUGAHARA et al. 1980). In regard to the functional consequences, suffice it to say that qualitative changes in the cell-associated heparan sulfate may have profound effects on the cellular ability to interact correctly with other cells or matrix constituents. In fact, this molecule is a highly charged polymer strategically located at the cell surface where most of the cellular interactions take place. Furthermore, the structural changes may have an effect on the molecular interactions normally occuring during the assembly of the basal lamina and may be directly implicated in the abnormal basal lamina produced by neoplastic cells (DAVID and VAN DEN BERGHE 1983). It is possible, therefore, that the lack of an ordered microenviroment around tumor cells could be essential for the full expression of the neoplastic phenotype (GALLAGER and HAMPSON 1984). If these reported alterations do indeed reflect the in vivo situation, these qualitative changes in molecular structure may then contribute to the abnormal growth behavior of the tumor cells.

Another example of tumor-associated abnormality in heparan sulfate proteoglycan has been reported occuring in patients with plasma cell disorders, i.e. multiple myeloma (Khoory et al. 1980; Palmer et al. 1984). In this case the abnormality appears to be a quantitative rather than a qualitative one; that is an increased level of circulating heparan sulfate with anticoagulant activity. In fact, functional coagulation studies showed that the circulating heparan sulfate had antithrombin III cofactor activity that could be neutralized by protamine or platelet factor 4 (Palmer et al. 1984). It is well established that the anticoagulant activity in heparin, a closely related molecule, depends on the negative charge density of sulfate and carboxyl groups, and on the amount of L-iduronic acid which are all present in heparan sulfate, though to a lesser extent than in heparin (Lindahl et al. 1986). Therefore, it is not surprising that abnormally high levels of circulating heparan sulfate may inhibit the coagulation cascade and induce fatal bleeding. Interestingly, a similar process has been recently described in an infant with acute monoblastic leukemia (Russel et al. 1984). Although the specific cellular source of this heparan sulfate has not been definitively established, it is likely that neoplastic cells do contribute in a significant way to its biosynthesis and release. Alternatively, it has been proposed (Russel et al. 1984) that the circulating heparan sulfate proteoglycan derives from the endothelial cells damaged by the chemotherapy. In any event, it is quite clear that overproduction and release of an otherwise normal heparan sulfate proteoglycan can have catastrophic consequences because of its potential anticoagulant activity.

It is apparent from the examples described above that the process of transformation markedly affects the mechanisms involved in the control of proteoglycan biosynthesis and secretion. Both qualitative and quantitative abnormalities may thus significantly affect the life of the host, and may contribute to the abnormal growth behavior or other oncogenic properties of the transformed cells.

5 Matrix Degradation by Neoplastic Cells

Malignant cells are confronted with several natural barriers during the invasion of adjacent structures and distal colonization of tissues (Woolley 1984). The extracellular matrix and the basement membranes represent the most important biological barrier and their active degradation is thought to facilitate invasive growth and metastatic potential (Liotta et al. 1983). These features are two of the most important prognostic indicators in cancer. There is evidence to indicate that malignant cells can digest matrix constituents either directly or indirectly by inducing the production and release of degradative enzymes in host mesenchymal cells (Liotta et al. 1982, 1983). The collagenases are probably the most important extracellular enzymes since they are capable of degrading cross-linked collagen fibers at neutral pH (Woolley 1984). The end result is either complete or partial collagenolysis with generation of small fragments which are eventually phagocyzed and degraded completely within lysosomes by the cathepsins, usually operating at a more acidic pH (Woolley 1984). Liotta et al. (1979) have identified a metalloproteinase from a metastatic tumor that preferentially digests type IV collagen, the major constituent of basement membranes (Kefalides et al. 1979). Subsequently, a correlation between the metastatic potential of malignant cells and the secretion of various collagenases has been demonstrated (LIOTTA et al. 1980; TARIN et al. 1982; TURPEENNIEMI-HUJANEN et al. 1985), thus supporting the concept that collagenolysis facilitates tumor cell invasion. It is notable that in the transformation of a benign tumor into a frankly invasive carcinoma of the breast, the immunoreactivity of the basement membrane progressively declines to eventually become undetectable at the invasive front and in the metastatic foci (BARSKY et al. 1983a). Concurrently, type IV collagenase immunoreactivity has been detected in invasive breast carcinoma (BARSKY et al. 1983b). Furthermore, immunohistochemical studies have shown the presence of collagenases in a variety of neoplasms, including metastatic melanoma and cancers of the skin, stomach and breast (for a review see WOOLLEY 1984). Another well studied enzyme that is associated with the malignant phenotype is cathepsin B, which is capable of digesting both collagen and proteoglycans. This enzyme has been localized at the invasive front of rabbit V2 carcinoma (GRAF et al. 1981), and increases several fold in breast carcinoma as compared to fibroadenoma (POOLE et al. 1978) and its activity has been correlated with enhanced metastatic potential (SLOANE et al. 1981).

Another important enzymatic step that may be operational in invasion is offered by endoglycosidase synthesized by melanoma (KRAMER et al. 1982) or lymphoma (VLODAVSKY et al. 1983) cells. This enzyme is capable of cleaving the heparan sulfate present in the extracellular matrix and basement membrane of endothelial cells (KRAMER et al. 1982; KRAMER and VOGEL 1984; NAKAJIMA et al. 1983, 1984). Specifically, B16 melanoma cells degrade the heparan sulfate chains into fragments approximately one-third of their original size. This enzymatic cleavage is observed only when the prelabeled matrix is incubated with the tumor cells, indicating that the tumor may express this trait in vivo. Additonally, melanoma cells with high propensity for lung metastasis degrade purified heparan sulfate from the mammalian lung at significantly higher rates than cells with lower metastatic potential (NAKAJIMA et al. 1983, 1984). Therefore, it is possible that the ability to extravasate and colonize a distant organ such as the lung may reside, at least in part, in the expression of this endoglucuronidase.

One of the most important concepts in modern cancer biology is the recognition that primary neoplasms are composed of heterogeneous cell populations (POSTE and FIDLER 1980). The expression of degradative enzymes is likely to be important for tumor progression, a process that is due to the sequential appearance of genetically altered subpopulations of cells with new characteristics within the neoplasms (NOWELL 1986). The expression of degradative enzymes would thus confer a selective advantage to a specific subpopulation. An open question, however, is whether these enzymatic activities reflect a continuous phenotypic trait or just a transient phase of enzyme production which depends on local tissue factors (WOOLLEY 1984) and cell interactions.

6 Neoplastic Induction of Matrix Biosynthesis

Desmoplasia, the excessive production of connective tissue around invasive neoplasms, has long been recognized as a phenomenon associated with certain types of human tumors. However, little has been done to answer some of the important questions about this response, such as: a) is there any specific qualitative or quantitative change in matrix constituents, b) what is the cellular source of matrix production, c) what is the role played by the cancer cells and d) is desmoplasia a specific response? In the past few years, we and others have attempted to answer some of these questions. For instance, Barsky et al. (1982) have demonstrated that human desmoplastic breast carcinomas exhibit a tenfold increase in the concentration of type V collagen, a product present only in minute amounts in normal breast or fibrocystic disease. Interestingly, an elevated number of host myofibroblasts or activated smooth muscle cells has been recognized in desmoplastic breast carcinomas (Seemayer et al. 1979) suggesting that these cells may be responsible for the excessive production of matrix (Schürch et al. 1981). Furthermore, we have shown (Iozzo et al. 1982) that human colon carcinoma tissue contains markedly elevated concentrations of chondroitin sulfate when compared with the normal counterpart, an observation also reported in a transplantable colon carcinoma of the rat (Isemura et al. 1982). In subsequent studies, (Iozzo and Wight 1982) we have demonstrated that human colon carcinoma in organ culture synthesizes primarily a small chondroitin sulfate proteoglycan, whereas normal colon synthesizes primarily heparan sulfate and only trace amounts of chondroitin sulfate proteoglycan. Several lines of evidence indicate that the stromal elements around the tumor represent the main cellular source for desmoplasia. First, the matrix around colon carcinoma incorporates the majority of the radiosulfate and retains affinity for alcian blue even at high magnesium chloride concentrations (Fig. 4), indicating that the stromal cells actively synthesize and accumulate proteoglycans. This phenotypic trait can also be retained in metastatic foci (Fig. 4e), suggesting that it may be a specific process closely related to the type of neoplastic cells. Second, human colon carcinoma cells in vitro synthesize a large heparan sulfate proteoglycan but no detectable amounts of chondroitin sulfate (Iozzo 1984b; Iozzo et al. 1986). Third, human colon fibroblasts are stimulated to synthesize chondroitin sulfate proteoglycan when cultured in the presence of medium conditioned by colon carcinoma cells. Partial characterization of the conditioned medium (Iozzo 1985b) showed that the stimulatory activity comprises a family of polypeptides that act on proteglycan metabolism without significantly stimulating cell replication. The work of Knudson et al. (1984) and Merrilees and Finlay (1985), supports these observations. They have demonstrated that the co-culture of carcinoma and mesenchymal cells or media conditioned by tumor cells can stimulate the production of both hyaluronic acid and sulfated glycosaminoglycans in normal fibroblasts. In agreement with our studies, Merrilees and Finlay (1985) have found that human colon carcinoma cells, when cultured alone, synthesize primarily heparan sulfate, but are capable of stimulating the synthesis of both hyaluronic acid and chondroitin sulfate in human fibroblasts. Further support for the neoplastic modulation of extracellular matrix is provided by the discovery that mammary carcinoma contains collagen-synthesis stimulating polypeptides (Bano et al. 1983). It has been proposed that these stimulatory activities may favor tumor growth, since proline analogues, which selectively block collagen deposition, can also block

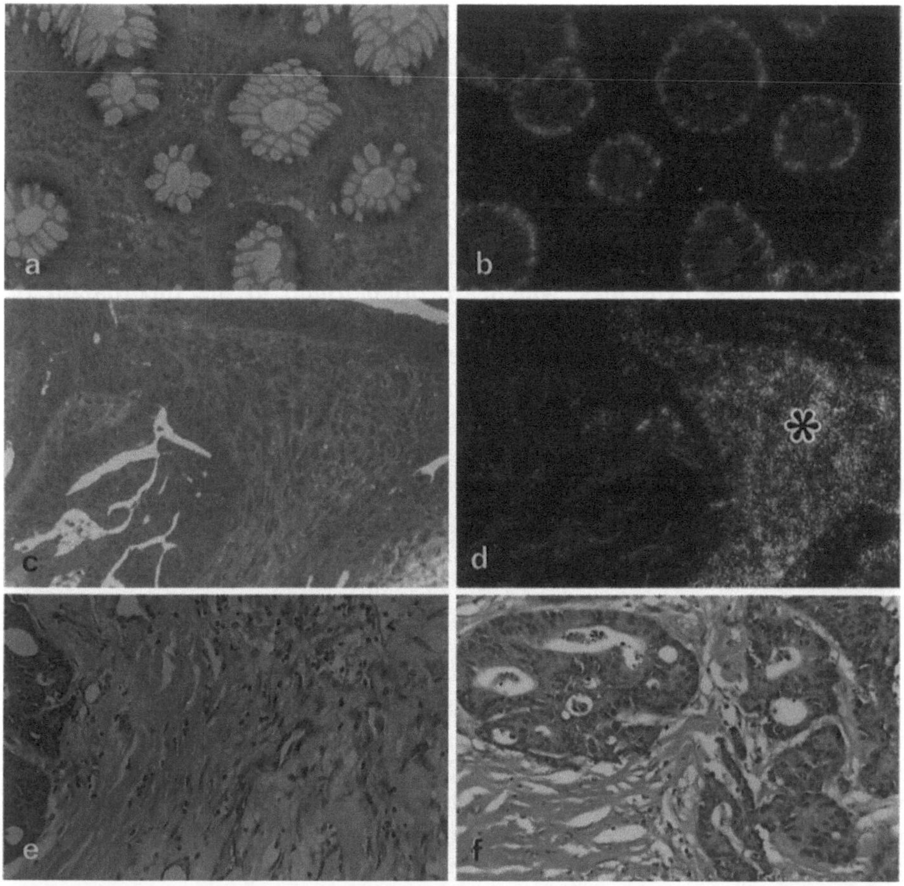

Fig. 4. Gallery of light and dark field micrographs of normal colon (**a**, **b**) and colon carcinoma (**c–f**). The top four panels represent autoradiographs of tissue incubated in an organ culture system with radiosulfate. Notice that in control (**a**, **b**) the radiosulfate is incorporated nearly exclusively by the epithelium, whereas in the tumor (**c**, **d**) the radiosulfate is incorporated primarily by the connective tissue stroma (*asterisk*). The accumulation of sulfated material in the peritumoral extracellular matrix is further shown by staining of the primary (**e**) or metastatic (**f**) colon carcinoma matrix with alcian blue at high (0.3 M) magnesium chloride concentrations. Methacrylate sections stained with H&E were subjected to autoradiography (Iozzo et al. 1982) and studied by light (**a**, **c**) and dark field (**b**, **d**) microscopy. Panels **e** and **f** represent paraffin sections of alcian blue-stained colon carcinoma at the primary site and at a metastatic pulmonary focus, respectively. **a–f** ×140

the growth of mammary carcinoma (Lewko et al. 1981). In a recent study, we (Iozzo and Müller-Glauser 1985) have demonstrated that the connective tissue of rabbit mesentery, following intraperitoneal implantation of V2 carcinoma cells, was markedly increased. Specifically the mesentery increased up to eight fold in thickness and several times in weight. This increase was not due to proliferation of tumor cells but was rather due to the marked increase in hyaluronic acid, chondroitin sulfate and collagen. In addition, the fibroblasts were also increased in number, suggesting that these mesenchymal cells were stimulated by the tumor cells to proliferate and synthesize matrix constituents. In support of this argument is the fact that intraperitoneal implantation of liver homogeneates did not cause any structural change in the rabbit mesentery (Iozzo and Müller-Glauser 1985).

From the above studies it appears that, at least in some malignant neoplasms, the cancer cells do modulate their extracellular environment either directly by cell-cell interactions, possibly via cell surface molecules, or indirectly via the elaboration of tumor cytokines that exert a stimulatory activity on the mesenchymal cells. These cytokines, which are often found in the supernatant of cancer cells in vitro, may function as local hormones in vivo, and exist in small quantities, traveling short distances between host cells. The expression of such cytokines may be the key to the specificity and appearance of desmoplasia in a given neoplasm. Their presence would also explain why only certain types of tumors are capable of inducing this response in host cells, whereas other tumors that either lack or are incapable of expressing these gene products fail to induce it.

7 Conclusions and Perspectives

This chapter has described quantitative and qualitative alterations in proteoglycans occuring in human tumors and has discussed some of the possible regulatory mechanisms. It is apparent from the evidence presented above that proteoglycans do play a role in the growth and evolution of malignant neoplasms. They function as a kind of molecular sieve, and as such regulate the filtration of a variety of "signals" that reach the cells and eventually influence their function. Proteoglycans are responsible for retaining water, can act as growth promoting agents, and favor cell adhesion and migration along defined routes. It is conceivable that, in analogy with developmental systems, the well-hydrated, proteoglycan-rich matrix may favor the evolution of malignant cells by providing a growth-supporting environment. It is also apparent that neoplastic cells with a capacity to invade neighboring structures have the potential to alter the adjacent connective tissue and to modulate the metabolism of host mesenchymal cells. This ability can be manifested by either the abnormal synthesis or degradation of matrix macromolecules or by the release of tumor cytokines that trigger the synthesis or breakdown of matrix constituents. The proposed concept that tumor cells "condition" their pericellular environment offers a novel perspective in investigating this complex phenomenon. If one accepts this hypothesis, then a number of new and provocative questions emerge. For example, are the tumor cells the sole protagonists or does the full expression of the tumor extracellular matrix depend on rather complex host-tumor cell interactions and if this were the case, should we not consider the tumor extracellular matrix as a "specific" response rather than as a non-specific repair as previously regarded? What are the molecular mechanisms that dictate the lack or overproduction of matrix in a given neoplasm? Is the presence of specific tumor cytokines the key to desmoplasia? Is there any tissue specificity? The point to be stressed is the difficulty in making generalizations about the mechanisms regulating matrix metabolism in cancer. It is probable that one type of tumor favors one mechanism over the others, but even that may exhibit subtle variations at various locations within the neoplasm itself. For the past several years our laboratory has been interested in investigating some of the above questions and the search for their answers is what makes research in this field challenging and worthwhile.

Acknowledgements. The author is deeply indebted to D.J. Slaymaker and J. Minda for their skilful technical assistance. This work was supported in part by Grant CA-39481 from the National Institutes of Health and Grant JFRA-95 from the American Cancer Society.

References

Bano M, Zwiebel JA, Salomon DS, Kidwell WR (1983) Detection and partial characterization of collagen synthesis stimulating activities in rat mammary adenocarcinomas. J Biol Chem 258:2729–2735

Barsky SH, Rao CN, Grotendorst GR, Liotta LA (1982) Increased content of type V collagen in desmoplasia of human breast carcinoma. Am J Pathol 108:276–283

Barsky SH, Siegal GP, Jannotta F, Liotta LA (1983a) Loss of basement membrane components by invasive tumors but not by their benign counterparts. Lab Invest 49:140–147

Barsky SH, Togo S, Garbisa S, Liotta LA (1983b) Type IV collagenase immunoreactivity in invasive breast carcinoma. Lancet 1:296–297

Castor CW, Naylor B (1977) Acid mucopolysaccharide composition of serous effusions. Study of 100 patients with neoplastic and non-neoplastic conditions. Cancer 20:462–466

Chiu B, Churg A, Tengblad A, Pearce R, McCaughey WTE (1984) Analysis of hyaluronic acid in the diagnosis of malignant mesothelioma. Cancer 54:2195–2199

David G, Van Den Berghe H (1983) Transformed mouse mammary epithelial cells synthesize undersulfated basement membrane proteoglycan. J Biol Chem 258:7338–7344

Gallager JT, Hampson IN (1984) Proteoglycans in cellular differentiation and neoplasia. Biochem Soc Trans 12:541–543

Graf M, Baici A, Sträuli P (1981) Histochemical localization of cathepsin B at the invasion front of the rabbit V2 carcinoma. Lab Invest 45:587–596

Hascall VC, Hascall GK (1981) Proteoglycans. In: Hay ED (ed) Cell biology of extracellular matrix. Plenum Press, New York, p 39

Heinegård DK, Paulsson M (1984) Structure and metabolism of proteoglycans. In: Piez DA, Reddi AH (eds) Extracellular biochemistry. Elsevier Publishing, New York, p 277

Hopwood JJ, Dorfman A (1978) Glycosaminoglycan synthesis by Wilms' tumor. Pediat Res 12:52–56

Hurst RE, Parmley RT, Nakamura N, West SS, Denys FR (1981) Heparan sulfate of AH-130 ascites hepatoma cells: a cell-surface glycosaminoglycan not displaced by heparin. J Histochem Cytochem 29:731–737

Iozzo RV (1984a) Proteoglycans and neoplastic-mesenchymal cell interactions. Hum Pathol 15:2–10

Iozzo RV (1984b) Biosynthesis of heparan sulfate proteoglycan by human colon carcinoma cells and its localization at the cell surface. J Cell Biol 99:403–417

Iozzo RV (1985a) Proteoglycans: structure, function, and role in neoplasia. Lab Invest 53:373–396

Iozzo RV (1985b) Neoplastic modulation of extracellular matrix. Colon carcinoma cells release polypeptides that alter proteoglycan metabolism in colon fibroblasts. J Biol Chem 260:7464–7473

Iozzo RV, Müller-Glauser W (1985) Neoplastic modulation of extracellular matrix: proteoglycan changes in the rabbit mesentery induced by V2 carcinoma cells. Cancer Res 45:5677–5687

Iozzo RV, Wight TN (1982) Isolation and characterization of proteoglycans synthesized by human colon and colon carcinoma. J Biol Chem 257:11135–11144

Iozzo RV, Goldes JA, Chen W-J, Wight TN (1981) Glycosaminoglycans of pleural mesothelioma: a possible biochemical variant containing chondroitin sulfate. Cancer 48:89–97

Iozzo RV, Bolender RP, Wight TN (1982) Proteoglycan changes in the intercellular matrix of human colon carcinoma. An integrated biochemical and stereologic analysis. Lab Invest 47:124–138

Iozzo RV, Ketterer CL, Slaymaker DJ (1986) Evidence of a small hydrophobic domain in the core protein of heparan sulfate proteoglycan from human colon carcinoma cells. FEBS Lett 206:304–308

Isemura M, Munakata H, Ototani N, Goto K, Yosizawa Z (1982) Glycosaminoglycans of rat colorectal adenocarcinoma. Gann 73:721–727

Kawai T, Suzuki M, Shinmei M, Maenaka Y, Kageyama K (1985) Glycosaminoglycans in malignant diffuse mesothelioma. Cancer 56:567–574

Kefalides NA, Alper R, Clark CC (1979) Biochemistry and metabolism of basement membranes. Int Rev Cytol 61:167–228

Keller KL, Keller JM, Moy JN (1980) Heparan sulfates from Swiss mouse 3T3 and SV3T3 Cells: 0-sulfate difference. Biochemistry 19:2529–2536

Khoory MS, Nesheim ME, Bowie EJW, Mann KG (1980) Circulating heparan sulfate proteoglycan anticoagulant from a patient with a plasma cell disorder. J Clin Invest 65:666–674

Knudson W, Biswas C, Toole BP (1984) Interactions between human tumor cells and fibroblasts stimulate hyaluronate synthesis. Proc Natl Acad Sci USA 81:6767–6771

Kramer RH, Vogel KG (1984) Selective degradation of basement membrane macromolecules by metastatic melanoma cells. J Natl Cancer Inst 72:889–897

Kramer RH, Vogel KG, Nicolson GL (1982) Solubilization and degradation of subendothelial matrix glycoproteins and proteoglycans by metastatic tumor cells. J Biol Chem 257:2678–2686

Lewko WM, Liotta LA, Wicha MS, Vonderhaar BK, Kidwell WR (1981) Sensitivity of N-nitrosomethylurea-induced rat mammary tumors to cis-hydroxyproline, an inhibitor of collagen production. Cancer Res 41:2855–2862

Lindahl U, Feingold DS, Rodén L (1986) Biosynthesis of heparin. Trends Biochem Sci 11:221–225

Liotta LA, Abe S, Gehron-Robey P, Martin GR (1979) Preferential digestion of basement membrane collagen by an enzyme derived from metastatic murine tumor. Proc Natl Acad Sci USA 76:2268–2272

Liotta LA, Tryggvason K, Garbisa S, Hart I, Foltz CM, Shafie S (1980) Metastatic potential correlates with enzymatic degradation of basement membrane collagens. Nature 284:67–68

Liotta LA, Thorgeirsson UP, Garbisa S (1982) Role of collagenases in tumor cell invasion. Cancer Metastasis Rev 1:277–288

Liotta LA, Rao CN, Barsky SH (1983) Tumor invasion and the extracellular matrix. Lab Invest 49:636–649

Merrilees MJ, Finlay GJ (1985) Human tumor cells in culture stimulate glycosaminoglycan synthesis by human skin fibroblasts. Lab Invest 53:30–36

Meyer K, Chaffee E (1939) Hyaluronic acid in pleural fluid associated with malignant tumor involving pleura and peritoneum. Proc Soc Exp Biol Med 42:797–800

Morse BS, Nussbaum M (1967) The detection of hyaluronic acid in the serum and urine of a patient with nephroblastoma. Am J Med 42:996–1002

Nakajima M, Irimura T, Di Ferrante DT, Di Ferrante N, Nicolson GL (1983) Heparan sulfate degradation: relation to tumor invasive and metastatic properties of mouse B16 melanoma sublines. Science 220:611–613

Nakajima M, Irimura T, Di Ferrante N, Nicolson GL (1984) Metastatic melanoma cell heparanase. Characterization of heparan sulfate degradation fragments produced by B16 melanoma endoglucuronidase. J Biol Chem 259:2283–2290

Nakamura N, Kojima J (1981) Changes in charge density of heparan sulfate isolated from cancerous human liver tissue. Cancer Res 41:278–283

Nakamura N, Hurst RE, West SS (1978) Biochemical composition and heterogeneity of heparan sulfates isolated from AH-130 ascites hepatoma cells and fluid. Biochim Biophys Acta 538:445–457

Nakano T, Fuji J, Tamura S, Amuro Y, Nabeshima K, Horai T, Hada T, Higashino K (1986) Glycosaminoglycans in malignant pleural mesothelioma. Cancer 57:106–110

Nowell PC (1986) Mechanisms of tumor progression. Cancer Res 46:2203–2207

Pal S, Strider W, Margolis R, Gallo G, Lee-Huang S, Rosenberg L (1978) Isolation and characterization of proteoglycans from human chondrosarcoma. J Biol Chem 253:1279–1289
Palmer RN, Rick ME, Rick PD, Zeller JA, Gralnick HR (1984) Circulating heparan sulfate anticoagulant in a patient with a fatal bleeding disorder. N Engl J Med 310:1696–1699
Poole AR, Tiltman KJ, Recklies AD, Stoker TAM (1978) Differences in secretion of the proteinase cathepsin B at the edges of human breast carcinomas and fibroadenomas. Nature 273:545–547
Poste G, Fidler IJ (1980) The pathogenesis of cancer metastasis. Nature 283:139–146
Powars DR, Allerton SE, Beierle J, Butler BB (1972) Wilms' tumor. Clinical correlation with circulating mucin in three cases. Cancer 29:1597–1605
Robinson J, Viti M, Höök M (1984) Structure and properties of an under-sulfated heparan sulfate proteoglycan synthesized by a rat hepatoma cell line. J Cell Biol 98:946–953
Rosenberg L, Tang L-H, Pal S (1979) Biochemical assessment of malignancy in human chondrosarcoma. In: Gregory JD, Jeanloz RW (eds) Glycoconjugate research, vol I. Academic Press, New York, p 393
Russel JB, Steinherz PG, Miller Dr, Hilgartner MW (1984) A heparin-like anticoaglant in an 8-month-old boy with acute monoblastic leukemia. Am J Hematol 16:83–90
Schürch W, Seemayer TA, Lagace R (1981) Stromal myofibroblasts in primary invasive and metastatic carcinomas. Virchows Arch (Pathol Anat) 391:125–139
Seemayer TA, Schürch W, Lagace R, Tremblay G (1979) Myofibroblasts in the stroma of invasive and meteastatic carcinoma. Am J Surg Pathol 3:525–533
Sloane BF, Dunn JR, Honn KV (1981) Lysosomal cathepsin B: correlation with metastatic potential. Science 212:1151–1153
Sugahara K, Schwartz NB, Dorfman A (1980) Defect in 3′-phosphoadenosine-5′-phosphosulfate formation in brachymorphic mice. Proc Natl Acad Sci USA 76:6615–6619
Tarin D, Hoyt BJ, Evans DJ (1982) Correlation of collagenase secretion with metastatic-colonization potential in naturally occurring murine mammary tumors. Br J Cancer 46:266–278
Thompson ME, Bromberg PA, Amenta JS (1969) Acid mucopolysaccharide determination: a useful adjunct for the diagnosis of malignant mesothelioma with effusions. Am J Clin Pathol 52:335–339
Thonar EJ-MA, Sweet MBE, Immelman AR, Lyons G (1979) Structural study on proteoglycan from human chondrosarcoma. Arch Biochem Biophys 199:179–189
Toole BP (1981) Glycosaminoglycans in morphogenesis. In: Hay ED (ed) Cell biology of extracellular matrix. Plenum Press, New York, p 259
Turpeeniemi-Hujanen T, Thorgeirsson UP, Hart IR, Grant SS, Liotta LA (1985) Expression of collagenase IV (basement membrane collagenase) activity in murine tumor cell hybrids that differ in metastatic potential. J Natl Cancer Inst 75:99–103
Underhill CB, Keller JM (1975) A transformation-dependent difference in the heparan sulfate associated with the cell surface. Biochem Biophys Res Commun 63:448–454
Vlodavsky I, Fuks Z, Bar-Ner M, Ariav Y, Schirrmacher V (1983) Lymphoma cell-mediated degradation of sulfated proteoglycans in the subendothelial extracellular matrix: relationship to tumor cell metastasis. Cancer Res 43:2704–2711
Waxler B, Eisenstein R, Battifora H (1979) Electrophoresis of tissue glycosaminoglycans as an aid in the diagnosis of mesotheliomas. Cancer 44:221–227
Wewer UM, Albrechtsen R, Rao CN, Liotta LA (1986) The extracellular matrix in malignancy. In: Kuhn K (ed) Rheumatology, vol 10. Karger, Basel, p 451
Winterbourne DJ, Mora PT (1981) Cells selected for high tumorigenicity or transformed by simian virus 40 synthesize heparan sulfate with reduced degree of sulfation. J Biol Chem 256:4310–4320
Woolley DE (1984) Collagenolytic mechanisms in tumor cell invasion. Cancer Metastasis Rev 3:361–372

Basal Membrane Antigens as Tumor Markers*

J. CASELITZ

1 Introduction

In multicellular organisms the basal membrane, a complex extracellular matrix, plays a leading role in maintenance of the structure and function of the organism

* Supported by the *Deutsche Forschungsgemeinschaft* and by the *Hamburger Stiftung zur Förderung der Krebsbekämpfung*

(VRACKO 1974, 1982; HAY 1978; FURTHMAYR 1982; STANLEY et al. 1982a, b; FOIDART 1984; ABRAHAMSON 1986; INGBER et al. 1986). The basal lamina (basement membrane, basement lamina) is primarily defined by its topography and structure, as revealed by light and electron microscopy. The electron microscopical analysis reveals three different zones: *lamina densa, lamina rara (sive lucida), lamina densa*. It is well known that the connective tissue matrix is composed of collagens, proteoglycans and glycoproteins, which influence the differentiation of the cells, their morphogenesis and the tendency to proliferation. MARTINEZ-HERNANDEZ and AMENTA (1983) have divided the substances composing the basal membrane into those which are *intrinsic* (collagen type IV, laminin, heparan sulfate proteoglycan, entactin) and those which are *extrinsic* (fibronectin and type V collagen).

2 Methodology of Morphological Analysis of the Basal Membrane

2.1 Histochemical Methods

For the detection of basal lamina in light microscopy, several histochemical techniques can be applied (VACCA 1985). These techniques are based partly on the classical *PAS method.* Lillie's PAS Allochrome method can be used for the distinction of reticulum fibers and basal lamina. The basal lamina can be visualized by the periodic acid-sodium bisulfite-resorcin fuchsin technique (VACCA 1985). *Reticulum fibers* and the *basal lamina* are decorated by Wilder silver method, whereas Weigert's haematoxylin-picro-sirius red stain labels collagen fibers, reticulum fibers and basement membrane (VACCA 1985). Other techniques have been developed for the analysis of the intercellular matrix.

The advantage of these techniques, especially that of PAS staining, is their application in daily routine work (FLOTTE et al. 1980). They allow detection of the structure of the basal membrane which may be sufficient for practical diagnostic application. For detailed questions, however, other methods have to be applied.

In *electron microscopical histochemistry*, the basal lamina is demonstated by uranium and lead staining, and by the use of ruthenium red.

2.2 Immunohistochemistry

The principles of *immunohistochemistry* have been reviewed in many good monographs where details of techniques and protocols can be found (e.g. ELIAS 1982; BULLOCK and PETRUCZ 1982, 1983; TAYLOR 1986). Currently the triple layer PAP technique, the avidin-biotin technique and the alkaline phosphatase technique are used, parallel to the still valuable technique of immunofluorescence. The important advantage of immunohistochemical investigation is an exact localization of virtually any substance in the basal membrane, provided there is the availability of appropriate antibodies.

Experimentally (and diagnostically in the analysis of autoaggressive diseases) human antibodies against reticulin fibers and against basal membrane substances can be applied (Goodpasture's syndrome and others) (ELIAS 1982).

In most investigations, experimentally produced antibodies against basal membrane substances, either polyclonal or monoclonal, have been used.

Antibodies against collagen IV (SCHUPPAN et al. 1980; FOIDART and YAAR 1981; KUEHN et al. 1981; WEBER et al. 1984a, b; MARTIN et al. 1985), against laminin (TIMPL et al. 1979; FOIDART et al. 1980; ENGEL et al. 1981; ALBRECHTSEN et al. 1981; TIMPL et al. 1982) and against fibronectin (LINDER et al. 1985; ENGVALL et al. 1978; KUUSELA et al. 1978; NERI et al. 1981; RUOSLAHTI et al. 1980, 1981, 1982; RUOSLAHTI 1984) have been used.

2.3 Immunoelectronmicroscopy

The classical analysis of the basal membrane has been done using conventional electron microscopy (e.g. BRIGGAMAN and WHEELER 1975; VRACKO 1974, 1982). Further information is provided by the combination of electron microscopy and immunohistochemistry. Immunoelectronmicroscopy follows the same principles as light microscopical immunohistochemical methods. In practice, many problems of appropriate preservation and appropriate reaction of the antibodies exist (e.g. TOKUYASU 1980; WILLINGHAM 1980; POLAK and VAN NOORDEN 1983). Since information on electronmicroscopy is provided by differences in electron density, the labelling is generally done by means of electron dense particles, such as ferritin, colloidal gold or others.

SANO et al. (1981) analysed collagen type IV in bovine lung parenchyma by an immunoelectronmicroscopical approach. FOIDART et al. (1980) studied the distribution of laminin with immunoelectronmicroscopical methods. FURCHT et al. (1978a) analysed the localization of fibronectin on the surface of myoblasts by lightmicroscopic and electronmicroscopic methods. HEDMAN (1980) presented an immunoelectronmicroscopical study of fibronectin. Some authors have analysed elastin at the ultrastructural level (DAMIANO et al. 1978, 1981, 1984; FUKUDA et al. 1984).

3 Overview of Basal Membrane and Its Associated Substances

3.1 Collagen Types

To date, about 10 different collagens have been identified (BORNSTEIN and SAGE 1980; GAY and MILLER 1983; MARTIN et al. 1985).

The *general structural principle* of the *collagens* is a composition from three polypeptide chains called alpha chains. The different *subtypes* of the collagens are characterized by the different aminosequences of the alpha chains. The chains are alpha helical, are characterized by the amino acid sequence glycine-X-Y and contain hydroxyproline or hydroxylysine as a major component.

Type I, II and III belong to the interstitial collagens (MARTIN et al. 1985). The extracellular collagens are identified as type I and III. Around the mesenchymal cells collagen type V is identified. This type is associated with the pericellular exocytoskeleton and collagen type IV (GAY and MILLER 1983). These collagens are associated with other proteins/glycoproteins, such as *fibronectin, chondroitin sulfate*, and *heparan sulfate, elastin* (RUOSLAHTI et al. 1981; URRY 1983; IOZZO 1984).

Collagen type III, originally detected in the fetal skin, is present in a variety of tissues like type I collagen. It is thought to comprise the reticulin fibers detected by histochemical methods (MARTIN et al. 1985).

The aggregation of *collagen type IV* forms a stable network for the basement membrane where other components are included (TIMPL et al. 1981). Type IV collagen is regularly and extensively localized in the *basal membranes* (MARTINEZ-HERNANDEZ et al. 1981a, b). *Collagen type IV* has an alpha helical structure which is interrupted by protein sequences which are different from Glycine-X-Y. It has relatively large amounts of 3-hydroxyproline and is richer in carbohydrate side chains than the interstitial collagens. Collagen type IV has, as a special feature, the 7S region. This is the cross-linking site for four collagen type IV molecules. All antibodies against collagen type IV decorate exclusively the basal membrane, the labelling being localized in the lamina rara and densa on the ultrastructural level (ROLL et al. 1980; MARTINEZ-HERNANDEZ et al. 1981a, b; SANO et al. 1981; FARQUHAR et al. 1982).

Collagen type V, a stromal and pericellular collagen, belongs to the *extrinsic components* of the basal membrane (MARTINEZ-HERNANDEZ and AMENTA 1983). It is related to collagen IV due to its amino acid content and chemical characteristics (CHUNG et al. 1976). Collagen V has been found in the basal membrane of the glomerulus, in other basal membranes it was not found (GAY et al. 1981; SANO et al. 1981; MARTINEZ-HERNANDEZ et al. 1982). Collagen type V is now accepted as being present in the basal membrane of some vessels and in the stroma surrounding collagen type I bundles (MARTINEZ-HERNANDEZ and AMENTA 1983). Type V collagen is increased in desmoplastic reactions of carcinomas (BARSKY et al. 1982).

3.2 Laminin

Laminin is one of the best characterized *non-collagenous* glycoproteins of the basal membranes which are associated with collagen type IV (TIMPL et al. 1979, 1982). Laminin has a cross like shape with one long and two short arms (ENGEL et al. 1981; TIMPL et al. 1982). The amino acid sequence of laminin is different from that of fibronectin (TIMPL et al. 1982).

Using antibodies against laminin, one finds a specific staining of the *basal lamina*, the reaction being more intense in the lamina rara (FARQUHAR et al. 1982; FOIDART et al. 1980; MADRI et al. 1980; MARTINEZ-HERNANDEZ et al. 1981a, b, 1982). Laminin is strongly involved in *cell adhesion* and *attachment* (TERRANOVA et al. 1980).

3.3 Fibronectin

Fibronectin (for review of the biochemical aspects: RUOSLAHTI et al. 1981; HYNES and YAMADA 1982) is a *dimeric protein* with two polypeptide chains (molecular weight *220,000 Dalton*). The fibronectins from different sources, one of which is *fibroblasts*, are chemically closely related and cannot be distinguished on the immunological level. Fibronectin being highly conserved during evolution, antibodies against this protein display an extensive interspecies cross reactivity. Fibronectins interact with many other macromolecules (collagens, fibrin, actin, heparin, DNA, hyaluronic acid, cell surfaces and bacteria). The interaction with collagens interferes with the basal membrane and fibronectin, although not specifically restricted to the basal membrane, may be regarded as an important factor (*extrinsic factor* according to MARTINEZ-HERNANDEZ and AMENTA 1983; see also COURTOY et al. 1980; D'ARDENNE et al. 1983a, b; D'ARDENNE and MCGEE 1984).

The biological activities of fibronectin are characterized as *mediation* of cell *adhesion* and *spreading*, promoting of cell *motility*, *prevention of fusion* of myoblasts and differentiation of chondrocytes, non-specific *opsonin* in plasma (KESKI-OJA et al. 1980; RUOSLAHTI et al. 1981; TARIN et al. 1982; TAKASHIMA and GRINNELL 1984). Fibronectin is produced by various cell types (*epithelial cells:* liver cells, kidney cells, cells of the gut, cells of the breast and of the amniotic membranes; *mesenchymal cells:* fibroblasts, myoblasts, chondrocytes, Schwann cells, endothelial cells, melanoma cells, macrophages and astroglial cells; see also HEDMAN et al. 1978; RUOSLAHTI 1981; HANTAI et al. 1983). Fibronectin is distributed in three different compartments: intracellular, at the cell surface and in the extracellular matrix (FURCHT et al. 1978a, b).

The *tissue distribution* of *fibronectin* follows the distribution of loose connective tissue, fibronectin being augmented in the limiting membranes such as basement membranes (MOSHER et al. 1977; MADRI et al. 1980; FOIDART and YAAR 1981; LAURIE et al. 1982). Fibronectin is located at the *interface* of the *basal membrane* and the abutting cells (RUOSLAHTI et al. 1981). As it plays a role in cell attachment and orientation of cells, fibronectin decreases towards the tips of the colonic epithelium, thus allowing a shedding of epithelial cells (QUARONI et al. 1978).

Apart from its importance in *neoplastic development*, fibronectin plays an interesting role in other pathological conditions. It is increased in sclerodermia and other diseases of the *skin* (COOPER et al. 1979; FYRAND 1980), in diseases of the *kidney* (e.g. diabetic nephropathy, glomerulonephritis, transplanted kidneys) and in *diabetes mellitus* (SCHEINMAN et al. 1978; WEISS et al. 1979; RUOSLAHTI 1981).

Lack of fibronectin, commonly reported for transformed cells in vitro and in some neoplastic lesions (see below), is not always due to the fact that malignant cells do not produce fibronectin, but that it seems to be released into the culture medium and is not found on the cell surface (RUOSLAHTI 1981).

The lack of fibronectin may be one of the features that identify cells with a capacity to metastasize.

3.4 Others

3.4.1 Entactin

Entactin is a highly sulfated protein with a molecular weight of approximately 150,000 Dalton. It is distinct from laminin. Although it has been localized in the rat kidney by immunoelectronmicroscopy, its definite localization in the basal membrane has still to be elucidated (MARTINEZ-HERNANDEZ and AMENTA 1983). Its distribution seems to be closely related to that of laminin, but on the immunoelectronmicroscopical level, the basal membranes may differ in their content and distribution of laminin and entactin (MCCARTHY et al. 1985).

3.4.2 Heparan Sulfate

Heparan sulfate proteoglycan is another component of the basal membrane which is dealt with in another chapter of this volume (Iozzo, Chapter 9). In immunohistochemical studies, the labelling of heparan sulfate seems to follow that of laminin and collagen type IV (BIREMBAUT et al. 1985).

3.4.3 Bullous Pemphigoid Antigen

The basal membrane antigen *"bullous pemphigoid antigen"* is characterized as a protein of 220,000 Dalton which is distinct from laminin and collagen IV (STANLEY et al. 1982a, b). It is found in the lamina lucida of the basal membrane zone, and in the basal membrane of squamous epithelia (WEBER et al. 1984a, b).

New antigens in basal membrane (e.g. nidogen) have been reported by TIMPL et al. (1983).

4 Change of Basal Membrane Associated Substances During Neoplastic Disorders

4.1 General Rules

The changes of the structure and synthesis of the basal membrane can be considered on the following levels:

4.1.1 Synthesis of Basal Membrane Associated Material by Neoplastic Cells

The substance which has especially been analysed with respect to the production site is *fibronectin*. Fibronectin has been thought to interfere with the *metastatic behaviour* of tumor cells (NERI et al. 1981). Primary carcinomas display a dispersed pattern of fibronectin which is also present in the surrounding tissue (STAMPFER et al. 1981). MARTIN et al. (1984) found a reciprocal expression of laminin and fibronectin receptors in the different tumor cells.

On the level of *collagen*, a difference between the different subtypes and neoplastic growth has not been shown (GAY and GAY 1985), although DAVID and VAN DEN BERGHE (1983) report the production of less sulfated proteoglycans by transformed mammary cells. KAO et al. (1984) showed that the matrix of human breast tumor cells is mitogenic for fibroblasts.

4.1.2 Synthesis of Basal Membrane Associated Substances by Surrounding Tissues as a Distinct Response of the Organism

Some human tumors are accompanied by a fibrous response. This reaction, often designated as scirrhous or desmoplastic, is well documented in mammary carcinomas (GAY and RHODES 1986). BARSKY et al. (1982) demonstrated that in the desmoplastic part of human mammary carcinoma collagen type V is augmented. Although former papers suggested tumor cells as the source of the desmoplastic material, it now generally accepted that cells of the surrounding stroma, especially the myofibroblasts (SEEMAYER et al. 1980), are the source (BARSKY et al. 1984). It is collagen type V which is synthesized by fibroblasts (GAY and GAY 1985).

4.1.3 Degradation of Basal Membrane Substances by Tumor Cells

In normal tissue and benign lesions, the basal membrane is generally intact as has been shown on the *electron microscopical level* (OZZELO 1959; FISHER 1976; STEGNER 1986) and by *immunohistochemistry* (GAY and GAY 1985). In preneoplastic conditions, disturbances of the basal membrane substances may be observed. In true invasive carcinomas, parts of the basal membrane are destroyed. This may be due to the influence of proteolytic enzymes (LIOTTA 1982; LIOTTA et al. 1983; LIOTTA 1984). Besides this "active" principle, destruction of the basal membrane may follow the increased pressure of proliferating cells (BARSKY et al. 1983; LIOTTA et al. 1983).

4.2 Changes in Tumors

An overview over the distribution of basal membrane substances is given in Table 1.

Table 1. Distribution pattern of basal membrane associated substances in neoplastic diseases. (Modified after BIREMBAUT et al. 1985)

Tumor	Intrinsic components (collagen type IV, laminin, heparansulfate)
Preinvasive carcinomas	Basal membrane labelling with interruptions
Invasive carcinomas	Irregular staining around well-differentiated (glandular) structures and adjacent to the basal cells of solid/epidermoid carcinomas
Sarcomas	
Leiomyosarcoma Neurofibrosarcoma Liposarcoma	Irregular pericellular staining in well differentiated areas
Malignant fibrous histiocytoma Osteosarcoma Malignant hemangiopericytoma undifferentiated sarcomas	Vascular basal membrane labelling No pericellular staining

Table 1 (continued)

Tumor	Extrinsic components (fibronectin)
Preinvasive carcinomas	Basal membrane labelling with interruptions, present in the *stroma*
Invasive carcinomas	present in the *stroma* intensely in *desmoplastic reaction*
Sarcomas	
Leiomyosarcoma	irregular pericellular labelling
Neurofibrosarcoma	
Liposarcoma	
Malignant fibrous histiocytoma	irregular peri- and intercellular labelling
Osteosarcoma	and vascular basal membrane labelling
Malignant hemangiopericytoma	
undifferentiated sarcomas	

4.2.1 Skin and Mucosa (Figs. 1, 2)

The changes in the multilayered epithelium of skin and mucosa are similar with respect to the basal membrane. In invasive and metastatic squamous carcinomas of the head and neck, collagen type IV displayed focal thickening, reduplication, aggregation, attenuation and segmental defects (CAM et al. 1984; CARTER et al. 1985; VISSER et al. 1986). Focal interruptions of laminin and collagen type IV in the basal membranes are found in *dysplasias of the laryngeal mucosa* (VISSER et al. 1986). However, intact basal membrane and reduplications may be found in some squamous cell carcinomas of the skin by the analysis of collagen type IV (GUSTERSON et al. 1984; GUSTERSON et al. 1986). A production of collagen type IV is found in tumors with a higher degree of differentiation (GUSTERSON et al. 1986).

In *solid basal cell carcinoma* the cells still appear to preserve their characteristic production of basal membrane material, but seem to have lost the ability of polar distribution of this material (WEBER et al. 1982).

In *cylindromas* of the skin, collagen type IV, laminin and bullous pemphigoid antigen were localized in the PAS positive hyaline zone, the staining was confined to a zone encircling the tumor nests (WICK and TIMPL 1980; WEBER et al. 1984a, b; KALLIOINEN et al. 1984; KALLIOINEN 1985). Immunoelectronmicroscopic observations revealed a positive staining for collagen type IV and laminin throughout the basal membrane material of the tumor (KALLIOINEN 1985). Fibronectin was uniformly distributed over the tumor stroma and at the periphery of the cell islands (WEBER et al. 1984a, b). Eccrine spiradenomas produce an excessive amount of basal membrane associated material, whereas the basal membrane is absent in many areas of trichofolliculomas and related tumors (KALLIOINEN et al. 1984). Malignant tumors of the skin (hidradenocarcinomas, sebaceous carcinomas) display narrow strips between infiltrating tumor cell clusters (KALLIOINEN et al. 1984).

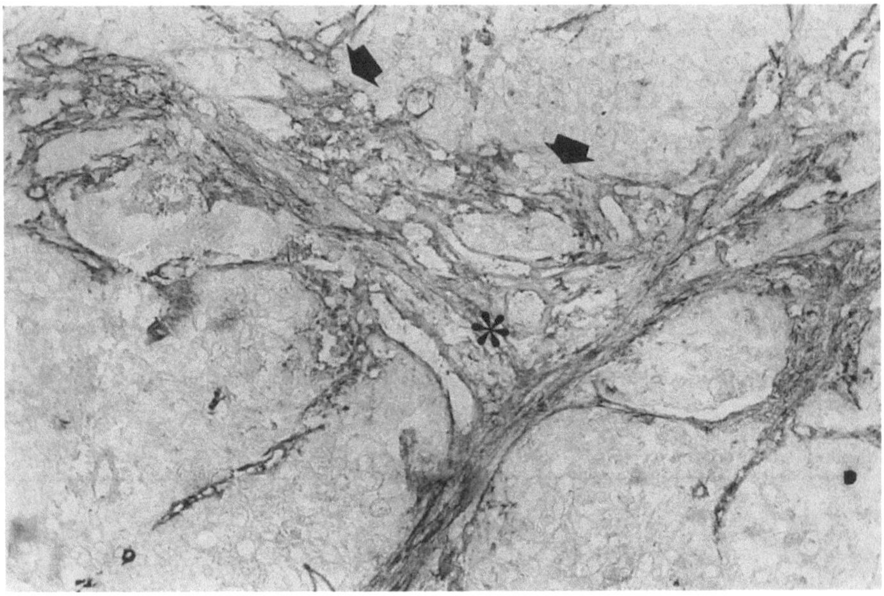

Fig. 1. Fibronectin in stroma of invasive squamous cell carcinoma. Intense staining in the stromal part *(asterisk)*. Disrupted fibronectin staining in the part of the basal membrane *(arrows)*. Cryostate section. × 300

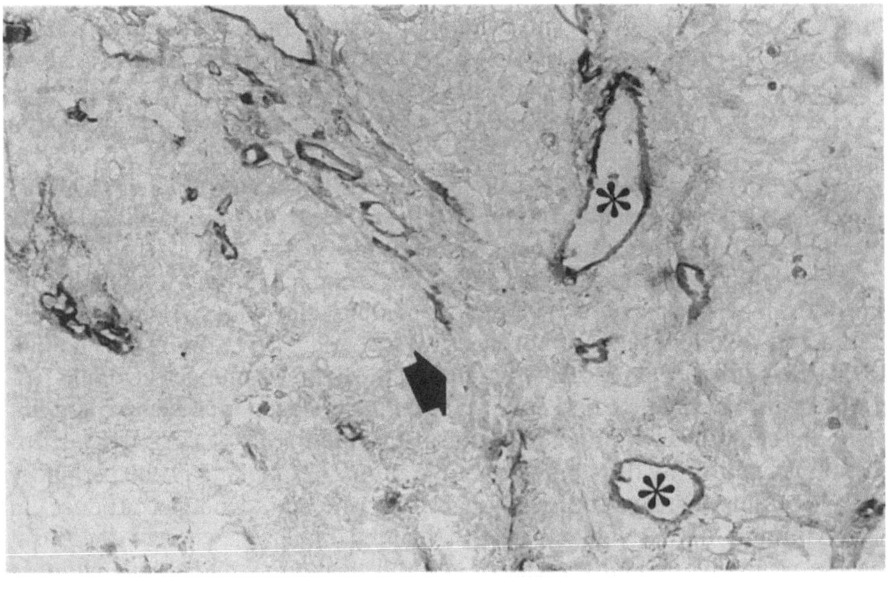

Fig. 2. Collagen type IV staining in invasive squamous cell carcinoma (identical cases as in Fig. 1). Intact basal membrane around the blood vessels *(asterisk)*. Disrupted basal membrane in the infiltrating part of the tumor *(arrow)*. Cryostate section. × 300

4.2.2 *Mammary Carcinoma*

In agreement with ultrastructural observations (OZZELLO 1959; FISHER 1976) laminin and collagen IV in *mammary carcinomas* are generally reduced (PITELKA et al. 1980; ALBRECHTSEN et al. 1981; GAY and GAY 1985; REMBERGER and NERLICH 1985; SCHAUER et al. 1985). The stages of neoplastic transformation of human breast tissue can be monitored by dissolution of basement membrane components (SIEGAL et al. 1981). Mouse mammary epithelial cells synthesize an undersulfated basement membrane proteoglycan (DAVID and VAN DEN BERGHE 1983). The synthesis of basement proteins is enhanced during the differentiation of rat mammary tumor epithelial cells (WARBURTON et al. 1982a, b).

The source of the substances appearing during the desmoplastic reaction in mammary carcinomas is the (myo)fibroblast (BARSKY et al. 1984). Collagen type V (BARSKY et al. 1982) and fibronectin (STAMPFER et al. 1981) are augmented in the desmoplastic stromal reaction of mammary carcinomas (CASELITZ et al. 1985a, b) and is found in the stromal part of medullary carcinomas in an irregular pattern (LABAT-ROBERT et al. 1981). In culture, fibronectin is reduced in malignant human mammary epithelial cells (YANG et al. 1980).

Myoepithelial cells are closely related to the basal membrane. The disturbances of the basal membrane during neoplastic invasion is accompanied by both an interruption of the myoepithelial cell layer and the staining for laminin and collagen IV (GUSTERSON et al. 1982; WARBURTON et al. 1982a, b). Myoepithelial cells have a close relationsship to basal membrane associated substances, collagen type IV, laminin and, in an indirect way, to fibronectin (GUSTERSON et al. 1982).

4.2.3 *Salivary Gland Tumors* (Figs. 3, 4, 5)

In *pleomorphic adenomas* (Figs. 3, 4, 5) a considerable amount of basal membrane associated material, collagen type IV and laminin, is localized in the stroma and around the tumor cell groups (D'ARDENNE et al. 1983a, b; CASELITZ 1984; TOTO and HSU 1985). The labelling for fibronectin is especially intense at the border of cell groups and duct-like structures (CASELITZ 1984). At the morphological level it cannot be decided whether fibronectin is "epithelial" or "mesenchymal" by origin (D'ARDENNE et al. 1983a, b). A possible source of part of the basal membrane associated substances is the *myoepithelial cell* (for observations in Sjögren's disease: PALMER 1986). The staining of the basal membrane substances seems to be parallel to, but not always identical with, the staining for elastic fibers. In *adenoid cystic carcinomas* laminin and collagen were found in the stromal trabeculae and pseudocysts, fibronectin displayed a similar, but sometimes more diffuse, staining pattern (D'ARDENNE et al. 1983a, b; CASELITZ 1984; CASELITZ et al. 1986; TOIDA et al. 1984; for mammary adenoid cystic carcinoma: WELLS et al. 1986). Elastic fibers were found in the neighbourhood (AZZOPARDI and ZAYID 1971; ADKINS and DALEY 1974; DAVID and BUCHNER 1980).

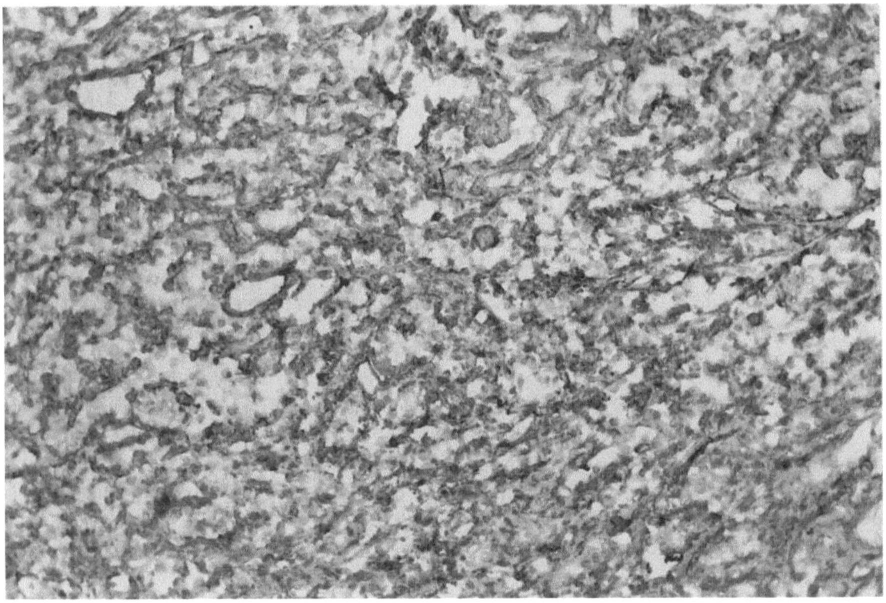

Fig. 3. Collagen type IV in pleomorphic adenoma. Strong labelling in stromal (mucoid) part of pleomorphic adenoma. Cryostate section. ×300

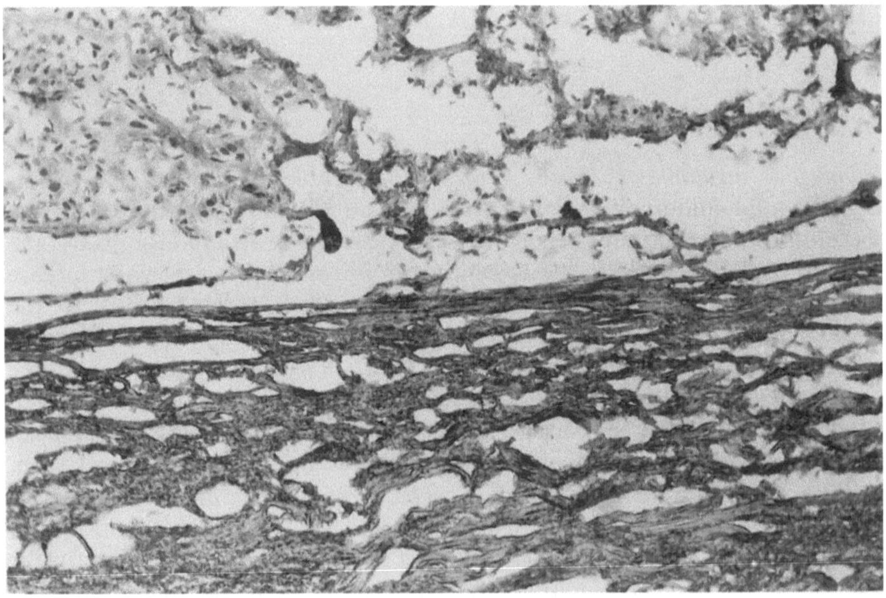

Fig. 4. Collagen type III in capsule of pleomorphic adenoma. Only weak staining in the tumor stroma. Cryostate section. ×300. (We gratefully acknowledge the kind gift of collagen type III antibody from Dr. SCHUPPAN, Berlin)

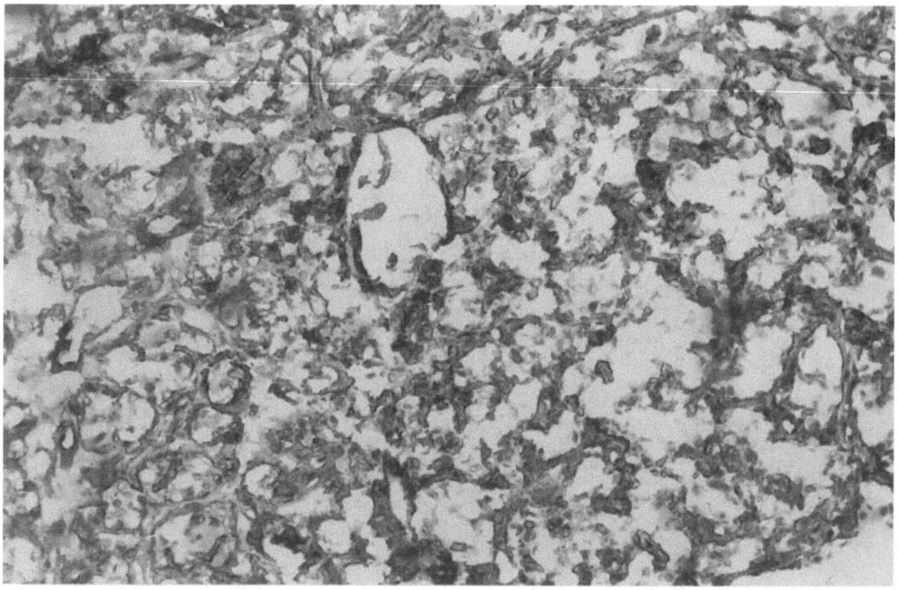

Fig. 5. Fibronectin in pleomorphic adenoma. Strong labelling in stromal (mucoid) part of pleomorphic adenoma. Cryostate section. × 300

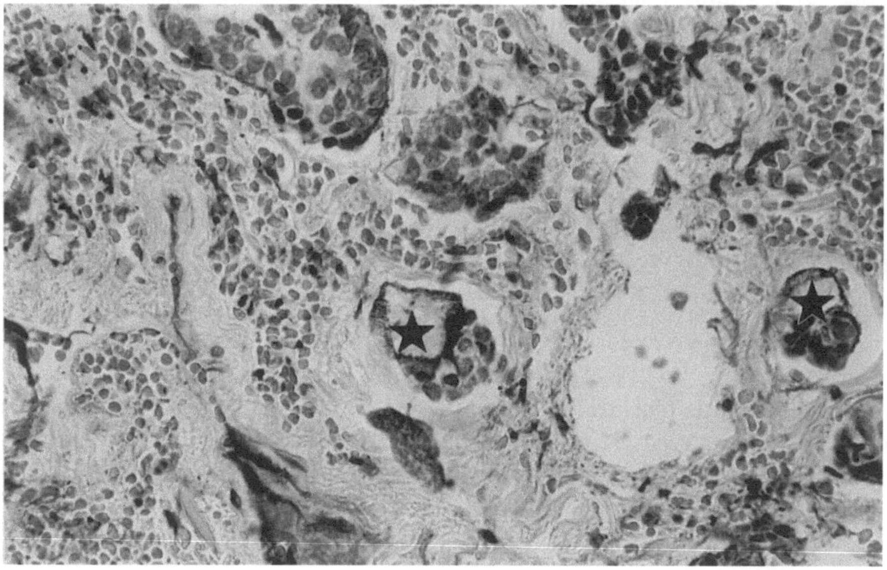

Fig. 6. Collagen type IV in adenoid cystic carcinoma. Strong staining of the material *(asterisks)* in the pseudocysts (especially at the border of the lining myoepithelial-like cells). Cryostate section. × 300

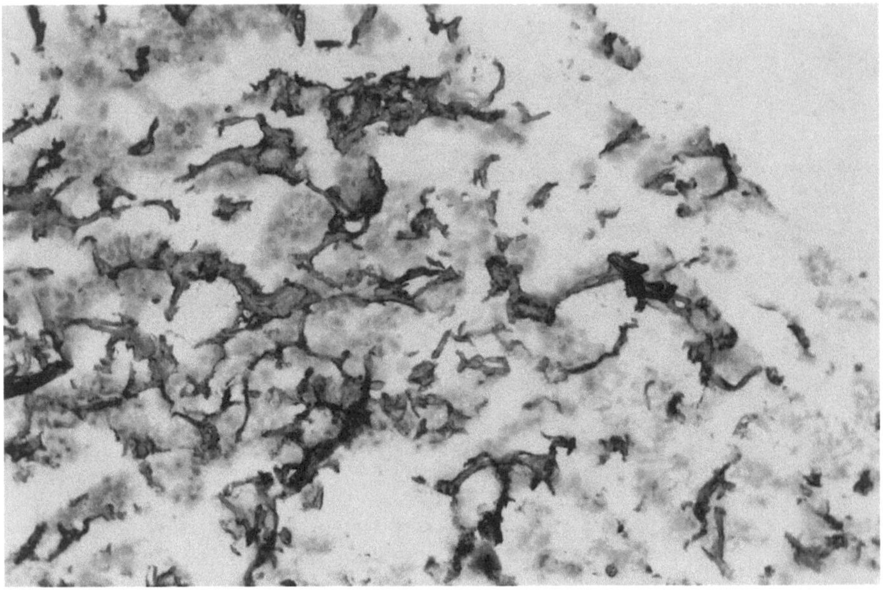

Fig. 7. Fibronectin in acinic cell tumor. Although the tumor is highly differentiated, there is a clear interruption of the fibronectin staining in the region of the basal membrane. Cryostate section. ×300. (We gratefully acknowledge the help of Dr. WUSTROW, Kiel, in providing fresh material)

In other types of malignant carcinomas of the salivary glands (Fig. 7), the staining for laminin, collagen type IV, and for fibronectin is generally interrupted in the region of the basal membrane (CASELITZ 1984; D'ARDENNE et al. 1983a, b).

4.2.4 Pancreas

The staining for *laminin* and *collagen type IV* is intact in pancreatitis. There is a loss of these substances in pancreatic carcinoma (BARSKY et al. 1983; HAGLUND et al. 1984).

4.2.5 Stomach

The so-called "linitis plastica" is accompanied by an intense reaction for *fibronectin* and *collagen type III* (D'ARDENNE et al. 1983a, b). These observations are similar to that of the desmoplastic carcinoma in mammary carcinoma.

4.2.6 Colon

Laminin was present in an *uninterrupted* linear staining in the basal membranes of normal, hyperplastic and adenomatous colorectal mucosa (KELLOKUMPU et al. 1985). The *early invasion* of colonic carcinomas was generally accompanied by a desmoplastic reaction with intense, but somewhat irregular staining for fibronectin and collagen IV (D'ARDENNE et al. 1983a, b). The basal lamina and the staining for *laminin* was disrupted in poorly differentiated colonic adenocarcinomas (BURTIN et al. 1982), whereas slight staining was still observed in the better differentiated tumors (KELLOKUMPU et al. 1985).

4.2.7 Lung and Bronchial Tree

In intraepithelial lesions, the basal membrane shows a disrupted staining for *collagen type IV* and *laminin*. These substances, although irregularly distributed in true invasive lesions, may be found around some well-differentiated solid tumor groups. *Fibronectin* is augmented in the stroma and displays no special augmentation in the region of the basal membrane (LABAT-ROBERT et al. 1981; BIREMBAUT et al. 1985).

4.2.8 Urinary Tract

Premalignant lesions display an intact basal membrane. Staining for *collagen type IV* and *laminin* is interrupted in invasive urothelial carcinomas. *Fibronectin* is still preserved in the stroma of the urinary bladder, but it is lost around infiltrating cells (LABAT-ROBERT et al. 1981).

4.2.9 Prostate Gland

Fibronectin is found pericellularly on the surface of smooth muscle cells in normal and hyperplastic prostate glands (LABAT-ROBERT et al. 1981). Athough still present at the surfaces of the muscle cells, fibronectin is lost around infiltrating malignant cells (LABAT-ROBERT et al. 1981).

4.2.10 Female Genital Tract

In normal exocervix and most dysplastic and malignant preinvasive lesions, the staining for *laminin* and *collagen type IV* is linear (BIREMBAUT et al. 1985). *Fibronectin* is visualized in the basal membrane of uterine glands and around the cells of the endometrial stroma (LABAT-ROBERT et al. 1981). It is lost at the basal membrane of invasive carcinomas, but may be present in the stroma in large quantities (LABAT-ROBERT et al. 1981).

4.2.11 Endocrine Tissue

BIREMBAUT et al. (1985) and LABAT-ROBERT et al. (1981) report a loss of basal membrane substances in follicular carcinomas of the thyroid gland.

4.2.12 Connective Tissue

In *benign proliferations*, *laminin*, *collagen type IV* and *fibronectin* are found in all vascular basal membranes and around all muscular and Schwann cells and adipocytes in a linear pericellular layer.

Staining for *laminin* and *collagen type IV* is found around the vessels in malignant mesenchymal tumors. However, there is a decrease in the pericellular staining of muscle cells, Schwann cells and adipocytes (Labat-Robert et al. 1981; Birembaut et al. 1985). Fibronectin, however, persists as intercellular and pericellular material in most sarcomas, especially in cases with fibroblastic differentiation (Birembaut et al. 1985). Stenman and Vaheri (1981) found single cells in sarcomas which were surrounded by fibronectin. It was abundant in the connective tissue (Stenman and Vaheri 1981) but Labat-Robert et al. (1981), found a loss of the pericellular labelling and a decreased staining of fibronectin in the extracellular matrix. Biphasic synovial cell sarcoma is reported to exhibit extracellular linear collagen type IV and laminin (Barsky et al. 1983). In the basal membrane of neurofibromas, collagen type IV and laminin were localized (Weber et al. 1984a).

5 Conclusion and Outlook

From the viewpoint of diagnostic pathology, *basal membranes* can be regarded as *structural barriers* to the invasion and metastatic procedure of malignant cells. Separating epithelial and other cells from the surrounding stroma, they are obstacles to invasion from tumor cells. For many substances which are directly and indirectly associated with the basal membrane, there is some evidence that the invasive progression of a tumor is accompanied by a loss of these substances.

In *surgical pathology*, antibodies against laminin and collagen type IV turn out to be very specific and valuable if one is interested in studying the structure and integrity of the basal membrane. Consequently, most studies agree that the immunohistochemical labelling for laminin and collagen type IV is intact and linear in normal tissues (Labat-Robert et al. 1981; Barsky et al. 1983; Carter et al. 1985; Gusterson et al. 1986). As far as can be demonstrated at the light microscopic level, there is a co-distribution of collagen type IV and laminin. In contrast, there is *interruption* of the basal membrane in many invasive tumors, immunohistologically visualized by the destruction of the decoration for collagen type IV and laminin (Albrechtsen et al. 1981; Labat-Robert et al. 1981; Birembaut et al. 1985; Gay and Gay 1985). This destruction has been shown in invasive carcinomas of the skin (Gusterson et al. 1986), of the breast (Albrechtsen et al. 1981; Gay and Gay 1985), of the lung (Labat-Robert et al. 1981), of the pancreas (Barsky et al. 1983; Haglund et al. 1984) and of other tissues (Birembaut et al. 1985). Thus, staining for laminin and collagen can be used to analyse the process of invasion.

Gusterson et al. (1986), however, observed that in some invasive carcinomas the immunohistological lining for laminin and collagen type IV is intact. This is in particular found in *well differentiated carcinomas*, where some tumor cells seem to preserve the capacity of forming a basal membrane. Therefore, there may be some exceptions to the general rule of interruption of the basal membrane in malignant tumors.

In *premalignant lesions*, many observations show disturbances of the basal membrane. This phenomenon has been interpreted as being due to the onset of infiltration. Other authors report an intact basal membrane in premalignant lesions. The differences may be explained by the different phases of a "premalignant" lesion and by the different sensitivity of immunohistochemistry due to technical details like fixation, different antibodies and others.

Fibronectin is generally reduced on the surface of neoplastic cells in comparison with their normal counterparts. Since the reduction at the surface of the cells is not always accompanied by a reduction of the production of fibronectin, its decrease may be due to the influence of *proteases* (HYNES 1981; HYNES and YAMADA 1982; SALO et al. 1982). McCARTHY et al. (1985) review the literature indicating that there is a general association between the loss in cell surface fibronectin and transformation. The details of this relationship may be more complicated, since many experimental studies have not always demonstrated a clear cut correlation between the metastatic potential of a cell and the lack of cell surface fibronectin. Different mechanisms may lead to the decrease in fibronectin (McCARTHY et al. 1985). Transformation may initiate a decrease of cell surface associated fibronectin due to the inhibition of production. Alternatively, the cell may degrade or alter this protein and there may be a change in the distribution of the cell surface receptors for fibronectin. Other authors offer the model of an inappropriate secretion of fibronectin to the surface (RUOSLAHTI 1984). From the experimental data, a choice between the different models cannot be made at the moment. Given the function of fibronectin in cell movement, its absence would perhaps explain the increased tendency to movement and invasion by malignant cells. For cell lines a *correlation* between the *decrease* of the cell surface fibronectin and their *tumorigenicity* has been found (KAHN and SHIN 1979).

Tumor cells display a *receptor* for fibronectin and can interact with fibronectin coated surfaces (RUOSLAHTI 1984). The future study of this receptor and its interaction with fibronectin will be of special interest for certain aspects of tumor biology.

In *metastatic processes* (SCHIRRMACHER 1984), attachment of tumor cells to *laminin* may play an important role. Tumor cells with a high affinity for laminin attachment display a higher pulmonary metastatic potential than those tumor cells with a low affinity for laminin (TERRANOVA et al. 1982). Antibodies against laminin can prevent the metastatic behaviour (McCARTHY et al. 1985). A specific receptor for laminin has been isolated in malignant cells (TERRANOVA et al. 1983). There are some indications of an association between the laminin molecule and the interaction with the contractile cytoskeleton (McCARTHY et al. 1985).

Future investigations will deal with a number of additional *attachment proteins*, such as gp 140, entactin, epinectin (McCARTHY et al. 1985) and others.

An interesting aspect is the *interaction* of the *basal membrane* associated substances with the *cytoskeleton* of the cell. Immunhistological and immunoelectronmicroscopic data point to the fact that fibronectin has an influence on binding to the orientation of actin filaments (D'ARDENNE and McGEE 1984). These studies would be interesting for the interpretation of a number of physiological and pathological aspects of the cell.

Basal membrane associated substances have been analysed in the *serum* of patients with neoplastic diseases (D'ARDENNE and McGEE 1984). A number of authors did not find differences in the plasma levels of fibronectin of normal and cancer patients. Intravascular coagulation produced a marked reduction in plasma fibronectin; liver metastasis were accompanied by low levels of fibronectin (D'Ardenne and McGee 1984).

Another point is the differences in the distribution pattern of the various basal membrane substances. It is necessary to establish the precise distribution pattern at the *ultrastructural level.* This would give new answers to several questions. Differences at the conventional immunohistogical level should be analysed on the basis of standardized techniques.

Basement membrane loss, however, is not always accompanied with malignancy. Some invasive tumors, especially the well-differentiated ones, exhibit focal extracellular reaction for both collagen type IV and laminin (Gusterson et al. 1986). However, basal membrane loss may be observed in some *reparative and inflammatory processes* (e.g. dermatitis herpetiformis).

Liotta (1984) stressed not only the importance of the composition of the basal membrane in neoplastic disease, but also the role of certain receptor proteins against the basal membrane substances. Experimental data suggest that there are receptors against this basal membrane component which may be altered in number or degree of occupancy in human neoplastic disease. Breast carcinoma tissue for example, contains a higher number of receptors and unoccupied receptors compared with benign breast tissue. The laminin receptors of normal breast epithelium seems to be polarized at the basal surface. In contrast, in invasive carcinoma cells this receptor may be distributed over the entire surface of the cell. The laminin receptor can stimulate hematogenous metastases.

There is evidence, however, that metastatic cells adhere specifically to basal membranes, essentially through an increase in plasma membrane receptors for laminin (Abrahamson 1986). When attached to basal membranes, these may be destroyed following the influence of proteolytic enzymes. This would permit the neoplastic cells to transmigrate across the altered basal membrane and metastasize. In contrast to malignant tumors, little apparent degradation takes place in benign tumors.

A particular point has to be made about certain special differentiated tumors like *mixed salivary gland tumors* which exhibit a focal augmentation of basal membrane associated material (e.g. laminin and fibronectin as "extrinsic" factors). Perhaps this increase in basal-membrane-like material may be related to the appearance of *myoepithelial cells* (and also to the clinical behavior). Other tumors with myoepithelial like cells display an augmentation of basal membrane material, although arranged in quite a different manner (e.g. pseudocysts in adenoid cystic carcinomas). This special arrangement may be a possible marker for the insiduous behaviour of the adenoid cystic carcinoma in the head and neck region.

References

Abrahamson DR (1986) Recent studies on the structure and pathology of basement membranes. J Pathol 149:257–278

Adkins KF, Daley TJ (1974) Elastic tissues in adenoid cystic carcinomas. Oral Surg 38:562–569

Albrechtsen R, Nielsen M, Wewer U, Engvall E, Ruoslahti E (1981) Basement membrane changes in breast cancer detected by immunohistochemical staining for laminin. Cancer Res 41:5076–5081

Azzopardi JG, Zayid I (1971) Elastic tissue in tumors of salivary glands. J Path 107:149–156

Barsky SH, Rao CN, Grotendorst GR, Liotta LA (1982) Increased content of type V collagen in desmoplasia of human breast carcinoma. Amer J Pathol 108:276–283

Barsky SH, Siegal GP, Jannotta F, Liotta LA (1983) Loss of basement membrane components by invasive tumors but not by their benign counterparts. Lab Invest 49:140–147

Barsky SH, Green WR, Grotendorst GR, Liotta LA (1984) Desmoplastic breast carcinoma as a source of human myofibroblasts. Amer J Path 115:329–333
Birembaut P, Caron Y, Adnet JJ, Foidart JM (1985) Usefulness of basement membrane markers in tumoral pathology. J Pathol 145:283–296
Bornstein P, Sage H (1980) Structurally distinct collagen types. Annu Rev Biochem 49:957–1003
Briggaman RA, Wheeler CE (1975) The epidermal-dermal junction. J Invest Dermatol 65:71–84
Bullock GR, Petrucz J (1982) Techniques in immunocytochemistry, vol 1. Academic Press, London New York
Bullock GR, Petrucz J (1983) Techniques in immunocytochemistry, vol 2. Academic Press, London New York
Burtin P, Chavanel G, Foidart JM, Martin E (1982) Antigens of basement membrane and the peritumoral stroma in human colonic adenocarcinomas. Immunofluorescence study. Int J Cancer 30:13–20
Cam Y, Caulet T, Bellon G, Poulin G, Legros M, Pytlinska M (1984) Immunohistochemical localization of macromolecules of the basement membrane and the peritumoral stroma in human laryngeal carcinomas. J Pathol 144:34–44
Carter RL, Burman JF, Barr L, Gusterson BA (1985) Immunohistochemical localization of basement membrane type IV collagen in invasive and metastatic squamous carcinomas of the head and neck. J Pathol 147:159–164
Caselitz J (1984) Das pleomorphe Adenom. Histogenese, zelluläre Differenzierung, Tumormarker. Habilitationsschrift, Hamburg
Caselitz J, Stegner HE, Osborn M, Walther B, Weber K (1985a) Die Rolle der Myoepithelzellen bei normalem und neoplastisch verändertem Mammagewebe. Verh Dtsch Ges Pathol 69:119–122
Caselitz J, Stegner HE, Spiegel A (1985b) Die Stromareaktion beim Mammacarcinom. Verh Dtsch Ges Pathol 69:254–257
Caselitz J, Schulze I, Seifert G (1986) Adenoid cystic carcinoma of the salivary glands. An immunohistochemical study. J Oral Pathol 15:308–318
Chung E, Rhodes RK, Miller E (1976) Isolation of three collagenous components of probable basement origin from several tissues. Biochem Biophys Res Commun 71:1167–1172
Cooper SM, Keyser AJ, Beaulieu AD, Ruoslahti E, Nimni ME, Quismorio FP (1979) Increase in fibronectin in the deep dermis of involved skin in progressive systemic sclerosis. Arthritis and Rheumatism 22:983–987
Courtoy PJ, Kanwar YS, Hynes RO, Farquhar MG (1980) Fibronectin localization in the rat glomerulus. J Cell Biol 87:691–696
D'Ardenne AJ, McGee JO (1984) Fibronectin in disease. J Pathol 142:235–251
D'Ardenne AJ, Burns J, Sykes BC, Bennett MK (1983a) Fibronectin and type III collagen in epithelial neoplasms of gastrointestinal tract and salivary glands. J Clin Pathol 36:756–763
D'Ardenne AJ, Burns J, Sykes BC, Kirkpatrick P (1983b) Comparative distribution of fibronectin and type III collagen in normal human tissues. J Pathol 141:55–69
Damiano VV, Tsang A, Christner P, Rosenbloom J, Weinbaum G (1978) Immunologic localization of elastin by electron microscopy. Am J Pathol 96:439–456
Damiano VV, Tsang A, Kucich U, Weinbaum G, Rosenbloom J (1981) Immunoelectronmicroscopic studies on cells synthesizing elastin. Conn Tiss Res 8:185–188
Damiano VV, Tsang A, Kucich U, Weinbaum G, Christner P, Rosenbloom J (1984) Secretion of elastin in the embryonic chick aorta as visualized by immunoelectron microscopy. Conn Tiss Res 4:153–164
David G, Van den Berghe H (1983) Transformed mouse mammary epithelial cells synthesize undersulfated basement membrane proteoglycan. J Biol Chem 258:7338–7344
David R, Buchner A (1980) Elastosis in benign and malignant salivary gland tumors. Cancer 45:2301–2310
Elias JM (1982) Principles and techniques in diagnostic histopathology. Noyes Publications, Park Ridge, USA
Engel J, Odermatt E, Angel A, Madri JA, Furthmayr H, Rohde H, Timpl R (1981) Shapes, domain organizations anf flexibility of laminin and fibronectin, two multifunctional proteins of the extracellular matrix. J Mol Biol 150:97–120
Engvall E, Ruoslahti E, Miller EJ (1978) Affinity of fibronectin to collagen of different genetic types and to fibrinogen. J Exp Med 147:1584–1595

Farquhar MG, Courtoy PJ, Lemkin MC, Kanwar YS (1982) Current knowledge of the functional architecture of the glomerula basement membrane. In: Kuehn R, Schoene HH, Timpl R (eds) New trends in basement membrane research. Raven Press, New York, pp 9–29

Fisher ER (1976) Ultrastructure of the human breast and its disorders. Am J Clin Pathol 66:291–375

Flotte TJ, Bell DA, Greco MA (1980) Tubular carcinoma and sclerosing adenosis. Am J Surg Pathol 4:75–77

Foidart JM (1984) Les membranes basales. Arch Anat Cytol Path 32:153–155

Foidart JM, Yaar M (1981) Type IV collagen, laminin and fibronectin at the dermo-epidermal junction. In: Prunieras M, Robert L (eds) Epidermal keratinocyte. Differentiation and fibrillogenesis. Karger, Basel, pp 175–188

Foidart M, Bere EW, Yaar M, Rennard SI, Gullino M, Martin GR, Katz SI (1980) Distribution and immunoelectron microscopic localization of laminin, a noncollagenous basement membrane glycoprotein. Lab Invest 42:336–342

Fukuda Y, Ferrans VJ, Crystal RG (1984) Development of elastic fibers of nuchal ligament, aorta and lung of fetal and postnatal sheep: An ultrastructural and electron microscopic immunohistochemical study. Am J Anat 170:597–629

Furcht LT, Mosher DF, Wendelshafer-Crabb G (1978a) Immunocytochemical localization of fibronectin (LETS protein) on the surface of L6 myoblasts: Light and electron microscopic studies. Cell 13:263–271

Furcht LT, Mosher DG, Wendelschafer-Crabb G (1978b) Effects of cell density and transformation on the formation of a fibronectin extracellular filamentous matrix on human fibroblasts. Cancer Res 38:4618–4623

Furthmayr H (ed) (1982) Immunohistochemistry of the extracellular matrix, vol I, II. CRC Press, Boca Raton, Florida

Fyrand O (1980) Studies on fibronectin in the skin. IV. Indirect immunofluorescence studies in lupus erythematodes. Br J Dermatol 102:167–171

Gay S, Gay RE (1985) Pathobiochemistry of connective tissue in breast cancer. Verh Dtsch Ges Pathol 69:62–73

Gay S, Miller EJ (1983) What is collagen, and what is not. Ultrastruct Pathol 4:365–377

Gay S, Rhodes RK (1986) Immunolocalization of the genetically-distinct collagen types in the pathology of connective tissue. In: Spicer SS, Garvin AJ, Hennigar GR (eds) Applications of histochemisty of pathologic diagnosis. Marcel Decker, New York

Gay S, Martinez-Hernandez A, Rhodes RK, Milkler EJ (1981) The collagenous exocytoskeleton of smooth muscle cells. Collagen Rel Res 1:377–386

Gusterson BA, Warbuton MJ, Mitchell D, Ellison M, Neville AM, Rudland PS (1982) Distribution of myoepithelial cells and basement membrane proteins in the normal breast and benign and malignant breast disease. Cancer Res 42:4763–4770

Gusterson BA, Warburton MJ, Mitchell D, Kraft N, Hancock WW (1984) Invading squamous cell carcinoma can retain a basal lamina. An immunohistochemical study using a monoclonal antibody to type IV collagen. Lab Invest 51:82–87

Gusterson BA, Clinton SS, Gough G (1986) Studies of early invasive and intraepithelial squamous cell carcinomas using an antibody to type IV collagen. Histopathology 10:161–169

Haglund C, Nordling S, Roberts PJ, Ekblom P (1984) Expression of laminin in pancreatic neoplasms and in chronic pancreatitis. Am J Surg Pathol 8:669–676

Hantai D, Gautron J, Labat-Robert J (1983) Immunolocalization of fibronectin and other macromolecules of the intercellular matrix in the striated muscle fiber of the adult rat. Collagen Rel Res 3:381–391

Hay ED (1978) Role of basement membranes in development and differentiation. In: Kefalides NA (ed) Biology and chemistry of basement membranes. Academic Press, New York, pp 118–136

Hedman K (1980) Intracellular localization of fibronectin using immunoperoxidase cytochemistry in light and electron microscopy. J Histochem Cytochem 28:1233–1241

Hedman K, Vaheri A, Wartivaara J (1978) External fibronectin of cultured human fibroblasts is predominantly a matrix protein. J Cell Biol 76:748–760

Hynes RO (1981) Fibronectin and its relation to cellular structure and behavior. In: Hay ED (ed) Cell biology of extracellular matrix. Plenum Press, New York, pp 295–33

Hynes RO, Yamada KM (1982) Fibronectins: Multifunctional modular glycoproteins. J Cell Biol 95:369–377

Ingber DE, Madri JA, Jamieson JD (1986) Basement membrane reorients pancreatic epithelial tumor cells in vitro. Am J Pathol 122:129–139

Iozzo RV (1984) Proteoglycans and neoplastic-mesenchymal cell interactions. Hum Pathol 15:2–10

Kahn P, Shin SI (1979) Cellular tumorgenicity in nude mice: Test of associations among loss of cell-surface fibronectin, anchorage independence, and tumor-forming ability. J Cell Biol 82:1–16

Kallioinen M (1985) Immunoelectron microscopic demonstration of the basement membrane components laminin and type IV collagen in the dermal cylindroma. J Pathol 147:97–192

Kallioinen M, Autio-Harmainen H, Dammert K, Risteli J, Risteli L (1984) Basement membrane laminin and type IV collagen in various benign and malignent adnexal tumors of the skin: An immunohistochemical study. J Invest Dermatol 83:276–280

Kao RT, Hall J, Engel L, Stern R (1984) The matrix of human breast tumor cells is mitogenic for fibroblasts. Am J Pathol 115:109–116

Kellokumpu I, Ekblom P, Scheinin TM, Andersson LC (1985) Malignant transformation in human colorectal mucosa as monitored by distribution of laminin, a basement membrane glycoprotein. Acta Pathol Microbiol Immunol Scand Sect A 93:285–291

Keski-Oja J, Sen A, Todaro GJ (1980) Direct association of fibronectin and actin molecules in vitro. J Cell Biol 85:527–533

Kuehn K, Wiedemann H, Timpl R, Ristelli J, Dieringer H, Voss T, Glanville RW (1981) Macromolecular structure of basement membrane collagens. Identification of 7S collagen as a crosslinking domain of type IV collagen. FEBS Lett 125:123–128

Kuusela P, Vaheri A, Palo J, Ruoslahti E (1978) Demonstration of fibronectin in human cerebrospinal fluid. J Lab Clin Med 92:595–601

Labat-Robert J, Birembaut P, Robert L, Adnet JJ (1981) Modification of fibronectin distribution pattern in solid human tumors. Diag Histopathol 4:299–306

Laurie GW, Leblond CP, Martin GR (1982) Localization of type IV collagen, laminin, heparan sulfate proteoglycan, and fibronectin to the basal lamina of basement membranes. J Cell Biol 95:340–344

Linder E, Vaheri A, Ruoslahti, Wartiovaara J (1975) Distribution of fibroblast surface antigen in the developing chick embryo. J Exp Med 142:42–49

Liotta LA (1982) Tumor extracellular matrix. Lab Invest 47:112–113

Liotta LA (1984) Tumor invasion and metastases: Role of the basement membrane. Am J Pathol 117:339–348

Liotta LA, Rao CN, Barsky SH (1983) Tumor invasion and the extracellular matrix. Lab Invest 49:636–649

Madri JA, Roll FJ, Furthmayr H, Foidart JM (1980) Ultrastructural localization of fibronectin and laminin in the basement membranes of the murine kidney. J Cell Biol 86:682–687

Martin GR, Kleinman HK, Terranova VP, Ledbetter S, Hassell JR (1984) The regulation of basement membrane formation and cell-matrix interactions by defined supramolecular complexes. Ciba Found Symp 108:197–212

Martin GR, Timpl R, Müller PK, Kühn K (1985) The genetically distinct collagens. Trends Biochem Sci 10:285–287

Martinez-Hernandez A, Amenta PS (1983) The basement membrane in pathology. Lab Invest 48:656–677

Martinez-Hernandez A, Marsh CA, Clark CC, Macarak EJ, Brownell AG (1981a) Fibronectin: its relationship to basement membranes. II. Ultrastructural studies in rat kidney. Collagen Rel Res 1:405–418

Martinez-Hernandez A, Marsh CA, Horn JF, Munoz E (1981b) Glomerular basement membrane: lamina rara, lamina densa. Renal Physiol 4:137–147

Martinez-Hernandez A, Gay S, Miller EJ (1982) Ultrastructural localization of type V collagen in rat kidney. J Cell Biol 92:343–351

McCarthy JB, Basara ML, Palm SL, Sas DF, Furcht LT (1985) The role of cell adhesions proteins – laminin and fibronectin – in the movement of malignant and metastatic cells. Cancer Metast Rev 4:125–152

Mosher DF, Saksela O, Keski-Oja J, Vaheri A (1977) Distribution of a major surface-associated glycoprotein, fibronectin, in cultures of adherent cells. J Supramol Struct 6:551–557

Neri A, Ruoslahti E, Nicolson GL (1981) Distribution of fibronectin on clonal cell lines of a rat

mammary adenocarcinoma growing in vitro and in vivo at primary and metastatic sites. Cancer Res 41:5082–5095
Ozzello L (1959) The behaviour of basement membranes in intraductal carcinoma of the breast. Am J Pathol 35:887–891
Palmer RM (1986) The identification of myoepithelial cells in human salivary glands. A review and comparison of light microscopical methods. J Oral Pathol 15:221–229
Pitelka D, Hamamoto ST, Taggart BN (1980) Basal lamina and tissue recognition in malignant mammary tumors. Cancer Res 40:1600–1611
Polak JM, Van Noorden S (1983) Immunocytochemistry, practical applications in pathology and biology. John Wright and Sons, Bristol
Quaroni A, Isselbacher KJ, Rouslahti E (1978) Fibronectin synthesis by epithelial crypt cells of rat small intestine. Proc Natl Acad Sci USA 75:5548–5552
Remberger K, Nerlich A (1985) Diagnostischer Wert der Darstellung von Basalmembranproteinen in benignen und malignen Mammaveränderungen. Verh Dtsch Ges Pathol 69:123–130
Roll FJ, Madri JA, Albert J, Furthmayr H (1980) Codistribution of collagen types IV and AB 2 in basement membranes and mesangium of the kidney. J Cell Biol 85:597–616
Ruoslahti E (1981) Fibronectin. J Oral Pathol 10:3–13
Ruoslahti E (1984) Fibronectin in cell adhesion and invasion. Cancer Metast Rev 3:43–51
Ruoslahti E, Hayman EG, Engvall E (1980) Fibronectin. In: Sell S (ed) Cancer markers. Humana Press, Clifton, New Jersey, pp 485–505
Ruoslahti E, Engvall E, Hayman EG (1981) Fibronectin: Current concepts of its structure and functions. Coll Rel Res 1:95–128
Ruoslahti E, Hayman EG, Pierschbacher M, Engvall E (1982) Fibronectin: Purification, immunochemical properties, and biological properties. Meth Enzymol 82:803–831
Salo T, Liotta LA, Keski-Oia J, Turpeenniemi-Hujanen T, Tryggvason K (1982) Secretion of basement membrane collagen degrading enzyme and plasminogen activator by transformed cells – role in metastasis. Int J Cancer 30:669–673
Sano J, Fujiwara S, Sato S, Ishizaki M, Su I (1981) AB (type V) and basement membrane (type IV) collagens in the bovine lung parenchyma: Electron microscopic localization by peroxidase-labeled antibody method. Biomed Res 2:20–29
Schauer A, Brehler R, Bergholz M (1985) Immunhistochemische Brustkrebsdiagnostik. Verh Dtsch Ges Pathol 69:171–186
Scheinman JI, Fisch AJ, Matas A, Michael AF (1978) The immunohistopathology of glomerular antigens. Am J Pathol 90:71–84
Schirrmacher V (1984) Eigenschaften von Tumorzellen als Voraussetzung der Metastasierung: Untersuchungen zum metastatischen Phänotyp. Verh Dtsch Ges Pathol 68:12–17
Schuppan D, Timpl R, Glanville RW (1980) Discontinuities in the triple helical sequence Gly-X-Y of basement membrane (type IV) collagen. FEBS Lett 115:297–300
Seemayer TA, Lagac'e R, Schürch W, Thelmo WL (1980) The myofibroblast: Biologic, pathologic, and theoretical considerations. Pathol Annu 15:443–470
Siegal GP, Barsky SH, Terranova VP, Liotta LA (1981) Stages of neoplastic transformation of human breast tissue as monitored by dissolution of basement membrane components. Invasion Metastasis 1:54–70
Stampfer MR, Vlodavsky I, Smith MS, Ford R, Becker FF, Riggs J (1981) Fibronectin production in human mammary cells. J Natl Cancer Inst 67:253–261
Stanley JR, Beckwith JB, Fuller RP, Katz SI (1982a) A specific antigenic defect of basement membrane is found in basal cell carcinoma but not in other epidermal tumors. Cancer 50:1486–1490
Stanley JR, Woodley DT, Katz SI, Martin GR (1982b) Structure and function of basement membrane. J Invest Dermatol 79, 69s–72s
Stegner HE (1986) Histopathologie der Mammatumoren. Licht- und elektronenmikroskopischer Atlas. Enke, Stuttgart (English edition in preparation)
Stenman S, Vaheri A (1981) Fibronectin human solid tumors. Int J Cancer 27:427–435
Takashima A, Grinnell F (1984) Human keratinocyte adhesion and phagocytosis promoted by fibronectin. J Invest Dermatol 83:352–358
Tarin D, Hoyt BJ, Evans DJ (1982) Correlation of collagenase secretion with metastic-colonization potential in naturally occurring murine mammary tumors. Br J Cancer 23:417–425
Taylor CR (1986) Immunomicroscopy: A diagnostic tool for the surgical pathologist. Saunders, Philadelphia London Toronto

Terranova VP, Rohrbach DH, Martin GR (1980) Role of laminin in the attachment of PAM 212 (epithelial) cells to basement membrane collagen. Cell 22:719–726

Terranova VP, Liotta LA, Russo RG, Martin GR (1982) Role of laminin in the attachment and metastasis of murine tumor cells. Cancer Res 42:2265–2269

Terranova VP, Rao CN, Kalebic T, Margulies IM, Goldstein IJ, Liotta LA (1983) Laminin receptor on human breast carcinoma cells. Proc Natl Acad Sci 444–448

Timpl R, Rohde H, Robey PG, Rennard SI, Foidart JM, Martin GR (1979) Laminin- a glycoprotein from basement membranes. J Biol Chem 254:9933–9937

Timpl R, Wiedemann H, Van Delden V, Furthmayr H, Kühn K (1981) A network model for the organization of type IV collagen molecules in basement membranes. Eur J Biochem 120:203–211

Timpl R, Rohde H, Ristelli L, Ott U, Robey PG, Martin GR (1982) Laminin. In: Cunningham L, Fredericksen DW (eds) Structural and contractile proteins. Meth Enzymol 82:831–848

Timpl R, Dziadek M, Fujiwara S, Nowack H, Wick G (1983) Nidogen: A new, self-aggregating basement membrane protein. Eur J Biochem 137:455–465

Toida M, Takeuchi J, Hara K, Sobue M, Tsukidata K, Goto K, Nakashima K (1984) Histochemical studies of intercellular components of salivary gland tumors with special reference to glycosaminoglycan, laminin and vascular elements. Virchows Arch (Pathol Anat) 403:15–26

Tokuyasu KT (1980) Immunochemistry on ultrathin frozen sections. Histochem J 12:381–403

Toto PD, Hsu DJ (1985) Product definition of pleomorphic adenoma of minor salivary glands. J Oral Pathol 14:818–832

Urry DW (1983) What is elastin; what is not. Ultrastruct Pathol 4:227–251

Vacca LL (1985) Laboratory manual of histochemistry. Raven Press, New York

Visser R, Van der Beek JMH, Havenith MG, Cleutjens JPM, Bosman FT (1986) Immunocytochemical detection of basement membrane antigens in the histopathological evaluation of laryngeal dysplasia and neoplasia. Histopathology 10:171–180

Vracko R (1974) Basal lamina scaffold-Anatomy and significance for maintenance of orderly tissue structure. Am J Pathol 77:313–346

Vracko R (1982) The role of basal lamina in maintenance of orderly tissue structure. In: Kuehn K, Schoene H, Timpl R (eds) New trends in basement membrane research. Raven Press, New York, pp 1–7

Warburton MJ, Ferns SA, Rudland PS (1982a) Enhanced synthesis of basement membrane proteins during the differentiation of rat mammary tumor epithelial cells into myoepithelial-like cells in vitro. Exp Cell Res 137:373–380

Warburton MJ, Mitchell D, Ormerod EJ, Rudland P (1982b) Distribution of myoepithelial cells and basement membrane proteins in the resting, pregnant, lactating, and involuting rat mammary gland. J Histochem Cytochem 10:667–676

Weber L, Krieg T, Müller PK, Kirsch E, Timpl R (1982) Immunofluorescent localization of type IV collagen and laminin in human skin and its application in junctional zone pathology. Br J Dermatol 106:267–273

Weber L, Krieg T, Timpl R (1984a) Übersicht: Basalmembranen – Struktur, Funktion, Pathologie. Hautarzt 35:279–286

Weber L, Wick G, Gebhart W, Krieg T, Timpl R (1984b) Basement membrane components outline the tumor islands in cylindroma. Br J Dermatol 111:45–51

Weiss MA, Ooi BS, Ooi YM, Engvall E, Ruoslahti E (1979) Immunofluorescent localization of fibronectin in the human kidney. Lab Invest 41:340–347

Wells CA, Nicoll S, Ferguson DJP (1986) Adenoid cystic carcinoma of the breast: a case with axillary lymph node metastasis. Histopathology 10:415–424

Wick G, Timpl R (1980) Study on the nature of the Good-Pasture antigen using a basement-producing mouse tumor. Clin Exp Immunol 39:733–741

Willingham MC (1980) Electron microscopic immunocytochemical localization of intracellular antigens in cultured cells: the EGF and ferritin bridge procedure. Histochem J 12:419–434

Yang NS, Kirkland W, Jorgenson T, Furmanski P (1980) Absence of fibronectin and presence of plasminogen activator in both normal and malignant human mammary epithelial cells in culture. J Cell Biol 84:120–130

Lectins and Blood Group Substances as "Tumor Markers" *

J. CASELITZ

* Supported by the *Sonderforschungsbereich 232* and by the *Hamburger Stiftung zur Förderung der Krebsbekämpfung*

List of abbreviations

Abbreviation	Name	Abbreviation	Name
BSA I, II	Bandeira (Griffonia) simplicifolia I, II	PHA	Phaseolus vulgaris (red kidney bean)
Con A	Concanavalia A (Canavalia ensiformis)	PNA	Arachis hypogaea (pea nut)
DBA	Dolichos biflorus (horse gram)	PWA	Phytolacca americana (Pokeweed)
HPA	Helix pomatia	PSA	Pisum sativum
LCA	Lens culinaris	RCA	Ricinus communis
LPA	Limulus polyphemus	SBA	Glycine max (Soy bean)
LTA	Lotus tetragonolobus	STA	Solanum tuberosum
MPA	Maclura pomifera	TGP	Tetragonolobus purpurea
PLA	Phaseolus lunatus (lima bean)	UEA	Ulex europaeus
		VVA	Vicia villosa
PHA	Phaseolus vulgaris (black kidney bean)	WGA	Triticum vulgaris (Wheat germ)

1 Introduction

In recent decades, research efforts have been focused on demonstrating changes in cell surface membrane during neoplastic transformation. This emphasis stems from the observation that the cell surface membrane may determine several characteristics of neoplastic cells such as decreased adhesion, loss of contact inhibition, increased growth rate, prolonged survival, increased invasiveness and motility, expression of "new" antigens and escape from immunological surveillance (review for glycolipids and glycoproteins in cancer: EMMELOT 1973; KÖTTGEN et al. 1979; YOGEESWARAN 1983; HAKOMORI 1984; BAENZINGER 1986).

A recent advance in the histochemistry of complex carbohydrates resulted from the use of lectins which bind to specific carbohydrates and localize them in tissue sections (RAPIN and BURGER 1974). AB isoantigens are widely distributed in human tissues and loss of AB isoantigen has been shown to be an early marker for neoplastic transformation in some tissues. The relation of ABH blood groups and development has been reviewed by SZULMAN (1980).

Lectins are defined as proteins or glycoproteins of non-immune origin which bind specifically to carbohydrate residues. The term "lectin" is derived from the latin "legere" and refers the specificity to "pick out" special carbohydrate chains (GOLDSTEIN and HAYES 1978; ROTH 1978). Lectins are often, but not exclusively, derived from plants and from animals (invertebrates and in lower vertebrates) (GOLDSTEIN 1981; LEATHEM and ATKINS 1983). The ubiquitous existence of lectins has now been established (LIS and SHARON 1986). More than 100 lectins have been purified. All possess at least two binding sites for carbohydrate chains. Through their specific interaction, they exhibit many important biological functions such as distinguishing human blood groups, preferential agglutination of immature or malignant cells and mitogenic stimulation of lymphocytes.

The extensive literature on lectins has been summarized in several reviews (UHLENBRUCK 1971, 1980, 1981; NICOLSON 1974, 1976, 1978; SHARON and LIS 1975; SCHREVEL et al. 1981; BOG-HANSEN 1982; SCHAUER 1982a, b; BOG-HANSEN and SPENGLER 1983; LEATHEM et al. 1983; BOG-HANSEN and VAN DRIESSCHE 1986; LIS and SHARON 1986).

The lectins may be classified on the basis of their carbohydrate binding specificity (usually agglutination of erythrocytes or precipitation of carbohydrate containing polymers) (GOLDSTEIN 1981; LIS and SHARON 1986).

1. D-Mannose (D-Glucose) binding lectins:
Canavalia ensiformis, Lens culinaris, Pisum sativum, Vicia faba.

This group is mitogenic for lymphocytes. According to their amino acid sequences, these lectins seem to be closely related.

2. N-Acetyl-D-Galactosamine binding lectins:
Dolichos biflorus, Phaseolus vulgaris, Soy bean agglutinin, Helix pomatia, Saphora japonica.

These lectins bind and agglutinate type A red blood cells. Phaseolus vulgaris (lima bean) is a good mitogen.

3. N-Acetyl-D-Glucosamine binding lectins:
Wheat germ, Ulex europaeus I, Solanum tuberosum, Phytolacca americana (Poke weed), Datura stramonium, Griffonia (sive Bandeiraea) simplicifolia I.

These lectins bind to chitin and are best inhibited by oligomeres of N acetyl D glucosamine.

Pokeweed is a potent mitogen for T and B lymphocytes, whereas Wheat germ agglutinin and Datura stramonium agglutinin are only weakly mitogenic towards T cells.

4. D-Galactose binding lectins:
Abrus precatorius, Arachis hypogaea (Pea nut), Griffonia (sive Bandeiraea) simplicifolia I, Maclura pomifera, Ricinus communis (Castor bean).

The GSA (BSA) and maclura pomifera lectins agglutinate type B erythrocytes.

5. L-fucose binding lectins:
Lotus tetragonolobus agglutinin, Ulex europaeus I, Anguilla anguilla.

These non-mitogenic lectins are useful reagents for typing of human O erythrocytes.

6. Sialic acid binding lectins:
Sialic acid binding lectins have been isolated from the horseshoe crab (Limulus polyphemus) and other types of crabs.

The importance of sialic acid containing glycoconjugates on cell surfaces should make these lectins most interesting for biology and other related scientific disciplines.

7. Lectins with complex carbohydrate binding sites:
Phaseolus vulgaris (red kidney bean)
Agaricus bisporus (mushroom)

The lectin of the red kidney bean has a complex binding site.

Neoplastic cell transformation is associated with an *altered carbohydrate composition* of plasma membranes (NICOLSON 1976; WARREN et al. 1978). Demonstration of altered lectin binding by tumor cells has been used to predict the likelihood of recurrence, the invasive potential of the neoplasm and to distinguish benign from malignant tumors. Lectin binding by tumor cells in tissue sections has been used for the study of various normal, dysplastic, and neoplastic tissues. The aim of this presentation is to elucidate the use of lectins as *morphological tumor markers*, and the literature reviewed is influenced by this purpose. Other aspects of lectins are dealt with in some excellent reviews (UHLENBRUCK 1971; NICOLSON 1974; SHARON and LIS 1975; ROTH 1978; SCHREVEL et al. 1981; SCHAUER 1982a, b; LEATHEM et al. 1983; LIS and SHARON 1986).

Among the large number of lectins, only some of these substances have been analysed on human materials. Those which are commonly used are: Peanut agglutinin (Arachis hypopaea), Wheatgerm agglutinin (Tritium vulgaris), Ulex europaeus, Helix pomatia, Dolichos biflorus, Ricinus communis, Concanavalia ensiformis (COOPER 1984). The relation of some lectins to the blood group substances is given in Table 1 (for details: NICOLSON 1974; ROTH 1978; SCHREVEL et al. 1981; LEATHEM et al. 1983; LIS and SHARON 1986).

Table 1. Lectins and their relation to blood group substances

Blood group agglutinated	Origin of lectin
A	Phaseolus limensis Viccia cracca Dolichos biflorus Crotalaria aegyptiaca
B	Bandeiraea simplicifolia I
A+B	Sophora japonica Calpurinea aurea
H	Cytisus sessiliformis Laburnum alpinum Lotus tetragonolobus Ulex europaeus
M	Iris amara
N	Vivia graminea Bauhinia purpurea
Disaccharide of the Thomsen-Friedenreich antigen	Arachis hypopeae

2 Methodology of Detection of Carbohydrate Residues in the Cellular Membrane

With regard to application in cytochemistry (SCHREVEL et al. 1979), the methods for the detection of cell glycoconjugates are summarized by SCHREVEL et al. (1981) and CULLING and REID (1982). Four main methods can be distinguished:

1. Detection of *anionic sites*
   a) Use of cationic dyes (Alcian red, toluidine blue, ruthenium red)
   b) Use of cationic colloid (colloidal iron)
   c) Use of cationized ferritin or proteins.
2. Detection of *alcohol residues*
3. Detection of *carbohydrate groups by use of lectins*
4. Application of *antibodies* (esp. monoclonal antibodies).

Methods 1 and 2 have to be regarded as techniques of broad specificity, whereas 3 and 4 are techniques of rather narrow specificity.

This paper concentrates on the application of lectins.

2.1 Affinity Histochemistry by Lectins

Lectins have been used in connection with various techniques for the identification of specific carbohydrate residues in tissue sections (ROTH and BINDER 1978; SCHULTE and SPICER 1983a, b, c, 1984; ALROY et al. 1984; HORISBERGER 1984; ROTH and TAATJES 1985). Although for some purposes frozen sections may be preferred (study of glycolipid receptors), formaldehyde fixed and paraffin embedded material is suitable for the investigation of many lectin binding sites.

At the *light microscopic level,* lectins have been applied with the following modifications (LEATHEM and ATKINS 1983):

1. *conjugated lectins* (coupled to fluorochromes or chromagens) have been used directly on tissue sections, e.g. lectin-fluorescein isothiocyanate (FITC) and lectin-horseradish peroxidase,
2. in *indirect immunological techniques,* lectins can be used in modifications like lectin-antilectin-FITC (or horseradish peroxidase).
3. lectins can be used as the primary step in the *peroxidase-antiperoxidase technique* (PAP technique)
4. after being coupled to *biotin,* lectins can be used in the avidin-biotin-peroxidase technique.

Although most papers present data for lectin use in light microscopy, elegant studies are being presented on the *ultrastructural* level, by using lectins coupled to gold (ROTH et al. 1983; ROTH and TAATJES 1985; ROTH 1986). The principles of the reactions are analogous to those in light microscopy.

Other applications are the use of lectin binding in *radioautography,* which may be quantified (SAVERIANO et al. 1981).

2.2 Methods for the Detection of Blood Group Antigens

Several methods can be applied to detect carbohydrate residues and blood group antigens on the cellular membrane (DENK et al. 1974; WEINSTEIN et al. 1981). The red cell adherence technique has been applied on histological sections for some time (WEINSTEIN et al. 1981). Immunohistochemical techniques may be used as well: PAP technique (WEINSTEIN et al. 1981) and the avidin-biotin-system on the basis of polyclonal and monoclonal antibodies. An overview of the distribution of blood grup substances is given in Tables 2 and 3.

Table 2. ABH antigens in human tumors. (According to WEINSTEIN et al. 1981)

Organ	Tumor
Blood vessels	Angioma
	Angiosarcoma
Breast	Adenocarcinoma
Colon	Adenocarcinoma
Fallopian tube	Adenocarcinoma
Larynx	Squamous carcinoma
Lung	Squamous carcinoma
	adenocarcinoma
Oral epithelium	Squamous carcinoma
Ovary	Cystadenoma
	Cystadenocarcinoma
Pancreas	Adenocarcinoma
Prostate	Adenocarcinoma
Skin	Epidermal appendage tumours
Stomach	Adenocarcinoma
Urinary bladder	Transitional carcinoma
	Inverted papilloma
Uterine cervix	Squamous carcinoma

Table 3. Presence of Isoantigens A, B, H in normal adult human tissues. (According to WEINSTEIN et al. 1981)

Location	Positive	Negative
Adrenal		Cortex, medulla
Blood	Erythrocytes	White blood cells
Blood vessels	Endothelial cells	
Breast	Glands Ducts	
Bronchus	Pseudostratified columnar epithelia Mucous glands	Serous glands
Central nervous system		Nerve cells Glial cells
Esophagus	Squamous epithelium	Basal cell layer
Duodenum	Epithelial cells Brunner's glands	
Colon	Proximal colon	Distal colon
Gallbladder	Columnar epithelia	
Endometrium	Epithelia (proliferative and secretory)	
Exocervix	Squamous epithelia	
Fallopian tube	Columnar epithelia	
Kidney	Glomeruli collecting tubes convoluted tubes	
Larynx	Squamous epithelia mucous glands	Serous glands
Liver	Kupffer cells	Hepatocytes
Lung	Mucous glands	Serous glands, Hepatocytes
Muscles		smooth and striated muscle cells
Ovary	Cyst fluid	
Pancreas	exocrine glands	islets of Langerhans
Pituitary	colloid pars intermedia	
Placenta	Trophoblast (partly)	Trophoblast Cytotrophoblast
Prostate	Glands	
Epidermis	Horny layer of squamous epithelium Sweat glands	Cells of basal and Malphigian layers Sebaceous glands
Small intestine	Columnar epithelia Brunner's glands Paneth cells	
Stomach	Columnar epithelia	
Testis		Tubules
Tongue	Squamous epithelia Mucous glands	Basal layer
Trachea	Pseudostratified Columnar epithelia Mucous glands	serous glands
Urinary system	Urinary bladder, ureter	
Vagina	Squamous epithelia mucous glands	basal layer
Connective tissue		Collagen, bone, cartilage

3 Distribution Pattern in Human Organs and Tumors

3.1 Epidermis and Mucosa

The human epidermis (and related structures, like oral mucosa) represents a model of differentiation, the expression of which may be analysed on fundamental and applied levels (ELIAS et al. 1983; RITTMAN and MACKENZIE 1983). REANO et al. (1982) have divided the lectins into 3 groups:

Group 1 (BSA I, DBA, Limulus polyphemus, TGP, UEA, WGA) did not stain the normal epidermis.
Group 2 (Canavalia ensiformis, Maclura pomifera, Phaseolus vulgaris, RCA 1) exhibited a "pemphigus-like" intercellular labelling of the whole epidermis with exception of the stratum corneum.
Group 3 (PNA, Glycine max, Helix pomatia, Sophora japonica) displayed a selective intercellular labelling of the stratum spinosum and the stratum granulosum. In psoriatic epidermis, not only the basal layer, but also cells from the adjacent lower stratum spinosum were found to be negative for the lectins in group 3 (REANO et al. 1982).

PNA (Peanut Agglutinin) (Fig. 1)

PNA binding sites in human epidermis have been analysed by REANO et al. (1982), HYUN et al. (1984a–d) and LÖNING (1984). Its expression in squamous cell carcinomas is variable. Most Merkel cell tumors were positive for PNA (PAJOR et al. 1986), this lectin being considered as a valuable marker for this tumor.

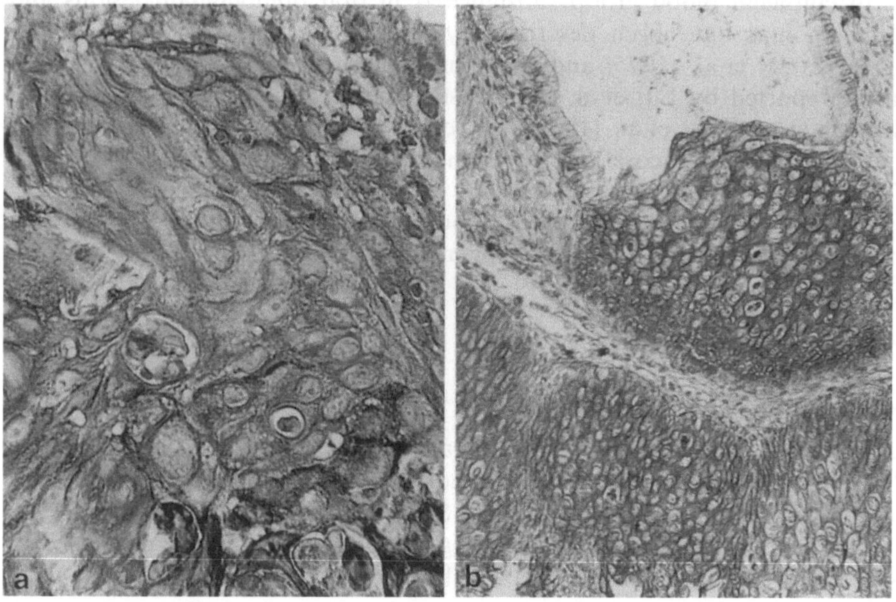

Fig. 1. a PNA binding sites in squamous cell carcinoma of the skin. Intense staining of tumor cells after sialidase (neuraminidase) digestion. ×300. **b** PNA binding sites in carcinoma in situ of the cervix. Intense staining of tumor cells after sialidase (neuraminidase) digestion. ×300

WGA (Wheat Germ Agglutinin)

WGA receptors do not play an important role in carcinomas of the epidermis and the mucosa (REANO et al. 1982, HYUN et al. 1984a–d). WGA receptors display a zonal arrangement in the oral mucosa (HOSAKA et al. 1985).

UEA (Ulex Europaeus Agglutinin)

UEA1 receptors do not play an important role in carcinomas of the epidermis and oral mucosa (REANO et al. 1982, HOSAKA et al. 1985).

HPA (Helix Pomatia Agglutinin)

The binding of HPA to squamous carcinomas of the epidermis and the mucosa is variable (LÖNING 1984).

Basal cell carcinomas are negative for DBA, whereas normal basal cells are in about 50% positive (COGGI et al. 1983). RCA 1 binding was slightly decreased in *actinic keratoses* and cases of *Bowen's disease*. In epidermal outgrowths from the edges of healing wound and in squamous and basal cell carcinomas, the binding of RCA 1 was diminished (GRAEM 1982). Receptors in oral carcinomas could not be demonstrated in invading islands of the tumors (DABELSTEEN and MACKENZIE 1978; DABELSTEEN et al. 1978). There is a shift in carbohydrate composition in *inflammatory* and, more important, in *neoplastic* lesions (DABELSTEEN and MACKENZIE 1978).

Other lectins (Con A, SBA, BSA, SJA) have been analysed in the human epidermis and oral mucosa and its neoplastic counterparts (REANO et al. 1982; GRAEM and DABELSTEEN 1984; HOSAKA et al. 1985).

Blood Group Antigens

Loss of epithelial blood group substance A in oral premalignant lesions and oral carcinomas have been described by DABELSTEEN and PINDBORG (1973), by DABELSTEEN et al. (1975) and by DABELSTEEN and DANIELS (1983). Similar data are reported by LIU et al. (1974) using the red cell adherence technique, and by PRENDERGAST et al. (1968). LIN et al. (1977) compared isoantigens A, B and H in normal mucosa and in carcinoma of the larynx. DABELSTEEN et al. (1974) analysed blood group substance A in carcinoma of the larynx.

ENGLAND et al. (1979) analysed blood group antigens A, B, H in normal skin and in tumors of the epidermal appendages.

3.2 Salivary Glands

PNA (Peanut Agglutinin) (Fig. 2)

PNA receptors are found in many salivary gland tumours, especially in the malignant ones like adenoid cystic carcinomas (CASELITZ et al. 1986), but also in pleomorphic adenomas (CASELITZ and SEIFERT 1981; CASELITZ 1987). DALEY et al. (1985) report a weak PNA staining in various salivary gland tumours, mostly apically localized, in the ductal cells or in the cystic areas.

WGA (Wheat Germ Agglutinin) (Fig. 3)

Corresponding to the normal tissue, WGA generally decorates the luminal membrane of the duct cells in the tubular parts of pleomorphic adenomas and ade-

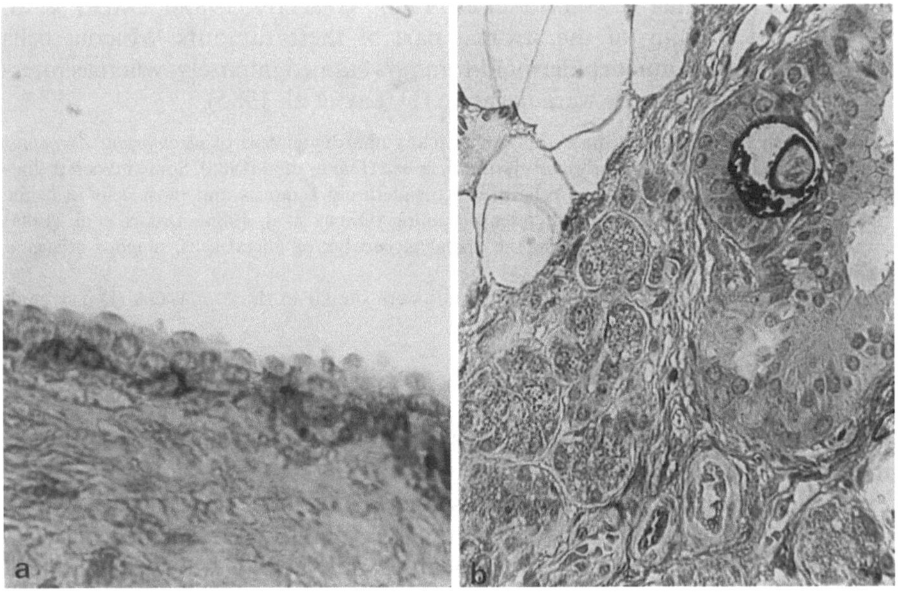

Fig. 2. **a** PNA binding sites in excretory duct of human parotid gland. Staining of basal cell fraction after sialidase digestion. ×300. **b** PNA binding sites in duct cells of human parotid gland. Staining of the apical part of the cells. ×300

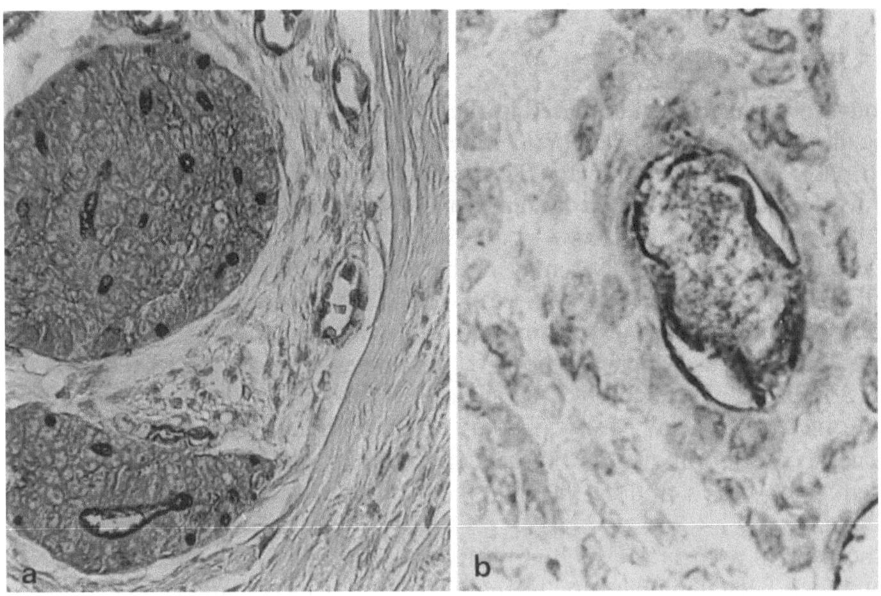

Fig. 3. **a** PNA binding sites in tubular part of adenoid cystic carcinoma. Staining of the luminal surface of the tumor cells. ×300. **b** WGA binding sites in pleomorphic adenoma. Strong staining of the apical part in the tubules. ×300

noid cystic carcinomas (CASELITZ et al. 1986; CASELITZ 1987). DALEY et al. (1985) report a staining of the stromal part of these tumours. Mucous cells and cystic contents in mucoepidermoid tumours stained intensely, whereas intermediate and squamous cells were negative (DALEY et al. 1985).

Con A binding was present in the cytoplasm of many non-ductal cells of *pleomorphic adenomas* and in most of the basaloid cells of *adenoid cystic carcinoma* (DALEY et al. 1985). Some myoepithelial-like cells were positive. Most mucous cells in mucoepidermoid tumours and most cells in acinic cell carcinomas displayed a moderate to intense binding (DALEY et al. 1985). DALEY et al. (1985) report a weak, erratic binding of *SBA* in the apical glycocalyx of ductal cells, mucous cells and cystic contents.

TGP (Tetragonolobus purpurea) binding features were similar to those of WGA (DALEY et al. 1985).

Blood Group Antigens

Blood group substances are found in the saliva. They can be identified in salivary gland tissue and salivary gland tumours (HAMPER et al. 1986). In normal tissue the expression of blood group substances corresponded to the serological blood group. Blood group substance H was found in almost every gland. In some glands, Le-b was rather selective for mucous acini. In salivary gland tumours, blood group substances were expressed in benign and highly differentiated tumors (HAMPER et al. 1986). WOLTERING et al. (1983) analysed parotid neoplasms by the specific red cell adherence technique. There was no reaction with red cell agglutination with the ductal epithelium. In malignant tumors there seems to be a loss of ABH substances as shown by this technique.

3.3 Mammary Gland and Mammary Carcinoma

PNA (Peanut Agglutinin) (Fig. 4)

Various lectins have been analysed in the mammary gland. The most important one is the *Peanut Agglutinin (PNA)* which detects the disaccharide Gal-GalNac which plays a leading role in the Thomsen-Friedenreich-phenomenon. It is related to the MN blood group system (SPRINGER et al. 1974, 1975; KLEIN et al. 1979; FRANKLIN 1983; HAGEMAN et al. 1983).

Despite the fact that PNA may be helpful in detecting carcinoma cells in lymph node metastases (STEGNER et al. 1981) its primary importance lies in the observation that the expression of lectin binding sites are demonstrated in those carcinomas which are positive for *estrogen receptors* (LEATHEM et al. 1983; BÖCKER et al. 1984, 1985; KLEIN and WÜRZ 1985; TOTOVIC et al. 1985; VIERBUCHEN et al. 1985; REMMELE et al. 1986). Although these studies confirm the general correlation of PNA binding sites and estrogen receptors, the practical problem often lies in finding the right "cut off level" for a given method and collective.

The general association of high differentiation and the expression of PNA receptors has been demonstrated by many groups (HOWARD and BATSAKIS 1980; HOWARD et al. 1981). Well differentiated mammary carcinomas will localize

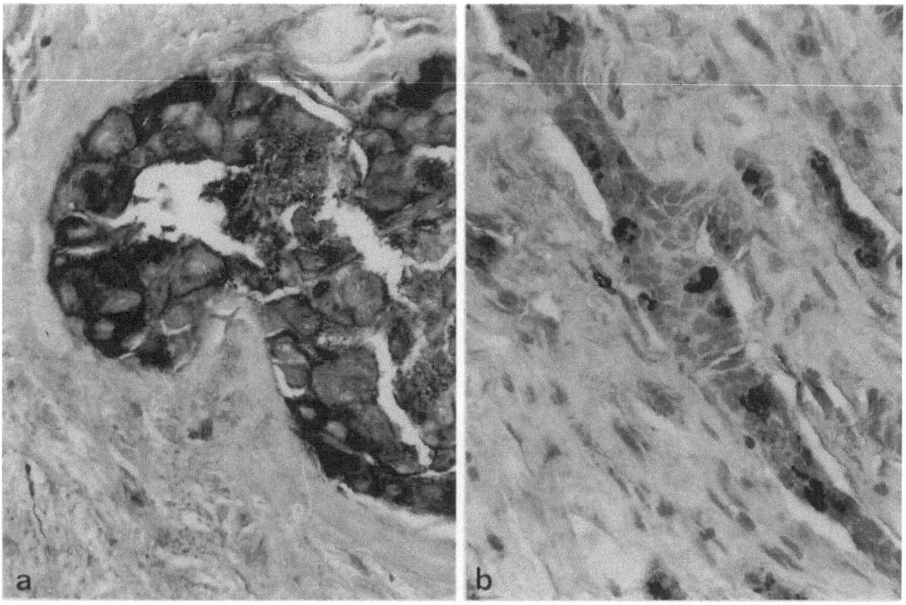

Fig. 4. a PNA binding sites (after sialidase digestion) in solid mammary carcinoma (NOS). Strong cytoplasmic staining of the tumor cells. ×300. **b** PNA binding sites (after sialidase digestion) in infiltrating lobular carcinoma of the mammary gland. Vacuolar and globular staining of the cytoplasm of the tumor cells. ×300

PNA to the apical portion of epithelial cells. Solid or poorly differentiated carcinomas of the breast will either localize PNA intracytoplasmically or not at all. KAHN and BAUMAL (1985) studied the differences in lectin binding in tissue sections of human and murine malignant tumors and their metastases. There was a weaker staining for PNA (without sialidase (neuraminidase) digestion) in those primary breast carcinomas which had no metastases. The staining was strong in metastatic tumours of breast carcinomas.

New aspects are introduced by the combination of affinity histochemistry with PNA and cytophotometry, and by the use of cell culture systems (DAXENBICHLER et al. 1985; ZIPPEL et al. 1985).

WGA (Wheat Germ Agglutinin) (Fig. 5)

In cancer cells there is generally a shift from luminal towards cytoplasmic staining (LEATHEM et al. 1983; FRANKLIN 1983; WALKER 1984).

Primary and metastatic breast carcinoma cells are equally positive for WGA (KAHN and BAUMAL 1985). WGA binding to tumour cells suggests that a large proportion of the exposed carbohydrates in mammary cancers consists of *N-acetyl-D-glucosamine (GlucNAc) and/or sialic acid.* There are some data suggesting that the carbohydrate stained in the current study may include CEA, which can be extracted from breast carcinomas by Sepharose columns coupled to WGA (SANTEN et al. 1980).

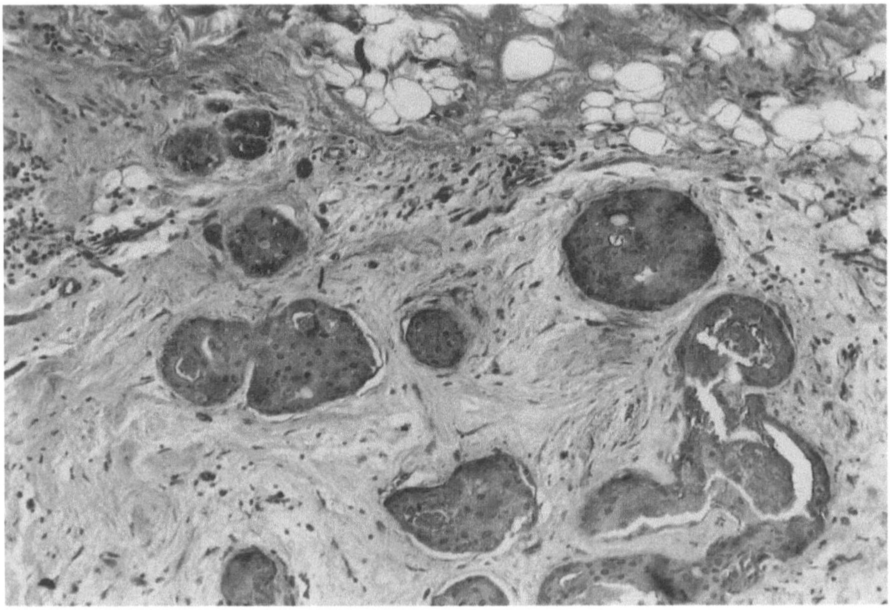

Fig. 5. WGA binding sites (after sialidase digestion) in solid carcinoma of the mammary gland. Strong cytoplasmic staining of the tumor cells. ×150

HPA (Helix Pomatia Agglutinin)

Helix pomatia agglutinin was observed in the luminal surface and the secretion of normal cell in the mammary gland (LEATHEM and ATKINS 1983). It is found in a subpopulation of cancer cells (LEATHEM et al. 1983).

DBA, RCA, Con A, PWA, LTA and SBA receptors have been studied in human mammary cancer cells (FRANKLIN 1983; LEATHEM and ATKINS 1983; LEATHEM et al. 1983; WALKER 1985b).

Blood Group Antigens

AB isoantigen expression in normal breast tissue was found to be largely confined to the mammary duct system, whereas loss of AB isoantigen expression was a feature of intraductal carcinoma (STRAUCHEN et al. 1980). Varying degrees of isoantigen loss were observed in *proliferative lesions* associated with fibrocystic disease. In contrast to other tissues, loss of AB isoantigen expression in proliferative breast lesions is not necessarily evidence of malignancy. GUPTA and SCHUSTER (1973) compared isoantigens A, B, H in benign and malignant lesions of the breast.

3.4 Lung and Bronchial Tree

PNA receptors have been analysed in *cell lines* of small cell lung cancer and squamous cell carcinoma (RAEDLER et al. 1985). 80 to 100% of the cells in the cell line were positive for PNA (RAEDLER et al. 1985). MIYAUCHI et al. (1982a) have identified PNA receptors in human lung carcinomas. RAEDLER et al. (1985) analysed the distribution of WGA, UEA, HPA, and RCA1 in cell cultures of human lung cancers (squamous cell carcinoma and oat cell carcinoma). 80 to 100% of the tumor cells were positive for RCA1. Cell cultures from *oat cell carcinomas* were negative for *Con A* (RAEDLER et al. 1985). *LCA* and *SBA* receptors were present in cell lines of human oat cell carcinoma and squamous cell carcinoma of the lung (RAEDLER et al. 1985). The same group analysed various other lectins in human lung carcinomas.

Blood Group Antigens

DAVIDSOHN and NI (1969) found a loss of AB isoantigens in human lung carcinomas.

3.5 Gastrointestinal Tract

3.5.1 Stomach

PNA (Peanut Agglutinin)

There was some change in the PNA binding pattern during metaplasia of the gastric mucosa (COGGI et al. 1983; KUHLMANN et al. 1983; BUR and FRANKLIN 1985). *Carcinomas* of the intestine and diffuse types contained both *PNA-positive* and *PNA-negative* cells of which PNA-positive cells occurred preferentially in the tumour periphery adjacent to normal mucosa (KUHLMANN et al. 1983). Extracellular mucus and cell borders are intensely stained by *PNA* after sialidase (neuraminidase) treatment, 50% of the cases being positive even without enzyme digestion (COGGI et al. 1983).

UEA (Ulex Europaeus Agglutinin)

Gastric carcinomas revealed a heterogenous picture for UEA with positive and negative tumor cells (KUHLMANN et al. 1983; BUR and FRANKLIN 1985).

Some normal cells of the gastric mucosa and some tumors were positive for *WGA* (FISCHER et al. 1983, 1984). About half of carcinomas of the intestinal type were positive for *DBA* (KUHLMANN et al. 1983). Tumour cells were sometimes strongly stained for BSA (BUR and FRANKLIN 1985), and occasionally for *SBA* receptors (COGGI et al. 1983; BUR and FRANKLIN 1985).

Blood Group Antigens

Generally, there is a loss of blood group substances in human gastric cancer (COWAN 1962; DAVIDSOHN et al. 1971a; DENK et al. 1974; PICARD et al. 1978; FINAN et al. 1983), but not in the surrounding normal mucosa (FINAN et al. 1983). SLOCOMBE et al. (1980) studied the distribution of blood group antigens in gastric carcinoma by the immunoperoxidase technique. No clear relationship between the tumor differentiation and preservation of blood group substances was found. In some, even well differentiated, tumours there was a loss of blood

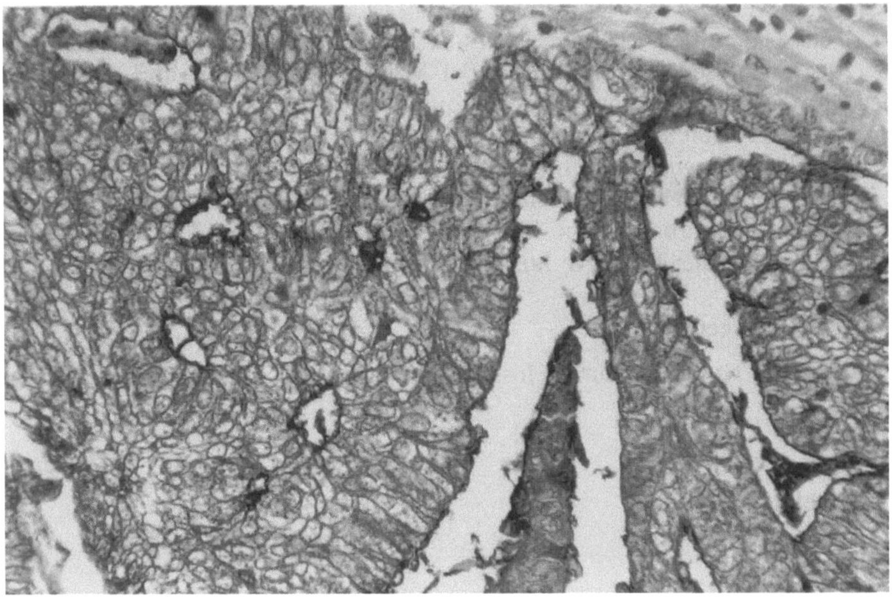

Fig. 6. WGA binding sites in adenocarcinoma of the stomach (intestinal type). ×300

group substances (A, B, H), obviously due to the fact that in these cells there is a blocking of the normal elongation of the carbohydrate portion of blood group substances (SLOCOMBE et al. 1980). Gastric carcinomas presented a loss of ABH antigens and an augmentation of Lewis a and I (Ma) antigens, the neoplastic tissue showing a marked heterogeneity of expression. Ca 19-9, corresponding to sialylated Lewis a antigen, was detected in 62% of gastric carcinomas (HIROHASHI et al. 1984). ERNST et al. (1984) detected Lewis a antigen in gastric carcinoma (80%) more often than in normal gastric mucosa (40%).

3.5.2 Colon

PNA (Peanut Agglutinin) (Fig. 7)

In adenomas, PNA reacts with the supranuclear region of adenoma cells that produces mucin goblets. On contrast, neoplastic non-mucinous columnar cells react with PNA in the apical and glycocalyx region (COOPER and REUTER 1983). PNA receptors are found in the apical and glycocalyx region of the colon *cancer cells*, indicating that the synthesis of MN blood group related glycoproteins is disturbed in these cells (BOLAND et al. 1982; COOPER 1982; HSU and RAINE 1982; COGGI et al. 1983). There was a strong labelling of the mucin in the lumen (BOLAND et al. 1982), especially in colloid carcinoma.

UEA (Ulex Europaeus Agglutinin)

The distribution pattern of UEA receptors in the normal human gastrointestinal tract, including the colon, and in adenomas and carcinomas of the colon have

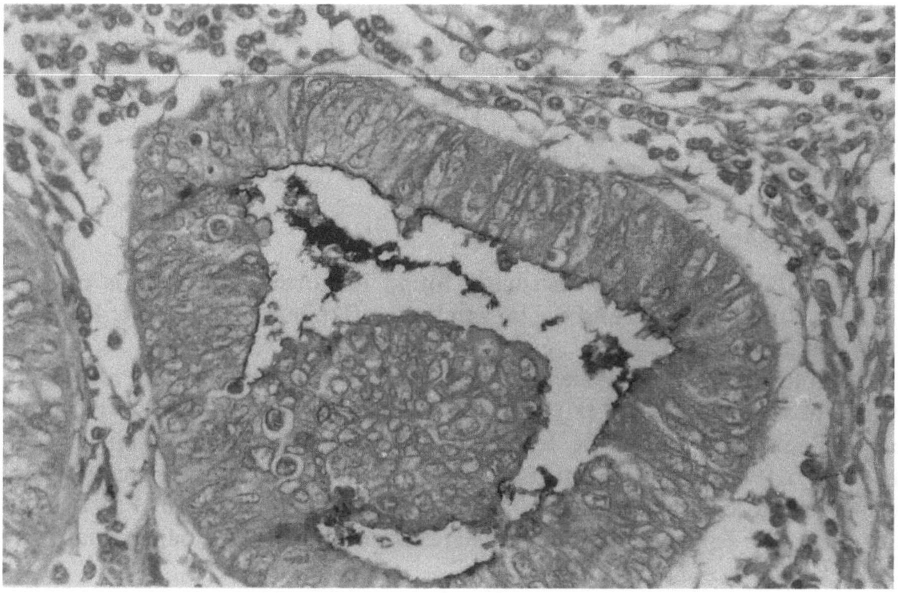

Fig. 7. PNA binding sites (after sialidase digestion) in adenocarcinoma of the colon. ×300

been analysed by YONEZAWA et al. (1982) and by biochemical methods (MATSUSHITA et al. 1985).

In contrast to the finding in the normal mucosa (COGGI et al. 1983), adenomas of the colonic mucosa are strongly positive for *DBA* (COGGI et al. 1983). Although dysplastic epithelium is positive for DBA1, this pattern seems to be lost in parts of the true carcinoma. *RCA* (RCA1) receptors are found in epithelial cells of the large bowel (FISCHER et al. 1983). In contrast to the findings with HPA, *Con A* receptors are strongly expressed in epithelial cells of the large bowel (FISCHER et al. 1983). According to NAKAYAMA et al. (1985) the *Bandeiraea (Groffonia) simplicifolia agglutinin II* displayed a strong affinity to colonic carcinoma cells. The staining was enhanced after prior digestion with neuraminidase.

Blood Group Substances

DAVIDSOHN et al. (1966) studied the distribution of A, B, H substances in gastrointestinal carcinoma. ABDELFATTAH-GAD and DENK (1980) analysed epithelial blood group antigens in human carcinoma of the distal colon. Other groups analysed the glycoproteins metabolism in inflammatory and neoplastic diseases of the human colon (KIM et al. 1974; KIM and ISAACS 1975). DENK et al. (1975) found epithelial blood group antigens in the secretory part of goblet cells with slight-to-moderate atypia and in the whole cell in cases of pronounced dedifferentiation.

3.6 Liver

Lectin receptors have been especially analysed in rodent tissue. CHERQUI et al. (1982) report the use of lectins for the characterization of the insulin receptor glycosidic moiety in rat adipocytes and hepatocytes. Virtanen et al. (1978) analysed the distribution of various lectins (Con A, WGA, RCA, SBA, Lotus tetragonolus agglutinin) in rat liver cell surface fractions.

3.7 Pancreas

PNA (Peanut Agglutinin)

Binding of *PNA* is only inconsistently found in normal tissue. It is obvious in hyperplastic conditions and ductal carcinoma (Klöppel et al. 1984; Klöppel and Fitzgerald 1986).

Hyperplastic lesions are strongly positive and duct carcinomas are inconsistently positive for *WGA* (Klöppel and Fitzgerald 1986). *UEA* receptors are prominent in hyperplastic lesions of the pancreatic gland (Klöppel and Fitzgerald 1986). *HPA* receptors are present in the normal pancreatic gland and in pancreatic carcinomas (Klöppel et al. 1984; Klöppel and Fitzgerald 1986), they are augmented in hyperplastic conditions of the pancreatic tissue (Klöppel and Fitzgerald 1986).

Other Lectins

DBA receptors present a similar distribution pattern as WGA receptors (Klöppel et al. 1984). Normal pancreatic tissue is negative, hyperplastic lesions are strongly positive, pancreatic carcinomas are positive (Klöppel and Fitzgerald 1986).

Blood Group Antigens

Davidsohn et al. (1971b) analysed the distribution of isoantigens A, B, H in carcinoma of the pancreas.

3.8 Urogenital Tract

3.8.1 Kidney

PNA receptors were localized in nephrogenic adenomas (Devine et al. 1984). The presence of PNA in *renal carcinoma* is analysed by Leathem and Atkins (1983). Nephroblastomas are positive for PNA (Wick et al. 1986).

There is a granular staining with renal oncocytomas with *WGA* (Holthöfer 1986). Epithelial parts of nephroblastomas are positive for WGA (Wick et al. 1986). *UEA* only binds to vascular endothelium (Ulrich et al. 1985). In some tumors, the degree of vascularization can be evaluated by use of this lectin. HPA is present in *renal carcinomas* (Raedler et al. 1982).

Other Lectins

Renal oncocytomas are reported to have a high affinity for *DBA* (Ulrich et al. 1985). In renal oncocytomas, there is a granular staining of the cytoplasm (Holthöfer 1986). Nephroblastomas are positive for SBA (Wick et al. 1986).

Blood Group Antigens

PEDAL and BAEDEKER (1985) analysed the distribution of A, B, H and secretor substances in kidneys. They found a reliable staining with the PAP technique, by using monoclonal antibodies.

3.8.2 Urinary Bladder

The expression and the loss of blood group antigens in bladder antigens have been analysed by several groups (for review: WEINSTEIN et al. 1981; ALROY et al. 1982a, b; LIMAS and LANGE 1982). Expression of the ABH blood group antigens has been related to tumor behaviour and patient's prognosis in transitional-cell carcinoma of the urinary bladder.

PNA (Peanut Agglutinin)

COON et al. (1982) found that the Thomsen-Friedenreich Antigen as detected by the PNA affinity histochemistry is absent in *normal bladder mucosa* and can only be detected after digestion with sialidase (neuraminidase). Three types of PNA binding in urothelial carcinomas were established (COOPER 1984, see also Table 4):

1. *cryptic* (+) *neoplasms* with PNA binding sites only after neuraminidase digestion
2. *Thomsen-Friedenreich antigen positive carcinoma cells* with PNA binding sites prior to neuraminidase digestion
3. *cryptic* (−) *neoplasms* that were negative for PNA binding sites even after neuraminidase treatment.

The worst prognosis is found in the group which were cryptic T antigen negative (COON et al. 1982). In this context, the expression of the Thomsen-Friedenreich antigen (without and after neuraminidase treatment) may be a prognostically useful immunohistochemical marker, together with the expression of ABH substances.

A new approach by dual parameter flow cytometric measurements of DNA content and PNA binding is presented by ÖRNTOFT et al. (1986). The cellular lectin binding seems to be correlated to the degree of aneuploidy.

Table 4. Diagrammatic presentation of probable structures corresponding to different T-antigen reactivities of bladder carcinomas (similar findings in other tumours)

Cryptic T antigen

```
NANA———Gal —beta 1–3— GalNac —alpha— Ser/Thr
        |
       NANA
```

T Antigen

```
Gal —beta 1–3— GalNAc —alpha— Ser/Thr
```

Cryptic antigen undetectable

```
?——— Ser/Thr
```

WGA (Wheat Germ Agglutinin) and UEA (Ulex Europaeus Agglutinin)

WGA receptors which are present in normal urothelial epithelium are lost in urinary bladder carcinomas (NIELSEN et al. 1986). A positive staining with *UEA* was observed in most blood group 0 patients with a urothelial carcinoma (79%) and in most tumours of patients with blood group A or B (90%). In blood group A patients, NIELSEN et al. (1986) found staining of the cytoplasm of the luminal urothelial cells. In situ dedifferentiation was associated with staining of cytoplasm of all layers, the truly invasive cells being negative for UEA (NIELSEN et al. 1986).

Blood Group Antigens

The behaviour of blood group substances in carcinomas of the urothelium is well documented in many papers (among others: KAY and WALLACE 1961; KATO 1977; JOHNSON and LAMM 1980; LIMAS and LANGE 1982; for review: WEINSTEIN et al. 1981). There is a loss of blood group substances in poorly differentiated carcinomas.

3.8.3 Male Genital Tract

3.8.3.1 Prostate Gland

PNA (Peanut Agglutinin)

A strong expression of PNA receptors is observed in many prostatic carcinomas. The pattern is not changed after antiandrogenic treatment (WERNERT and DHOM 1986).

Other Lectins

UCCI et al. (1983) and LEATHEM and ATKINS (1983) analysed the distribution of lectins in the normal prostate gland and in carcinomas of this organ.

Blood Group Antigens

GUPTA et al. (1973) report a loss of AB isoantigens in prostate cancer.

3.8.3.2 Testis

In *human germ cell tumors,* most embryonal carcinomas in adult testis were positive, all yolk sac tumours were positive for *PNA*. Most seminomas and choriocarcinomas were negative (TESHIMA et al. 1984). Squamous epithelium in teratomas were positive for PNA (TESHIMA et al. 1984). *WGA* receptors were found in *spermatoza* of man (DE CEREZO et al. 1982). In seminomas, all tumor cells were positive for this lectin (LEE et al. 1985). All yolk sac tumours and 7 of 16 embryonal carcinomas were positive for *UEA* whereas seminomas and choriocarcinomas were negative (TESHIMA et al. 1984). Squamous epithelium in teratomas was positive for UEA (TESHIMA et al. 1984).

Other Lectins

LEE et al. (1985) and TESHIMA et al. (1984) report the different binding patterns of lectins in the various tumors of the testis.

3.9 Female Genital Tract

3.9.1 Endometrium

More than 70% of endometrial carcinomas displayed a binding for PNA with different structural patterns (ZIPPEL et al. 1986). *WGA receptors* were localized at the cell luminal border of endometrial carcinoma (KLUSKENS et al. 1984). *Con A* binding is increased in cells in endometrial carcinoma (KLUSKENS et al. 1984). LEE and DAMJANOV (1985) analysed different lectin binding pattern with respect to the functional status of the endometrium.

3.9.2 Ovary

There are few reports on the distribution of lectin binding sites in ovarian tissue. DIETEL et al. (1986) found that the staining pattern of various lectins (*PNA*, *WGA*, *UEA I*, *RCA*, *Con A*, *SBA*) was not correlated with the cellular differentiation of normal tissue and various neoplastic disorders.

Concerning the *blood group substances*, blood group substances of the ABO system and the Lewis system were found in normal tissues and benign tumors. In borderline lesions the over all expression of blood group substances was reduced by 15%. The reduction of blood group substance expression was even more distinct malignant ovarian tumors (DIETEL et al. 1986).

3.9.3 Placenta

PNA and WGA binding sites are found at the basement membrane of the human placenta (LEE and DAMJANOV 1984). *UEA 1* receptors have not been found in the trophoblast (LEE and DAMJANOV 1984). LEE and DAMJANOV (1984) analysed various lectins and their stage related binding to human placental tissues (*HPA*, *DBA*, *Con A*, *BPA*, *BSA I*, *BSA II*, *HAA*, *LBA*, *LCA*, *MPA PHA E*, *PHA L*, *RCA 1*, *RCA 2*, *SBA*, *STA*, *VVA*).

Blood Group Antigens

LOKE and BALLARD (1973) localized blood group A antigens in human trophoblast cells. STEJSKAL et al. (1973) analysed the distribution of isoantigens in the human fetus.

3.10 Haematopoietic and Lymphoreticular System

3.10.1 Haematopoietic and Related System

The different blood group substances can be identified on the surface of red cells, either by the use of polyclonal and monoclonal antibodies or by appropriate lectins (for review: UHLENBRUCK

1971 and others). STODDART et al. (1980) have analysed lectin staining of carbohydrate of haemic cells. BAYER et al. (1976) localized lectin and antibody receptors on erythrocytes.

Untreated erythrocytes do not bind *PNA*. Owing to the removal of terminal sialic acid from its protein (glycophorin), erythrocytes react with PNA after digestion with neuraminidase (POSCHMANN et al. 1976 and others). The removal of neuraminic (sialic) acid exposes the normally cryptic T blood group antigen. *WGA* labelling for leucocytes is generally stronger than for erythrocytes and platelets (ZINSMEISTER and ACKERMAN 1983). *RCA* reacted with the plasmalemma and cytoplasm of normal human bone marrow cells and blood cells. There was no detectable difference between normal and leukaemic cells (STODDART et al. 1980). *Con A* reacted with the plasmalemma and cytoplasm of normal human bone marrow cells and blood cells.

3.10.2 Lymphoid Tissue

PNA (Peanut Agglutinin)

Different groups obtained different results by use of PNA, probably due to the various techniques of detecting lectin receptors (COOPER 1984). PNA has been shown to react with activated T cells (CHERVENAK and COHEN 1982), it binds to follicular center lymphocytes (STRAUCHEN 1985), to human B lymphocytes (COOPER 1984), and reacts with immature human thymocytes (REISNER et al. 1979). Centroblasts and centrocytes are said to react with PNA (COGGI et al. 1983). HOWARD and BATSAKIS (1982) found that PNA was a marker for tissue histiocytes. There is an intracytoplasmic binding in histiocytes of reactive germinal centers (REE and HSU 1983). Nodular lymphomas display a diffuse staining with PNA (REE and HSU 1983). Cells in Burkitt's lymphoma are positive for PNA (MIYAUCHI et al. 1982). REE et al. (1983) demonstrated that stromal macrophage-histiocytes were positive in diffuse large cell lymphomas. MÖLLER (1982) found that Hodgkin cells are positive for PNA. This kind of reaction may be useful in paraffin embedded material. HSU et al. (1985) did not find a constant reaction with Hodgkin and Sternberg-Reed cells (cell line studies in Hodgkin's disease: UHLENBRUCK et al. 1986).

WGA binding sites found in in macrophages and tumor cells of diffuse large cell lymphoma (REE et al. 1983). STRAUCHEN (1985) has analysed UEA receptors in lymphomas. Lymphatic vessels, in contrast to blood vessels, are negative for UEA (WALKER 1985a).

Con A binding sites have been identified in lymphoma cells (RAPIN and BURGER 1974) and in stromal macrophages-histiocytes of diffuse large cell lymphoma (REE 1983). STRAUCHEN (1985) reports binding of Con A to the follicular center macrophages and a positive population of dendritic cells in follicular lymphomas. Con A labelling to stromal histiocytes in paraffin sections may indicate a high risk of relapse in Hodgkin's disease (REE 1986). *SBA* binding sites have been found in lymphoma cells by RAPIN and BURGER (1974). RAEDLER et al. (1986a, b) found that the majority of SBA positive cells were T 11 positive and had to be regarded as T cells. The SBA positive cells were interpreted as immature.

3.10.3 Blood Vessels and Lymphatic Vessels UEA (Ulex Europaeus Agglutinin)

The value of UEA as a marker for vascular epithelium has been demonstrated by various papers (YONEZAWA et al. 1982; HOLTHÖFER et al. 1982; BORISCH et al. 1983; MIETTINEN et al. 1983; MÖLLER and LENNERT 1984; WALKER 1985a). Kaposi sarcoma, haemangioma and angiosarcoma are UEA 1 and factor VIII positive (ANGERVALL et al. 1985; BENDELAC et al. 1985). Another lectin with fucose binding, Lotus agglutinin, failed to react with endothelium, suggesting a rather specific type of fucose expression in human endothelial cells (HOLTHÖFER et al. 1982).

Blood Group Antigens

FEIGL et al. (1976) analysed blood group isoantigens in benign and malignant vascular tumors.

3.11 Central Nervous System and Pituitary Gland

PNA (Peanut Agglutinin)

The main targets of PNA binding in the normal central nervous system and pituitary gland are the luminal membrane of the *ependymal cells*, the brush border of the *chorioid plexus* after neuraminidase digestion and a subpopulation of cells in the *anterior pituitary gland. Astrocytes* are positive for PNA, as well (SCHWECHHEIMER et al. 1984). MÜLLER et al. 1980 demonstrated receptors for PNA in normal choroid plexus and of plexus papillomas, and draw some implications for the role of Thomsen Friedenreich antigen in bacterial and viral infections of the brain. Nerve fibers, perineurinum and blood vessels are negative for PNA (ESTRUCH and DAMJANOV 1986). *Granular cells* and *granular cell tumors* are strongly positive for PNA (SCHWECHHEIMER et al. 1983).

UEA (Ulex Europaeus Agglutinin)

UEA binding sites were only found in *vascular endothelial cells* and in some cells of the pituitary gland (SCHWECHHEIMER et al. 1984). Nerve fibers and perineurinum are negative (ESTRUCH and DAMJANOV 1986).

Blood Group Antigens

JELLINGER and DENK (1974) analysed blood group isoantigens in angioblastic meningiomas and haemangioblastomas of the central nervous system. They found that the prominent cell type in haemangioblastomas is the endothelial cell.

3.12 Other Tissues

Lectin-binding sites have been analysed in *melanin-producing* cells by UNO et al. (1985). Melanoma cells, dermal melanocytes, epidermal melanocytes, and pigmented nevus cells were studied on the electron microscopical level by use of horseradish peroxidase coupled lectins (*PNA*, *WGA*, *RCA*, *CON A*). *Melanoma cells* had a higher number of WGA binding sites and a lower number of PNA binding sites (UNO et al. 1985).

In *ameloblastomas* (VEDTOFTE and DABELSTEEN 1981), cell membrane bound WGA receptors were shown in the epithelial cells. The distribution of RCA 1 and SBA receptors was related to the morphology of peripheral cells in ameloblastomas.

SOBRINHO-SIMOES and DAMJANOV (1986) present data on the lectin binding of papillary and follicular carcinoma of the thyroid gland. *PNA* receptors are not found in normal thyroid tissue (follicular cells, colloid, stroma). Only one poorly differentiated carcinoma and one papillary carcinoma were positive for this lectin (SOBRINHO-SIMOES and DAMJANOV 1986). *WGA* reacted strongly with follicular cells, colloid and stroma. Receptors for this lectin were positive in all thyroid carcinomas (SOBRINHO-SIMOES and DAMJANOV 1986). *UEA* only reacted with blood vessels in normal and neoplastic thyroid cancers (SOBRINHO-SIMOES and DAMJANOV 1986).

HPA receptors were found in stromal blood vessels of tissue from blood group A donor patients. HPA receptors were localized in all 7 papillary carcinomas of the study (SOBRINHO-SIMOES and DAMJANOV 1986). Benign adenomatous lesions and follicular carcinomas were negative for this receptor. *DBA* reacted with stroma of normal thyroid tissue. Only one poorly differentiated follicular carcinoma was positive (Sobrinho-Simoes and Damjanov 1986). *RCA* I and II receptors are found in follicular cells and colloid to a varying degree. They are regularly present in the stroma. RCA I and II receptors were detected in benign adenomatous lesions, as well as in follicular and papillary carcinomas (Sobrinho-Simoes and Damjanov 1986). *Succinyl Con A* binding sites were demonstrated in follicular cells, colloid and stroma. They are present in all carcinomas (Sobrinho-Simoes and Damjanov 1986).

4 Discussion

Carbohydrate residues of membrane glycoproteins and glycolipids have a major influence on the chemical and physical properties of the cell membrane and can serve to regulate cell metabolism, membrane transport, growth and various functional activities of the cell. Surface carbohydrates are known to be involved in cell recognition, cell interaction and serve as receptor sites for a wide variety of substances (Köttgen et al. 1979; Hakomori 1984; Baenzinger 1986). The type and the spatial orientation of these carbohydrates are variable and form branching chains on the external surface of the cell.

The use of the multifaceted properties of lectins has led to their application in the investigation of various cellular phenomena. The primary event in these applications involves the interaction of the lectin with specific carbohydrates residues on the surface of the cell membrane. Lectins coupled to fluorescent tags, ferritin and radiotracers have been used to study the nature, number, and distribution of sugars on cells and tissues (Uhlenbruck 1971; Nicolson 1974; Sharon and Lis 1975; Schrevel et al. 1981; Schauer 1982a, b; Yogeeswaran 1983; Leathem et al. 1983; Lis and Sharon 1986). A large number of biologically important glycoproteins (e.g. membrane components, plasma components, enzymes and antigens) have been isolated by means of immobilized lectin columns. An effective use of lectins has been the development of techniques for the fractionation of animal cells. Lectins have been employed as carriers for the delivery of chemotherapeutic agents.

The distribution of lectin-binding carbohydrates is highly ordered in benign tissues and the disorganization of the carbohydrates in neoplasms may be exploited for diagnostic and therapeutic purposes. From morphological studies of the application of lectins, the following general observations can be made:

1. During neoplastic dedifferentiation there is a *change of membrane sugars*.
2. There are some lectins which are associated with *malignant differentiation* (PNA).
3. In tumors, there is often a *loss* of *"higher" blood group substances*, when these substances belong to the normal "background" of the patient.

A good example for the type of changes occurring in tumours is given by a *model of PNA* related change (Table 4).

The T antigen is believed to be the precursor carbohydrate of a variety of glycoproteins and glycolipids, including M and N blood group determinants. Being normally masked by *N-acetylneuraminic (sialic) acid residues*, this antigen is only accessible for binding to PNA after neuraminidase digestion. It has been postulated that T antigen expression in malignancy is a result of accelerated degradation or incomplete synthesis of normal cell surface glycoproteins (SPRINGER et al. 1975; see Table 4). Different patterns of distribution are found by PNA affinity histochemistry: *granular globular pattern, vacuolar pattern, diffuse pattern* (BÖCKER et al. 1985). The data show that the lectin binding sites are the expression of functional differentiation. Obviously, the disaccharid Gal–Gal–Nac which is detected by the peanut agglutinin is not bound to the same protein components as in erythrocytes (HAMPER et al. 1984; BÖCKER et al. 1985). The expression of PNA receptors is correlated with the appearance of estrogen receptors in mammary carcinomas (BÖCKER et al. 1984; KLEIN and WÜRZ 1985; TOTOVIC et al. 1985).

PNA receptors are found in various types of salivary tumors. It seems to be expressed more intensely in malignant salivary gland tumors (e.g. carcinomas in pleomorphic adenomas) than in the benign counterparts (CASELITZ 1987). The expression of PNA receptors alone cannot be taken as a sign of malignancy, but as a marker displaying a certain tendence of differentiation. The PNA-receptor specific disaccharide is not bound to the same protein component as in erythrocytes (HAMPER et al. 1984).

The presence of PNA receptors in *colonic* epithelium has been identified bound to the Golgi apparatus in normal cells (SATO and SPICER 1982a, b; COOPER 1984). Taking the normal route of glycoprotein synthesis, the supranuclear distribution corresponds to the detection of PNA of nascent oligosaccharides prior to the addition of terminal neuraminic acid (COOPER 1984). This orderly arrangement is distributed during neoplastic development (COOPER 1984). 50% of the adenocarcinomas are positive for PNA (COGGI et al. 1983). Other authors found a labelling of mucin in all carcinomas of the large bowel. The divergence of the results may be due to technical reasons (fixation, embedding procedure, different types of affinity histochemistry).

In urothelial carcinomas, the presence of TF antigen is generally detected in the low differentiated tumors (COON et al. 1982). Interestingly, there are neoplasms with direct (without digestion) PNA receptors, neoplasms with cryptic PNA receptors and those without any PNA receptors. The worst prognosis was found in those carcinomas which were directly positive for Thomsen Friedenreich receptors and the cryptic Thomsen Friedenreich negative tumors.

In lymphoid tissue, PNA receptors have been found in activated T cells (CHERVENAK and COHEN 1982), follicular center cells (STRAUCHEN 1985), in B cells and in histiocytes (REE et al. 1983). MÖLLER (1982) claimed that Hodgkin cells are positive for PNA. Thus, PNA may be used in identifying several cell types in a given tissue. SCHWECHHEIMER et al. (1983) showed that this lectin was positive for granular cells. In this context, it can be used as an histochemical marker.

A lectin which is valuable in differential diagnosis is the UEA lectin. Many authors (HOLTHÖFER et al. 1982; BORISCH et al. 1983; MIETTINEN et al. 1983; MÖLLER and LENNERT 1984) have shown that this fucose specific lectin binds to endothelial cells under normal and neoplastic conditions. It may be used to analyse the normal architecture in different tissues (e.g. MÖLLER and LENNERT 1984), and to identify the vascular nature of some tumors (ANGERVALL et al. 1985; BENDELAC et al. 1985).

The AB isoantigens, in addition to their presence on erythrocytes and in secretions of certain individuals, are expressed by a variety of body tissues. AB isoantigens are demonstrable on the cell surface of epithelial cells of the skin, oral cavity, bronchopulmonary tract, gastrointestinal tract, genitouronary tract, and vascular endothelium. The expression of AB isoantigens by epithelial lining type cells and general absence of these antigens in mesenchymal cells may imply that these antigens are related to epithelial differentiation. In stratified epithelia, AB isoantigen expression is absent from the least differentiated cell type, namely the basal layer. AB antigens become present in the more differentiated superficial layers. This gives support to the idea that AB isoantigen expression is a marker of differentiated epithelial cell function.

The disappearance of AB isoantigens in malignant epithelial tumours has been demonstrated for carcinomas of the lung, the gastrointestinal tract, bladder, cervix uteri, and prostate. The loss of blood group A and B determinants has been documented in a large variety of human tumours.

On the basis of large studies Hakomori (1984, 1986) found the following changes:

1. Tumor tissues *delete A and B* determinants due to blocked activity of A and B transferases.
2. The *type 2 chain* is fucosylated by enhanced transferases and *accumulates Le x (or X) hapten* in some tumors.
3. The *type 1 chain* is fucosylated by enhanced transferases, Le a and Le b hapten being accumulated regardless of the Lewis status of the host (Hakomori 1984).

Many monoclonal antibodies with a distinctive reactivity towards human lung, gastric and colonic cancer and towards leukaemias, are identified as being directed against Le x structure (Hakomori and Kannagi 1983). *Le x* structure is present in many normal human tissues (gastrointestinal epithelial cells of kidney tubules) and its accumulation in human neoplasms (e.g. colonic cancer) is noteworthy (Fukushi et al. 1984; Hakomori et al. 1983).

Some monoclonal antibodies reacting with tumour cells were specific for *Le b* or *sialosyl Le a* (Koprowski et al. 1979, 1981). Accumulation of both Le a and Le b antigens in human adenocarcinoma may be the basis for an association of sialosyl Le a or Le b.

Blocked synthesis of ABH antigens may go together with the accumulation of precursor carbohydrates type 1 or type 2 chain (Hakomori 1984). A-like antigens may be expressed in tumours of 0 or B blood group. Other antigens playing a role in human tumours are P-like and P1-like antigens, P K antigen in Burkitt's lymphoma and the Forssman and related antigens (Hakomori 1984).

The following conclusions may be drawn from the studies with human blood groups in tumours (Hakomori 1984).

1. Human blood group ABH determinants in erythrocytes are carried by unbranched and branched type 2 chains. Fetal and newborn erythrocytes contain mainly unbranched type 2 chain and adult erythrocytes contained branched type 2 chain.

2. In the majority of human carcinomas, blood group ABH determinants are lost due to incomplete synthesis, consequently the precursor substances increase, either type 1 chain (Le c type) or type 2 chain (Ii type).

3. Various types of human adenocarcinoma (lung, gastric, colonic, liver, pancreatic) accumulate a large quantity of fucolipids with Le a, X (Le x) and their sialylated derivatives. Various monoclonal antibodies have been identified as being directed to these structures.

4. Some tumours derived form gastrointestinal tissue of blood group 0 or B individuals present a neosynthesis of the incompatible A-like antigen, different from the Forssman antigen. P-like structures may play a special role in individual cases (Hakomori 1984).

References

Abdelfattah-Gad M, Denk H (1980) Epithelial blood group antigens in human carcinoma of the distal colon: Further studies on their pathologic significance. J Natl Cancer Inst 64:1025–1028

Alroy J, Gavris V, Heaney J, Szoka F, Ucci A (1982a) Alterations in specific carbohydrate residues in non-neoplastic epithelium (i.e. urothelium of human urinary bladder). Lab Invest 46:3A

Alroy J, Szoka FC, Heaney JA, Ucci AA (1982b) Lectin as a proble for carbohydrate residues in non-neoplastic urothelium of human urinary bladder. J Urol 128:189–193

Alroy J, Ucci AA, Pereira EA (1984) Lectins: Histochemical probes for specific carbohydrate residues. In: Dellelis RA (ed) Advances in immunohistochemistry. Masson, New York, pp 67–88

Angervall L, Kindblom LG, Karlsson K, Stener B (1985) Atypical hemangioendothelioma of venous origin. A clinicopathologic, angiographic, immunohistochemical, and ultrastructural study of two endothelial tumors within the concept of histocytoid hemangioma. Am J Surg Pathol 9:504–516

Baenzinger JU (1986) The role of glycosylation in protein recognition. Am J Pathol 124:382–391

Bayer EA, Wilchek M, Skutelsky E (1976) Affinity cytochemistry: The localization of lectin and antibody receptors on erythrocytes via the avidin-biotin complex. FEBS Lett 68:240–244

Bendelac A, Kanitakis I, Chouvet B, Viac I, Thivolet J (1985) Sarcome de Kaposi: Etude immunohistochemique comparative et intérêt histogénétique de marqueurs endothéliaux. Ann Pathol 5:45–52

Böcker W, Klaubert A, Bahnsen J, Schweikhart G, Pollow K, Mitze M, Kreienberg R, Beck T, Stegner HE (1984) Peanut lectin histochemistry of 120 mammary carcinomas and its relation to tumor type, grading, staging, and receptor status. Virchows Arch (Pathol Anat) 403:149–161

Böcker W, Bein G, Poschmann A, Seitz C, Pollow K (1985) Beziehungen zwischen Erdnuß-Kektinhistochemie, Thomsen-Friedenreich-Antigen (TFA) und Östrogen-/Progesteronrezeptoren in Mammakarzinomen. In: Bellmann O, Krebs D, Totovic V (eds) Hormonabhängigkeit des Mammacarcinoms. Aktuelle diagnostische Möglichkeiten. Zukschwerdt, München Bern Wien, pp 67–74

Bog-Hansen TC (ed) (1982) Lectins. Biology, biochemistry, clinical biochemistry, vol 2. de Gruyter, Berlin New York

Bog-Hansen TC, Spengler GA (eds) (1983) Lectins. Biology, biochemistry, clinical biochemistry, vol 3. de Gruyter, Berlin New York

Bog-Hansen TC, Van Driessche E (eds) (1986) Lectins. Biology, biochemistry, clinical biochemistry, vol 5. de Gruyter, Berlin New York

Boland CR, Montgomery CK, Kim YS (1982) Alterations in human colonic mucin occurring with cellular differentiation and malignant transformation. Proc Natl Acad Sci USA 79:2051–2055

Borisch B, Möller P, Harms D (1983) Lektin Ulex Europaeus I als Marker in der Differentialdiagnose von Gefäßtumoren. Pathologe 4:241–243

Bur M, Franklin WA (1985) Lectin binding to human gastric adenocarcinomas and adjacent tissues. Am J Pathol 119:279–287

Caselitz J (1987) Das Pleomorphe Adenom der Speicheldrüsen. Histogenese, zelluläre Differenzierung, Tumormarker. Fischer, Stuttgart New York

Caselitz J, Seifert G (1981) Immunhistologische Untersuchungen zum Verteilungsmuster von intrazytoplasmatischen Antigenen und Lektinrezeptoren bei Parotiskarzinomen. Ver Dtsch Ges Pathol 65:378

Caselitz J, Schultze I, Seifert G (1986) Adenoid cystic carcinoma of the salivary glands: An immunohistochemical study. J Oral Pathol 73:347–359

Cherqui G, Capeau J, Caron M, Picard J (1982) Use of lectins for the characterization of the insulin receptor glycosidic moiety in rat adipocytes and hepatocytes. In: Bog-Hansen TC (ed) Lectins. Biology, biochemistry, clinical biochemistry, vol 2. de Gruyter, Berlin New York, pp 257–272

Chervenak R, Cohen JJ (1982) Peanut lectins' binding as a marker for activated T-lineage lymphocytes. Thymus 4:61–67

Coggi G, Dell' Orto P, Bonoldi E, Doi P, Viale G (1983) Lectins in diagnostic pathology. In: Bog-Hansen TC, Spenler GA (eds) Lectins. Biology, biochemistry, clinical biochemistry, vol 3. de Gruyter, Berlin New York, pp 87–103

Coon JS, Weinstein RS, Summers JL (1982) Blood group precursor T-antigen expression in human urinary bladder carcinoma. Am J Clin Pathol 77:692–699

Cooper HS (1982) Peanut lectin-binding sites in large bowel carcinoma. Lab Invest 47:383–390

Cooper HS (1984) Lectins as probes in histochemistry and immunohistochemistry: The peanut (Arachis hypogaea) lectin. Hum Pathol 15:904–906

Cooper HS, Reuter VE (1983) Peanut lectin-binding sites in polyps of the colon and rectum. Lab Invest 49:655–661

Cowan WK (1962) Blood group antigens on human gastrointestinal carcinoma cells. Br J Cancer 16:535–540

Culling CFA, Reid PE (1982) Histochemistry of sialic acids. In: Schauer R (ed) Sialic acids. Chemistry, metabolism and function. Springer, Wien New York, pp 173–193

Dabelsteen E, Daniels TE (1983) Tumor associated antigens in oral cancer and premalignant lesions. Dtsch Z Mund Kiefer Gesichts Chir 7:87–92

Dabelsteen E, Mackenzie IC (1978) Expression of Ricinus communis receptors on epithelial cells in oral carcinomas and oral wound. Cancer Res 38:4676–4680

Dabelsteen E, Pindborg JJ (1973) Loss of epithelial blood group substance A in oral carcinomas. Acta Pathol Microbiol Scand (Sect A) 81:435–444

Dabelsteen E, Mygind N, Henriksen B (1974) Blood group substance A in carcinoma of the larynx. Acta Otorhinolaryngol 77:360–367

Dabelsteen E, Roed-Petersen B, Pindborg JJ (1975) Loss of blood group epithelial antigens A and B in oral premalignant lesions. Acta Pathol Microbiol Scand (Sect A) 83:292–300

Dabelsteen E, Fejerskov O, Noren O, MacKenzie JC (1978) Concanavalin A and Ricinus communis receptor sites in normal human oral mucosa. Invest Dermatol 70:11–15

Daley TD, Tolson ND, Wysocki GP (1985) Lectin probes of glycoconjugates in human salivary gland neoplasms: 2. J Oral Pathol 14:531–538

Davidsohn I, Kowarik S, Lee CL (1966) A, B and O substances in gastrointestinal carcinoma. Arch Pathol 81:381–390

Davidsohn I, Ni LY (1969) Loss of antigen ABH in carcinomas of the lung. Am J Pathol 57:307–334

Davidsohn I, Ni LY, Stejskal R (1971a) Tissue isoantigens A, B, H in carcinoma of the stomach. Arch Pathol 92:456–464

Davidsohn I, Ni LY, Stejskal R (1971b) Tissue isoantigens A, B, H in carcinoma of the pancreas. Cancer Res 31:1244–1250

Daxenbichler G, Dürken M, Marth CH, Böck G, Dapunt O (1985) Bindung des Erdnuß-Lektins an menschliche Mammacarcinomzellen in Kultur. In: Bellmann O, Krebs D, Totovic V (eds) Hormonabhängigkeit des Mammacarcinoms. Aktuelle diagnostische Möglichkeiten. Zukschwerdt, München Bern Wien, pp 105–113

De Cerezo JMS, Bueno MP, Skowronski B, Cerezo AS (1982) Immunohistochemical localization of concanavalin A and wheat germ receptors in the normal human spermatozoa. Am J Reprod J Immunol 2:246–249

Denk H, Tappeiner G, Davidovits A, Eckerstorfer R, Holzner JH (1974) Carcinoembryonic antigen and blood group substances in carcinomas of the stomach and colon. J Natl Cancer Inst 53:933–942

Denk H, Holzner JH, Obiditisch-Mayr I (1975) Epithelial blood group antigens in colon polyps. I. Morphologic distribution and relationship to differentiation. J Natl Cancer Inst 54:1313–1317

Devine P, Ucci AA, Krain H, Gavris VE, Bhagavan BS, Heaney JA, Alroy J (1984) Nephrogenic adenoma and embryonic kidney tubules share PNA receptor sites. Am J Clin Pathol 81:728–732

Dietel M, Arps H, Hölzel F, Viale G, Dell' Orto P, Kröger A, Niendorf A (1986) Blood group substances, CEA, and lectins in ovarian tumors. Cancer Detect Prevent 9:511–520

Elias PM, Chung JC, Orozco-Topete R, Nemanic MK (1983) Membrane glycoconjugate visualization and biosynthesis in normal and retinoid-treated epidermis. J Invest Dermatol 81:81s–85s

Emmelot P (1973) Biochemical properties of normal and neoplastic cell surfaces; a review. Eur J Cancer 9:319–333

England D, Solie B, Winkelmann R (1979) Isoantigens A, B, H in normal skin and tumors of the epidermal appendages. Arch Pathol Lab Med 103:586–590

Ernst C, Atkinson B, Wysocka M, Blaszczyk M, Herlyn M, Sears H, Steplewski Z, Koprowski H (1984) Monoclonal antibody localization of Lewis antigens in fixed tissue. Lab Invest 50:394–400

Estruch R, Damjanov I (1986) Lectin histochemistry applied to human nerves. Arch Pathol Lab Med 110:730–735

Feigl W, Denk H, Davidovits A, Holzner JH (1976) Blood group isoantigenes in human benign and malignant vascular tumors. Virchows Arch (A) 370:323–332

Finan PJ, Wight DG, Lennox ES, Sacks SH, Bleehen NM (1983) Human blood group isoantigen expression on normal and malignant gastric epithelium studied with anti-A and anti-B monoclonal antibodies. J Natl Cancer Inst 70:679–685

Fischer J, Klein PJ, Vierbuchen M, Fischer R (1983) Histochemical and biochemical characterization of glycoprotein components in normal gastric mucosa, intestinal metaplasia and gastric cancers with lectins. In: Bog-Hansen TC, Spengler GA (eds) Lectins. Biology, biochemistry, clinical biochemistry, vol 3. de Gruyter, Berlin New York, pp 167–178

Fischer J, Klein PJ, Vierbuchen M, Skutta B, Uhlenbruck G, Fischer R (1984) Characterization of glycoconjugates of human gastrointestinal mucosa by lectins. I. Histochemical distribution of lectin binding sites in normal alimentary tract as well as in benign and malignant gastric neoplasms. J Histochem Cytochem 32:681–689

Franklin, WA (1983) Tissue binding of lectins in disorders of the breast. Cancer 51:295–300

Fukushi Y, Nudelman E, Levery SB, Hakomori S (1984) A hybridoma antibody (FH6) directed to a human cancer-associated fucogangliosided (VI 3 Neu AC V 3 III Fuc 2nLC6); novel fucolipids accumulating in human cancer. J Biol Chem 259:10511–10517

Goldstein IJ (1981) Lectins: An overview. Biotechs. News about advanced biochemicals for life science research. EY Laboratories, San Matea, USA

Goldstein IJ, Hayes CE (1978) The lectins: Carbohydrate-binding proteins of plants and animals. Adv Carbohydr Chem Biochem 34:127–340

Graem N (1982) Ricinus communis agglutinin A binding to the cell membrane in benign, premalignant and malignant epidermal lesions. Acta Pathol Microbiol Immunol Scand Sect A 90:455–462

Graem N, Dabelsteen E (1984) Species identification of epidermal cells in the human skin/nude mouse model with lectins. Acta Pathol Microbiol Immunol Scand Sect A 92:317–322

Gupta RK, Schuster R (1973) Isoantigens A, B, H in benign and malignant lesions of breast. Am J Pathol 72:253–257

Gupta RK, Schuster R, Christian WD (1973) Loss of isoantigens ABH in prostate. Am J Pathol 70:439–445

Hageman Ph, Bobrow L, Van der Valk MA, Misdorp W, Hilkens J (1983) Reactions with lectins on benign and malignant lesions of the human breast. In: Bog-Hansen TC, Spengler GA (eds) Lectins. Biology, biochemistry, clinical biochemistry, vol 3. de Gruyter, Berlin New York, pp 105–118

Hakomori S (1984) Philip Levine Award Lecture: Blood group glycolipid antigens and their modifications as human cancer antigens. Am J Clin Pathol 82:635–648

Hakomori S (1986) Glycosphingolipide. In: Schirrmacher V (ed) Krebs – Tumoren, Zellen, Gene. Spektrum, Heidelberg, pp 178–188

Hakomori S, Kannagi R (1983) Glycosphingolipids as tumor-associated and differentiation markers. J Natl Cancer Inst 71:231–251

Hakomori S, Nudelman E, Levery SB, Patterson CM (1983) Human cancer-associated gangliosides defined by a monoclonal antibody (iB9) directed to sialosyl-alpha-2-galactosyl residue: A preliminary note. Biochem Biophys Res Comm 113:791–798

Hamper K, Caselitz J, Seifert G, Seitz R, Poschmann A (1984) Die Verteilung des Thomsen-Friedenreich-Antigens in Speicheldrüsentumoren. Verh Dtsch Ges Pathol 68:562

Hamper K, Caselitz J, Seifert G, Seitz R, Poschmann A (1986) The occurrence of blood group substances (A, B, H, Le-a, Leb) in salivary gland tumors. An immunohistochemical investigation. J Oral Pathol 15:334–338

Hirohashi S, Shimosato Y, Ino Y, Tome Y, Watanabe M, Hirota T, Itabashi M (1984) Distribution of blood group antigens and CA 19-9 in gastric cancers and non-neoplastic gastric mucosa. Gann 75:540–547

Holthöfer H (1986) Histochemical features in renal oncocytomas as studied with labeled lectins. In: Bog-Hansen TC, Van Driessche E (eds) Lectins. Biology, biochemistry, clinical biochemistry, vol 5. de Gruyter, Berlin New York, pp 457–461

Holthöfer H, Virtanen I, Kariniemi AK, Hormia M, Linder E, Miettinen A (1982) Ulex europaeus I lectin as a marker for vascular endothelium in human tissues. Lab Invest 47:60–66

Horisberger M (1984) Lectin cytochemistry. In: Polak JM, Varndell IM (eds) Immunolabelling for electron microscopy. Elsevier, Amsterdam New York Oxford, pp 249–258

Hosaka M, Murase N, Takai Y, Fukui S, Mori M (1985) Lectin binding in oral mucosa of mammals. Acta Histochem Cytochem 18:33–48

Howard DR, Batsakis JG (1980) Cytostructural localization of a tumor-associated antigen. Science 210:201–203

Howard DR, Batsakis JG (1982) Peanut agglutinin: A new marker for tissue histiocytes. Am J Clin Pathol 77:401–408

Howard DR, Ferguson P, Batsakis JG (1981) Carcinoma-associated cytostructural antigenic alterations: Detection by lectin binding. Cancer 47:2872–2877

Hsu SM, Raine L (1982) Versatility of biotin-labeled lectins and avidin-biotin-peroxidase complex for localization of carbohydrate in tissue sections. J Histochem Cytochem 30:157–161

Hsu SM, Ree HJ (1980) Histochemical studies on lectin binding in reactive lymphoid tissues. J Histochem Cytochem 31:538–543

Hsu SM, Yang K, Jaffe ES (1985) Phenotypic expression of Hodgkin's and Reed-Sternberg cells in Hodgkin's disease. Am J Pathol 118:209–217

Hyun KH, Hosaka M, Fukui S, Sugimoto A, Mori M (1984a) Lectin binding sites in basal cell epitheliomas and carcinomas. Acta Histochem Cytochem 17:507–523

Hyun KH, Murase N, Mitani H, Mori M (1984b) Histochemical expression of lectin binding in papillomas. Acta Histochem Cytochem 17:631–647

Hyun KH, Nakai M, Kawamura K, Mori M (1984c) Histochemical studies of lectin binding patterns in keratinized lesions, including malignancy. Virchows Arch A 402:337–351

Hyun KH, Sumimoto S, Fukui S, Mori M (1984d) Lectin-binding histochemistry of verruca vulgaris and seborrheic keratosis. Acta Histochem Cytochem 17:399–413

Jellinger K, Denk H (1974) Blood group isoantigens in angioblastic meningiomas and hemangioblastomas of the central nervous system. Virchows Arch A 364:137–144

Johnson JD, Lamm DL (1980) Prediction of bladder tumor invasion with the mixed cell agglutination test. J Urol 13:23–28

Kahn HJ, Baumal R (1985) Differences in lectin binding in tissue sections of human and murine malignant tumors and their metastases. Am J Pathol 119:420–429

Kato T (1977) Detection of A, B and H (0) antigens in normal and neoplastic epithelium of the urinary bladder by the specific red cell adherence test (SRCA). Tohoku J Exp Med 121:239–246

Kay HEM, Wallace DM (1961) A and B antigens of tumors arising from urinary epithelium. J Natl Cancer Inst 26:1349–1365

Kellokumpu I, Karhi K, Andersson LC (1986) Lectin-binding sites in normal, hyperplastic, adenomatous and carcinomatous human colorectal mucosa. Acta Pathol Microbiol Immunol Scand Sect A 94:271–280

Kim YS, Isaacs R (1975) Glycoprotein metabolism in inflammatory neoplastic disease of human colon. Cancer Res 35:2092–2097

Kim Y, Isaacs R, Perdomo J (1974) Alterations of membrane glycoproteins in human colonic adenocarcinoma. Proc Natl Acad Sci 71:4869–4873

Klein PJ, Würz H (1985) Korrelation zwischen Lektinhistochemie und Hormonsensibilität von Mammacarcinomen. In: Bellmann O, Krebs D, Totovic V (eds) Hormonabhängigkeit des Mammacarcinoms. Aktuelle diagnostische Möglichkeiten. Zukschwerdt, München Bern Wien, pp 55–66

Klein PJ, Newman RA, Müller P, Uhlenbruck G, Citoler P, Schaefer HE, Lennartz KJ, Fischer R (1979) The presence and significance of the Thomsen Friedenreich antigen in mammary gland: II. Its topochemistry in normal hyperlastic and carcinoma tissue of the breast. J Cancer Res Clin Oncol 93:205–214

Klöppel G, Fitzgerald PJ (1986) Pathology of nonendocrine pancreatic tumors. In: Go V, Gardener JD, Brooks FP, Lebenthal E, Dimagno E, Scheele G (eds) The exocrine pancreas. Biology, pathobiology, and diseases. Raven Press, New York, pp 649–674

Klöppel G, Dreyer T, Lampe V, Kalthoff H, Schmiegel WH, Von Bülow M, Kern HF, Morohoshi T, Heitz PU (1984) Immunzytochemische Typisierung von exokrinen Pankreastumoren und ihren Metastasen. Verh Dtsch Ges Pathol 68:104–107

Kluskens LF, Kluskens JL, Bibbo M (1984) Lectin binding in endometrial adenocarcinoma. Am J Clin Pathol 82:259–266

Koprowski H, Steplewski Z, Mitchell K, Herlyn M, Herlyn D, Fuhrer P (1979) Colorectal carcinoma antigens detected by hybridoma antibodies. Somat Cell Genet 5:957–972

Koprowski H, Herlyn M, Steplewski Z, Sears HF (1981) Specific antigen in serum of patients with colon carcinoma. Science 212:53–54

Köttgen E, Bauer C, Reutter W, Gerok W (1979) Neue Ergebnisse zur biologischen und medizinischen Bedeutung von Glykoproteinen. Klin Wochenschr 57:151–159

Kuhlmann WD, Peschke P, Wurster K (1983) Lectin-peroxidase conjugates in histopathology of gastrointestinal mucosa. Virchows Arch (Pathol Anat) 398:319–328

Leathem AJC, Atkins NJ (1983) Lectin binding to paraffin sections. In: Bullock GR, Petrusz P (eds) Techniques in immunocytochemistry, vol 2. Academic Press, London New York Paris San Diego San Francisco, pp 41–70

Leathem A, Dokai J, Atkins (1983) Lectin binding to normal and malignant breast tissue. Diagn Histopathol 6:171–180

Lee MC, Damjanov I (1984) Lectin histochemistry of human placenta. Differentiation 28:123–128

Lee MC, Damjanov I (1985) Pregnancy-related changes in the human endometrium revealed by lectin histochemistry. Histochemistry 82:275–280

Lee MC, Talerman A, Oosterhuis, JW, Damjanov I (1985) Lectin histochemistry of classic and spermatocytic seminoma. Arch Pathol Lab Med 109:938–942

Limas C, Lange P (1982) A, B, H antigen detectability in normal and neoplastic urothelium. Cancer 49:2476–2484

Lin F, Liu PI, McGregor DH (1977) Isoantigens A, B and H in morphologically normal mucosa and in carcinoma of the larynx. Am J Clin Pathol 68:372–376

Lis H, Sharon N (1986) Lectins as molecules and as tools. Ann Rev Biochem 55:35–67

Liu PI, McGregor DH, Liu JG, Dunlap DL, Jinks WI, Lin F, Przybylski C, Miller LA (1974) Carcinoma of the oral cavity evaluated by specific red cell adherance test. Oral Surg 38:56–64

Loke YW, Ballard AC (1973) Blood group A antigens on human trophoblast cells. Nature 245:329–330

Löning T (1984) Immunpathologie der Mundschleimhaut. Fischer, Stuttgart New York

Matsushita Y, Yonezawa S, Nakamura T, Shimizu S, Ozawa M, Muramatsu T, Sato E (1985) Carcinoma-specific Ulex europaeus Agglutinin-I binding glycoproteins of human colorectal carcinoma and its relation to carcinoembryonic antigen. J Natl Canc Inst 75:219–226

Miettinen M, Holthöfer H, Lehto VP, Miettinen A, Virtanen I (1983) Ulex europaeus I lectin as a marker for tumors derived from endothelial cells. Am J Clin Pathol 79:32–36

Miyauchi T, Muramatsu H, Ozawa M, Mizuta T, Suzuki T, Muramatsu T (1982a) Receptors for peanut agglutinin isolated from cell lines of human Burkitt lymphoma, lung squamous cell carcinoma and gastric signet ring cell carcinoma. Gann 73:581–587

Miyauchi T, Yonezawa S, Takamura T, Chiba T, Tejima S, Ozawa M, Sato E, Muramatsu T (1982b) A new fucosyl antigen expressed on colon adenocarcinoma and embryonal carcinoma cells. Nature 299:168–169

Möller P (1982) Peanut lectin: A useful tool for detecting Hodgkin's cells in paraffin sections. Virchows Arch (Pathol Anat) 396:313–317

Möller P, Lennert K (1984) On the angiostructure of lymph nodes in Hodgkin's disease. Virchows Arch (Abt A) 403:257–270

Müller W, Klein PJ, Vierbuchen MJ, Uhlenbruck G (1980) Lectin binding sites in the choroid plexus and choroid plexus papillomas. Neurosurg Rev 3:57–65

Nakayama J, Katsuyama T, Ono K, Akamatsu T, Hattori H (1985) Large bowel carcinoma-specific antigens detected by the lectin, Griffonia simplicifolia agglutinin-II. Jpn J Cancer Res 76:1078–1084

Nicolson GL (1974) The interaction of lectins with animal cell surfaces. Int Rev Cytol 39:89–190
Nicolson GL (1976) Trans-membrane control of the receptors on normal and tumor cells: II. Surface changes associated with transformation and malignancy. Biochem Biophys Acta 458:1–72
Nicolson GL (1978) Ultrastructural localization of lectin receptors. In: Koehler JK (ed) Advanced techniques in biological electron microscopy. II. Specific Ultrastructural probes. Springer, Berlin Heidelberg New York, pp 1–38
Nielsen MJS, Örntoft TF, Bendtsen L, Wolf H (1986) Lectin binding to neoplastic human epithelium. In: Bog-Hansen TC, Van Driessche E (eds) Lectins. Biology, biochemistry, clinical biochemistry, vol 5. de Gruyter, Berlin New York, pp 463–468
Örntoft TF, Bendtsen L, Poulsen HS, Petersen SE, Wolf H, Pedersen S, Bolund L (1986) Dual parameter flowcytometric measurement of DNA-content and lectin binding in human bladder tumors. In: Bog-Hansen, Van Driessche (eds) Lectins. Biology, biochemistry, clinical biochemistry, vol 5. de Gruyter, Berlin New York, pp 545–550
Pajor L, Bal'azs M, Balogh J, Brittig F, Jo'os L, Linse R, Scholz M, Suba Zs, T'oth P (1986) Ultrastructural, lectin histochemical and immunohistological observations on Merkel cell tumors. Pathol Res Pract 181:45–49
Pedal I, Baedeker C (1985) Immunoenzymatische Darstellung der Isoantigene A, B und H in fäulnisverändertem Nierengewebe. Z Rechtsmed 94:9–20
Picard J, Edvard E, Feizi T (1978) Changes in the expression of the blood groups A, B, H, Le a and Le b antigens in glycoprotein-rich of gastric carcinomas. J Clin Lab Immunol 1:119–128
Poschmann A, Fischer K, Grundmann A, Vongjirad A (1976) Neuraminidaseinduzierte Hämolyse. Monatsschr Kinderheilkd 124:15–24
Prendergast RC, Tot PD, Gargiulo AW (1968) Reactivity of blood group substances of neoplastic oral epithelium. J Dent Res 47:306–310
Raedler A, Boehle A, Otto U, Raedler E (1982) Differences of glycoconjugates exposed on hypernephroma and normal kidney cells. J Urol 128:1109–1113
Raedler A, Böttcher S, Stalling F, Kaukel E (1985) Lectin binding sites on bronchial carcinoma cells. In: Bog-Hansen TC (ed) Lectins. Biology, biochemistry, clinical biochemistry, vol 4. de Gruyter, Berlin New York, pp 137–142
Raedler A, Hachmann E, Buhl M, Greten H, Thiele HG (1986a) Phenotype of SBA+T cells in human bone marrow. In Bog-Hansen TC, Van Driessche (eds) Lectins. Biology, biochemistry, clinical biochemistry, vol 5. de Gruyter, Berlin New York, pp 557–564
Raedler A, Kelm R, Thiele HG, Greten H (1986b) Activated PNA+peripheral immunocytes in inflammatory bowel diseases. In: Bog-Hansen TC, Van Driessche E (eds) Lectins. Biology, biochemistry, clinical biochemistry, vol 5. de Gruyter, Berlin New York, pp 399–408
Rapin AMC, Burger MM (1974) Tumor cell surfaces: General alterations detected by agglutinins. Adv Cancer Res 20:1–91
Reano A, Faure M, Jacques Y, Reichert U, Thivolet J (1982) Lectins as markers of human epidermal cells. Differentiation 22:205–210
Reano A, Ledger PW, Bonnefoy JY, Thivolet J (1984) Con A- and PNA-binding glycoproteins of human epidermis. J Invest Dermatol 83:202–205
Ree HJ (1983) Lectin histochemistry of malignant tumors. II. Concanavalin A: a new histochemical marker for macrophage-histiocytes in follicular lymphoma. Cancer 51:1639–1646
Ree HJ (1986) Concanavalin A-binding histiocytes in Hodgkin's disease. A predictor of early relapse. Cancer 58:87–95
Ree HJ, Hsu SM (1983) Lectin histochemistry of malignant tumors. I. Peanut agglutinin (PNA) receptors in follicular lymphoma and follicular hyperplasia: An immunohistochemical study. Cancer 51:1631–1638
Ree HJ, Raine L, Crowley JP (1983) Lectin binding patterns in diffuse large cell lymphoma. Cancer 52:2089–2099
Reisner Y, Biniaminov M, Rosenthal E (1979) Interaction of peanut agglutinin with normal human lymphocytes and with leukemic cells. Proc Natl Acad Sci 76:447–452
Remmele W, Hildebrand U, Hienz HA, Klein PJ, Vierbuchen M, Behnken LJ, Heicke B, Scheidt E (1986) Comparative histological, histochemical, immunohistochemical and biochemical studies on oestrogen receptors, lectin receptors, and Barr bodies in human breast cancer. Virchows Arch (A) 409:127–147

Rittmann BR, Mackenzie IC (1983) Effects of histological processing on lectin binding on oral mucosa and submucosa. Histochem J 15:467–474

Roth J (1978) The lectins – molecular probes in cell biology and membrane research. Exp Pathol 3 (Suppl): 1–186

Roth J (1986) Light and electron microscopic localization of cellular glycoconjugates with lectin-gold complexes: Demonstration of glycocalyx specialization and intracellular glycosylation compartments. In: Bog-Hansen TC, Van Driessche E (eds) Lectins. Biology, biochemistry, clinical biochemistry, vol 5. de Gruyter, Berlin New York, pp 419–431

Roth J, Binder M (1978) Colloidal gold, ferritin, and peroxidase as markers for electron microscopic double labeling lectins techniques. J Histochem Cytochem 26:163–169

Roth J, Taatjes DJ (1985) Glycocalyx heterogeneity of rat kidney urinary tubule: demonstration with a lectin-gold technique specific for sialic acid. Eur J Cell Biol 39:449–457

Roth J, Brown D, Orci L (1983) Regional distribution of N-acetyl-D-galactosamine residues in the glycocalyx of glomerular podocytes. J Cell Biol 96:1189–1196

Salfelder A, Caselitz J (1983) Immunhistologische Aspekte des Cystadenolymphoms im Gewebeschnitt und Heterotransplantat. Verh Dtsch Ges Pathol 67:586

Santen RJ, Collette J, Franchimont P (1980) Partial purification of carcinoembryonic-reactive antigen from breast neoplasms using lectin and antibody affinity chromatography. Cancer Res 40:1181–1188

Sato A, Spicer SS (1982a) Ultrastructural visualization of galactose in the glycoprotein of gastric surface cells with a peanut lectin conjugate. Histochem J 14:125–138

Sato A, Spicer SS (1982b) Ultrastructural visualization of glycosyl residues in various alimentary epithelial cells with the peanut lectin-horseradish peroxidase. Histochemistry 73:607–624

Saveriano N, Drinnan M, Santer V, Osmond DG (1981) Lectin-binding patterns of small lymphocytes in bone marrow, thymus and spleen: demonstration of lymphocyte subsets by quantitative radioautography. Eur J Immunol 11:870–876

Schauer R (1982a) Chemistry, metabolism, and biological functions of sialic acids. Adv Carbohydr Chem Biochem 40:131–234

Schauer R (1982b) Sialic acids. Chemistry, metabolism and function. Springer, Wien New York

Schrevel J, Kieda C, Caigneaux E, Gros D, Delmotte F, Monsigny M (1979) Visualization of cell-surface carbohydrates by a general two-step lectin technique: lectins and glycosylated cytochemical markers. Biol Cell 36:259–266

Schrevel J, Gros D, Monsigny M (1981) Cytochemistry of cell glycoconjugates. Progress in histochemistry and cytochemistry, vol 14, Nr 2. Fischer, Stuttgart New York

Schulte BA, Spicer SS (1983a) Light microscopic detection of sugar residues in glycoconjugates of salivary glands and the pancreas with lectin-horseradish peroxidase conjugates. I. Mouse. Histochem J 15:1217–1238

Schulte BA, Spicer SS (1983b) Histochemical evaluation of mouse and rat kidneys with lectin-horseradish peroxidase conjugates. Am J Anat 168:345–362

Schulte BA, Spicer SS (1983c) Light microscopic histochemical detection of sugar residues in secretory glycoproteins of rodent and human tracheal glands with lectin-horseradish peroxidase conjugates and the galactose oxidase-Schiff sequence. J Histochem Cytochem 31:391–403

Schulte BA, Spicer SS (1984) Light microscopic detection of sugar residues in glycoconjugates of salivary glands and the pancreas with lectin-horseradish peroxidase conjugates. II. Rat. Histochem J 16:3–20

Schwechheimer K, Möller P, Schnabel P, Waldherr R (1983) Emphasis on peanut lectin as a marker for granular cells. Virchows Arch (Pathol Anat) 399:289–297

Schwechheimer K, Weiss G, Schnabel P, Möller P (1984) Lectin target cells in human central nervous system and the pituitary gland. Histochemistry 80:165–169

Sharon N, Lis H (1975) Use of lectin for the study of membranes. Methods Membr Biol 3:143–200

Slocombe GW, Berry CL, Swettenham KV (1980) The variability of blood group antigens in gastric carcinoma as demonstrated by the immunoperoxidase technique. Virchows Arch (Pathol Anat) 387:289–300

Sobrinho-Simoes M, Damjanov I (1986) Lectin histochemistry of papillary and follicular carcinoma of the thyroid gland. Arch Pathol Lab Med 110:722–729

Springer GF, Desai PR, Banatwala I (1974) Blood group MN specific substances and precursors in normal and malignant human breast tissues. Naturwissenschaften 61:457–458

Springer GF, Desai PR, Banatwala I (1975) Blood group MN antigens and precursors in normal and malignant human breast glandular tissue. J Natl Cancer Inst 54:335–339

Stegner HE, Fischer K, Poschmann A (1981) Immunohistochemical localization of Thomsen-Friedenreich antigen in normal and malignant breast tissue using peroxidase-antiperoxidase technique. Tumor Diagnostik 3:127–130

Stejskal R, Lill PH, Davidsohn I (1973) A, B, and H isoantigens in the human fetus. Dev Biol 34:274–281

Stoddart RW, Collins RD, Jacobson W (1980) Lectin staining of carbohydrates of haemic cells: The cells of normal blood and bone marrow and of the myeloid leukemias. J Pathol 131:321–332

Strauchen JA (1985) Lectin binding properties of human follicular center lymphocytes and follicular (nodular) lymphomas. Hematol Oncol 3:11–23

Strauchen JA, Bergman SM, Hanson TAS (1980) Expression of A and B tissue isoantigens in benign and malignant lesions of the breast. Cancer 45:2149–2155

Szulman AE (1980) The ABH blood groups and development. Curr Top Dev Biol 14:127–144

Teshima S, Hirohashi S, Shimosato Y, Kishi K, Ino Y, Matsumoto K, Yamada T (1984) Histochemically demonstrable changes in cell surface carbohdydrates of human germ cell tumors. Lab Invest 50:271–277

Thatcher N, Hashimi K, Chang J, Swindell R, Crowther D (1980) Anti-T antibody in malignant melanoma patients. Cancer 46:1378–1382

Tolson ND, Daley TD, Wysocki GP (1985) Lectin probes of glycoconjugates in human salivary glands: 1. J Oral Pathol 14:423–430

Totovic V, Dvorak O, Bellmann O, Sigmund M (1985) Die Lektinhistochemie zur Erfassung der Hormonabhängigkeit des Mammacarcinoms – Problematik der Quantifizierung. In: Bellmann O, Krebs D, Totovic V (eds) Hormonabhängigkeit des Mammacarcinoms. Aktuelle diagnostische Möglichkeiten. Zukschwerdt, München Bern Wien, pp 46–54

Ucci AA, Alroy J, Orgad U, Goyal V, Gavris V (1983) Distribution of lectin binding sites in normal and pathologic human prostates. Lab Invest 48:87 A–88 A

Uhlenbruck G (1971) Immunbiologie. Goldmann, München

Uhlenbruck G (1980) Lektine: Eine Einführung, Entwicklungen und Aussichten. J Clin Chem Clin Biochem 18:373–374

Uhlenbruck G (1981) Lektine, Toxine und Immunotoxine. Naturwissenschaften 68:2949–2953

Uhlenbruck G, Schwonzen M, Pissors W, Diehl V (1986) Lectins and cancer: Characterization of Hodgkin-derived cell lines by various lectins. In: Bog-Hansen TC, Van Driessche (eds) Lectins. Biology, biochemistry, clinical biochemistry, vol 5. de Gruyter, Berlin New York, pp 433–440

Ulrich W, Horvat R, Krisch K (1985) Lectin histochemistry of kidney tumours and its pathomorphological relevance. Histpathology 9:1037–1050

Uno A, Hori Y, Saida T, Seki Y, Oohara K, Kukita A (1985) Lectin-binding sites and patterns of melanin-producing cells. In: Pigment Cell 1985. Biological, molecular and clinical aspects of pigmentation. Proceedings of the XIIth International Pigment Cell Conference, Tokyo

Vedtofte P, Dabelsteen E (1981) Receptors for the lectins wheat germ, ricinus communis I and Soybean in ameloblastomas and normal oral mucosa. Acta Pathol Microbiol Scand Sect A 89:439–449

Vierbuchen M, Klein PJ, Würz H, Kreuder J, Fischer R (1985) Experimentelle Grundlagen der diagnostischen PNA (peanut-agglutinin) Histochemie des Mammacarcinoms. In: Bellmann O, Krebs D, Totovic V (eds) Hormonabhängigkeit des Mammacarcinoms. Aktuelle diagnostische Möglichkeiten. Zukschwerdt München Bern Wien, pp 21–45

Virtanen I, Miettinen A, Wartiovaara J (1978) Lectin-binding sites are found in rat liver cell plasma membrane only on its extracellular surface. J Cell Sci 29:287–296

Walker RA (1984) The binding of peroxidase-labelled lectins to human breast epithelium. II. The reactivity of breast carcinomas to wheat germ agglutinin. J Pathol 144:101–108

Walker RA (1985a) Ulex europeus 1-peroxidase as a marker of vascular endothelium: its application in routine histopathology. J Pathol 146:123–127

Walker RA (1985b) The binding of peroxidase-labelled lectins to human breast epithelium. IV. The reactivity of breast carcinomas to peanut, soy bean and dolichos biflorus agglutinins. J Pathol 145:269–277

Warren L, Buck CA, Tuczynski GP (1978) Glycopeptide changes and malignant transformation: A possible role for carbohydrate in malignant behavior. Biochem Biophys acta 516:97–127

Weinstein RS, Coon J, Alroy J, Davidsohn I (1981) Detection and tissue-associated blood group antigens ABH (0) in human carcinomas. In: DeLellis, RA (ed) Diagnostic immunohistochemistry. Masson, New York, pp 239–261

Wernert N, Dhom G (1986) Immunohistochemical demonstration of peanut agglutinin (PNA) binding sites in the prostate and prostatic carcinoma before and after antiandrogenic treatment. XXVIII. Symposion der Gesellschaft für Histochemie, Gargellen

Wick MR, Manivel C, O'Leary TP, Cherwitz DL (1986) Nephroblastoma: A comparative immunocytochemical and lectin-histochemical study. Arch Pathol Lab Med 110:630–635

Woltering EA, Tuttle SE, James AG, Sharma HM (1983) ABO (H) cell surface antigens in benign and malignant parotid neoplasms. J Surg Oncol 24:177–179

Wurster K, Peschke P, Kuhlmann WD (1983) Cellular localization of lectin-affinity in tissue sections of normal human duodenum. Virchows Arch Abt A 402:1–9

Yogeeswaran G (1983) Cell surface glycolipids and glycoproteins in malignant transformation. Adv Cancer Res 38:289–350

Yonezawa S, Nakamura T, Tanaka S, Sato E (1982) Glycoconjugate with Ulex europaeus agglutinin-I-binding sites in normal mucosa, adenoma, and carcinoma of the human large bowel. J Natl Cancer Inst 69:777–785

Zinsmeister VAP, Ackermann GA (1983) Surface localization of wheat germ agglutinin and concanavalin A binding sites on normal human blood cells: An ultrastructural histochemical study. Tissue and Cell 15:673–683

Zippel HH, Benzenberg F, Hackenberg R, Hölzel F, Schulz KD, Klein PJ (1985) Untersuchungen in Zellkulturen von Mammakarzinomen: Lektinhistochemische PNA-Bindungen in Abhängigkeit von den Zellpassagen. In: Bellmann O, Krebs D, Totovic V (eds) Hormonabhängigkeit des Mammacarcinoms. Aktuelle diagnostische Möglichkeiten. Zukschwert, München Bern Wien, pp 114–121

Zippel HH, Hackenberg R, Benzenberg F, Hölzel F, Schulz KD, Klein PJ (1986) Histochemical investigations on lectin binding in normal and neoplastic endometrial tissue and in cell cultures of endometrial carcinomas. In: Bog-Hansen TC, Van Driessche E (eds), vol 5. Lectins. Biology, biochemistry, clinical biochemistry, vol 5. de Gruyter, Berlin New York, pp 441–448

Neuroendocrine Tumor Markers

PH.U. HEITZ

1 Neuroendocrine Tumors

1.1 Definition and Markers

Neuroendocrine tumors arise from organs known to secrete neuroendocrine messengers, e.g. hormones or substances with paracrine activity, and/or to contain cells displaying neuroendocrine characteristics or activity. The term "neuroendocrine" is used here to define the secretory products of the cells and tumors rather than their nature and embryological derivation (see SOLCIA et al. 1984). Although individual tumor types do not occur frequently, neuroendocrine tumors as a group are common.

There are several markers characteristic of neuroendocrine cells, some of which have been used for tumor typing (see below). One group of markers occurs independently of production of regulatory peptides by normal or tumorous neuroendocrine cells. This group of *broad-spectrum markers* (Sect. 1.3) includes 1) the presence of voltage-dependent Na^{+} and/or Ca^{++} channels in the cell membrane, 2) receptors for specific ligands, e.g. nerve growth factor,

3) cytoskeletal proteins, e.g. intermediate filaments, neurofilaments (Sect. 1.3.2), 4) neuron-specific enolase (Sect. 1.3.3), 5) granule matrix constituents, e.g. chromogranins and Leu-7 (Sects. 1.3.4 and 1.3.5), 6) granule membrane constituents, e.g. cytochrome B 561 or synaptophysin (Sect. 1.3.5), and 7) amine biosynthetic enzymes.

The second group of markers represents *specific markers*, including regulatory peptides (Sect. 1.4).

1.2 Role of Tumor Markers in the Diagnosis and Differential Diagnosis

The purpose of the use of tumor markers depends on the investigated tumor types. In general, the following questions can often be answered, at least partially:
- localization of the tumor,
- precise functional diagnosis,
- differential diagnosis,
- grade of differentiation,
- assessment of biological behavior,
- effects of therapy,
- morphologico-functional correlations.

Conventional histology using tinctorial stains is still a prerequisite for the diagnosis of tumors. However, it is not possible to reach a precise, functionally significant diagnosis of neuroendocrine tumors using conventional histology. These tumors often cause characteristic symptoms, or well defined syndromes caused by inappropriate hormone secretion. The concentration in the patient's serum of many secretory products can be determined. It is necessary that the pathologist provide a specific and functionally significant diagnosis and determine the morphological-functional correlations. The final evidence for the production and secretion of a substance can be provided only by the combined effort of the clinician, the biochemist, and the morphologist.

In order to define posttranslational products or to localize mRNA at the cellular and/or subcellular levels, the morphologist must use immunocytochemical methods or in-situ hybridization techniques. The use of these techniques is of particular interest in the differentiation of metastases of neuroendocrine tumors. Some types of tumors, e.g., pancreatic endocrine tumors, endocrine tumors of the gastrointestinal tract (carcinoids) are of low grade malignancy, and can be treated, using drugs with a specific action directed against specific types of tumor cells. The precise definition of a metastasis and/or the localization of the primary is therefore important for the patient.

1.3 Broad-Spectrum Markers (see Fig. 12)

1.3.1 Significance

The importance of broad-spectrum markers is to provide the pathologist with a clue to the neuroendocrine differentiation of a tumor, if conventional histology does not reveal it. In addition, these markers are independent of hormone production by tumor cells, e.g. they are valuable in the diagnosis of tumors producing messengers, as well as in the diagnosis of inactive or poorly differentiated tumors.

1.3.2 Cytokeratins

With a few exceptions, cells of neuroendocrine tumors contain cytokeratins, in particular cytokeratins 8 and 18 (HOEFLER et al. 1986). Using a monoclonal antibody directed against an epitope of a sequence common to all presently known cytokeratins (FRANKE et al. 1987), and reacting with the cytoskeleton of all epithelial and mesothelial cells (monoclonal antibody lu-5; VON OVERBECK et al. 1985), virtually all cells of neuroendocrine tumors yield a reaction. The reaction pattern may be pancytoplasmic, "dot-like" paranuclear, or cell membrane-associated (Fig. 1a, 2a, 5a). Exceptions, e.g. tumor cells not containing cytokeratins, are cells of tumors of the adrenal medulla and paragangliomas, steroid-hormone-producing tumors of the adrenal cortex, testis, and ovary, and melanocytes. An immunocytochemical reaction for cytokeratins used as the first step in a diagnostic algorithm (Fig. 12) may therefore give an important clue to the differential diagnosis of neuroendocrine tumors.

1.3.3 Neuron-Specific Enolase

The soluble, cytoplasmic metallo-enzyme enolase (E.C. 4.2.1.11) catalyzes the interconversion of 2-phospho-D-glycerate and phosphoenolpyruvate (PEP) in the glycolytic pathway, and is present in every cell. Enolase is a dimer with a molecular weight of 87,000 daltons. The following dimers occur in various cell types: dimer alpha-alpha in the hepatocyte and other non-neural tissues, dimer beta-beta in muscle cells, and dimer gamma-gamma in neurones and neuroendocrine cells ("neuron-specific enolase"). A heterodimer alpha-gamma has also been found to occur in neurons and neuroendocrine cells (SCHMECHEL et al. 1978a, b, 1980).

The occurrence of various dimers in a given cell is not mutually exclusive, but there is a markedly higher concentration of enolase gamma-gamma than of other dimers in neurons and neuroendocrine cells. The term "neuron-specific enolase" is therefore too restrictive because this dimer is not restricted to neuroendocrine cells. Nevertheless, using antisera of high specificity for dimer gamma-gamma, it is possible to localize it at the light and electron microscopic level in the cytoplasm of neurons and neuroendocrine cells (Fig. 1b, 2b), but only to a very limited extent in non-neuroendocrine cells. This reaction has been used extensively and successfully in morphological analysis (TAPIA et al. 1981; DHILLON et al. 1982; OSKAM et al. 1985; IWASE et al. 1986; OSBORN et al. 1986), and as a serum marker of neuroendocrine tumors (CARNEY et al. 1982; PRINZ and MARANGOS 1982; ZELTZER et al. 1983).

In our experience, neuron-specific enolase is a useful marker for neuroendocrine tumors in the second step of a diagnostic flow-chart (Fig. 12), provided that the tissue was rapidly and thoroughly fixed. A non-specific reaction often occurs in necrotic or autolytic areas. It is rare that all cells of a given tumor display a clear-cut reaction. There is no detectable reaction for neuron-specific enolase in tumors of the parathyroid glands, tumors of the thyroid with follicular differentiation, and in tumors producing steroid hormones.

1.3.4 Chromogranins

Chromogranins are a group of acidic monomeric proteins of various sizes, which form a major part of the soluble proteins in the secretory granules of the adrenal medulla (WILSON and LLOYD 1984). The largest of the polypeptides is chromogranin A (MW 68,000 daltons), which makes up approx. 40 to 50% of the total soluble granule proteins of the adrenal medulla. A monoclonal antibody to human chromogranin A has been shown to react with normal (endocrine) cells and

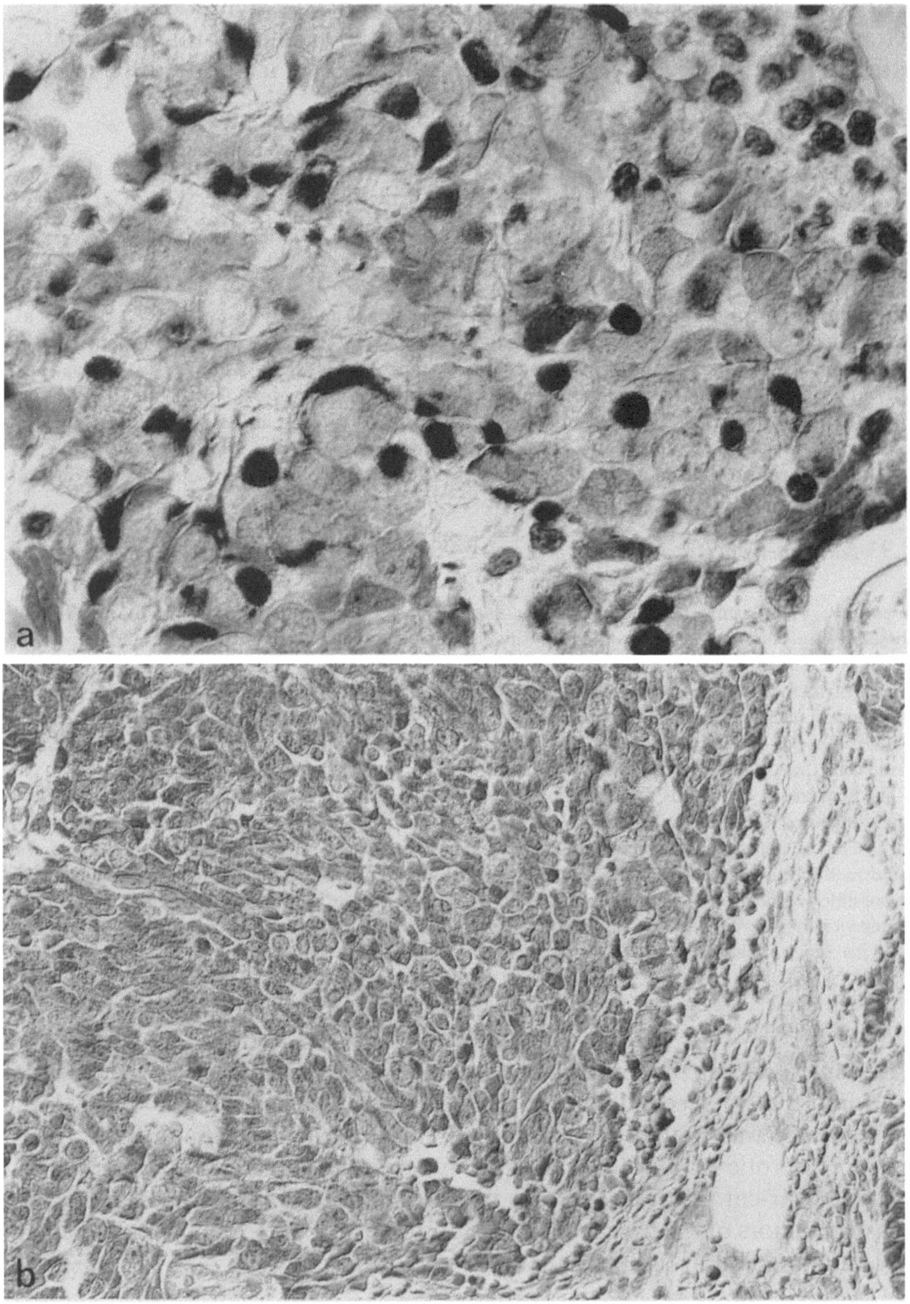

Fig. 1a, b. Merkel cell tumor (neuroendocrine carcinoma of the skin). Avidin-biotin complex technique, differential interference contrast optics. **a** "Dot-like" reaction for cytokeratins in many tumor cells. Monoclonal antibody to a panepithelial antigen, ×1000. **b** Neuron-specific enolase in the cytoplasm of the majority of tumor cells, ×200

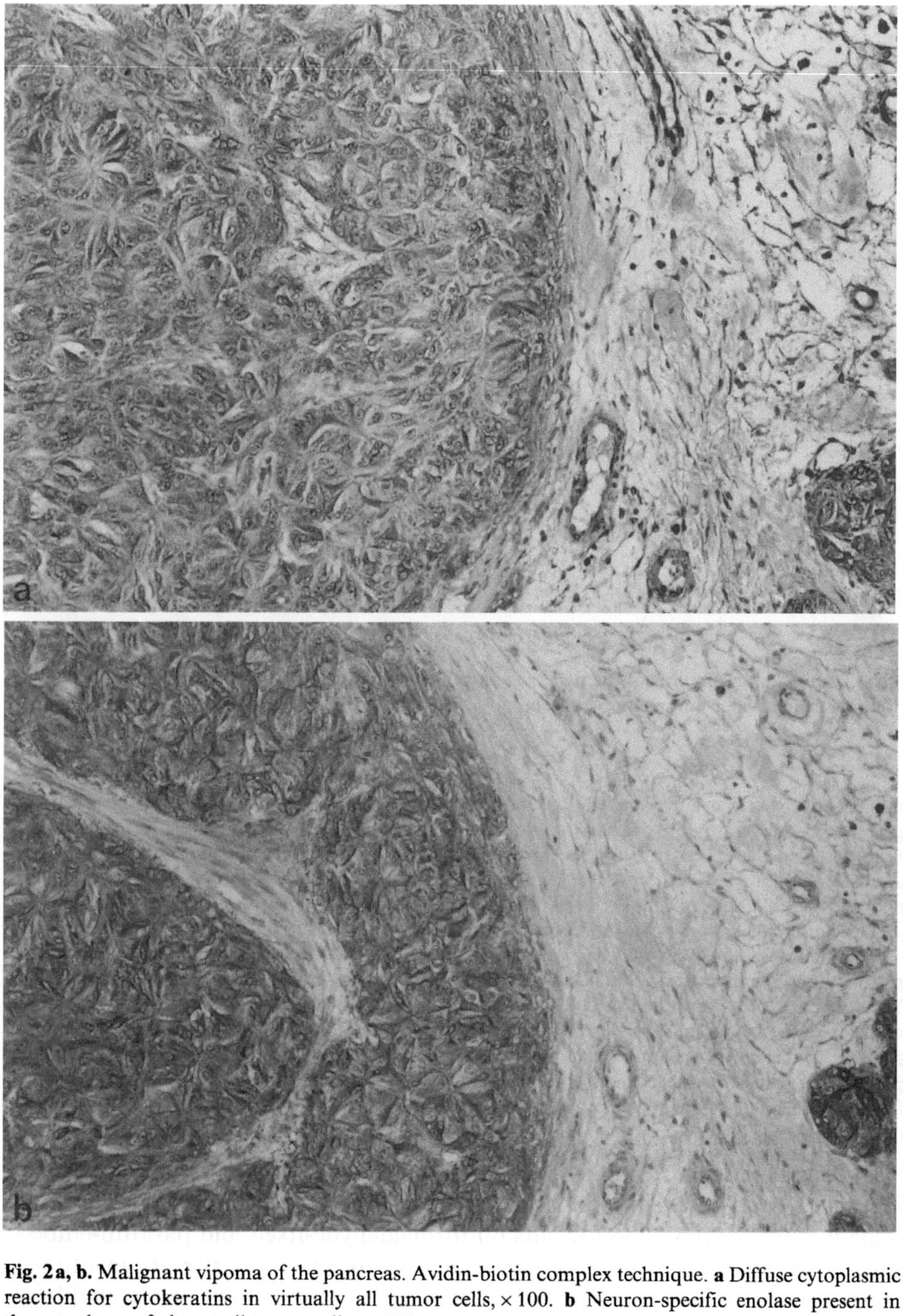

Fig. 2a, b. Malignant vipoma of the pancreas. Avidin-biotin complex technique. **a** Diffuse cytoplasmic reaction for cytokeratins in virtually all tumor cells, ×100. **b** Neuron-specific enolase present in the cytoplasm of almost all tumor cells, ×100

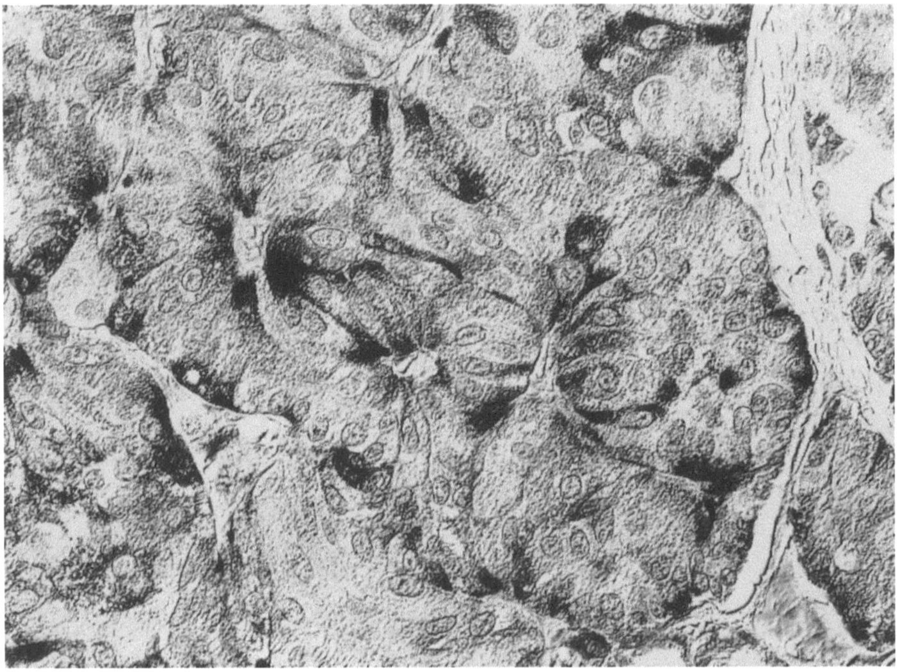

Fig. 3. Malignant insulinoma of the pancreas. Strong reaction for chromogranin A in a large number of tumor cells. Avidin-biotin complex technique, ×400, differential interference contrast optics

tumors arising in the adrenal medulla, endocrine pancreas (Fig. 3), parathyroid glands, anterior pituitary gland, the thyroid (C-cells), and medullary carcinomas. In addition, a focal staining has been observed in Merkel cell tumors, neuroblastoma, and small-cell carcinoma of the lung. The antibody did not react with melanomas, naevi, the posterior pituitary gland, and neurones (LLOYD and WILSON 1983; EHRHART et al. 1986). Chromogranin A could be localized in cytoplasmic secretory granules of paragangliomas by immunoelectron microscopy (WILSON and LLOYD 1984). In addition, chromogranin A was shown to be secreted by the aforementioned tumors. The measurement of plasma chromogranin A may therefore be a useful diagnostic procedure in patients presumed to be suffering from neuroendocrine tumors (O'CONNOR and DEFTOS 1986). Recently, the sequence of bovine chromogranin A has been described, using its complementary DNA. In addition, the presence of the mRNA of prechromogranin A in adrenal medulla, brain, anterior pituitary and parathyroid glands has been demonstrated. It is conceivable that chromogranin A is a precursor for smaller peptides or that it may be important in the binding process of intragranular calcium (IACANGELO et al. 1986).

In our experience, using the aforementioned monoclonal antibody the specific reaction for chromogranin A is a useful adjunct in the diagnosis of neuroendocrine tumors, yielding good results on formaldehyde-fixed and paraffin-embedded tissue. A large number, but not all, cells react in the tumors mentioned above. However, we found the following exceptions: 1) We were unable to obtain a reaction of chromogranin A in small cell carcinoma of the bronchus, and 2) in our series the reaction was not reliable in endocrine tumors of the gastrointestinal tract (carcinoids), and it was often negative in carcinoids of the colon and rectum. In contrast to tumors a large number of cells were regularly visualized in the epithelium of the entire gastrointestinal tract. In contrast to neuron-specific enolase, the reaction for chromogranin A is useful to differentiate melanocytic tumors from other neuroendocrine neoplasms. It can be used as a second-step marker (Fig. 12).

1.3.5 Other Broad-Spectrum Markers

A new monoclonal antibody (Phe-5) has recently been commercialized. It can be used on formaldehyde-fixed paraffin-embedded tissue. In our test series, the reactions were almost identical to the antibody directed against chromogranin A. A reaction was obtained in some carcinoids of the colon, but there was no reaction in small cell carcinomas of the lung. At present, no literature concerning this antibody and the recognized antigen is yet available.

One potentially useful marker for neuroendocrine cells is synaptophysin. Synaptophysin is a specific component isolated from the membranes of presynaptic vesicles of bovine neurons. It is a glycosylated polypeptide (MW 38,000 daltons), and it has been reported to be visualized regularly in almost all neurons and in a majority of neuroendocrine tumors. A prerequisite for its demonstration is good preservation of conventionally fixed and paraffin-embedded tissue or incubation of freshly frozen tissue (GOULD et al. 1986; WIEDENMANN et al. 1986; GOULD et al. 1987).

Recently, it has been reported that anti-lymphocyte antibody Leu-7 (HNK), directed against an epitope present on the cell surface of natural killer (NK) lymphocytes, and with myelin-associated glycoprotein (MAG), yields an intense reaction with normal and neoplastic adrenal medullary cells. Immunocytochemical studies and immunoblot analyses showed that this antibody apparently reacts with the carbohydrate moiety of a protein (molecular weight 75,000 daltons) localized in the matrix of chromaffin granules. The antibody was shown to react with a small percentage of normal pancreatic islet cells, cells of the anterior pituitary, and endocrine cells of the gastrointestinal tract. We found a reaction in tumor cells of formaldehyde-fixed, paraffin-embedded tissue of medullary thyroid carcinomas, pheochromocytomas, paragangliomas, endocrine pancreatic tumors, and carcinoids of the stomach, duodenum, ileum, and rectum. No reaction was obtained in carcinoids of the colon. It was suggested that Leu-7 immunoreactivity might be a marker of specific subsets of secretory granules. It is also important to realize that antibody Leu-7 reacts with the surface of cells of small-cell carcinoma of the lung (TISCHLER et al. 1986).

1.4 Specific Markers (see Fig. 12)

Peptides, glycoproteins, biogenic amines, catecholamines, and, to a limited extent, steroid hormones may be used as markers in a second or third step of an algorithm designed for the precise typing of neuroendocrine tumors. The immunocytochemical localization of secretory products of neurons and of endocrine cells producing peptides, proteins, glycoproteins and/or amines, and of chromogranins is based on the fact that the intracellular pathway of these substances is regulated as well as polarized. In a steady state of the cell metabolism, a post-Golgi pool of secretory products, precursors thereof, and substances contained in the matrix of secretory granules exists. The secretory products can be stored for several hours in granules (KELLY 1985; LAUFFER et al. 1985; WICKNER and LODISH 1985). In this post-Golgi compartment, a high concentration of secretory products and other substances contained in the granular matrix is present. The granules can be analyzed morphologically by electron microscopy. Secretory products as well as other substances in the granular matrix can be identified and localized at light and electron microscopic levels by immunological reactions.

The extensive use of these markers has provided the cell biologist and the pathologist with new insights into the process of hormone biosynthesis and the secretion and the regulation thereof, and into the phenotypic heterogeneity of tumor cells.

2 Tumors of the Pituitary Gland

Tumors of the adenohypophysis leading to symptoms account for 6–10% of all intracranial neoplasms; microadenomas can be detected in 8–22.5% of all autopsied patients. Their prevalence increases with age. By using tinctorial stains (HE and PAS-orange G stains) the following variants of pituitary tumors can be differentiated: acidophil, mucoid, chromophobe, and oncocytic tumors (SAEGER 1981; for the complete classification of the World Health Organization see WILLIAMS et al. 1980). By far the most frequent tumors are chromophobe adenomas, followed by acidophil and mucoid tumors. The delimitation of these tumors from the adjacent pituitary parenchyma is often difficult because a tumor capsule is most often lacking. In addition, it is unfortunately impossible to correlate reliably the properties of tumor cells based on tinctorial stains, with the production of hormones (Fig. 4a and b).

The precise diagnosis of tumors of the pituitary gland was based for a long time on electron microscopy (LANDOLT 1975; EZRIN et al. 1982; KOVACS and HORVATH 1986). At present, the diagnosis is most often made by using immunocytochemistry, which is more reliable and faster than electron microscopy (HEITZ and OBERHOLZER 1984). Electron microscopy and immunoelectron microscopy are important tools for research.

The purpose of localizing hormonal tumor markers in the pituitary gland is to provide a precise diagnosis, to establish morphologico-functional correlations, and to observe the effects of therapy. Assessment of the biological behaviour of pituitary tumors can still be based on conventional histological observation because metastasizing carcinomas are extremely uncommon, and the invasion of bony structures surrounding the sella can be readily seen at conventional histology. The localization of the tumor is possible using tinctorial stains because the alveolar structure is in general much coarser or may be lacking in the tumor, in contrast to the adjacent pituitary parenchyma.

The most important immunocytochemical reactions are those for prolactin, growth hormone (Fig. 4b), and corticotropin. In addition, it is very useful to localize the alpha-chain, which is virtually identical in all glycoprotein hormones (LANDOLT and HEITZ 1986), the hormone specific beta-chains of follicle stimulating hormone (FSH), luteinizing hormone (LH), and thyrotropin (TSH).

The ratio of chromophobe tumors is 39.1–79.5%, of acidophil tumors 15–59%, and of mucoid tumors 0–19% based on conventional histology (BAKAY

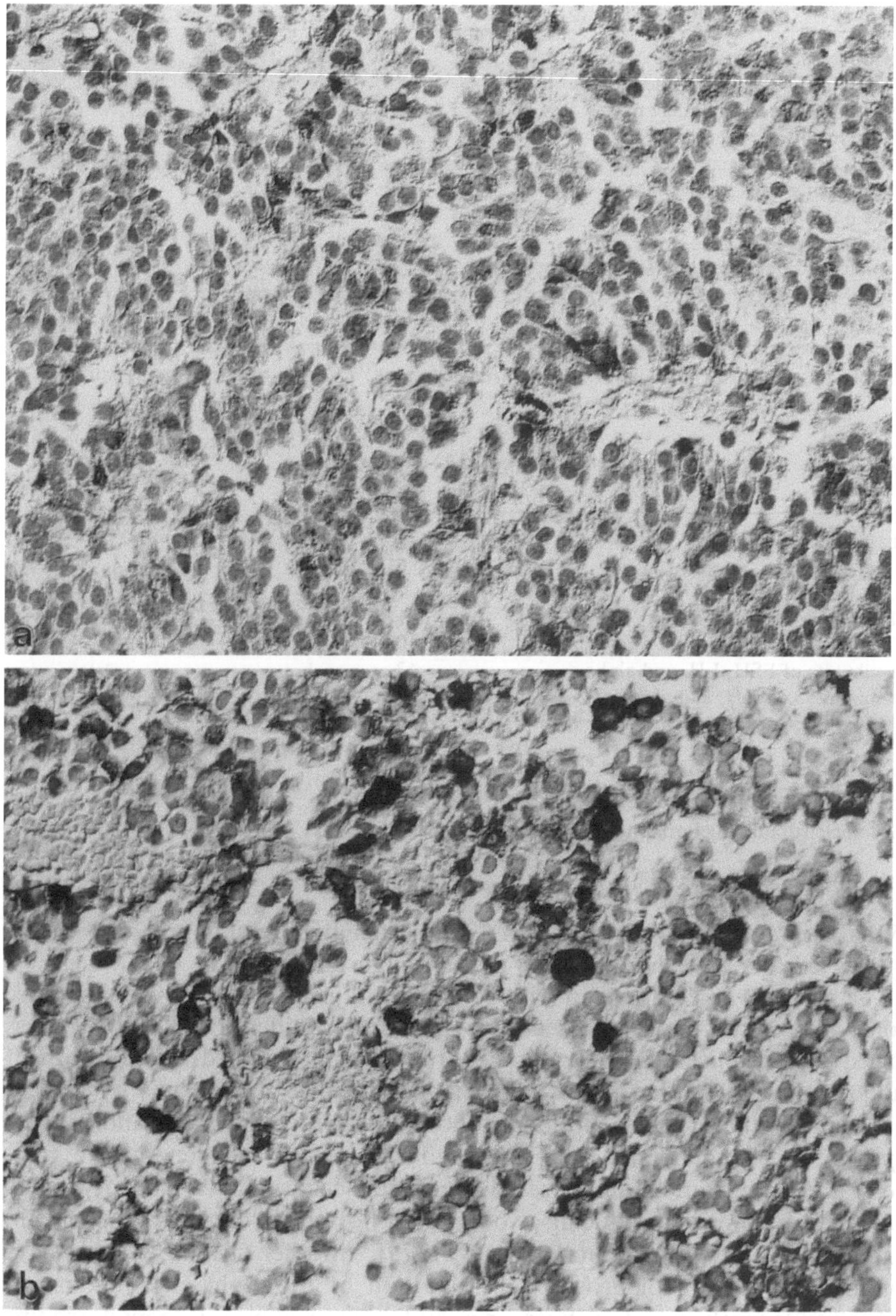

Fig. 4 a, b. Pituitary adenoma causing acromegaly. Differential interference contrast optics. **a** Absence of staining: chromophobe adenoma, PAS-orange G, ×200. **b** Presence of growth hormone in the cytoplasm of many tumor cells: growth hormone secreting adenoma. Avidin-biotin complex technique, ×400

Table 1. Type of tumors of the pituitary gland

Prolactinomas	194	51 %
Growth hormone-producing adenomas	93	24.1%
Corticotropin-producing adenomas	26	7 %
FSH-producing adenomas	3	0.8%
LH-producing adenomas	5	1.3%
TSH-producing adenomas	4	1 %
"Alpha-only" adenomas	13	3.4%
Multihormonal adenomas	13	3.4%
Adenomas without detectable secretory products	30	8 %
	381	100 %

FSH, follicle-stimulating hormone; *LH*, luteinizing hormone; *TSH*, thyroid-stimulating hormone

Table 2. Alpha-subunit of glycoprotein hormones in 92 pituitary adenomas

Alpha-only	13
GH and alpha	6
PRL and alpha	1
ACTH and alpha	1
GH and PRL and alpha	10
Beta-subunits of FSH, LH, and alpha	12
Beta-subunits and GH, and alpha	3
	46

GH, growth hormone; *PRL*, prolactin; *ACTH*, corticotropin; *FSH*, follicle-stimulating hormone; *LH*, luteinizing hormone

1950; GIROD et al. 1980; SAEGER 1981). The systematic use of immunocytochemistry has led to a considerably more precise diagnosis (Table 1).

The progress in comparison with conventional histology is obvious because in only 8% of the tumors no secretory product was identified. This progress allowed the establishment of a modern, functionally-oriented classification of pituitary tumors (Table 1). It is striking that the prolactinoma is by far the most common tumor of the pituitary, followed by growth hormone and corticotropin producing adenomas. FSH-, LH- and TSH-producing adenomas are rare. One of the important findings was the common occurrence of phenotypically heterogeneous tumors producing two or more hormones, most often prolactin combined with growth hormone (HEITZ 1979). Another striking result was a combined production of growth hormone and of the alpha-chain (common to the pituitary glycoprotein hormones and human chorionic gonadotropin; Table 2). An additional finding was the occurrence of "alpha-only adenomas", which was found by radioimmunoassay of the patient's sera (MACFARLANE et al. 1980; RIDGWAY et al. 1981; KLIBANSKY et al. 1983), and by immunocytochemistry (LANDOLT and HEITZ 1986). These tumors do not cause hormonally-induced symptoms.

The results obtained with immunocytochemical techniques on tumor tissue correlate well with clinical and biochemical findings. In addition, the immunocytochemical localization of hormones can be correlated with the localization of receptors for hypothalamic hormones on tumor cells, which opens new avenues for future research (REUBI et al. 1987).

3 Tumors of the Thyroid Gland

Diagnosis and treatment of tumors of the thyroid gland are good examples of the use of functional criteria in oncology, e.g. determination of iodine-trapping by tumor tissue, or of the production of thyroglobulin. The two most important problems in the diagnosis of these tumors are 1) the assessment of the biological behaviour of highly differentiated follicular adenomas and carcinomas respectively, and 2) the classification of poorly differentiated tumors. At present, the first problem can be solved only by a very precise and thorough histological inspection of the tumor, carefully excluding or confirming penetration and perforation of the tumor capsule, and the invasion of blood vessels. Immunocytochemistry and electron microscopy are of no help in solving this problem.

Significant progress has been made in the differentiation of poorly differentiated tumors. Information on the cellular characteristics of these tumors is essential because therapy and prognosis of the various tumor types are very different (MEISSNER 1984). Tumor cells with epithelial differentiation can be defined by immunocytochemical localization of cytokeratins. There is a coexpression of cytokeratins and vimentin in approx. 50% of follicular, and in virtually all papillary carcinomas (SCHROEDER 1986). The analysis of large series of poorly differentiated tumors has yielded the following information: 1) The presence of cytokeratins (Fig. 5a), often combined with vimentin (Fig. 5b) can be observed in most anaplastic carcinomas (CARCANGIU et al. 1985; HEITZ 1986). 2) "Undifferentiated carcinomas" of the small cell type according to the WHO classification (HEDINGER and SOBIN 1974) are often malignant non-Hodgkin lymphomas (TOBLER et al. 1984; ANSCOMBE and WRIGHT 1985; RALFKIAER et al. 1985). They are most often negative for cytokeratins, but give reactions for various markers of lymphocytes. 3) Tumors with characteristics of malignant hemangioendotheliomas are uncommon but they apparently exist (EGLOFF 1983; PFALTZ et al. 1983; RUCHTI et al. 1984). Their reaction pattern can be summarized as follows: factor VIII-related antigen positive, Ulex europaeus agglutinin 1 positive, cytokeratins negative. 4) The histological aspect of medullary carcinomas can be solid and poorly differentiated. These carcinomas may be identified by a positive reaction for calcitonin, calcitonin-gene-related peptide and carcinoembryonic antigen.

The differential diagnosis of tumors with follicular or C-cell differentiation is most often possible by visualizing markers such as triiodothyronin (T3), thyroxin (T4), and thyroglobulin (Fig. 6) in tumors with follicular differentiation. Calcitonin, neuron-specific enolase, chromogranin A, and carcinoembryonic anti-

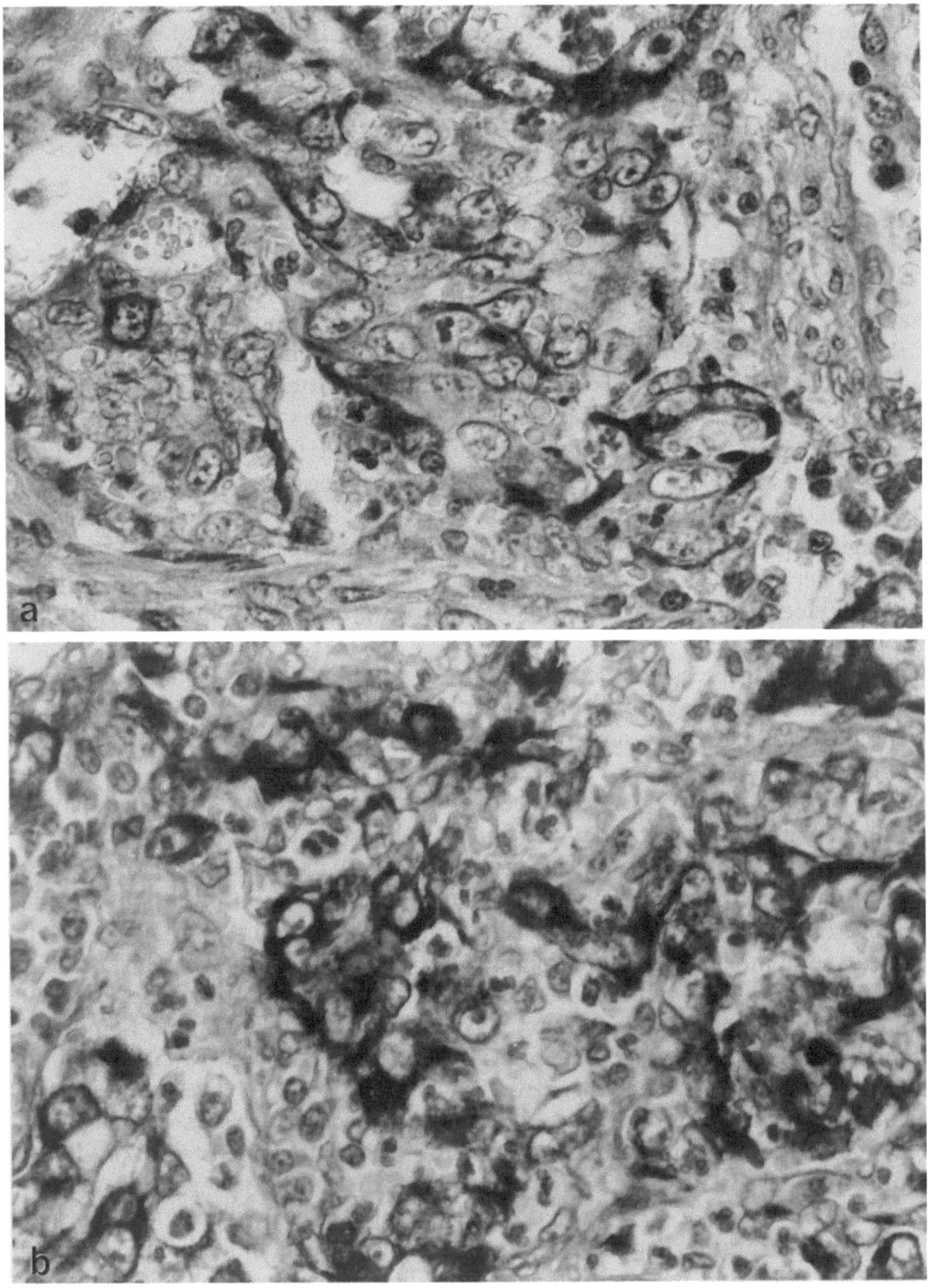

Fig. 5a, b. Undifferentiated (anaplastic) carcinoma of the thyroid. Avidin-biotin complex technique. **a** Presence of cytokeratins in the cytoplasm of many tumor cells, × 500. **b** Co-expression of vimentin with cytokeratin, × 500

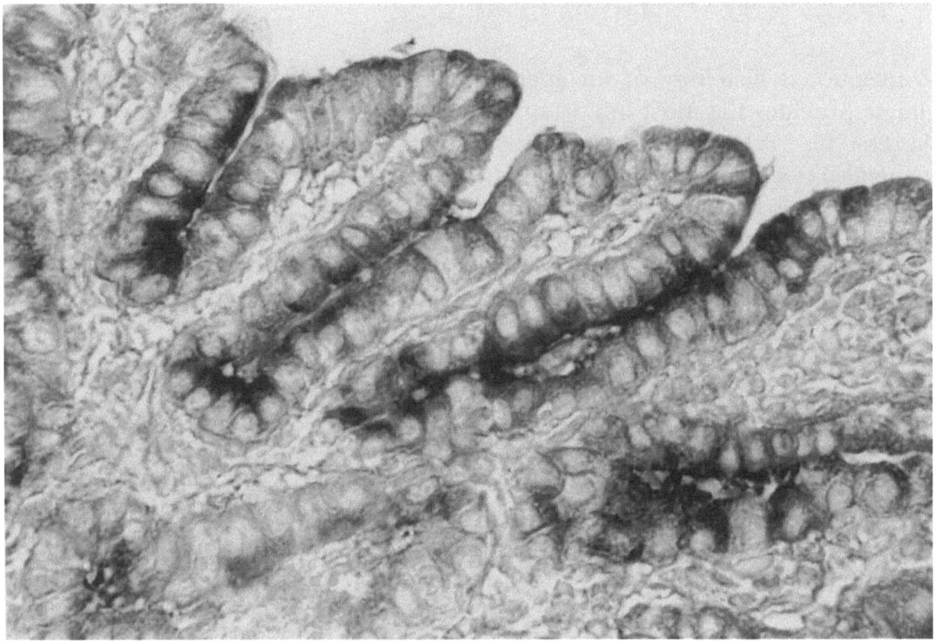

Fig. 6. Papillary thyroid carcinoma producing thyroglobulin. Unlabeled antibody enzyme technique, ×500

gen are typical markers of medullary carcinomas. Recently the production of several other peptides by medullary thyroid carcinomas has been reported. These findings offer an additional possibility of recognizing tumors with C-cell characteristics (KRISCH et al. 1985; SIKRI et al. 1985). Lately, some authors have described tumors with combined follicular and C-cell differentiation in a limited number of tumors (LJUNGBERG et al. 1984).

An important aid in the diagnosis and therapeutic management of the patient is the determination of production of thyroglobulin by thyroid tumors with follicular or papillary differentiation. Most authors agree that the majority of tumors with histologically recognizable follicular or papillary differentiation produce thyroglobulin (92–98%). Production of thyroglobulin can be shown by immunocytochemistry on formaldehyde-fixed and paraffin-embedded tissue (BOECKER et al. 1980; RYFF-DE LECHE et al. 1986). After total thyroidectomy the determination of the serum concentration of thyroglobulin can serve as a guide in the search for metastases without carrying out a total body scan. Levels of serum thyroglobulin rising above the normal range provide an indication of recurrency or metastases. On the other hand, tumors lacking follicular differentiation (solid tumors), and poorly differentiated carcinomas do not produce significant amounts of thyroglobulin. The immunohistochemical findings in the tumor tissue correlate well with the findings of serum levels of thyroglobulin (RYFF-DE LECHE et al. 1986).

4 Tumors of the Parathyroid Gland

Conventional histology of an adenoma or a hyperplasia of the parathyroid glands provides the most important clues to the diagnosis (AKERSTROEM 1980; ALTENAEHR 1981). It is possible to localize parathyroid hormone by immunocytochemical techniques (DIETEL et al. 1980). In addition, adenomas of the parathyroid glands have been reported to contain chromogranin A. The localization of these markers is rarely necessary for the diagnosis of these tumors, but they have played an important role in the research into the pathways of intracellular transport of hormones.

5 Endocrine Tumors of the Pancreas

The classification of pancreatic endocrine tumors has been revolutionized by the systematic use of radioimmunoassay for measuring hormones in the patient's plasma, and by electron microscopy and immunocytochemistry for visualizing the site of hormone production in the cells. Prior to the use of these techniques the classification was simple, as only B- and non-B cell tumors could be differentiated by conventional histology with special tinctorial stains. No reliable criterion for typing cells, or for hormone production or secretion is possible by means of histological observation. The growth pattern of a tumor, e.g. trabecular, glandular, or solid, cannot be reliably correlated with the various types of tumors (HEITZ 1984).

In our opinion, immunocytochemical and/or electron microscopical analysis of these tumors is a must at present for functional classification (HEITZ et al. 1982).

Virtually all endocrine tumors of the pancreas contain broad-spectrum markers, e.g. cytokeratins, neuron-specific enolase (HEITZ and KLOEPPEL 1984), and chromogranins (Fig. 2b, 3), and yield positive reactions to the antibody Phe-5. By these reactions they can be differentiated reliably from exocrine tumors, which sometimes resemble endocrine tumors, e.g. the solid cystic (papillary-cystic) tumor (MOROHOSHI et al. 1987). The markers of the second or third steps in the diagnostic algorithm are eutopic pancreatic hormones, e.g. insulin, glucagon, somatostatin, and pancreatic polypeptide. In addition, hormones "ectopic" to the adult human pancreas can be visualized, e.g. gastrin, vasoactive intestinal polypeptide (VIP), or, if appropriate, corticotropin, calcitonin and others. The occurrence of growth hormone-releasing hormone immunoreactivity has recently been reported in approx. 25% of pancreatic endocrine tumors, even in the absence of associated acromegaly (DAYAL et al. 1986).

The problem of the definition of the biological behavior in pancreatic endocrine tumors is particularly difficult to solve. Many authors agree that histological observation does not allow a prediction of the course of the disease in the absence of massive invasion of vessels or of an adjacent organ, which are rare findings. The only reliable criterion is metastases in the regional lymph

Table 3. Pancreatic endocrine tumors

Tumor	Approx. ratio of all secreting pancreatic endocrine tumors	Biological behavior
Insulinoma	70–75%	90% benign
Glucagonoma	1%	50% malignant
Somatostatinoma	rare	?
PP-oma	rare	50% benign
Gastrinoma	20–25%	60% malignant
VIP-oma	3–5%	60% malignant
Non-secreting tumors	20–25% [a]	50% malignant

[a] of the total of pancreatic endocrine tumors
PP, pancreatic polypeptide; *VIP*, vasoactive intestinal polypeptide

nodes or the liver. It is therefore of interest that the production, and sometimes the secretion, of the alpha-subunit of glycoprotein hormones (and possibly the rare production of the beta-subunit of human chorionic gonadotropin) could be found in up to 75% of tumors with proven malignancy (KAHN et al. 1977; BLACKMAN et al. 1980; OEBERG and WIDE 1981; HEITZ et al. 1983). These findings mean that the tumor producing the alpha-subunit should be considered malignant, while the absence of alpha by no means proves the benign behavior of the tumor. In some patients the course of the illness, and the presence of metastases could be monitored by measuring serum levels of pancreatic hormones and of the alpha-subunit over several years (KOELZ et al. 1987).

The present widely accepted classilfication of pancreatic endocrine tumors, their prevalence, and their biological behavior are given in Table 3.

6 Endocrine Tumors (Carcinoids) of the Gastrointestinal Tract and Bronchus

The histological growth pattern of carcinoid tumors is quite uniform. It is therefore impossible to draw a conclusion concerning the site of the primary by analyzing a metastasis by histological observation. In addition, it is often impossible to differentiate carcinoids from pancreatic endocrine tumors, and it may be difficult to distinguish an atypical carcinoid of the bronchus from a small cell carcinoma. Reactions, which visualize markers specific to particular localizations of carcinoids, and which differentiate atypical bronchus carcinoids from small-cell carcinomas, are therefore useful.

Carcinoids consistently react to antibodies directed against "broad-spectrum" markers, with the exception of chromogranin A and Phe-5 (Sections 1.3.4, 1.3.5).

Several series of carcinoid tumors of the gastrointestinal tract have been analyzed by immunocytochemistry using antibodies to serotonin and a large

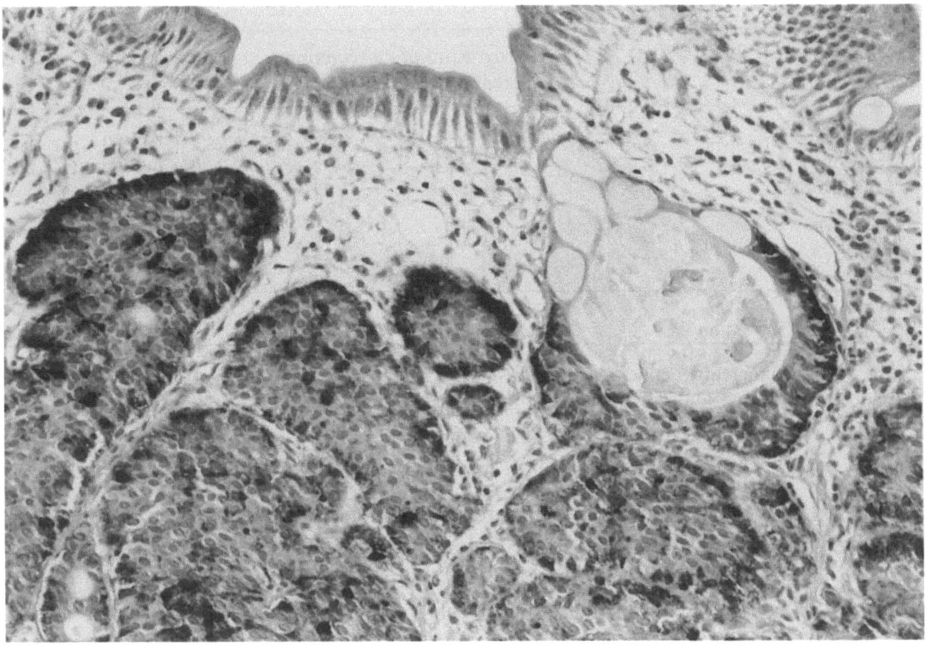

Fig. 7. Endocrine tumor of the ileum (carcinoid). Presence of substance P in many tumor cells. Unlabeled antibody enzyme method, ×200

series of peptides, including tachykinins (ALUMETS et al. 1977, 1980, 1981, 1983; YANG et al. 1983; DAYAL and WOLFE 1984; CONLON et al. 1985; MÄRTENSSON et al. 1985; IWAFUCHI et al. 1986; KOMMINOTH 1986; STAMM et al. 1986). The most important findings in gastrointestinal carcinoids can be summarized as follows: serotonin is the most important marker for the entire group of tumors, while tachykinins, especially substance P, occur almost exclusively in endocrine tumors of the ileum (Fig. 7). These two markers also allow the differentiation of carcinoids of the gastrointestinal tract from pancreatic endocrine tumors, in which they never occur. Gastrin, somatostatin, and pancreatic polypeptide can be localized predominantly in carcinoids of the duodenum. The other peptides investigated, including neurotensin, motilin, glicentin, peptide YY (TATEMOTO et al. 1982), calcitonin, corticotropin, and growth hormone-releasing hormone (DAYAL et al. 1986) occur much less frequently, and are not specific for any localization of the primary. In addition, the alpha-chain of glycoprotein hormones was found in carcinoids of the stomach, duodenum, and rectum, but not in tumors of the ileum. This marker is therefore of only limited use for the localization of the primary, or for the differentiation of a metastasis from a pancreatic endocrine tumor (HEITZ et al. 1987). A striking new finding is the occurrence of a substance identical to prostatic acid phosphatase, or cross-reacting with antisera and monoclonal antibodies raised against prostatic acid phosphatase (SOBIN et al. 1986). We can confirm this finding. It is also important to realize that there is a reaction for protein S-100-beta (VAN ELDIK et al. 1986) in the majority of tumors. This reaction is not localized in epithelial tumor cells, but in Schwann-cells included in the tumors (HOEFLER and AUBOECK 1984; KOMMINOTH 1986).

The substances localized in bronchial carcinoids, including atypical bronchial carcinoids, and in small-cell cancer include some of the broad-spectrum markers (Section 1.3). In addition, the alpha-chain of glycoprotein hormones was found in 27 out of 55 bronchial carcinoids, but not in small-cell cancer. Serotonin and other peptides could be localized less frequently in carcinoids, but not in 12 small-cell carcinomas. The immunocytochemical reactions for these substances may therefore be useful for the differentiation of atypical carcinoids from small-cell carcinomas (KOMMINOTH 1986).

7 Tumors of the Adrenal Medulla and the Chemoreceptor System

The diagnosis of these tumors often relies on histology. Cytokeratins are lacking in the tumor cells (VON OVERBECK et al. 1985) while virtually all tumors contain significant amounts of neuron-specific enolase (TAPIA et al. 1981; KOMMINOTH 1986; OSBORN et al. 1986), chromogranin A (WILSON and LLOYD 1984) and the neurofilament polypeptide with a molecular weight of 68 kD (OSBORN et al. 1986). Catecholamines may be visualized by using formaldehyde-induced fluorescence (FALCK et al. 1962) or by condensation with glyoxylic acid (DE LA TORRE and SURGEON 1976; LINDVALL et al. 1980). These reactions cannot be carried out in formaldehyde-fixed and paraffin-embedded tissue. Sometimes the production of neuropeptides, e.g. vasoactive intestinal polypeptide, somatostatin and others can be demonstrated.

Protein S-100 can be visualized in satellite, or sustentacular cells (LAURIOLA et al. 1985; PELC et al. 1986). This reaction yields sometimes very striking patterns (Fig. 8).

8 Steroid Hormone-Producing Tumors of the Adrenal Cortex, the Testis, and the Ovary

Steroid hormone-producing cells lack broad spectrum markers, such as cytokeratins, neuron-specific enolase, or chromogranin. It is difficult to localize steroid hormones specifically by immunocytochemistry because their intracellular concentration is low. This is due to the absence of a post-Golgi pool in steroid hormone-producing cells. Enzyme deficiencies or the activity of enzymes can sometimes be demonstrated by enzyme histochemistry, e.g. 3-beta-OH-steroid-dehydrogenase (Fig. 9). By this means, cells producing steroids or lacking this enzyme, e.g. virilizing tumors, can be localized. The diagnosis of this group of tumors can in general be made by conventional histology, and electron microscopy is often valuable for a precise diagnosis.

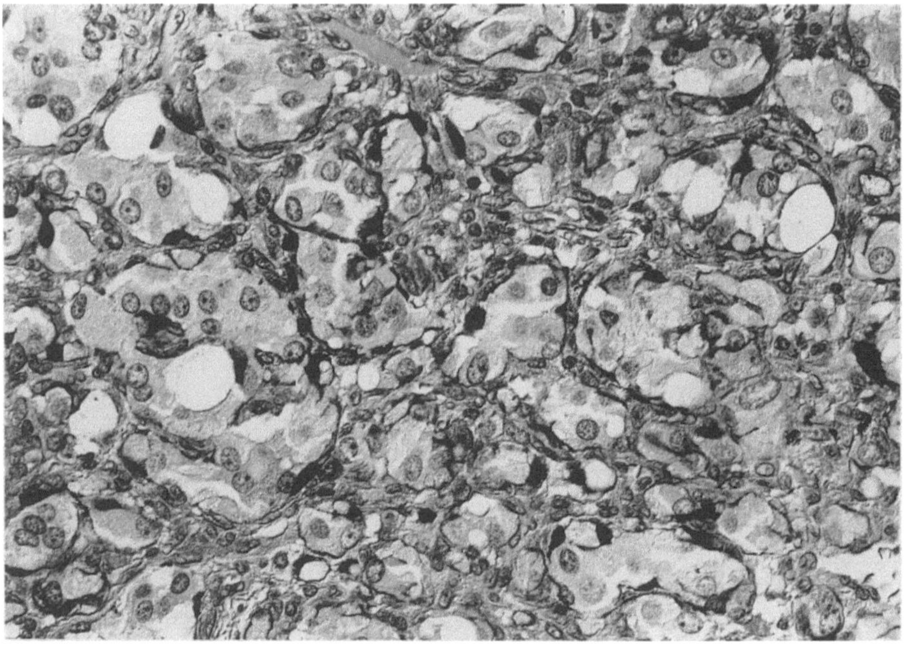

Fig. 8. Lymph node metastasis of a malignant paraganglioma. Localization of protein S-100 in satellite, but not in tumor cells. Avidin-biotin complex technique, ×400

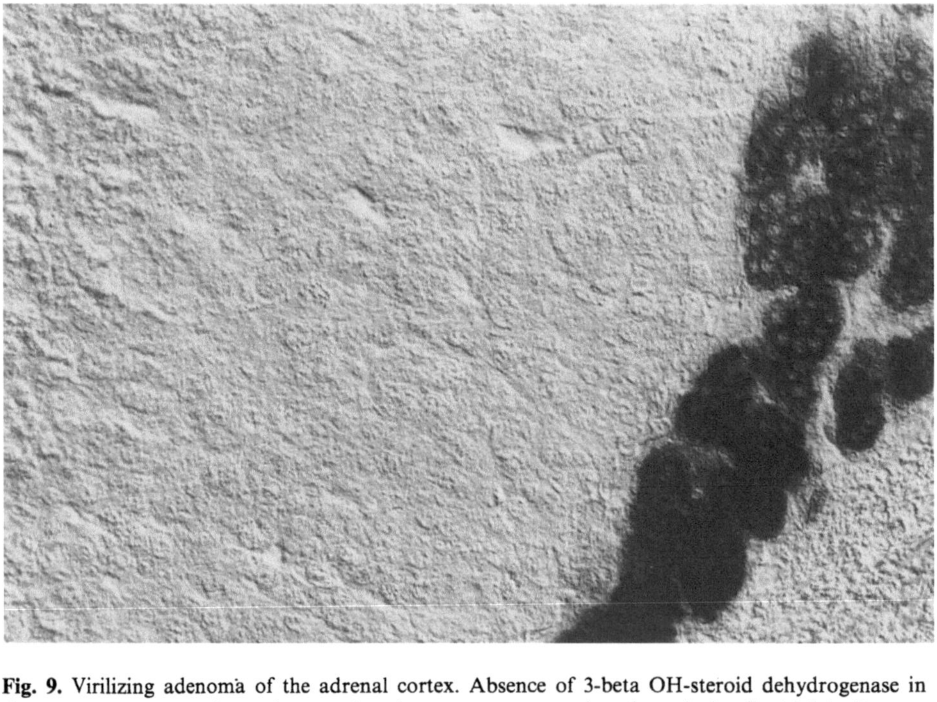

Fig. 9. Virilizing adenoma of the adrenal cortex. Absence of 3-beta OH-steroid dehydrogenase in the tumor, presence of reaction product in non-tumorous adrenal cortical cells (right). Enzyme histochemical reaction for 3-beta OH-steroid dehydrogenase, ×200, differential interference contrast optics

9 Melanoma

The diagnosis of metastases of melanoma is a difficult problem to solve histologically if small amounts of melanin are present, or if pigment is lacking. The precise diagnosis can often be made by electron microscopy based on the presence of cytoplastic (pre-)melanosomes. Immunocytochemistry can also be helpful: Tumorous melanocytes and their normal counterparts fail to react to antibodies directed against cytokeratins and chromogranin A, but contain vimentin, neuron-specific enolase, and protein S-100 (LOEFFEL et al. 1985; KOMMINOTH 1986; VANSTAPEL et al. 1986). Using one or more of these markers, melanocytic cells can be identified reliably and differentiated from other tumor cells. The reactions for vimentin and S-100 (Fig. 10a and b) are in general intense and present in virtually all tumor cells, while the reaction for neuron-specific enolase is often only focally positive and of lesser intensity. CHERESH et al. (1984) have reported an alkali-labile O-acetylated product of the neurectoderm-associated disialoganglioside GD3 to be expressed specifically by melanoma cells. It will perhaps be possible in the near future to diagnose melanocytic lesions specifically by using single markers. At present, the electron microscopic diagnosis and/or the findings of the aforementioned combination of markers is the most reliable way to diagnose melanocytic tumors.

It is sometimes very difficult to differentiate benign from malignant melanocytic lesions in the skin. A reliable marker should be found for malignancy expressed exclusively by melanocytic tumor cells displaying malignant biological behavior. It has been shown that the expression of neuron-specific enolase increases progressively from benign to malignant (DHILLON and RODE 1982), but this quantitative change has not proved to be a useful criterion in our hands, because differences in staining intensity of neuron-specific enolase may be caused by a large number of technical factors, not related to the biological behavior of the tumor cells. Recently the expression of a high molecular weight chondroitin sulphate proteoglycan consisting of a 250 kD core glycoprotein and a number of glycosaminoglycan side chains by human malignant, but not by normal melanocytes, has been reported (RETTIG et al. 1986). It is therefore conceivable that specific markers for malignancy may be used for melanocytic lesions.

10 Merkel Cell Tumors (Neuroendocrine Carcinoma of the Skin)

TOKER described 5 tumors of the skin with a trabecular growth pattern in 1972, and subsequently TANG and TOKER (1978) reported the ultrastructural features of these tumors, including cytoplasmic granules. On the basis of the structural resemblance of the tumor cells to cells of the epidermis first described by MERKEL (1875) the tumors are now known as Merkel cell tumors (neuroendocrine carcinomas) of the skin.

The differentiation of this tumor type from skin metastases of malignant non-Hodgkin lymphomas, melanomas, or of other tumors is very important

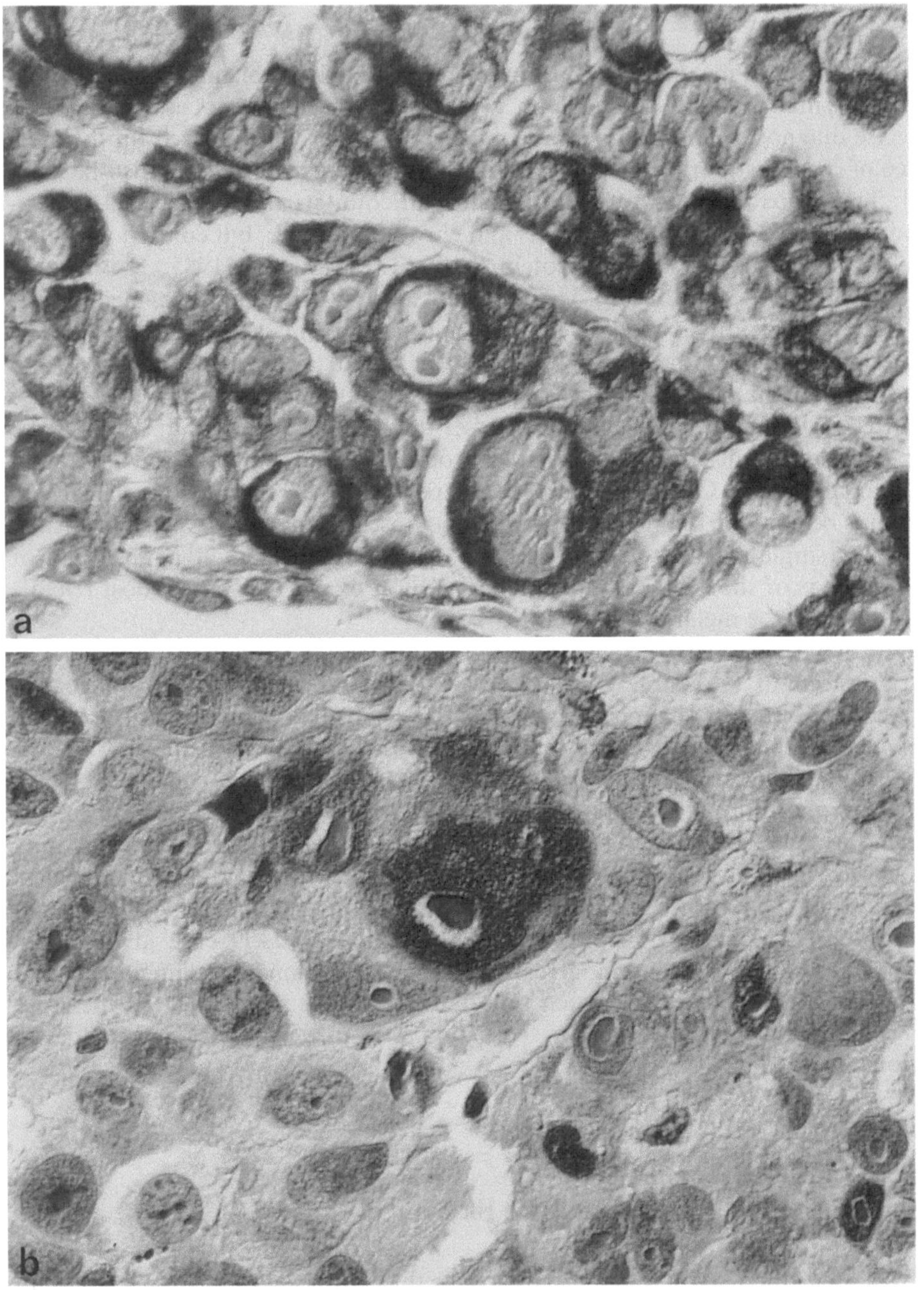

Fig. 10a, b. Malignant melanoma. Avidin-biotin complex technique, differential interference contrast optics. a Localization of vimentin in the cytoplasm of the majority of tumor cells, ×1000. b Localization of protein S-100 in several tumor cells, ×1000

in view of its biological behavior of low grade malignancy and, therefore, of the therapeutic implications.

Most authors agree that Merkel cell tumors display characteristics of epithelial and neuroendocrine cells. They yield a strong reaction to cytokeratins 8 and 18, often a dot-like paranuclear reaction is present (Fig. 1a). By this reaction, a metastasis of a malignant lymphoma, of melanoma, and of all types of sarcomas can be excluded. In addition, a generally conspicuous reaction for neuron-specific enolase (Fig. 1b) in at least some of the tumor cells, and for chromogranin A in the large majority of the tumor cells and neurofilaments, permit the differentiation of Merkel cell tumors from all non-neuroendocrine neoplasms (GLOOR et al. 1982; WICK et al. 1983; GOULD et al. 1985; BATTIFORA and SILVA 1986; KOMMINOTH 1986). Additional characteristics of a Merkel cell tumor are the presence of a narrow rim of connective tissue between the dermal tumor and the epidermis, and the presence of cytoplasmic granules (diameter 100 to 200 nm; SILVA and MACKAY 1981; PILOTTI et al. 1982; FRIGERIO et al. 1983) and bundles of intermediate filaments (thickness approx. 10 nm). In a small number of tumors the production and secretion of peptides has been reported (GOULD et al. 1980; IWASAKI et al. 1981; DRIJKONINGEN et al. 1986), but this is apparently uncommon. In a recent series the absence of an immunocytochemical reaction for bombesin, leu-enkephalin, met-enkephalin, and beta-endorphin has been reported (WICK et al. 1985). In one report the presence of protein S-100 in Merkel cell tumors has been described (VOIGT et al. 1985). This finding is at variance with those of other authors.

11 Tumors Producing "Ectopic Hormones"

Rapidly accumulating evidence indicates that many tumors produce a variety of substances, including molecules with hormonal or hormone-like biological activity. Ectopic hormone production is one particular form of paraneoplasia. The term ectopic hormone production refers to production of tumor cells causing remote effects resembling known biological actions of hormones (LIDDLE et al. 1969). By definition, ectopic hormones are produced by tumors arising in an organ not secreting the substance under normal physiological conditions. This definition implies that hormone production by metastases of a tumor arising in an endocrine gland is eutopic rather than ectopic. However, it becomes increasingly difficult to apply the given definitions strictly, since new substances are currently found to be produced by hitherto unexpected tissues (HEITZ and STAUB 1981; HEITZ and OBERHOLZER 1984). The proof of ectopic hormone production is difficult. It includes, among other techniques, histology, electron microscopy, immunocytochemistry and immunoelectron microscopy (HEITZ et al. 1981; HEITZ 1986). It is impossible to correlate morphology as analyzed by conventional histology with the production of any hormones by a tumor.

A large number of symptoms induced by ectopic hormone production have been described. The most common tumors associated with ectopic hormone production are carcinomas of the bronchus and the pancreas, carcinoids in

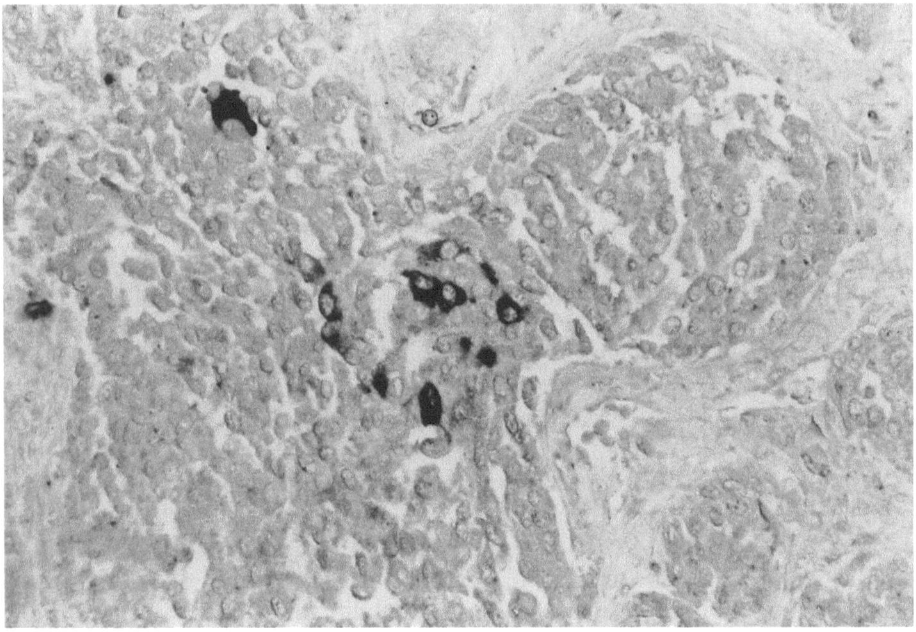

Fig. 11. Malignant paraganglioma of the mediastinum associated with Cushing's disease: production of opiocortins (lipotropin) in a group of tumor cells. Unlabeled antibody enzyme method, ×100

various organs, and thymomas. In many organs the tumors are of the so-called small-cell type. The substances most frequently secreted by these tumors excert effects identical or similar to those of opiocortins (lipotropin-corticotropin-endorphin; Fig. 11), vasopressin (antidiuretic hormone), and gastrin. In addition, elevated serum concentrations of calcitonin and the alpha-chain of the glycoprotein hormones can often be found. The hormones can also be localized in the tumor by immunocytochemistry. Immunocytochemical evidence for ectopic hormone production by tumors is important, because it cannot be provided by conventional histology.

12 Conclusions

The discovery of tumor markers for neuroendocrine tumor cells has permitted far more precise characterizations of many tumors than conventional histology. However, a careful histological observation of a given tumor is a prerequisite for a rational and precise immunocytochemical analysis. In some cases, electron microscopy may be more appropriate than immunocytochemistry for the typing of a tumor, or both techniques might be used in combination to solve such a problem. In the near future *in-situ* hybridization techniques visualizing mRNA of markers may become an important and sensitive tool in typing tumor cells.

The definition of marker production is at present feasible for many tumors on formaldehyde-fixed and paraffin-embedded tissue. The correlation with clini-

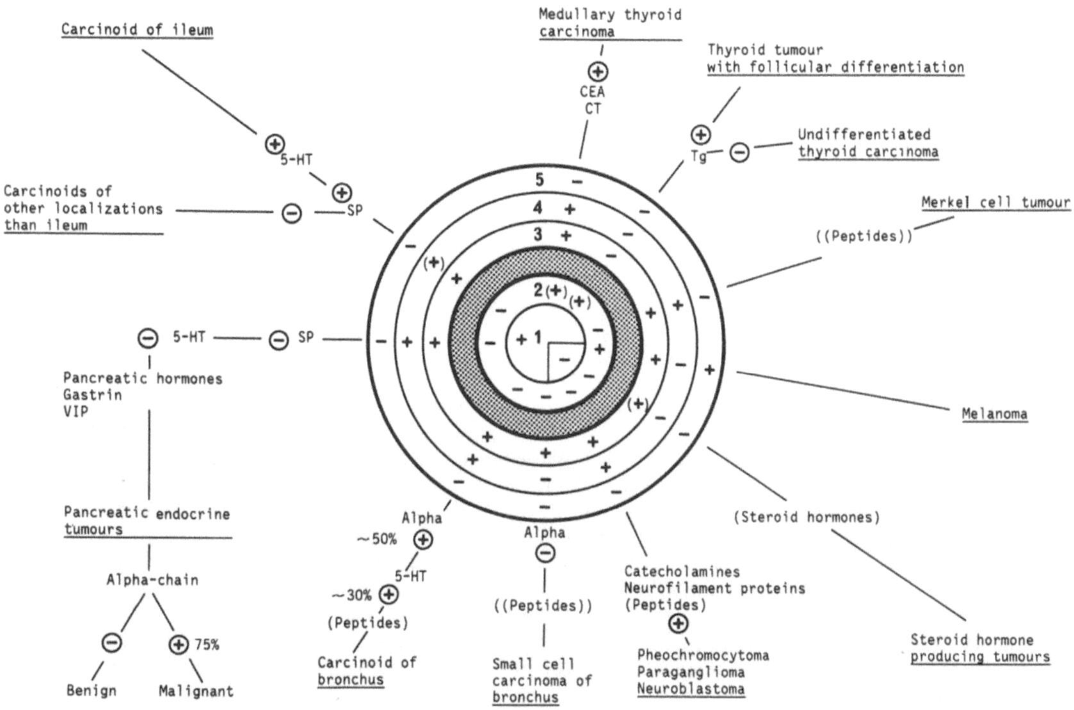

Fig. 12. Proposed systematic diagnostic flow-chart for neuroendocrine tumors. In general, only selected reactions are necessary to reach a final diagnosis. The indicated results of reactions refer exclusively to tumor cells. For details see text. First step of the diagnostic algorithm: Reactions 1 and 2, second step: reactions 3 to 5.

Key: First–step: *1*, cytokeratins; *2*, vimentin. Second step: *3*, neuron-specific enolase; *4*, chromogranin A; *5*, protein S-100. *CT*, calcitonin; *CEA*, carcino-embryonic antigen; *Tg*, thyroglobulin; *Alpha-chain*, alpha subunit of glycoprotein hormones; *5-HT*, 5-hydroxytryptamine (serotonin); *VIP*, vasoactive intestinal polypeptide; *SP*, substance P. ( ): Reactions only in part of the tumors, (( )): reactions uncommon

cal and biochemical findings can often be established. This represents great progress over conventional histology.

The extensive use of immunocytochemistry has revolutionized the morphological classification of some groups of neuroendocrine tumors. In order to make a classification useful it must be as simple as possible, and it must be meaningful to everybody working in the field. We therefore classify the tumors according to the leading clinical symptoms or to the predominantly secreted hormone. In addition, the probable biological behavior of the tumor (e.g. benign versus malignant), and additional secretory products are mentioned in the diagnosis.

It is furthermore important to realize that most tumors are not characterized by a single marker specific for the given type of tumor cells, but by a group of markers. A precise diagnosis can therefore often be reached by using a diagnostic flow-chart, using combinations of markers (Fig. 12). One important finding made by immunocytochemical observation of tumors was the frequently occurring phenotypic cell heterogeneity of neuroendocrine tumors (and probably of many other tumors), exemplified by the simultaneous or asynchronous production of more than one secretory product by various cells of the same tumor. With a few exceptions, reliable markers of malignancy are not known at present. The search for this type of tumor marker must be one of the goals of future research.

Acknowledgements. The author acknowledges the expert technical assistance of Mrs. M. KASPER, Mrs. E. OBRIST, and the excellent secretarial work of Mrs. C. STÖCKLI.

References

Akerström G (1980) Parenchymal cell mass determination in parathyroid diagnosis. Acta Univ Upsal 352

Altenähr L (1981) Nebenschilddrüsen. In: Dörr W, Seifert E, Uehlinger E (eds) Spezielle pathologische Anatomie, vol 14, Endokrine Organe. Springer, Berlin Heidelberg New York, pp 415–521

Alumets J, Hakanson R, Ingemansson S, Sundler F (1977) SP and 5-HT in granules isolated from an intestinal argentaffin carcinoid. Histochemistry 52:217–222

Alumets J, Falkmer S, Grimelius L, Hakanson R, Ljungberg O, Sundler F, Wilander E (1980) Immunocytochemical demonstration of enkephalin and beta-endorphin in endocrine tumors of the rectum. Acta Pathol Microbiol Scand Sect A 88:103–109

Alumets J, Alm P, Falkmer S, Hakanson R, Ljungberg O, Martensson H, Sundler F, Tibblin S (1981) Immunohistochemical evidence of peptide hormones in endocrine tumors of the rectum. Cancer 48:2409–2415

Alumets J, Falkmer S, Hakanson R, Martensson H, Nobin A, Lasson A (1983) Neurohormonal peptides in endocrine tumors of the pancreas, stomach and upper small intestine: an immunohistochemical study of 27 cases. Ultrastruct Path 5:55–72

Anscombe AM, Wright DH (1985) Primary malignant lymphoma of the thyroid – a tumor of mucosa-associated lymphoid tissue: Review of seventy-six cases. Histopathology 9:81–97

Bakay L (1950) The results of 300 pituitary adenoma operations. J Neurosurg 7:204–255

Battifora H, Silva EG (1986) The use of antikeratin antibodies in the immunohistochemical distinction between neuroendocrine (Merkel cell) carcinoma of the skin, lymphoma, and oat cell carcinoma. Cancer 58:1040–1046

Blackman NR, Weintraub BD, Rosen SW, Kourides IA, Steinwascher K, Gail MH (1980) Human placental and pituitary glycoprotein hormones and their subunits as tumor markers. A quantitative assessment. J Natl Cancer Inst 65:81–93

Böcker W, Dralle H, Hüsselmann H, Bay V, Brassow M (1980) Immunohistochemical analysis of thyroglobulin synthesis in thyroid carcinomas. Virchows Arch A (Pathol Anat) 385:187–200

Carcangiu ML, Steeper T, Zampi G, Rosai J (1985) Anaplastic thyroid carcinoma. A study of 70 cases. Am J Clin Pathol 83:135–158

Carney DN, Ihde DC, Cohen MH, Marangos PJ, Bunn PA Jr, Minna JD, Gazdar AF (1982) Serum neuron-specific enolase: A marker for disease extent and response to therapy of small-cell lung cancer. Lancet I:583–585

Cheresh DA, Reisfeld RA, Varki AP (1984) O-acetylation of disialoganglioside GD3 by human melanoma cells creates a unique antigenic determinant. Science 225:844–846

Conlon JM, Schäfer G, Schmidt WE, Lazarus LH, Becker HD, Creutzfeldt W (1985) Chemical and immunochemical characterisation of substance P-like immunoreactivity and physalaemin-like immunoreactivity in a carcinoid tumor. Regulatory Peptides 11:117–132

Dayal Y, Wolfe HJ (1984) Regulatory substances in clinically nonfunctioning gastrointestinal carcinoids. In: Falkmer S, Hakanson R, Sundler F (eds) Evolution and tumor pathology of the neuroendocrine system. Elsevier, Amsterdam New York Oxford, pp 497–517

Dayal Y, Lin HD, Tallberg K, Reichlin S, DeLellis RA, Wolfe HJ (1986) Immunocytochemical demonstration of growth hormone-releasing factor in gastrointestinal and pancreatic endocrine tumors. Am J Clin Pathol 85:13–20

De La Torre JC, Surgeon W (1976) A methodological approach to rapid and sensitive monoamine histofluorescence using a modified glyoxylic acid technique. The SGP method. Histochemistry 49:81–93

Dhillon AP, Rode J (1982) Patterns of staining for neurone specific enolase in benign and malignant melanocytic lesions of the skin. Diagnostic Histopathology 5:169–174

Dhillon AP, Rode J, Leathem A (1982) Neurone specific enolase: an aid to the diagnosis of melanoma and neuroblastoma. Histopathology 6:81–92

Dietel M, Dorn-Quint G (1980) By-pass secretion of human parathyroid adenomas. A particular intracellular form of rapid adaptation to external stimuli. Lab Invest 43:116–125

Drijkoningen M, DeWolf-Peeters C, Van Limbergen E, Desmet V (1986) Merkel cell tumor of the skin. An immunohistochemical study. Hum Pathol 17:301–307

Egloff B (1983) Hemangioendothelioma of the thyroid. Virchows Arch A (Pathol Anat) 400:119–142

Ehrhart M, Grube D, Bader MF, Aunis D, Gratzl M (1987) Chromogranin A in the pancreatic islet: Cellular and subcellular distribution. J Histochem Cytochem (to be published)

Ezrin C, Kovacs K, Horvath E (1982) Pathology of the adenohypophysis. In: Bloodworth JMB (ed) Endocrine pathology. General and surgical 2nd edn. Williams & Wilkins, Baltimore London, pp 101–132

Falck B, Hillarp NA, Thieme G, Torp A (1962) Fluorescence of catecholamines and related compounds condensed with formaldehyde. J Histochem Cytochem 10:348–354

Franke WW, Gudat F, Heitz PU (1987) Reaction of antibody lu-5 with epitopes of cytokeratins. Virchows Arch A 411:137–147

Frigerio B, Capella C, Eusebi V, Tenti P, Azzopardi JG (1983) Merkel cell carcinoma of the skin: the structure and origin of normal Merkel cells. Histopathology 7:229–249

Girod C, Mazzuca M, Trouillas J, Tramu G, Lheritier M, Beauvillain JC, Claustrat B, Dubois MP (1980) Light microscopy, fine structure and immunohistochemistry studies of 278 pituitary adenomas. In: Derome JP, Jedynak CP, Peillon F (eds) Pituitary adenomas. Biology, physiopathology and treatment. Asclepios, France, pp 3–18

Gloor F, Heitz PU, Hofmann E, Höfler H, Maurer R (1982) Das neuroendokrine Merkelzellkarzinom der Haut. Schweiz Med Wochenschr 112:141–148

Gould VE, Dardi LE, Memoli VA, Johannessen JV (1980) Neuroendocrine carcinomas of the skin: Light microscopic, ultrastructural, and immunhistochemical analysis. Ultrastruct Pathol 1:499–509

Gould VE, Moll R, Moll I, Lee I, Franke WW (1985) Biology of Disease. Neuroendocrine (Merkel) cells of the skin: Hyperplasias, dysplasias, and neoplasms. Lab Invest 52:334–353

Gould VE, Lee I, Wiedenmann B, Moll R, Chejfec G, Franke WW (1986) Synaptophysin: A novel marker for neurons, certain neuroendocrine cells and their neoplasms. Hum Pathol 17:979–983

Gould VE, Wiedenmann B, Lee I, Schwechheimer K, Dockhorn-Dworniczak B, Radosevich JA, Moll R, Franke WW (1987) Synaptophysin expression in neuroendocrine neoplasms as determined by immunocytochemistry. Am J Pathol 126:243–257

Hedinger C, Sobin LH (1974) Histological typing of thyroid tumors. International Histological Classification of Tumors. WHO, Geneva

Heitz PU (1979) Multihormonal pituitary adenomas. Hormone Res 10:1–13

Heitz PU (1984) Pancreatic endocrine tumors. In: Klöppel G, Heitz PU (eds) Pancreatic pathology. Churchill Livingstone, pp 206–232

Heitz PU (1986) Hormonale Tumormarker. Verh Dtsch Ges Pathol 70:64–81

Heitz PU, Klöppel G (1984) Pancreatic endocrine tumors. In: Falkmer S, Hakanson R, Sundler F (eds) Evolution and tumor pathology of the neuroendocrine system. Elsevier Biomedical Press, Amsterdam, pp 481–496

Heitz PU, Oberholzer M (1984) Electron microscopy as a tool in the diagnosis of endocrine tumors. Pathol Res Pract 178:427–430

Heitz PU, Staub JJ (1981) Paraneoplastische endokrine Syndrome. In: Dörr W, Seifert G, Uehlinger E (eds) Spezielle pathologische Anatomie, vol 14, Endokrine Organe. Springer, Berlin Heidelberg New York, pp 1205–1271

Heitz PU, Klöppel G, Polak JM, Staub JJ (1981) Ectopic hormone production by endocrine tumors: Localization of hormones at the cellular level by immunocytochemistry. Cancer 48:2029–2037

Heitz PU, Kasper M, Polak JM, Klöppel G (1982) Pancreatic endocrine tumors. Immunocytochemical analysis of 125 tumors. Hum Pathol 13:263–271

Heitz PU, Kasper M, Klöppel G, Polak JM, Vaitukaitis JL (1983) Alpha-chain production of pancreatic endocrine tumors – a specific marker for malignancy. Cancer 51:277–282
Heitz PU, Von Herbay G, Klöppel G, Komminoth P, Kasper M, Höfler H, Müller KM, Oberholzer M (1987) The expression of subunits of human chorionic gonadotropin (hCG) by nontrophoblastic, nonendocrine, and endocrine tumors. Am J Clin Pathol 88
Höfler H, Auböck L (1984) S-100 Protein in Karzinoiden. Verh Dtsch Ges Pathol 68:86–91
Höfler H, Denk H, Lackinger E, Helleis G, Polak JM, Heitz PU (1986) Immunocytochemical demonstration of intermediate filament cytoskeleton proteins in human endocrine tissues and (neuro-) endocrine tumors. Virchows Arch A (Pathol Anat) 409:609–626
Iacangelo A, Affolter HU, Eiden LE, Herbert E, Grimes M (1986) Bovine chromogranin A sequence and distribution of its messenger RNA in endocrine tissues. Nature 323:82–86
Iwafuchi M, Watanabe H, Ishihara N, Shimoda T, Iwashita A, Ito S (1986) Peptide YY immunoreactive cells in gastrointestinal carcinoids: Immunohistochemical and ultrastructural studies of 60 tumors. Hum Pathol 17:291–296
Iwasaki H, Mitsui T, Kikuchi M, Imai T, Fukushima K (1981) Neuroendocrine carcinoma (trabecular carcinoma) of the skin with ectopic ACTH production. Cancer 48:753–756
Iwase K, Nagasaka A, Kato K, Ohtani S, Nagatsu I, Ohyama T, Nakai A, Aono T, Nakagawa H, Shinoda S, Inagaki M, Tei T, Miyakawa S, Kawase K, Miura K, Nakamura T, Kanno T, Kuwayama A, Kageyama N (1986) Enolase subunits in patients with neuroendocrine tumors. Clin Endocrinol Metab 63:94–101
Kahn CR, Rosen SW, Weintraub BD, Fajans SS, Gorden P (1977) Ectopic production of chorionic gonadotropin and its subunits by islet cell tumors. N Engl J Med 297:565–569
Kelly B (1985) Pathways of protein secretion in eukaryotes. Science 230:25–32
Klibansky A, Ridgway EC, Zervas NT (1983) Pure alpha subunit-secreting pituitary tumors. J Neurosurg 59:585–589
Koelz A, Kränzlin M, Gyr K, Meier V, Bloom SR, Heitz PU, Stalder H (1987) Rebound effect and escape of diarrhea inhibition during treatment with a long-acting somatostatin analogue (SMS 201–995) in patients with vipoma. Gastroenterology 92:527–531
Komminoth P (1986) Immunzytochemie: Ein Fortschritt in der Differenzierung unklarer Tumoren. Immunzytochemische Studie an 224 neuroendokrinen Tumoren. Inaug. Diss., University of Basel, Switzerland
Kovacs K, Horvath E (1986) Atlas of tumor pathology, vol 21, 2nd series, Tumors of the pituitary gland. Armed Forces Institute of Pathology, Washington
Krisch K, Krisch I, Horvat G, Neuhold N, Ulrich W (1985) The value of immunohistochemistry in medullary thyroid carcinoma: A systematic study of 30 cases. Histopathology 9:1077–1089
Landolt AM (1975) Ultrastructure of human sella tumors. Correlations of clinical findings and morphology. Acta Neurochir Suppl 22
Landolt AM, Heitz PU (1986) Alpha-subunit-producing pituitary adenomas: Immunocytochemical and ultrastructural studies. Virchows Arch A (Pathol Anat) 409:417–431
Lauffer L, Garcia PD, Harkins RN, Coussens L, Ullrich A, Walter P (1985) Topology of signal recognition particle receptor in endoplasmic reticulum membrane. Nature 318:334–388
Lauriola L, Maggiano N, Sentinelli S, Michetti F, Cocchia D (1985) Satellite cells in the normal human adrenal gland and in pheochromocytomas. An immunohistochemical study. Virchows Arch B (Cell Pathol) 49:13–21
Liddle GW, Nicholson WE, Island DP, Orth DN, Abe K, Lowder SC (1969) Clinical and laboratory studies of ectopic humoral syndromes. Rec Progr Horm Res 25:283–314
Lindvall O, Björklund A, Falck B, Loren I (1980) New aspects on factors determining the sensitivity of the formaldehyde and glyoxylic acid fluorescence histochemical methods for monoamines. Histochemistry 68:169–181
Ljungberg O, Bondeson L, Bondeson AG (1984) Differentiated thyroid carcinoma, intermediate type. A new tumor entity with features of follicular and parafollicular cell carcinoma. Hum Pathol 15:218–228
Lloyd RV, Wilson BR (1983) Specific endocrine tissue marker defined by a monoclonal antibody. Science 222:628–630
Löffel SC, Gillespie Y, Mirmiran SA, Miller EW, Golden P, Askin FB, Siegal GP (1985) Cellular immunolocalization of S100 protein within fixed tissue sections by monoclonal antibodies. Arch Pathol Lab Med 109:117–122

MacFarlane IA, Beardwell CG, Shalet SM, Darbyshire PJ, Hayward E, Sutton ML (1980) Glycoprotein hormone alpha subunit secretion by pituitary adenomas: Influence of external irradiation. Clin Endocrinol 13:215–222

Märtensson H, Nobin A, Sundler F, Falkmer S (1985) Endocrine tumors of the ileum. Cytochemical and clinical aspects. Pathol Res Pract 180:356–363

Meissner WA (1984) Atlas of tumor pathology, 2nd series, Tumors of the thyroid gland, suppl. fasc. 4. Armed Forces Institute of Pathology, Washington

Merkel F (1875) Tastzellen und Tastkörperchen bei den Hausthieren. Arch Mikrosk Anat Entwicklungsmech 11:636–652

Morohoshi T, Kanda M, Horie A, Chott A, Dreyer T, Klöppel G, Heitz PU (1987) Immunocytochemical markers of uncommon pancreatic tumors: Acinar cell carcinoma, pancreatoblastoma and solid cystic (papillary-cystic) tumor. Cancer (to be published)

O'Connor DT, Deftos LJ (1986) Secretion of chromogranin A by peptide-producing endocrine neoplasms. N Engl J Med 314:1145–1151

Oeberg K, Wide L (1981) hCG and hCG subunits as tumor markers in patients with endocrine pancreatic tumors and carcinoids. Acta Endocrinol 98:256–260

Osborn M, Dirk T, Käser H, Weber K, Altmannsberger M (1986) Immunohistochemical localization of neurofilaments and neuronspecific enolase in 29 cases of neuroblastoma. Am J Pathol 122:433–442

Oskam R, Rijksen G, Lips CJM, Staal GEJ (1985) Enolase isozymes in differentiated and undifferentiated medullary thyroid carcinomas. Cancer 55:394–399

Pelc S, Gompel C, Simonet ML (1986) S-100 protein expression in satellite and Schwann cells in neuroblastoma. An immunohistochemical and ultrastructural study. Virchows Arch B (Cell Pathol) 51:487–495

Pfaltz M, Hedinger C, Saremaslani P, Egloff B (1983) Malignant hemangioendothelioma of the thyroid and factor XIII-related antigen. Virchows Arch A (Pathol Anat) 401:177–184

Pilotti S, Rilke F, Lombardi L (1982) Neuroendocrine (Merkel cell) carcinoma of the skin. Am J Surg Pathol 6:243–254

Prinz RA, Marangos PJ (1982) Use of neuron-specific enolase as a serum marker for neuroendocrine neoplasms. Surgery 92:887–889

Ralfkiaer N, Gatter KC, Alcock C, Heryet A, Ralfkiaer E, Mason DY (1985) The value of immunocytochemical methods in the differential diagnosis of anaplastic thyroid tumors. Br J Cancer 52:167–170

Rettig WJ, Real FX, Spengler BA, Biedler JL, Old LJ (1986) Human melanoma proteoglycan: Expression in hybrids controlled by intrinsic and extrinsic signals. Science 321:1281–1284

Reubi JC, Heitz PU, Landolt AM (1987) Visualization of somatostatin receptors and correlation with immuno-reactive GH and PRL in human pituitary adenoma. J Clin Endocrinol Metab 65:65–73

Ridgway EC, Klibansky A, Ladenson PW, Clemmons D, Beitins IZ, McArthur JW, Martorana MA, Zervas NT (1981) Pure alpha-secreting pituitary adenomas. N Engl J Med 304:1254–1259

Ruchti C, Gerber HA, Schaffner T (1984) Factor VIII-related antigen in malignant hemangioendothelioma of the thyroid: Additional evidence for the endothelial origin of this tumor. Am J Clin Pathol 82:474–480

Ryff-de Lèche A, Staub JJ, Kohler-Faden R, Müller-Brand J, Heitz PU (1986) Thyroglobulin production by malignant thyroid tumors: An immunocytochemical and radioimmunoassay study. Cancer 57:1045–1053

Säger W (1981) Hypophyse. In: Dörr W, Seifert G, Uehlinger E (eds) Spezielle pathologische Anatomie. Springer, Berlin Heidelberg New York, pp 1–226

Schmechel DE, Marangos PJ, Brightman M (1978a) Neurone-specific enolase is a molecular marker for peripheral and central neuroendocrine cells. Nature 276:834–836

Schmechel DE, Marangos PJ, Zis AP, Brightman M, Goodwin FK (1978b) Brain enolases as specific markers of neuronal and glial cells. Science 199:313–315

Schmechel DE, Brightman MW, Marangos PJ (1980) Neurons switch from non-neuronal enolase to neuron-specific enolase during differentiation. Brain Res 190:195–214

Schröder S (1986) Morphologische Prognosekriterien maligner Schilddrüsentumoren. Immunhistologie, Elektronenmikroskopie, Autoradiographie, DNS-Cytophotometrie. Habil. Schrift, Universität Hamburg

Sikri KL, Varndell IM, Qutayba A, Hamid MB, Wilson BS, Kamey T, Ponder BAJ, Lloyd RV, Bloom SR, Polak JM (1985) Medullary carcinoma of the thyroid. An immunocytochemical and histological study of 25 cases using eight separate markers. Cancer 56:2481–2491

Silva E, MacKay B (1981) Neuroendocrine (Merkel cell) carcinomas of the skin: An ultrastructural study of nine cases. Ultrastruct Pathol 2:1–9

Sobin LH, Hjermstad BM, Sesterhenn IA, Helwig EB (1986) Prostatic acid phosphatase activity in carcinoid tumors. Cancer 58:136–138

Solcia E, Capella C, Buffa R, Usellini L, Fiocca R, Sessa F, Tortora O (1984) The contribution of immunohistochemistry to the diagnosis of neuroendocrine tumors. Semin Diagn Pathol 1:285–296

Stamm B, Hedinger CE, Saremaslani P (1986) Duodenal and ampullary carcinoid tumors. A report of 12 cases with pathological characteristics, polypeptide content and relation to the MEN I syndrome and von Recklinghausen's disease (neurofibromatosis). Virchows Arch A (Pathol Anat) 408:475–489

Tang CK, Toker C (1978) Trabecular carcinoma of the skin: an ultrastructural study. Cancer 42:2311–2321

Tapia FJ, Barbosa AJA, Marangos PJ, Polak JM, Bloom SR, Dermody C, Pearse AGE (1981) Neuron-specific enolase is produced by neuroendocrine tumors. Lancet I:808–811

Tatemoto K, Carlquist M, Mutt V (1982) Neuropeptide Y – a novel brain peptide with structural similarities to peptide YY and pancreatic polypeptide. Nature 296:659–660

Tischler AS, Mobtaker H, Mann K, Nunnemacher G, Jason WJ, Dayal Y, DeLellis R, Adelman L, Wolfe HJ (1986) Anti-lymphocyte antibody Leu-7 (HNK-1) recognizes a constituent of neuroendocrine granule matrix. J Histochem Cytochem 34:1213–1216

Tobler A, Maurer R, Hedinger C (1984) Undifferentiated thyroid tumors of diffuse small cell type. Histological and immunohistochemical evidence for their lymphomatous nature. Virchows Arch A (Pathol Anat) 404:117–126

Toker C (1972) Trabecular carcinoma of the skin. Arch Dermatol 105:107–110

Van Eldik LJ, Jensen RA, Ehrenfried BA, Whetsell WO (1986) Immunohistochemical localization of S100 beta in human nervous system tumors by using monoclonal antibodies with specificity for the S100 beta polypeptide. J Histochem Cytochem 34:977–982

Vanstapel MJ, Gatter KC, DeWolf-Peeters C, Mason DY, Desmet VD (1986) New sites of human S-100 immunoreactivity detected with monoclonal antibodies. Am J Clin Pathol 85:160–168

Voigt JJ, Alsaati T, Gorguet B, Caverivière P, Scarna H, Bugat R, Delsol G (1985) Carcinome à cellules de Merkel de la peau. Etude anatomo-clinique, ultrastructurale et immunohistochimique de 14 cas. Ann Pathol 5:195–203

Von Overbeck J, Stähli C, Gudat F, Carmann H, Lautenschlager C, Dürrmüller U, Takacs B, Miggiano V, Stähelin T, Heitz PU (1985) Immunocytochemical characterization of an anti-epithelial monoclonal antibody (mAB lu-5). Virchows Arch A (Pathol Anat) 407:1–12

Wick MR, Goellner JR, Scheithauer BW, Thomas JR, Sanchez NP, Schroeter AL (1983) Primary neuroendocrine carcinomas of the skin (Merkel cell tumors). A clinical, histologic, and ultrastructural study of thirteen cases. Am J Clin Pathol 79:6–13

Wick MR, Millns JL, Sibley RK, Pittelkow MR, Winkelmann RK (1985) Secondary neuroendocrine carcinomas of the skin. An immunohistochemical comparison with primary neuroendocrine carcinoma of the skin ("Merkel cell" carcinoma). J Am Acad Dermatol 13:134–142

Wickner WT, Lodish HF (1985) Multiple mechanisms of protein insertion into and across membranes. Science 230:400–407

Wiedenmann B, Franke WW, Kuhn C, Moll R, Gould VE (1986) Synaptophysin: A marker protein for neuroendocrine cells and neoplasms. Proc Natl Acad Sci USA 83:3500–3504

Williams ED, Siebenmann RE, Sobin LH (1980) Histological typing of endocrine tumors. International Histological Classification of Tumors. WHO

Wilson BS, Lloyd RV (1984) Detection of chromogranin in neuroendocrine cells with a monoclonal antibody. Am J Pathol 115:458–468

Yang K, Ulrich T, Cheng L, Lewin KJ (1983) The neuroendocrine products of intestinal carcinoids. An immunoperoxidase study of 35 carcinoid tumors stained for serotonin and eight polypeptide hormones. Cancer 51:1918–1926

Zeltzer PM, Parma AM, Dalton A, Siegel SE, Marangos PJ, Sather H, Hammond D, Seeger RC (1983) Raised neuron-specific enolase in serum of children with metastatic neuroblastoma. Lancet II:361–363

Morphological Markers in Neuro-Oncology

P. KLEIHUES, M. KIESSLING and R.C. JANZER

1 Introduction

Like many other morphological disciplines, surgical neuropathology has been greatly advanced by the introduction of immunohistochemical methods. The assessment of antigenic marker proteins in nervous system tumors has generally led to a higher level of diagnostic accuracy. Although the spectrum of available antibodies with proven diagnostic usefulness is still limited, some previously difficult differential diagnoses have become less troublesome and ambiguous (BONNIN and RUBINSTEIN 1984). This is particularly true for the distinction of gliomas and embryonal CNS tumors from metastatic lesions of epithelial and mesenchymal origin, as well as from malignant lymphomas. In addition, immunocytochemistry has expanded our knowledge of the origin of some human brain tumors with a re-evaluation of several entities, the histogenesis and classification of which had been disputed for decades (ZÜLCH 1979) due to the lack of reliable histomorphological criteria. Thus, the identification of abundant glial fibrillary acidic protein (GFAP) in most giant cells of the 'monstrocellular sarcoma' has led to its re-classification as giant cell glioblastoma with a sarcomatous component, i.e., a variant of the glioblastoma. The presence of numerous GFAP positive cells in superficially located cerebral neoplasms of young adults previously classified as malignant mesenchymal tumors, has allowed the identification of a new and now widely acknowledged tumor type, the pleomorphic xanthoastrocytoma (Kepes et al. 1979; Grant and Gallagher 1986). The frequent lack of GFAP expression and the presence of NSE (Stefansson and Wollmann 1980, 1981) in the subependymal giant cell astrocytoma associated with tuberous sclerosis have caused speculations that these slowly growing, benign lesions may be partially of neuronal origin (Nakamura and Becker 1983; Bonnin et al. 1984). It should, however, be emphasized that the presence within tumor cells of the astrocytic intermediate filament protein GFAP is not necessarily proof of their astrocytic origin. In contrast to their normal counterparts, neoplastic cells have a tendency towards unpredictable changes in gene expression, which may lead to the suppression of a gene product typically present, or the synthesis of marker proteins normally absent, in the cell line of origin. In embryonal tumors, the focal expression of glial or neuronal markers may be indicative of a tendency towards a distinct line of differentiation. In some instances, the possibility cannot be dismissed that a marker protein such as GFAP has been taken up from neighbouring reactive astrocytes. This is exemplified by the dilemma of how to interpret the occasional presence of GFAP in stromal cells of the cerebellar hemangioblastoma (Deck and Rubinstein 1981; McComb et al. 1986). Future studies employing in situ hybridization techniques for the identification of specific mRNA sequences (Lewis and Cowan 1985) may aid in overcoming such ambiguities.

2 Biochemically Defined Neural Marker Proteins

2.1 Glial Fibrillary Acidic Protein (GFAP)

Since its discovery by Eng et al. in 1971, GFAP has been extensively investigated and is presently regarded as the most useful marker antigen for astroglial cells in health and disease. Its impact on surgical neuropathology is unrivalled by any other tumor marker. This is mainly due to the availability of a large panel of reliable antibodies, a high immunocytochemical "signal to noise" ratio and a low chance of cross reactivity with GFAP-like antigens in extraneural neoplasms.

Biochemical and immunological studies have revealed that GFAP is a cytoskeletal protein of 55 kD which was shown to polymerize *in vitro* to form smooth, curvilinear or rectilinear 10 nm intermediate filaments. When the *in vitro* assembly is carried out in the absence of reducing agents, approximately 50% of the total protein migrates as a dimer, indicating that the polymerization is accompanied by oxidation of SH-groups (Rueger et al. 1979). There is recent evidence that subtypes of GFAP exist which can be distinguished by specific monoclonal antibodies (Gheuens et al. 1984). It remains to be seen whether this polymorphism can be related to specific cell populations or disease entities.

The distribution of GFAP in the normal brain has been extensively reviewed (Eng 1980; Osborn and Weber 1983; Traub 1985) and is summarized in Table 1. It should be pointed out that the

Table 1. Expression of GFAP in normal and neoplastic tissues

Cell type/Tissue	Comment
Central nervous system	
Astrocytes	More accentuated in fibrous (reactive) than in protoplasmic astrocytes
Bergmann glia	In normal cerebellum and, more markedly, following the loss of Purkinje neurons (SCHACHNER et al. 1977)
Radial glia	In the developing fetal brain (CHOI and LAPHAM 1978; CHOI 1986)
Tanycytes	Subependymal tapering cells (ROESSMANN et al. 1980; SHAW et al. 1981)
Oligodendrocytes	Transient expression during ontogeny (CHOI and KIM 1984)
Ependymal cells	Only during brain development (ROESSMANN et al. 1980)
Rosenthal fibers	Variable staining; amorphous core often negative (JANZER and FRIEDE 1981; SMITH and LANTOS 1984)
Peripheral and autonomic nervous system	
Schwann and satellite cells	Limited expression in human Schwann cells. Antigen may differ from that present in CNS astrocytes (JESSEN et al. 1984; JESSEN and MIRSKY 1985)
Enteric glia	Positively identified in rats (JESSEN and MIRSKY 1980)
Neural tumors	
Astrocytomas	Reliable marker for this tumor type (see KUMAR and MARSDEN 1986)
Oligoastrocytomas	Mainly in neoplastic astrocytes (HERPERS and BUDKA 1984; NAKAGAWA et al. 1986)
Oligodendrogliomas	In neoplastic 'gliofibrillary oligodendrocytes' (HERPERS and BUDKA 1984; NAKAGAWA et al. 1986)
Choroid plexus papillomas	Only focal expression (RUBINSTEIN and BRUCHER 1981)
Ependymomas	Variable expression; often accentuated in perivascular pseudorosettes (DUFFY et al. 1979)
Glioblastoma multiforme	Patchy distribution of GFAP, accentuated in the giant cell subtype
Medulloblastoma	Focal expression in approx. 10% of cases (see Table 2)
Neurofibromas	Only occasionally expressed (MEMOLY et al. 1984)
Extraneural tissues	
Adenohypophysis	Follico-stellate cells (VELASCO et al. 1982; HÖFLER et al. 1984)
Epiglottis	Epiglottic cartilage (KEPES et al. 1984; BUDKA 1986)
Salivary gland	Myoepithelial cells (ACHTSTÄTTER et al. 1987)
Liver	Perisinusoidal stellate cells of rat liver (GARD et al. 1985)
Extraneural tumors	
Pituitary adenoma	Only one case reported (HÖFLER et al. 1984)
Pleomorphic salivary adenoma	Case report; only few cells were GFAP positive (NAKAZUTO et al. 1982)
Metastatic renal carcinoma	Case report (BUDKA 1986)
Papillary meningioma	Case report; triple expression of GFAP, vimentin and desmin (BUDKA 1986)

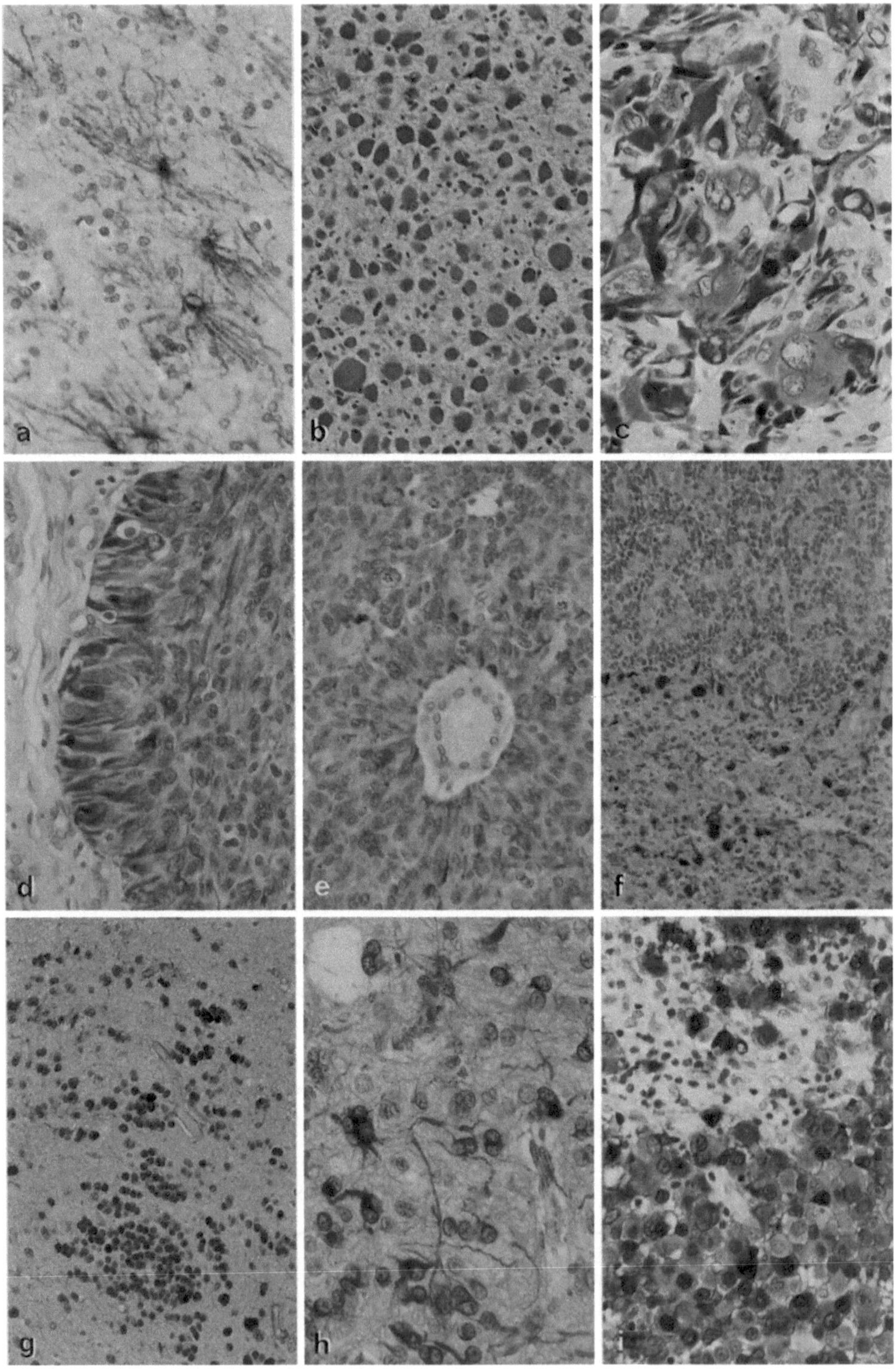
a
b
c
d
e
f
g
h
i

extent of GFAP expression is closely linked to the functional state of astroglia and greatly influenced by surrounding tissue elements. Generally, protoplasmic astrocytes contain little GFAP whereas reactive fibrous astrocytes show abundant GFAP expression (Fig. 1a). Exposure to cyclic AMP (SENSENBRENNER et al. 1980), wounding of the cerebral cortex with formation of a glial scar (BIGNAMI and DAHL 1976) and contact with mesodermal tissue (HERPERS et al. 1984) is accompanied by a marked increase in immunoreactive GFAP (Fig. 1d). Glial filament proteins are evenly distributed in the perikaryon and cytoplasmic processes, the cell nucleus usually being spared. Rosenthal fibers, i.e. conglomerates of glial filaments, present in a variety of chronic pathological conditions, show variable GFAP antigenicity but the surrounding rim of cytoplasm is usually strongly GFAP positive (JANZER and FRIEDE 1981; SMITH and LANTOS 1985). Co-expression of GFAP with vimentin is not uncommon and this has led to the suggestion that in response to functional differences, these intermediate filaments may act together in the formation of the cytoskeleton (WANG et al. 1984). There is, however, evidence that certain GFAP-positive cells do not express vimentin and these have tentatively been classified as protoplasmic astrocytes (SHAW et al. 1981). In addition, GFAP is frequently co-expressed with S-100 protein, particularly in reactive astrocytes and differentiated astrocytomas. In several neoplasms, however, immunoreactivities to these marker proteins dissociate and this may be a valuable diagnostic criterion (KIMURA et al. 1986).

Intermediate filament typing in diagnostic oncology is based on the concept that neoplastic cells retain the pattern of intermediate filament proteins typical for the tissue or cell type of origin (OSBORN and WEBER 1983; RAMAEKERS et al. 1983). For GFAP this has been largely confirmed, although the practicing pathologist should be aware of some notable exceptions (Table 1). The role of GFAP in the histopathological diagnosis of gliomas has been extensively reviewed (DECK et al. 1978; VELASCO et al. 1980; TASCOS et al. 1982; BONNIN and RUBINSTEIN 1984; COLLINS 1984; GULLOTTA et al. 1985; KUMAR and MARSDEN 1986; SCHWECHHEIMER 1986). Two major limitations to the application of GFAP should be emphasized. First, it does not allow a clear-cut distinction of neoplastic cells from reactive astrocytes in the adjacent brain or from entrapped astrocytes within an invading neoplasm. Although in tumor cells GFAP expression is often restricted to the perikaryon and a few, plump cell processes (Figs. 1f, 2b), it can be very difficult to unequivocally identify astrocytic differentiation in non-astroglial tumors, e.g. oligodendrogliomas (HERPERS and BUDKA 1984; NAKAGAWA et al. 1986) and medulloblastomas (MANNOJI et al. 1981; COFFIN et al. 1983). In hamartomatous lesions, astrocytes may show long, but irregularly formed cell processes (Fig. 1h). Secondly, the histopathological assessment of GFAP in astrocytic tumors is of little assistance in tumor grading.

◁ **Fig. 1. a** Infiltration zone of a glioblastoma multiforme. Expression of GFAP by reactive astrocytes. In contrast to neoplastic astrocytes (shown in **b**), reactive astrocytes characteristically have numerous delicate cell processes. **b** Gemistocytic astrocytoma. GFAP-immunoreactivity is accentuated in the perinuclear cell body of neoplastic cells. Note the virtual absence of celll processes. **c** Glioblastoma multiforme. Marked expression of GFAP by large, multinucleated giant cells. **d** Anaplastic ependymoma. Expression of GFAP is accentuated at the border towards surrounding connective tissue. **e** Anaplastic ependymoma. GFAP-immunoreactivity is largely restricted to perivascular pseudorosettes. **f** Malignant glioma with an anaplastic focus lacking GFAP and a sharp transition into a more differentiated tumor area with marked GFAP expression. **g** Glial hamartoma of the temporal lobe. Clusters of undifferentiated small cells show marked immunostaining with S-100 protein. **h** Glial hamartoma (same case as in **g**). GFAP immunostaining shows malformed astrocytes with long, but irregular cell processes. **i** Cerebral metastasis of a malignant melanoma. Most tumor cells are strongly positive for S-100 protein

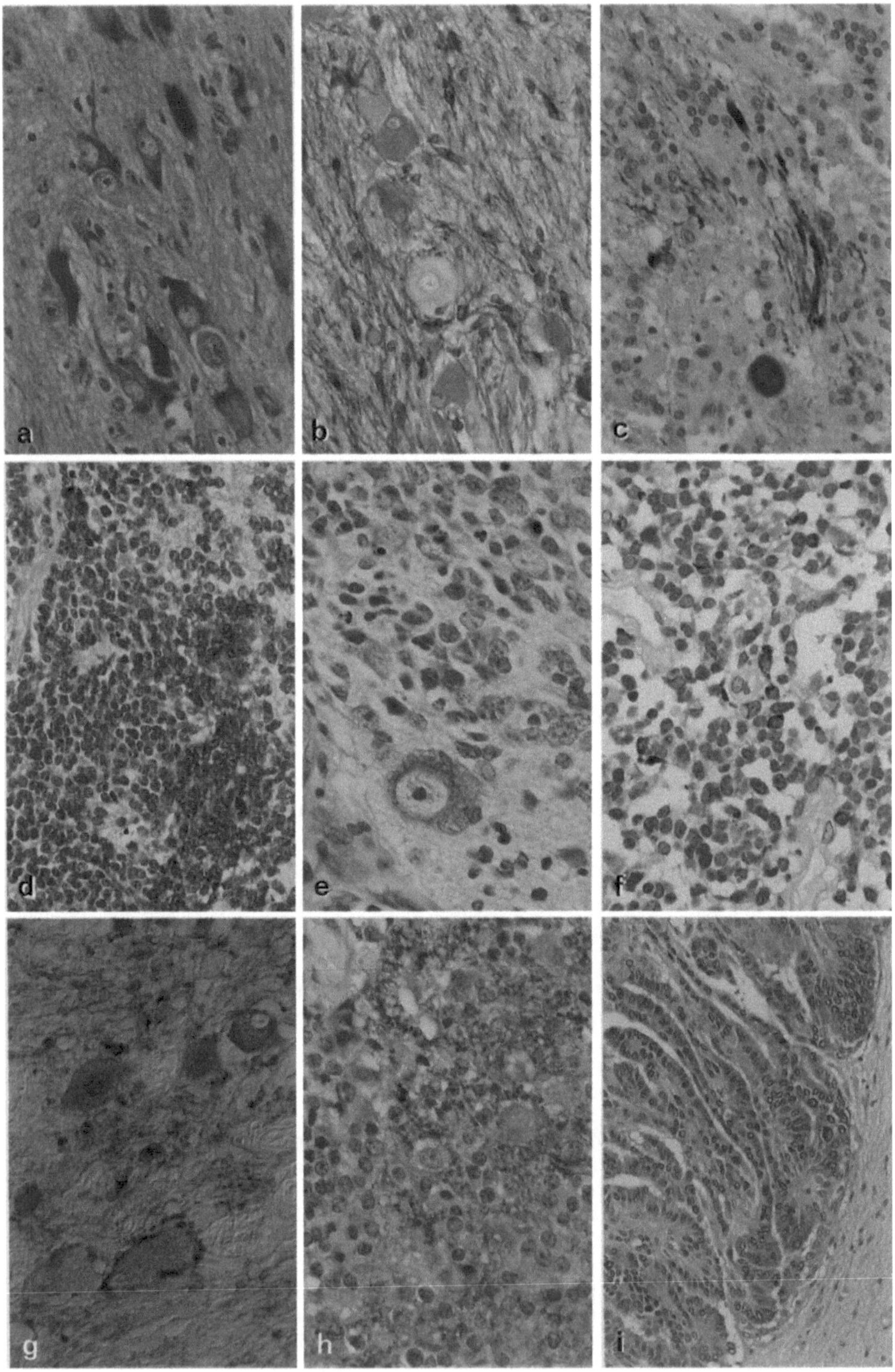
a
b
c
d
e
f
g
h
i

Although neoplastic progression with increasing anaplasia is usually paralleled by a decrease in GFAP expression (JACQUE et al. 1979; DUFFY et al. 1980), this inverse correlation is too variable to be of prognostic value.

Comparison of results obtained in different labaratories is often difficult since a steadily increasing range of commercially available polyclonal and monoclonal antibodies is in use, with variable specificity and sensitivity. In addition, the immunoreactivity of GFAP in paraffin-embedded sections depends on the dilution of the primary antibody (BIGBEE et al. 1977), the interval between tissue removal and fixation (DEARMOND et al. 1983), on the type of fixative and the length of fixation (DIXON and ENG 1981; JIE et al. 1986). With respect to the latter, it is generally agreed that rapid fixation in neutral buffered 4% formaldehyde (NBF) overnight or short fixation in NBF (1 hr) followed by 70% ethanol, give satisfactory results with most available antibodies (ZHENG et al. 1986). The quenching effect of formalin fixation is probably due to extensive protein crosslinking and can be partially reversed by trypsin digestion (MEPHAM et al. 1979; ENG 1985). Monoclonal antibodies tend to bind less consistently to GFAP in formalin-fixed tissues since each of them recognizes antigenic determinants which may be spatially closely related, but not identical. These epitopes may be unpredictably sensitive or resistant to fixation. It has, therefore, been proposed to use a 'cocktail' of monoclonal antibodies to GFAP. This makes a loss of immunoreactivity due to fixation-induced alteration of antigen structure less likely (MCLENDON et al. 1986). In contrast to polyclonal antisera, such MAB mixtures offer the advantage of continuous, unlimited supply.

2.2 S-100 Protein

A thermolabile, acidic protein of MW 20–25,000 found in bovine brain but not liver was first isolated in 1965 by MOORE. It was named S-100 because of its solubility in 100% ammonium sulphate solution at neutral pH. Later it was shown to consist of two polypeptide chains of 93 and 91 amino acids (alpha- and beta-subunit), sharing 54 identical amino acids. S-100-alpha and S-100-beta form homo- and heterodimers (alpha/alpha = S-100ao, alpha/beta = S-100a, beta/beta = S-100b) but the physiological role of these dimers has not been determined (ISOBE et al. 1983).

Based on biochemical structural similarity, it was suggested that S-100 was a member of the group of calcium binding proteins. There is a soluble and a membrane-bound form of S-100 (HAGLID et al. 1976). The latter is bound to the cytoplasmic side of mitochondria, endoplasmic reticulum

◁ **Fig. 2. a** Cerebral ganglioglioma. Marked GFAP expression by the neoplastic glial component. **b** Cerebral ganglioglioma. NSE immunoreactivity in numerous neoplastic neuronal cell bodies and processes. **c** Spinal ganglioneuroblastoma. NFP expression by ganglionic cells and their processes. **d** Focal expression of NSE in a medulloblastoma, indicating neuronal differentiation. **e** Medulloblastoma with neuronal differentiation. NFP expression is restricted to mature ganglionic cells. **f** Medulloblastoma with astrocytic differentiation. GFAP immunoreactivity is restricted to a small perinuclear cytoplasmic rim. **g** Ganglioglioma. Synaptophysin stains the neuronal tumor compartment. Note the punctuate accentuation along the plasma membrane, reflecting the distribution of synaptic junctions. **h** Supratentorial primitive neuroectodermal tumor (PNET) with focal neuronal differentiation. Synaptophysin immunoreactivity in these areas shows a granular, network-like pattern. **i** Diffuse expression of carbonic anhydrase C in the cerebral metastasis of an anaplastic choroid plexus papilloma and Golgi apparatus. It is found in free and membrane-bound ribosomes and can be secreted into the extracellular space (COSGRAVE et al. 1983). In addition to its cytoplasmic distribution, S-100 is expressed in the cell nucleus where a stimulatory effect on RNA polymerase I has been reported (MICHETTI et al. 1976). Its synthesis is cell cycle dependent and highest in quiescent cells (FAN 1982), but nothing is known about a possible gene-regulatory function. In the development of the CNS it promotes neurite extension (KLIGMAN and MARSHAK 1985), and has been implicated in microtubule disassembly (BAUDIER et al. 1982), in cation permeability regulation of lipid membranes (CALISSANO and BANGHAM 1971) and in the regulation of protein phosphorylation and dephosphorylation (KUO et al. 1986). Its detailed biological function, however, is still unknown (for review see DONATO 1986).

Originally thought to be nervous system-specific, S-100 has now been identified in over 20 different cell types of normal tissues and their neoplastic counterparts (DONATO 1986; NAKAJIMA et al. 1984; WEISS et al. 1983). Within the normal nervous system, it is present in astrocytes and oligodendrocytes (LUDWIN et al. 1976), ependymal cells (YAMAGUCHI 1980), in some neurons (HAAN et al. 1982; LOEFFEL et al. 1985; MOLNAR et al. 1985; VANSTAPEL et al. 1985), and in Schwann cells (VANSTAPEL et al. 1985). In contrast to GFAP, it is preferentially expressed in cell nuclei (Figs. 2g, h). The relative amount and the distribution of the S-100 subunits in different human tumors is very variable (TAKAHASHI et al. 1984). More than 90% of the cerebral S-100 belongs to the S-100b fraction (JENSEN et al. 1985). Accordingly, the immunohistochemical staining obtained by heterologous polyclonal antibodies recognizing both subunits is very similar, if not identical, to that seen with monospecific anti-S-100-beta antisera as far as the normal (MOLNAR et al. 1985; LOEFFEL et al. 1985; VANSTAPEL et al. 1985) and neoplastic (VAN ELDICK et al. 1986) nervous system is concerned. S-100 is expressed in astrocytomas, ependymomas, glioblastomas, schwannomas, glial hamartomas (Fig. 1g), and the adamantinomatous areas of craniopharyngiomas. In addition, it is strongly positive in melanocytic lesions and can be expressed in chordomas, chondrosarcomas, lipomas, liposarcomas, fibromatosis, histiocytosis X, eosinophilic granulomas, mixed sweat gland tumors, pleomorphic adenomas of the salivary gland, medullary carcinomas of the breast, bronchioalveolar carcinomas of the lung and teratomas of the ovary (KAHN et al. 1982; NAKAMURA et al. 1983; TAKAHASHI et al. 1984; WEISS et al. 1983). Its principal diagnostic application in neuro-oncology is the differential diagnosis of benign and malignant spindle cell tumors since it is consistently expressed in schwannomas and neurofibromas (WEISS et al. 1983; WEIDENHEIM and CAMPBELL 1986). The strong intracytoplasmic expression of S-100 in all melanocytic lesions, including amelanotic melanomas (Fig. 1i) has also proved very useful. It has recently been reported that the alpha-subunit of S-100 is present in human malignant melanomas, but not in Schwann cell-derived tumors of the peripheral nervous system (ISOBE et al. 1984).

2.3 Neuron-Specific Enolase

Neuron-specific enolase (NSE) was first isolated from brain tissue by MOORE and co-workers (1965) and initially referred to as 14.3.2 protein, based on chromatographic and electrophoretic criteria (MOORE 1975). Later, it was designated neuron-specific protein (MARANGOS et al. 1975) until BOCK and DISSING (1975) discovered that this protein was in fact an isoenzyme of the glycolytic enzyme enolase (2-phospho-D-glycerate hydrolase, E.C. 4.2.1.11). Like several other glycolytic enzymes, enolase exists as a set of isoenzymes. Each of the 5 dimeric isoenzymes known is composed of two out of three immunologically distinct subunits (α, β and γ) which are coded for by separate genes (PEARCE et al. 1976; CHEN and GIBLET 1976). Unlike liver and skeletal muscle which express mainly a single isoenzyme (αα or ββ, respectively) it has been established by chromatographic and immunoassay techniques (MARANGOS et al. 1979; SUZUKI et al. 1980) that nervous tissue contains a mixture of isoenzymes, two homodimers (αα and γγ) and one heterodimer (αγ). The γ subunit, present either in the hybrid or γγ form, accounts for approximately 2% of total soluble brain proteins (MARANGOS et al. 1979; HULLIN et al. 1980; KATO et al. 1982). Most other tissues (e.g. liver and skeletal muscle) have γ subunit levels three orders of magnitude lower than the brain. Compared to the αα homodimer (MW 87,000), γγ enolase is characterized by a lower molecular weight (MW 78,000), a more acidic isoelectric point and greater stability towards chloride inactivation (MARANGOS et al. 1978; SCHMECHEL 1985). Immunochemical studies of the subunit distribution in normal tissues have been performed employing purified antibodies specific for either the α or γ subunits (ROYDS et al. 1982; HAIMOTO et al. 1985). At present, immunohistochemical techniques do not allow discrimination between the neuron-specific (γγ) and the hybrid (αγ) enolase.

Within the CNS, γ enolase is almost exclusively localized in neurons whereas α enolase expression is found in glial, meningothelial, endothelial and Schwann cells and has thus been referred to as non-neuronal enolase (NNE) (SCHMECHEL et al. 1978). Neuroblasts also contain NNE, but show a switch of expression from α to γ enolase during the process of differentiation (SCHMECHEL et al. 1980). Immunohistochemical and the more sensitive enzyme-linked immunoassay techniques have demonstrated that γ-enolase (mainly in the hybrid form) is also widely distributed in extraneural tissues. High levels of enolase are present in the adrenal and pituitary gland, in cells of the diffuse neuroendocrine system (SCHMECHEL et al. 1978), and in a variety of non-neuronal and non-neuroendocrine cell populations (HAIMOTO et al. 1985).

In spite of the recognition that this protein is not neuron-specific, γ-enolase is currently employed as a broad-range marker for neoplastic neuronal and neuroendocrine cells (TAPIA et al. 1981; GHOBRIAL and ROSS 1986; TAYLOR et al. 1986). Within the CNS, it has been used as an indicator of neuronal differentiation in neuroblastomas (ODELSTAD et al. 1981), retinoblastomas (NAKAJIMA et al. 1984) and a variety of primitive neuroectodermal tumors, including the medulloblastoma (Fig. 2d). Reservation about its usefulness for the classification of CNS tumors, however, has arisen from the co-expression of GFAP and NSE in reactive astrocytes (VINORES et al. 1984; VINORES and RUBINSTEIN 1985) and the positive immunoreactivity of tumor cells occasionally found in various types of gliomas (HAGLID et al. 1973; NAKAJIMA et al. 1984; VINORES et al. 1984), in meningiomas (NAKAJIMA et al. 1984) and neurinomas (NAKAJIMA et al. 1984). This clearly demonstrates that the γ subunit of NSE can no longer be regarded as a neuron-specific tumor marker. NSE is further expressed in neuroblastomas, ganglioneuroblastomas, gangliogliomas (Fig. 2b), paragangliomas (IRONSIDE et al. 1985) and in the majority of neuroendocrine tumors, including intermediate and small cell carcinomas of the lung. In addition, there are numerous studies reporting NSE-like immunoreactivity in non-neuronal tumors including malignant melanomas, Schwann cell neoplasms, chordomas, fibroadenomas of the mammary gland and a variety of carcinomas of non-neuroendocrine origin (DHILLON et al. 1982; NAKAJIMA et al. 1983, 1984; VINORES et al. 1984; NESLAND et al. 1986; PÅHLMAN et al. 1986).

2.4 Neurofilament Proteins (NFP)

Neurofilaments constitute one of five subclasses of intermediate filaments. All intermediate filaments have a strongly conserved, rod-shaped, alpha-helical 40 kD domain, probably derived from an ancestral intermediate filament gene (Shaw et al. 1984; Pruss et al. 1981). NFP have a hypervariable amino-terminal head and carboxy-terminal tail on both sides of the 40 kD alpha-helix. The tail ends of the neurofilament proteins, which are unusually long for an intermediate filament, contain a domain with a high content of acidic residues (Liem et al. 1985a). This "b" domain represents the binding site for the Bodian silver stain, which specifically stains neurofilament proteins (Weber et al. 1983; Gambetti et al. 1981). Expression of neurofilament proteins has, so far, only been recorded in neuronal cells, predominantly in axons (Drake and Lasek 1984). Its function is not clear, but may not be essential since it is entirely lacking in the well functioning nervous system of arthropods. Neurofilaments are composed of 3 subclasses of polypeptide chains with molecular weights of 68–75, 140–175 and 180–212 kD (Shaw et al. 1984). Since there is a relatively great interspecies variation in molecular weight the terms H-, M- and L-subunit (for "high", "middle" and "low") are commonly used. These subunits are differentially expressed during brain development (Shaw and Weber 1982) and have a differential distribution within the neuronal cell (Hirokawa et al. 1984). Although all three subunits are found in the axon as well as in the cell body and dendrites, phosphorylated epitopes on the H- and M-subunits recognized by monoclonal antibodies have so far only been detected in the axon, but not in cell bodies or dendrites (Sternberger and Sternberger 1983; Goldstein et al. 1983). The H-subunit forms cross-bridges between neurofilaments (Hirokawa 1982) while the L-subunit represents the binding site for cross-links with microtubule associated proteins, including tau-protein (Liem et al. 1985b). In diagnostic neuro-oncology, commercially available monoclonal antibodies recognizing the H- and M-subunit have preferentially been used. In most reports, it is not specified whether these reacted with phosphorylated or non-phosphorylated epitopes. Despite these limitations, neurofilament protein expression in tumor cell bodies has only been found in neuronal tumors, e.g., phaeochromocytoma, ganglioneuroma, ganglioneuroblastoma (Fig. 2c), and neuroblastoma (Mukai et al. 1986; Osborn et al. 1986; Trojanowski et al. 1984). In our experience, the presence of neurofilaments is highly specific for neuronal tumors. However, NFP are most consistently expressed in neoplasms with advanced neuronal differentiation and may be difficult to detect in undifferentiated neuroblastomas.

2.5 Synaptophysin

Synaptophysin, an acidic homoligomeric glycoprotein of subunit MW 38,000 was first isolated from rat and bovine brain in 1985 (Jahn et al. 1985; Wiedenmann and Franke 1985). Its expression during brain development correlates with synaptogenesis and nerve terminal differentiation (Knaus et al. 1986). Recent biochemical characterization has identified this protein as an integral calcium-binding membrane protein of presynaptic vesicles in neurons and similar vesicles in neuroendocrine cells (WIEDENMANN and FRANKE 1985; REHM et al. 1986). Synaptophysin appears to span the membrane of these vesicles and has an epitope exposed on the cytoplasmic surface to which several polyclonal (JAHN et al. 1985) and monoclonal antibodies have been raised (WIEDENMANN and FRANKE 1985; WIEDENMANN et al. 1986). The epitope recognized by the antibody SY 38 is reliably demonstrable by immunofluorescence in frozen sections (WIEDENMANN et al. 1986; SCHWECHHEIMER et al. 1987) and in conventionally fixed, routinely processed paraffin embedded material (for details see GOULD et al. 1986).

In normal tissues, immunoreactivity is present in all types of neurons, at the neuromuscular junction of cross striated muscle, in the adrenal medulla and certain cells of the dispersed neuroendocrine system (WIEDENMANN and FRANKE 1985; GOULD et al. 1986; WIEDENMANN et al. 1986). Immunostaining characteristically reveals a fine punctuate cytoplasmic pattern which most probably reflects the distribution of synapses and/or neurosecretory vesicles. Accordingly, immunoreactivity in mature CNS neurons is accentuated at the cell periphery (Fig. 2g), whereas in neuroendocrine cells it is spread rather evenly throughout the cytoplasm (WIEDENMANN et al. 1986; GOULD et al. 1987). Within the neuroendocrine system, synaptophysin is present in both "neural" and "epithelial" cell types, indicating its independence from cytoskeletal characteristics or the neuropeptide composition of individual cells (GOULD et al. 1986). Thus, co-expression with the neurofilament triplet, cytokeratins, vimentin or a combination of intermediate filament markers has been observed.

Synaptophysin immunostaining was found to be reliable and specific for a variety of differentiated and undifferentiated neuronal tumors, including neuroblastomas, ganglioneuroblastomas, ganglioneuromas, glio-neuronal hamartomas as well as primitive neuroectodermal tumors (Fig. 2h), including the medulloblastomas (GOULD et al. 1986; SCHWECHHEIMER et al. 1987). Synaptophysin is further expressed in almost the entire spectrum of neural and epithelial neuroendocrine neoplasms, irrespective of their degree of differentiation, cytoskeletal constituents and hormone production. In contrast to NSE, immunohistochemical studies of malignant epithelial and mesenchymal tumors have so far failed to show synaptophysin expression. The consistent negativity of malignant melanomas is particularly remarkable, despite their positive immunoreactivity for NSE and several neuropeptide antigens. The absence of synaptophysin immunostaining in non-neural and non-neuroendocrine cells and in neoplasms derived therefrom indicates that synaptophysin may prove to be a specific and highly reliable marker for the identification of neuroblastic and neuroendocrine cells.

3 Human Brain Tumor-Associated Antigens Defined by Monoclonal Antibodies

Being directed towards a single antigenic determinant, MABs have the potential of recognizing proteins that are unique to certain tumor types and restricted to neoplastic cells. In addition, the continuous production of a homogeneous antibody could result in diagnostic and therapeutic applications *in vivo*. Over the past five years, a great number of murine MABs has been raised, using different approaches for immunization. Their potential use and limitations have recently been reviewed by BULLARD and BIGNER (1985). These authors concluded that the antibodies developed so far have greatly advanced our knowledge of the complexity of antigenic determinants in human brain and brain tumors, but that their practical use is still limited by a variety of inherent disadvantages. These include a marked phenotypic heterogeneity of human neuroectodermal tumors, limited sensitivity, inability to detect certain human histocompatibility complex (HLA) specificities and to evoke in mice the autochthonous human immunological response to tumor-associated antigens.

Following immunization with fetal human brain, KEMSHEAD and co-workers developed the murine antibody UJ13A which selectively binds to tissues of neuroectodermal origin and neoplasms derived therefrom. This pan-neuroectodermal antibody has been successfully used to distinguish neural tumors from lymphomas and epithelial metastatic lesions but the antigen is inactivated by fixation (ALLAN et al. 1983). This also applies to a panel of MABs to neuroectodermal oncofetal antigens developed by WIKSTRAND et al. (1982) and GARSON et al. (1985) which recognize neuroectodermal tumors as well as human fetal CNS and lymphoid tissues. A similar specificity for neuroectodermal tumors and oncofetal tissues was found in a hybridoma raised against melanoma cells (LIAO et al. 1981). In several studies, antibodies were produced by immunizing mice with cultured glioma cell lines. The MABs BF7 and GE7 demonstrated specific binding to glioma cells in a radioimmunoassay (SCHNEGG et al. 1981) but histologically detectable immunoreactivity was present in both normal astrocytes and a subset of neoplastic glial cells in human gliomas (KRAJEWSKI et al. 1986). Special fixation and embedding techniques were necessary to maintain antigenicity.

A detailed analysis of MABs developed by immunization with human glioma-derived cell lines (HGL) revealed a complex pattern of antigenic heterogeneity (WIKSTRAND et al. 1985). Human glioma lines which reacted most extensively with anti-HGL MABs also had the highest reactivity rate with anti-lymphoid marker antibodies. Antigenicity in frozen sections from human glioblastomas and astrocytomas was found to be even more complex, with immunoreactivity to anti-HGL MABs present in 66–100% of cases. None of the MABs tested was sufficiently specific and reliable for diagnostic purposes. This also holds true for most antibodies raised against neuroblastomas which usually cross-react with haematopoietic cells and neuroectoderm-derived tissues and tumors. The recently developed MAB 5A7 appears to be more selective, showing no reactivity with normal or neoplastic haematopoietic cells (GROSS et al. 1986).

In conclusion, the MABs raised in an attempt to selectively define glioma-associated antigens are at present not very helpful in the histopathological diagnosis of human nervous system tumors, mainly due to extensive cross-reactivity with normal CNS cells and extraneural tissues and their general sensitivity to formalin fixation.

4 Application of Tumor Markers to Diagnostic Neuro-Oncology

4.1 Neuronal Tumors

Highly differentiated neuronal neoplasms do not usually give rise to diagnostic problems. This is particularly true for gangliocytomas, gangliogliomas, ganglioneuromas and ganglioneuroblastomas in which mature ganglionic cells can be identified both by their histological appearance and immunoreactivity for NSE (Fig. 2b), NFP (Fig. 1c) and synaptophysin (Fig. 2g). However, these neuronal marker proteins can be of great assistance in the interpretation of neoplasms with less advanced neuronal differentiation. Thus, retroperitoneal and adrenal neuroblastomas can be difficult to distinguish from other small round-cell tumors (e.g., Ewing's sarcoma, malignant lymphoma, rhabdomyosarcoma, Wilms' tumor). *Neuroblastomas* can be positively identified by their immunoreactivity for NFP, but abundant neurofilaments are only present in the more differentiated neoplasms. In immature neuroblastomas, immunoreactivity for NFP is lower and can easily be abolished by fixation. A recent report by OSBORN et al. (1986) indicates that alcohol fixation and paraffin embedding causes a preferential inactivation of epitopes recognized by antibodies to the low molecular weight NFP. This is at variance with a study by MUKAI et al. (1986) who found that in formalin-fixed, paraffin-embedded neuroblastomas, a polyvalent rabbit antiserum to the 68-kD subcomponent was most reliable and sufficiently discriminating for diagnostic application. NSE is also consistently expressed in neuroblastomas, but is less specific than NFP since it has also been found in a variety of other tumors (VINORES et al. 1984; NAKAJIMA et al. 1984; OSBORN et al. 1986). Synaptophysin (GOULD et al. 1986) and microtubule-associated proteins (ARTLIEB et al. 1985) may soon be established as the most useful marker proteins for neuroblastomas and neoplasms of the diffuse neuro-endocrine system since their distribution appears to be highly specific and their immunoreactivity less affected by fixation. None of these markers has been shown to be of use in the grading of neuroblastomas. In contrast, immunoreactivity to S-100 protein has recently been claimed to be associated with favourable prognosis (SHIMADA et al. 1985).

In the *olfactory neuroblastoma*, both NSE (CHOI and ANDERSON 1985) and synaptophysin (Fig. 2h) are expressed in irregularly shaped cell clusters. Immunoreactive S-100 protein is restricted to cells that form an interconnecting network at the periphery of lobular tumor cell nests (CHOI and ANDERSON 1985). With progressive differentiation, olfactory neuroblastomas also express NFP (TROJANOWSKI et al. 1982). Due to their common origin from photoreceptor cells, retinoblastomas and pineal parenchymal tumors (pineocytomas) express the retinal S-antigen (DONOSO et al. 1985; KORF et al. 1986; PERENTES et al. 1986).

4.2 Astrocytomas

The most valuable marker for normal and neoplastic astrocytes is *GFAP* (RAFF et al. 1979; BONNIN and RUBINSTEIN 1984). In astrocytomas, there is a general tendency for decreasing GFAP expression to accompany increasing dedifferentiation (TASCOS et al. 1982; VELASCO et al. 1980; JACQUE et al. 1979). This is well illustrated (Fig. 1f) by the occasional finding in anaplastic gliomas of sharply delineated foci of small anaplastic subclones with complete lack of GFAP expression within more differentiated areas with correspondingly high levels of GFAP (SCHMITT 1983a, 1983b). However, even in the most anaplastic variants of astrocytoma, there are some tumor cell clusters still expressing GFAP. On the other hand, in highly differentiated astrocytomas there is never GFAP-expression by 100% of the tumor cells. Occasionally, low grade astrocytomas demonstrate a surprisingly meager GFAP-expression whereas frankly malignant lesions may contain a high incidence of GFAP-positive tumor cells (KIMURA et al. 1986). Two immunohistological staining patterns are usually observed. Cytoplasmic staining with perinuclear accentuation is mostly found in protoplasmic and gemistocytic astrocytomas, in the astrocytic component of gangliogliomas (Fig. 2a), and in the bizarre giant cells of anaplastic astrocytomas. A diffuse staining of the fibrillary matrix corresponds to neoplastic cell processes in pilocytic and fibrillary astrocytomas. Small round cells with scanty cytoplasm and processes tend not to express GFAP. Rosenthal fibres are typically found in pilocytic astrocytomas, but are occasionally encountered in other subtypes of astrocytic tumors (HWANG and BORIT 1982). They show a variable immunoreactivity for GFAP but a positively stained cytoplasmic rim is consistently visible (JANZER and FRIEDE 1981; SMITH and LANTOS 1985). Pale eosinophilic hyaline granules, structures of unknown origin and significance, are often present in pilocytic and fibrillary astrocytomas (Smith and Lantos 1985) and stain diffusely positive for GFAP.

The expression of *S-100 protein* in astrocytomas is very similar to that of GFAP. Co-expression has been reported (Kimura et al. 1986), with a similar staining pattern of perinuclear accentuation in gemistocytic lesions *versus* a preferential staining of cell processes in pilocytic and fibrillary astrocytomas. However, there is often an additional expression of S-100 protein in the nucleus of neoplastic astrocytes, especially in small tumor cells with scanty cytoplasm. A dissociation of immunoreactivity with preferential or selective expression of S-100 protein is found in the subependymal giant cell tumors associated with tuberous sclerosis (Kimura et al. 1986), in glial hamartomas of the temporal lobe (Fig. 1 g) and, occasionally, in malignant astrocytic tumors. As for GFAP, a low S-100 protein content correlates with advanced dedifferentiation (Jacque et al. 1979).

Another useful marker is *glutamine synthetase*, an enzyme abundant in extraneural tissues like liver, but within the CNS restricted to normal, reactive and neoplastic astrocytes (Pilkington and Lantos 1982). Co-expression with GFAP has been reported. The overall staining pattern is very similar to that found with GFAP and S-100 protein, including the tendency for decreasing expression with increasing dedifferentiation of the astrocytoma. However, there are some differences in the cellular staining pattern. Glutamine synthetase is expressed most heavily in the cytoplasm of protoplasmic astrocytes with a low content of glial intermediate filaments and to a lesser extent in the cell processes of fibrous astrocytes.

The following markers are not specific for astrocytomas. Expression of *NSE* (mostly the alpha-, but to a lesser extent also the gamma-subunit) has been observed in all histopathological variants. In most astrocytic neoplasms, a small proportion of cells co-expresses NSE and GFAP, while in the majority of cells either NSE or GFAP is exclusively present (Vinores and Rubinstein 1985). *Vimentin*, an intermediate filament protein of 57 kD, originally thought to be specifically expressed in mesenchymal cells (Altmannsberger et al. 1981), is consistently expressed in normal (Chiu et al. 1981) and neoplastic (Yung et al. 1985) astrocytes. Its cellular staining pattern is very similar to that of GFAP. GFAP-positive cells in astrocytomas may also be positive for vimentin, but some vimentin-positive cells in astrocytomas are GFAP-negative (Wai-Kwan et al. 1985). Since vimentin is expressed early in astrogliogenesis (Dahl 1981; Pixley and DeVellis 1984), one could assume that its presence in astrocytomas indicates a lower degree of differentiation. However, this could not be established in larger series comparing GFAP and vimentin expression in human gliomas (Schiffer et al. 1986).

Fibronectin expression in astrocytomas is controversial. *In vivo* studies have failed convincingly to demonstrate fibronectin in locations other than the extracellular matrix produced by leptomeningeal or blood vessel cells. A faint diffuse staining pattern with an intercellular distribution around neoplastic glial cells was interpreted as being due to vessel leakage (Paetau et al. 1980; Bellon et al. 1985). However, neoplastic cells derived from astrocytomas definitely express fibronectin *in vitro*, in addition to laminin and collagen (McKeever et al. 1986). Within individual neoplastic glial cells, co-expression of fibronectin with GFAP, S-100 or NSE has never been observed. On the other hand, non-neoplastic astrocytes are able to produce fibronectin and laminin *in vitro,* and co-expression of fibronectin and GFAP within the cytoplasm of embryonic and early postnatal rat brain cells has recently been reported (Liesi et al. 1986). *NFP* and *synaptophysin* are not expressed in neoplastic astrocytes. Staining of entrapped axons is, however, frequently observed.

4.3 Oligodendrogliomas

Several marker proteins have been identified in human and rodent oligodendrocytes, including galactocerebroside (Raff et al. 1978), cyclic nucleotide phosphodiesterase (Kim et al. 1984), myelin-specific P_2 (Trapp et al. 1983) and proteolipid proteins (Hartman et al. 1982). However, these antigens are not consistently expressed in neoplastic oligodendrocytes. This is similarly true for myelin basic protein (MBP), myelin-associated glycoprotein (MAG) and carbonic anhydrase C (Kumpulainen et al. 1982; Weller et al. 1985; Nakagawa et al. 1986). Although some oligodendrogliomas show patchy immunoreactivity for NSE and S-100 protein, these markers are neither reliable nor specific for this tumor type (Weller et al. 1985; Nakagawa et al. 1986).

GFAP immunoreactivity in oligodendrogliomas has been identified in three cell types. As in a variety of other CNS tumors, reactive astrocytes are found to be entrapped within the tumor or around the vascular mesenchymal stroma. These reactive astrocytes are characterized by their stellate morphology and abundant cytoplasm (Nakagawa et al. 1986). Neoplastic astrocytes with eccentric nuclei are mainly found in oligo-astrocytomas, often closely intermingled with GFAP-negative neoplastic oligodendroglial cells. In addition, about half of all oligodendrogliomas contain GFAP-positive neoplastic oligodendrocytes which may be scattered singly or in clusters throughout the tumor. These cells are categorized as neoplastic by their typical appearance, with centrally located nuclei surrounded by a small rim of immunoreactive cytoplasm (Herpers and Budka 1984; Nakagawa et al. 1986). Although various interpretations have been put forward to explain the presence of GFAP-positive neoplastic cells with oligodendroglial morphology, most authors favour the hypothesis of Herpers and Budka (1984), who related this feature to the transient ontogenetic GFAP expression by oligodendrocytes during myelinogenesis (Choi and Kim 1984). Alternatively, the possibility of a divergent differentiation based on the existence of a common bipotential glial pregenitor cell for astrocytes and oligodendrocytes (Raff et al. 1983) has been discussed.

The monoclonal antibody *anti-Leu 7* (Abo and Balch 1981), a marker for natural killer cells (HNK-1), recognizes a carbohydrate determinant present on a family of neural plasma membrane glycoproteins, including myelin-associated glycoprotein (MAG), the neural cell adhesion molecule (N-CAM), and the L1 and J1 glycoproteins (Kruse et al. 1985). It is expressed in various components of the nervous system, including oligodendroglia (Schuller-Pe-

TROVIC et al. 1983). Consequently, anti-Leu 7 has been suggested as a useful marker for oligodendrogliomas (MOTOI et al. 1985). This was confirmed by PERENTES and RUBINSTEIN (1986) and NAKAGAWA et al. (1986) who tested a large series of oligodendrogliomas, as well as a variety of other neuroepithelial tumors. To a variable extent, up to 90% of oligodendrogliomas were stained with anti-Leu 7. Although not specific, it was much more abundant (in terms of the number of immunoreactive tumor cells) in oligodendrogliomas than in other gliomas, primitive neuroectodermal tumors and schwannomas (PERENTES and RUBINSTEIN 1985). Immunoreactivity is restricted to cell membranes and processes, in contrast to the intracytoplasmic immunoreactivity for GFAP. Thus, the demonstration of abundant cell-surface anti-Leu 7 immunoreactivity in combination with interspersed GFAP-positive neoplastic cells may be helpful in the diagnosis of oligodendroglial neoplasms.

4.4 Ependymal and Choroid Plexus Tumors

No marker antigen specific for normal or neoplastic ependymal cells has yet been identified. In analogy to fetal ependymal lining cells (ROESSMANN et al. 1980), ependymomas focally express S-100 protein and GFAP (DUFFY et al. 1979; WELLER et al. 1985; KIMURA et al. 1986; MIETTINEN et al. 1986), preferentially in the perivascular pseudorosettes (Fig. 1e). In our experience, this distribution pattern, rather than the specificity of GFAP expression itself, is a helpful diagnostic feature. If perivascular pseudorosettes are not found, the multiplicity and variability of GFAP expression largely reduces its value as a single immunoprobe for ependymomas. In these cases, the additional expression of vimentin and the lack of immunoreactive cytokeratins may serve to discriminate ependymomas from choroid plexus papillomas and metastatic carcinomas (MIETTINEN et al. 1986).

Choroid plexus papillomas (CPP) are usually distinguishable from other intraventricular tumors (e.g., papillary ependymomas and metastatic adenocarcinomas) by their distinctive histoarchitecture with a layer of isomorphic cuboidal or columnar epithelium resting on a fibrovascular stroma. Rare anaplastic variants and metastatic tumors may, however, lack these typical morphologic features and require immunohistochemical analysis for classification and adequate therapy. In accordance with their histogenetic origin from neuroepithelial (ventricular) precursor cells, choroid plexus neoplasms frequently show immunoreactivity for S-100 protein (NAKAMURA et al. 1983; WELLER et al. 1985; COFFIN et al. 1986; KIMURA et al. 1986). In rare cases, focal NSE expression has also been reported (VINORES et al. 1984; COFFIN et al. 1986). A high proportion of CPP contain either clusters or single GFAP-positive cells which are morphologically indistinguishable from adjacent epithelial cells without GFAP expression (RUBINSTEIN and BRUCHER 1981; TARATUTO et al. 1983; MCCOMB and BURGER 1983; WELLER et al. 1985; COFFIN et al. 1986; MIETTINEN et al. 1986; GIANGASPERO et al. 1987). This finding has been interpreted as an indication of focal ependymal differentiation (RUBINSTEIN and BRUCHER 1981; TARATUTO et al. 1983). The epithelial character of choroid plexus cells is reflected by the expression of epithelial marker antigens. In contrast to ependymomas, most benign and malignant CPP contain cytokeratins (in particular the 'non-epidermal' keratin types 8, 18, and 19) and the desmosome plaque proteins desmoplakin I and II (COFFIN et al. 1986; MIETTINEN et al. 1986). In addition to these markers, anaplastic choroid plexus papillomas may also express the carcinoembryonic antigen (CEA) (COFFIN et al. 1986).

Due to their specialized function in CSF production, normal and neoplastic choroid plexus epithelia contain high concentrations of *carboanhydrase C* (CAC) (ROUSSEL et al. 1979; KUMPULAINEN and KORHONEN 1982; WELLER et al. 1986), an enzyme engaged in active transport processes (MASUZAWA and SATO 1983). Immunohistochemical studies of a large series of primary and metastatic brain tumors (WELLER et al. 1985) have demonstrated that CAC immunoreactivity is conserved in all benign and malignant choroid plexus papillomas (Fig. 2i), but absent in gliomas (including ependymomas) and metastatic carcinomas. CAC thus appears to be the most reliable and specific immunoprobe for choroid plexus neoplasms.

4.5 Glioblastoma Multiforme

All marker antigens detected in astrocytomas can also be expressed in the glioblastoma multiforme (GBM), although usually with a lower percentage of immunopositive cells and a reduced staining intensity. There is no evidence that the extent of GFAP expression in glioblastomas correlates with favorable prognosis. Although major portions of a GBM may not express any of the above mentioned markers, in our experience wider sampling will reveal some tumor cell clusters positive for GFAP or S-100. It is noteworthy that multinucleated giant cells often strongly express GFAP (Fig. 1c) and S-100 protein, whereas the occasional giant cells in anaplastic oligodendrogliomas very rarely do so. The presence of giant cells in glioblastomas has been shown to be correlated with a somewhat better prognosis (BURGER and VOLLMER 1980). Receptors for *epithelial growth factor (EGF)* are found in up to 40% of tumor cells of glioblastoma multiforme (LIBERMAN et al. 1984; NOBLE, personal communication), but the prognostic relevance of this finding is uncertain. In the gliosarcoma, the exact nature of the sarcomatous component is still controversial. There is agreement that its most likely cellular derivation is from the vascular wall of tumor vessels. Expression of endothelial cell markers *(factor VIII-related antigen and Ulex europaeus I agglutinin)* and of basement membrane components like laminin and fibronectin in the sarcomatous component is consistent with an origin from the endothelial hyperplasia typically found in anaplastic gliomas (SCHIFFER et al. 1984; SLOWIK et al. 1985). Ultrastructural findings and the expression of smooth muscle myosin (KISHIWAKA et al. 1986) suggest the possible participation of myofibroblasts, smooth-muscle cells, pericytes or fibroblasts in the genesis of the sarcomatous component of gliosarcomas.

4.6 Medulloblastomas and Related Primitive Neuroectodermal Tumors (PNET)

Although the cerebellar medulloblastoma has for more than 60 years constituted a well defined clinical and histopathological entity, its histogenesis has long been disputed (for review see RUBINSTEIN 1975). It is now generally agreed that this neoplasm originates from a pluripotent matrix cell population. Some authors maintain that this is most likely to be the external granular layer of the cerebellum. Others have proposed the hypothesis that this tumor type is not restricted to the cerebellum but occurs at all levels of the CNS. Accordingly, the subependymal matrix cells should be regarded as precursor cell population. There is indeed evidence that medulloblastoma-like tumors occur in the cerebral hemispheres and it has been proposed (HART and EARLE 1973) that this group of embryonal neoplasms should be named primitive neuroectodermal tumors (PNET). The conceptual basis for this nomenclature is the assumption that PNETs share a common progenitor population, i.e. the subependymal matrix cells, and that their neoplastic transformation at various levels of the CNS leads to tumors with similar morphology and biology (RORKE 1983). Analogous to non-transformed matrix cells, PNET cells have the potential for neuronal, glial and ependymal differentiation. Although a tendency for various lines of differentiation has been noted in the past, only recent immunohistochemical techniques have allowed us to estimate more reliably the incidence and extent to which this occurs.

Historically, *neuronal differentiation* has been most conspicuous, since neuroblastic (Homer Wright) rosettes can easily be detected in a significant proportion of medulloblastomas. Immunocytochemical studies indicate that NSE constitutes a valuable marker for incipient neuronal differentiation (GIORDANA et al. 1983; VELASCO et al. 1985; BURGER et al. 1987). In a recent evaluation of 133 medulloblastomas (Table 2), we found sharply delineated islands of NSE-positive cells (Fig. 2d) in approximately one third of the cases. Often, the nuclei of these cells had a more vesicular chromatin, occasionally with transition into ganglionic cell types. Medulloblastomas with more advanced neuronal differentiation showed clearly identifiable ganglionic cells and these often expressed neurofilament protein, predominantly in the cell processes but occasionally also in the perikaryon (Fig. 1e). This is in accordance with the findings of VELASCO et al. (1985) but at variance with those of ROESSMANN et al. (1983) and FRIEDMAN

Table 2. Differentiation in medulloblastomas

Neuroblastic rosettes		36%
– With ganglioid cells	13%	
Neuron-specific enolase (NSE) positive clusters		35%
– With neuroblastic rosettes	58%	
– Without neuroblastic rosettes	42%	
GFAP-positive neoplasic astrocytes		10%
Desmoplastic variant		11%

133 Cases of the ongoing SIOP/GPO MED 84 study. For details see BURGER et al. (1987)

et al. (1985) who reported positive NFP staining in small undifferentiated cells whose neuroblastic or neuronal nature was not obvious from their appearance in conventionally stained histological sections. In our experience, there is a continuous line of neuronal differentiation, the earliest characteristic being the focal expression of NSE and formation of neuroblastic rosettes, followed by the appearance of ganglioid cells and, eventually, the expression of NFP (BURGER et al. 1987).

Glial differentiation in the medulloblastoma is much more difficult to define. Since specific and reliable markers have not been identified for neoplastic oligodendrocytes and ependymal cells, GFAP is at present the only marker that allows the estimation of glial differentiation in primitive neuroectodermal tumors. Most medulloblastomas contain GFAP-positive cells but the distinction of neoplastic *versus* reactive cells can be very difficult, particularly when these are sparsely dispersed within the neoplasm. Accordingly, the reported incidence of glial differentiation ranges from absent or very uncommon (COFFIN et al. 1983; SCHINDLER and GULLOTTA 1983) to 50% (ROESSMANN et al. 1983) and more (PASQUIER et al. 1983; PALMER et al. 1981). In our view, the typing of GFAP-positive cells in medulloblastomas as proposed by MANNOJI et al. (1981) is quite useful. These authors define neoplastic GFAP-positive cells as morphologically identical to the ordinary medulloblastoma cell with a hyperchromatic nucleus and a very scanty cytoplasm (Fig. 2f). Reactive, entrapped astrocytes show a sparse chromatin and well-developed cytoplasmic processes. In addition, there is an intermediate type with fairly rich cytoplasm but short, plump cell processes and it is assumed that at least some of these cells are neoplastic. When we applied these stringent criteria to a large series of medulloblastomas (Table 2), we found unequivocal evidence of GFAP expression by neoplastic cells in only 10% of cases, and this figure agrees well with the findings of MANNOJI et al. (1981), SCHINDLER and GULLOTTA (1983) and KUMANISHI et al. (1985). We could not confirm the observation of HERPERS and BUDKA (1985) that the occurrence of GFAP-positive neoplastic cells is restricted to the desmoplastic variant. In contrast to neuronal differentiation, GFAP expression by medulloblastoma cells is usually restricted to a few foci of dispersed cells, with no evidence of a progressive differentiation into mature astrocytes.

4.7 Tumors of the Peripheral Nervous System

The only reliable tumor marker for Schwann cell-derived tumors is S-100 protein. Only the beta-subunit, but not the alpha-subunit is present in these neoplasms (ISOBE et al. 1984). The marker is expressed in all benign schwannomas and neurofibromas and in the majority of malignant Schwann cell tumors (WEISS et al. 1983; DAIMARU et al. 1985) in a diffuse cytoplasmic and nuclear distribution, generally with a stronger expression in Antoni A than in Antoni B areas. The same is true for the occasional Schwann cell tumors, which are NSE-positive (NAKAJIMA et al. 1984). Fibronectin and laminin can be demonstrated in the basement membrane surrounding the tumor cells (GIORDANA et al. 1985). Myelin basic protein and myelin-specific P2 are usually absent in Schwann cell tumors

(CLARK et al. 1985). The presence of axons, as demonstrated by neurofilament staining, may be of help in the differential diagnosis of spindle cell tumors. Rhabdomyoblastic differentiation in malignant schwannomas (Triton tumor) can be verified with anti-myoglobin antibodies.

4.8 Meningiomas

Meningiomas have been histogenetically linked to arachnoidal cap cells (KEPES 1986). Despite the existence of diverse morphological subtypes, all meningiomas have common immunohistochemical characteristics that parallel those of their putative non-neoplastic progenitor cells. Meningeal neoplasms share morphological features of mesenchymal and epithelial tissues and this is reflected by the co-expression of markers for epithelial and mesenchymal cells. All meningiomas so far tested showed cytoplasmic and membrane-bound immunoreactivity for *epithelial membrane antigen (EMA)* (THEAKER et al. 1986; KEPES 1986), a glycoprotein marker of epithelial and mesothelial differentiation (PINKUS and KURTIN 1985). Similarly, meningiomas consistently exhibit a strong positive reaction for the major antigenic constituents of desmosomal plaques, *desmoplakin* I and II, another distinctive diagnostic feature of epithelium-derived tumors (MOLL et al. 1986). Desmoplakin staining yields a fine punctuate pattern confined to cell borders, and ultrastructural immunolocalization has pinpointed reactivity to desmosomal junctions (KARTENBECK et al. 1984; SCHWECHHEIMER et al. 1984).

Although initial reports have emphasized the exclusive expression of *vimentin* in meningiomas (KARTENBECK et al. 1984; SCHWECHHEIMER et al. 1984; HALLIDAY et al. 1985), this concept has subsequently been challenged by an increasing number of observations suggesting a more complex pattern of intermediate filament expression. Although it remains undisputed that fibrillary cytoplasmic vimentin immunoreactivity is present in virtually all meningiomas and thus represents the principal cytoskeletal component (KARTENBECK et al. 1984; SCHWECHHEIMER et al. 1984; HALLIDAY et al. 1985; THEAKER et al. 1986), some authors have also detected focal immunostaining for cytokeratins, in particular adjacent to so-called pseudopsammoma bodies (YUNG et al. 1984; THEAKER et al. 1986). Double-labeling studies have localized cytokeratin and vimentin expression to different cells (THEAKER et al. 1986). Recently, even triple expression of vimentin, GFAP and cytokeratin was reported in a case of malignant papillary meningioma (BUDKA 1986). These studies, however, are outnumbered by those which failed to demonstrate expression of cytoskeletal components other than vimentin, including cytokeratins (SCHLEGEL et al. 1980; KARTENBECK et al. 1984; SCHWECHHEIMER et al. 1984; HALLIDAY et al. 1985), desmin (KARTENBECK et al. 1984; SCHWECHHEIMER et al. 1984), NFP (KARTENBECK et al. 1984; SCHWECHHEIMER et al. 1984) and glial filaments (TASCOS et al. 1982). Since some of these negative results have been substantiated by immunoblot analyses (KARTENBECK et al. 1984; SCHWECHHEIMER et al. 1984; HALLIDAY et al. 1985), the issue remains at present unresolved.

S-100 protein which is rather uniformly expressed in schwannomas, has also been focally detected in meningiomas (NAKAMURA et al. 1983; THEAKER et al.

1986) and thus does not contribute significantly to the differential diagnosis of these two most frequent types of extracerebral intracranial neoplasms. Basement membrane antigens (laminin, fibronectin, type IV and V collagen) which are also abdundant in schwannomas were found to be largely restricted to vessel walls, whorl formations and psammoma bodies in meningiomas (KOCHI et al. 1983; BELLON et al. 1985; GIORDANA et al. 1985; MCCOMB and BIGNER 1985).

In view of the consistent expression of mesenchymal (vimentin) and epithelial (epithelial membrane antigen, desmoplakin) markers in normal and neoplastic meningothelial cells, combined probing for these antigens probably represents the most valuable tool presently available to discriminate rare anaplastic or atypical meningiomas from other primary or secondary leptomeningeal neoplasms, such as metastatic carcinomas, gliomas and schwannomas. There are only two other examples of an apparent potential for epithelial differentiation in tumors of presumed mesenchymal origin, i.e., mesotheliomas and synovial sarcomas (CHURG 1985; SALISBURY and ISAACSON 1985).

5 Immunohistochemical Determination of Proliferating Cells

Since the most prominent feature of cells undergoing malignant transformation is a loss of growth control, pathologists have always been eager to estimate the rate of cell proliferation in biopsies from human neoplasms, in particular in slowly growing but ultimately recurring lesions of the CNS. However, the mitotic index cannot be regarded as a reliable parameter since the mitotic phase accounts for less than 10% of the total cell cycle time, and during the interval between tissue removal and fixation, cells may complete mitosis without new cells entering the mitotic phase. Alternative methods have, therefore, been used which give a more reliable estimation of tumor cell proliferation, e.g. tritiated thymidine incorporation in vivo (HOSHINO and WILSON 1979), flow cytophotometry (for review see SPAAR et al. 1986) and immunohistochemical localization of bromodeoxyuridine (YOSHII et al. 1986).

GERDES and co-workers (1983, 1984, 1985) have developed the monoclonal antibody Ki-67 which recognizes a nuclear antigen expressed in the G_1, G_2, S and M phases of the cell cycle. Only resting cells (G_0 phase) are consistently negative for Ki-67. The use of this antibody is at present restricted to frozen sections since immunoreactivity is lost during fixation. Estimation of the growth fraction of CNS tumors, as determined by Ki-67 immunoreactivity, can be easily performed and yields reliable data even in tumors with a very low cell turnover, e.g. pilocytic astrocytomas, schwannomas and pituitary adenomas. In a recent report on 40 neoplasms of the nervous system (BURGER et al. 1986), the highest incidence of stained nuclei was found in a metastatic carcinoma (57%). The percentage of stained cells in gliomas correlated with their histologic grading and known biological behavior, the lowest values being found for pilocytic astrocytomas and the highest for the glioblastoma multiforme. In fibrillary astrocytomas, there was a good correlation between the Ki-67 index and the

Table 3. Ki-67 Expression in intracranial tumors

Diagnosis	n	Range	Mean
Metastatic carcinoma	6	2.2–56.7	23.4
Medulloblastoma	1	10.2	10.2
Glioblastoma	32	1.3–27.6	10.9
Astrocytoma, anaplastic	3	1.6– 4.7	3.3
Astrocytoma	7	0.6– 3.2	1.7
Oligodendroglioma	11	0.7– 4.9	1.9
Oligo-astrocytoma	3	1.3– 2.9	2.1
Meningioma	11	0.2– 4.0	1.8
Ependymoma	4	0.2– 1.5	1.1
Astrocytoma, pilocytic	6	0.4– 1.4	0.9
Pituitary Adenoma	32	0.2– 3.7	1.1
Schwannoma	3	0.2– 2.7	1.0

Data are given as percentage of Ki-67 positive nuclei. Modified from BURGER et al. (1986)

degree of nuclear polymorphism. Since this report was published, additional cases have been examined (Table 3). The wide range of Ki-67 immunoreactivity in the glioblastoma multiforme was particularly striking. At the lower end, these lesions did not differ from benign astrocytomas. There was no apparent correlation with the histology of these tumors and it remains to be shown, whether these differences in the growth fraction of glioblastomas correlate with patient survival. Recent investigations by ZUBER et al. (1987) and GIANGASPERO et al. (1987) corroborate these findings, with Ki-67 indices for the glioblastoma ranging from less than 2% to 32%. As expected, pituitary adenomas were shown to have a low percentage of Ki-67-positive cells but a recent survey of 31 cases (LANDOLT et al. 1987) showed a considerable variation. Low values were present in 11 endocrine inactive adenomas (0.1–1%), higher values in 6 patients with Cushing syndrome (1.1–1.5%), whereas 12 prolactinomas covered the entire range from 0.2 to 3.7%. Invasive adenomas demonstrated significantly higher values than non-invasive neoplasms.

6 Concluding Remarks

The immunohistochemical assessment of neural marker antigens has become an integral part of the histopathological diagnosis of nervous system neoplasms. Antibodies presently available for routine application are directed towards cell constituents present in both normal and neoplastic neural cell lineages. They contribute to the determination of the histogenetic origin of neoplasms and their distinction from extraneural epithelial and mesenchymal tumors. However, there are no markers that assist in tumor grading and in the distinction of normal and reactive glial cells from highly differentiated, invading gliomas. Future efforts should be directed towards the development of antibodies that make these distinctions possible.

References

Abo T, Balch TA (1981) Differentiation antigen of human NK and K cells identified by a monoclonal antibody (HNK-1). J Immunol 127:1024–1029

Achtstätter T, Moll R, Anderson A, Schwechheimer K, Kuhn C, Franke WW (1986) Cell type specificity of the expression of glial filament protein as demonstrated by monoclonal antibody. Differentiation 31:206–227

Allan PM, Garson JA, Harper EI, Asser U, Coakham HB, Brownell B, Kemshead JT (1983) Biological characterization and clinical applications of a monoclonal antibody recognizing an antigen restricted to neuroectodermal tissues. Int J Cancer 31:591–598

Altmannsberger M, Osborn M, Schaver A, Weber K (1981) Antibodies to different intermediate filament proteins. Cell type-specific markers on paraffin-embedded human tissues. Lab Invest 45:427–434

Artlieb U, Krepler R, Wiche G (1985) Expression of microtubule-associated proteins, map-1 and map-2, in human neuroblastomas and differential diagnosis of immature neuroblasts. Lab Invest 53:684–691

Baudier J, Briving C, Deinum J, Haglid K, Soerskog L, Wallin M (1982) Effect of S-100 proteins and calmodulin on Ca-induced disassembly of brain microtubule proteins in vitro. FEBS Lett 147:165–167

Bellon G, Cavlet T, Cam Y, Pluot M, Poulin G, Pytlinska M, Bernard MH (1985) Immunohistochemical localization of macromolecules in the basement membrane and extracellular matrix of human gliomas and meningeomas. Acta Neuropathol (Berl) 66:245–252

Bigbee JW, Kosek JC, Eng LF (1977) Effects of primary antiserum dilution on staining of "antigen-rich" tissues with the peroxidase-antiperoxidase technique. J Histochem Cytochem 25:443–447

Bignami A, Dahl D (1976) The astroglial response to stabbing. Immunofluorescence studies with antibodies to astrocyte-specific protein (GFA) in mammalian and submammalian vertebrates. Neuropathol Appl Neurobiol 2:99–111

Bignami A, Dahl D (1979) The radial glia of Müller in the rat retina and their response protein. Exp Eye Res 28:63–69

Bock E, Dissing J (1975) Determination of enolase activity connected to the brain-specific-protein 14.3.2. Scand J Immunol 4:31–36

Bonnin JM, Rubinstein LJ (1984) Immunohistochemistry of central nervous system tumors. Its contributions to neurosurgical diagnosis. J Neurosurg 60:1121–1133

Bonnin JM, Rubinstein LJ, Papasozomenos SCh, Marangos PJ (1984) Subependymal giant cell astrocytoma. Significance and possible cytogenetic implications of an immunohistochemical study. Acta Neuropathol (Berl) 62:185–193

Budka H (1986) Non-glial specificities of immunocytochemistry for the glial fibrillary acidic protein (GFAP). Acta Neuropathol 72:43–54

Bullard DE, Bigner DD (1985) Applications of monoclonal antibodies in the diagnosis and treatment of primary brain tumors. J Neurosurg 63:2–16

Burger PC, Vollmer RT (1980) Histologic factors of prognostic significance in the glioblastoma multiforme. Cancer 46:1175–1186

Burger PC, Shibata T, Kleihues P (1986) The use of the monoclonal antibody Ki-67 in the identification of proliferating cells: Application to surgical neuropathology. Am J Surg Pathol 10:611–617

Burger PC, Grahmann FC, Bliestle A, Kleihues P (1987) Differentiation in the medulloblastoma. A histological and immunohistochemical study. Acta Neuropathol (Berl) 73:115–123

Calissano P, Bangham AD (1971) Effect of two brain specific proteins (S-100 and 14.3.2) on cation diffusion across artificial membranes. Biochem Biophys Res Commun 43:504–509

Chen S-H, Giblet ER (1976) Enolase: Human tissue distribution and evidence for three different loci. Ann Hum Genet 39:277–280

Chiu FC, Norton WT, Fields KL (1981) The cytoskeleton of primary astrocytes in culture contains actin, glial fibrillary acidic protein and the fibroblast-type filament protein vimentin. J Neurochem 37:147–155

Choi BH (1986) Glial fibrillary acidic protein in radial glia of early human fetal cerebrum: A light and electron microscopic immunoperoxidase study. J Neuropathol Exp Neurol 45:408–418

Choi HSH, Anderson PJ (1985) Immunohistochemical diagnosis of olfactory neuroblastoma. J Neuropathol Exp Neurol 44:18–31

Choi BH, Kim RC (1984) Expression of glial fibrillary acidic protein in immature oligodendroglia. Science 223:407–409
Churg A (1985) Immunohistochemical staining for vimentin and keratin in malignant mesothelioma. Am J Surg Pathol 9:360–365
Clark HB, Minesky JJ, Agrawal D, Pluot M (1985) Myelin basic protein and P2 protein are not immunohistochemical markers for Schwann cell neoplasms. A comparative study using antisera to S-100, P2 and myelin basic protein. Am J Pathol 121:96–101
Coffin CM, Mukai K, Dehner LP (1983) Glial differentiation in medulloblastomas. Histogenetic insight, glial reaction, or invasion of brain? Am J Surg Pathol 7:555–565
Coffin CM, Wick MR, Braun JT, Dehner LP (1986) Choroid plexus neoplasms. Clinicopathologic and immunohistochemical studies. Am J Surg Pathol 10:394–404
Collins VP (1984) Monoclonal antibodies to glial fibrillary acidic protein in the cytologic diagnosis of brain tumors. Acta Cytol 28:401–406
Cosgrave JW, Heikkila JG, Marks A, Brown IR (1983) Synthesis of S-100 protein on free and membrane-bound polysomes of the rabbit brain. J Neurochem 40:806–813
Dahl D (1981) The vimentin-GFAP protein transition in rat neuroglia cytoskeleton occurs at the time of myelination. J Neurosci Res 6:741–748
Daimaru Y, Hashimoto H, Enjoji M (1985) Malignant peripheral nerve-sheath tumors (malignant schwannomas). An immunohistochemical study of 29 cases. Am J Surg Pathol 9:434–444
DeArmond SJ, Fajardo M, Naughton SA, Eng LF (1983) Degradation of glial fibrillary acidic protein by a calcium dependent proteinase: an electroblot study. Brain Res 262:275–282
Deck JHN, Rubinstein LJ (1981) Glial fibrillary acidic protein in stromal cells of some capillary hemangioblastomas: Significance and possible implications of an immunoperoxidase study. Acta Neuropathol (Berl) 54:173–181
Deck JHN, Eng Lf, Bigbee J, Woodcock SM (1978) The role of glial fibrillary acidic protein in the diagnosis of central nervous system tumors. Acta Neuropathol (Berl) 42:183–190
Dhillon AP, Rode J (1982) Patterns of staining for neuron-specific enolase in benign and malignant melanocytic lesions of the skin. Diagn Histopathol 5:169–174
Dixon RG, Eng LF (1981) Glial fibrillary acidic protein in the retina of the developing albino rat: An immunoperoxidase study of paraffin-embedded tissue. J Comp Neurol 195:305–321
Donato R (1986) S-100 proteins. Cell Calcium 7:123–145
Donoso LA, Folberg R, Arbizo V (1985) Retinal S-antigen and retinoblastoma: A monoclonal antibody histopathologic study. Arch Ophthalmol 103:855–857
Drake PF, Lasek RJ (1984) Regional differences in the neuronal cytoskeleton. J Neurosci 5:1173–1186
Duffy PE, Graf L, Huang Y-Y, Rapport MM (1979) Glial fibrillary acidic protein in ependymomas and other brain tumors. J Neurol Sci 40:133–146
Duffy PE, Huang Y-Y, Rapport MM, Graf L (1980) Glial fibrillary acidic protein in giant cell tumors of brain and other gliomas: A possible relationship to malignancy, differentiation, and pleomorphism of glia. Acta Neuropathol (Berl) 52:51–57
Eng LF (1980) The glial fibrillary acidic (GFA) proteins of the nervous system. In: Bradshaw RA, Schneider DM (eds) Proteins of the Nervous System. 2nd Edition. Raven Press, New York, pp 85–117
Eng LF (1985) Glial fibrillary acidic protein (GFAP): The major protein of glial intermediate filaments in differentiated astrocytes. J Neuroimmunol 8:203–214
Eng LJ, Vanderhaeghen JJ, Bignami A, Gerstl B (1971) An acidic protein isolated from fibrous astrocytes. Brain Res 28:351–354
Fan K (1982) S-100 protein synthesis in cultured glioma cells is G_1-phase of cell cycle-dependent. Brain Res 237:498–503
Friedman HS, Burger PC, Bigner SH, Trojanowski JQ, Halperin EC, Bigner DD (1985) Establishment and characterization of the human medulloblastoma tumor cell line D283 MED. J Neuropathol Exp Neurol 44:592–605
Gambetti P, Autilio-Gambetti L, Papasozomenos SC (1981) Bodian's silver method stains neurofilament polypeptides. Science 213:1521–1522
Gard AL, White FP, Dutton GR (1985) Extra-neural glial fibrillary acidic protein (GFAP) immunoreactivity in perisinusoidal stellate cells of rat liver. J Neuroimmunol 8:359–375
Garson JA, Coakham HB, Kemshead JT, Brownell B, Harper EI, Allan P, Bourne S (1985) The

role of monoclonal antibodies in brain tumor diagnosis and cerebrospinal fluid (CSF) cytology. J Neuro-Oncol 3:165–171

Gerdes J (1985) An immunohistological method for estimating cell growth fractions in rapid histopathological diagnosis during surgery. Int J Cancer 35:169–171

Gerdes J, Schwab U, Lemke H, Stein H (1983) Production of a mouse monoclonal antibody reactive with a human nuclear antigen associated with cell proliferation. Int J Cancer 31:13–20

Gerdes J, Lemke H, Baisch H, Wacker H-H, Schwab U, Stein H (1984) Cell cycle analysis of a cell proliferation-associated human nuclear antigen defined by the monoclonal antibody Ki-67. J Immunol 133:1710–1715

Gheuens J, de Schutter E, Noppe M, Lowenthal A (1984) Identification of several forms of the glial fibrillary acidic protein, or alpha-albumin, by a specific monoclonal antibody. J Neurochem 43:964–970

Ghobrial M, Ross ER (1986) Immunocytochemistry of neuron-specific enolase: A reevaluation. In: Zimmerman HM (eds) Progress in Neuropathology, Vol. 6. Raven Press, New York, pp 199–221

Giangaspero F, Doglioni C, Rivano MT, Pileri S, Gerdes J, Stein H (1987a) Growth fraction in human brain tumors defined by the monoclonal antibody Ki-67. Acta Neuropathol (Berl) 74:179–182

Giangaspero F, Kleihues P, Yasargil MG (1987b) Metastasis of a choroid plexus papilloma 15 years after surgical resection and radiotherapy. J Neurosurg (submitted for publication)

Giordana MT, Mauro A, Migheli A, Schiffer D (1983) Contributions of immunohistochemistry to the problem of differentiation in medulloblastoma. Ital J Neurol Sci 4:411–415

Giordana MT, Germano I, Giaccone G, Mauro A, Migheli A, Schiffer D (1985) The distribution of laminin in human brain tumors: An immunohistochemical study. Acta Neuropathol (Berl) 67:51–57

Goldstein ME, Sternberger LA, Sternberger NH (1983) Microheterogeneity ("neurotypy") of neurofilament proteins. Proc Natl Acad Sci USA 80:3101–3105

Gould VE (1985) The coexpression of distinct classes of intermediate filaments in human neoplasms. Arch Pathol Lab Med 109:984–985

Gould VE, Lee I, Wiedenmann B, Moll R, Chejfec G, Franke WW (1986) Synaptophysin: A novel marker for neurons, certain neuroendocrine cells, and their neoplasms. Hum Pathol 17:979–983

Gould VE, Wiedenmann B, Lee I, Schwechheimer K, Dockhorn-Dworniczak B, Radosevich JA, Moll R, Franke WW (1987) Synaptophysin expression in neuroendocrine neoplasms as determined by immunocytochemistry. Am J Pathol 126:243–257

Grant JW, Gallagher PJ (1986) Pleomorphic xanthoastrocytoma. An immunohistochemical reappraisal. Am J Surg Pathol 10:336–341

Gross N, Beck D, Carrel S, Munoz M (1986) Highly selective recognition of human neuroblastoma cells by mouse monoclonal antibody to a cytoplasmic antigen. Cancer Res 46:2988–2994

Gullotta F, Schindler F, Schmutzler R, Weeks-Seifert A (1985) GFAP in brain tumor diagnosis: Possibilities and limitations. Pathol Res Pract 180:54–60

Haan EA, Boss BP, Cowan WM (1982) Production and characterization of monoclonal antibodies against "brain specific" proteins 14.3.2. and S-100. Proc Natl Acad Sci USA 79:7585–7589

Haglid K, Carlsson C-A, Stavrou D (1973) An immunological study of human brain tumors concerning the brain specific proteins S-100 and 14.3.2. Acta Neuropathol (Berl) 24:187–196

Haglid K, Hamberger A, Hansson H-A, Persson L, Rönnbäch L (1976) Cellular and subcellular distribution of the S-100 protein in rabbit and rat central nervous system. J Neurosci Res 2:175–192

Haimoto H, Takahashi Y, Koshikawa T, Nagura H, Kato K (1985) Immunohistochemical localization of gamma enolase in normal human tissues other than neurons and neuroendocrine tissues. Lab Invest 52(3):257–263

Halliday WC, Yeger H, Duwe GF, Phillips MJ (1985) Intermediate filaments in meningiomas. J Neuropathol Exp Neurol 44:617–623

Hart MN, Earle KM (1973) Primitive neuroectodermal tumors of the brain in children. Cancer 32:890–897

Hartman BK, Agrawal HC, Agrawal D, Kalmbach S (1982) Development and maturation of central nervous system myelin: Comparison of immunohistochemical localization of proteolipid protein and basic protein in myelin and oligodendrocytes. Proc Natl Acad Sci USA 79:4217–4220

Herpers MJHM, Budka H (1984) Glial fibrillary acidic protein (GFAP) in oligodendroglial tumors: Gliofibrillary oligodendroglioma and transitional oligoastrocytoma as subtypes of oligodendroglioma. Acta Neuropathol (Berl) 64:265–272
Herpers MJHM, Budka H (1985) Primitive neuroectodermal tumors including the medulloblastoma: Glial differentiation signaled by immunoreactivity for GFAP is restricted to the pure desmoplastic medulloblastoma ("arachnoidal sarcoma of the cerebellum"). Clin Neuropathol 4:12–18
Herpers MJHM, Budka H, McCormick D (1984) Production of glial fibrillary acidic protein (GFAP) by neoplastic cells: Adaptation to the microenvironment. Acta Neuropathol (Berl) 64:333–338
Hirokawa N (1982) Cross-linker system between neurofilaments, microtubules, and membrane organelles in frog axons revealed by the quick-freeze, deep-etching method. J Cell Biol 94:129–142
Hirokawa N, Glicksman MA, Willard MB (1984) Organization of mammalian neurofilament polypeptides within the neuronal cytoskeleton. J Cell Biol 98:1523–1536
Höfler H, Walter GF, Denk H (1984) Immunohistochemistry of folliculo-stellate cells in normal human adenohypophysis and in pituitary adenomas. Acta Neuropathol (Berl) 65:35–40
Hoshino T, Wilson CB (1979) Cell kinetic analyses of human malignant brain tumors (gliomas). Cancer 44:956–962
Hullin DA, Brown K, Kynoch PAM, Smith C, Thompson RJ (1980) Purification, radioimmunoassay, and distribution of human brain 14.3.2. protein (nervous-system specific enolase) in human tissues. Biochim Biophys Acta 628:98
Hwang TL, Borit A (1982) Rosenthal fibers in glioblastoma multiforme. Acta Neuropathol (Berl) 57:230–232
Ironside JW, Royds JA, Taylor CB, Timperley WR (1985) Paraganglioma of the cauda equina: A histological, ultrastructural and immunocytochemical study of two cases with a review of the literature. J Pathol 145:195–201
Isobe T, Ishioka N, Masupa T, Takahashi Y, Ganno S, Okuyama T (1983) A rapid separation of S-100 subunits by high performance liquid chromatography: The subunit composition of S-100 proteins. Biochem Int 6:419–426
Isobe T, Ichimori K, Nakajima T, Okuyama T (1984) The α subunit of S-100 protein is present in tumor cells of human malignant melanoma, but not in schwannoma. Brain Res 294:381–384
Jacque CM, Kujas M, Poreau A, Raoul M, Collier P, Racadot J, Baumann N (1979) GFA and S-100 protein levels as an index for malignancy in human gliomas and neurinomas. J Natl Cancer Inst 62:479–483
Jahn R, Schiebler W, Ouimet C, Greengard P (1985) A 38,000-dalton membrane protein (p38) present in synaptic vesicles. Proc Natl Acad Sci USA 82:4137–4141
Janzer RC, Friede RL (1981) Do Rosenthal fibers contain glial fibrillary acid protein? Acta Neuropathol (Berl) 55:75–76
Jensen, Marshak DR, Anderson C, Lukas TJ, Watterson DM (1985) Characterization of human brain S-100 fraction: Amino acid sequence of S-100 β. J Neurochem 45:700–705
Jessen KR, Mirsky R (1980) Glial cells in the enteric nervous system contain glial fibrillary acidic protein. Nature 286:736–737
Jessen KR, Mirsky R (1985) Glial fibrillary acidic polypeptides in peripheral glia. Molecular weight, heterogeneity and distribution. J Neuroimmunol 8:377–393
Jessen KR, Thorpe R, Mirsky R (1984) Molecular identity, distribution and heterogeneity of glial fibrillary acidic protein: An immunoblotting and immunohistochemical study of Schwann cells, satellite cells, enteric glia, and astrocytes. J Neurocytol 13:187–200
Kahn HJ, Marks A, Thom H, Baumal R (1982) Role of antibody to S-100 protein in diagnostic pathology. Am J Clin Pathol 79:341–347
Kartenbeck J, Schwechheimer K, Moll R, Franke WW (1984) Attachment of vimentin filaments to desmosomal plaques in human meningiomal cells and arachnoidal tissue. J Cell Biol 98:1072–1081
Kato K, Ishiguro Y, Suzuki F, Ito A, Semba R (1982) Distribution of nervous system-specific forms of enolase in peripheral tissues. Brain Res 237:441
Kepes JJ (1986) The histopathology of meningiomas. A reflection of origins and expected behavior? J Neuropathol Exp Neurol 45:95–107
Kepes JJ, Rubinstein LJ, Eng LF (1979) Pleomorphic xanthoastrocytoma: A distinctive meningocerebral glioma of young subjects with relatively favorable prognosis. A study of 12 cases. Cancer 44:1839–1852

Kepes JJ, Rubinstein LJ, Chiang H (1984) The role of astrocytes in the formation of cartilage in gliomas. An immunohistochemical study of four cases. Am J Pathol 117:471–483

Kim SU, McMorris FA, Sprinkle FJ (1984a) Immunofluorescence demonstration of 2′.3′-cyclic nucleotide 3′-phosphodiesterase in cultured oligodendrocytes of mouse, rat, calf and human. Brain Res 300:195–199

Kim SU, Moretto G, Ruff B, Shin DH (1984b) Culture and cryopreservation of adult human oligodendrocytes and astrocytes. Acta Neuropathol (Berl) 64:172–175

Kimura T, Budka H, Soler-Federspiel S (1986) An immunocytochemical comparison of the glia-associated proteins glial fibrillary acidic protein (GFAP) and S-100 protein in human brain tumors. Clin Neuropathol 5:21–27

Kishiwaka M, Tsuda N, Fujii H, Nishimori I, Yokoyama H, Kihara M (1986) Glioblastoma with sarcomatous component associated with myxoid change. A histochemical, immunohistochemical and electron microscopic study. Acta Neuropathol (Berl) 70:44–52

Kligman D, Marshak DR (1985) Purification and characterization of a neurite extension factor from bovine brain. Proc Natl Acad Sci USA 82:7136–7139

Knaus P, Betz H, Rehm H (1986) Expression of synaptophysin during postnatal development of the mouse brain. J Neurochem 47:1302–1304

Kochi X, Tani E, Morimura T, Itagaki T (1983) Immunohistochemical study of fibronectin in human glioma and meningioma. Acta Neuropathol (Berl) 59:119–126

Korf H-W, Klein DC, Zigler JS, Gery I, Schachenmayr W (1986) S-Antigen-like immunoreactivity in a human pineocytoma. Acta Neuropathol (Berl) 69:165–167

Krajewski S, Schwendemann G, Weizsäcker M, Wechsler W, de Tribolet N (1986) Binding specificity of two monoclonal antiglioma antibodies: Immunocytochemical studies using a new tissue embedding technique. Acta Neuropathol (Berl) 69:124–131

Kruse J, Keilhauer G, Faissner A, Timpl R, Schachner M (1985) The JI glycoprotein – a novel nervous system cell adhesion molecule of the L2/HNK-I family. Nature 316:146–148

Kumanishi T, Washiyama K, Watabe K, Sekiguchi K (1985) Glial fibrillary acidic protein in medulloblastomas. Acta Neuropathol (Berl) 67:1–5

Kumar S, Marsden HB (1986) Glial fibrillary acidic protein (GFAP) in human brain tumors. In: Staal EJ, van Weelen CWM (eds) Markers of human neuroectodermal tumors, chap 3. CRC Press, Boca Raton, Florida, pp 25–51

Kumpulainen T, Korhonen LK (1982) Immunohistochemical localization of carbonic anhydrase isoenzyme C in the central and peripheral nervous system of the mouse. J Histochem Cytochem 30:283–292

Kumpulainen T, Dahl D, Korhonen LK, Nyström SHM (1983) Immunolabeling of carbonic anhydrase isoenzyme C and glial fibrillary acidic protein in paraffin-embedded tissue sections of human brain and retina. J Histochem Cytochem 31:879–886

Kuo W-N, Blake T, Cheema IR, Dominguez J, Nicholson J, Puente K, Shells P, Lowery J (1986) Regulatory effects of S-100 protein and parvalbumin on protein kinases and phosphoprotein phosphatases from brain and skeletal muscle. Mol Cell Biochem 71:19–24

Landolt AM, Shibata T, Kleihues P (1987) Growth rate of human pituitary adenomas. J Neurosurg (to the published)

Lewis SA, Cowan NJ (1985) Temporal expression of mouse glial fibrillary acidic protein mRNA studied by a rapid in situ hybridization procedure. J Neurochem 45:913–919

Liao SK, Clarke BJ, Kwong PC (1981) Common neuroectodermal antigens on human melanoma, neuroblastoma, retinoblastoma, glioblastoma, and fetal brain revealed by hybridoma antibodies raised against melanoma cells. Eur J Immunol 11:450–454

Liberman TA, Bartal AD, Yarden Y, Schlessinger J, Soreq H (1984) Expression of EGF-receptors in human brain tumors. Cancer Res 44:753–760

Liem RKH, Chin SSM, Moraru E, Wang E (1985a) Monoclonal antibodies to epitopes on different regions of the 200,000 dalton neurofilament protein. Probes for the geometry of the filament. Exp Cell Res 156:419–428

Liem RKH, Pachter JS, Napolitano EW, Chin SSM, Moraru E, Heimann R (1985b) Associated proteins as possible cross-linkers in the neuronal cytoskeleton. Ann NY Acad Sci 455:492–509

Liesi P, Kirkwood T, Vaheri A (1986) Fibronectin is expressed by astrocytes cultured from embryonic and early postnatal rat brain. Exp Cell Res 163:175–185

Loeffel SC, Gillespie Gy, Mirmiran SA, Miller EW, Golden P, Askin FB, Siegal GB (1985) Cellular immunolocalization of S-100 protein within fixed tissue sections by monoclonal antibodies. Arch Pathol Lab Med 109:117–122

Ludwin SK, Kosek JC, Eng LF (1976) The topographical distribution of S-100 and GFAP proteins in the adult rat brain: An immunohistochemical study using horseradish peroxidase labelled antibodies. J Comp Neurol 165:197–208

Mannoji H, Takeshita I, Fukui M, Ohta M, Kitamura K (1981) Glial fibrillary acidic protein in medulloblastoma. Acta Neuropathol (Berl) 55:63–69

Marangos PJ, Zomzely-Neurath C, Luk DCM, York C (1975) Isolation and characterization of the nervous system protein 14.3.2. from rat brain: Purification, subunit composition, and comparison to the beef brain protein. J Biol Chem 250:1884–1891

Marangos PJ, Parma AM, Goodwin FK (1978) Functional properties of neuronal and glial isoenzymes of brain enolase. J Neurochem 31:727

Marangos PJ, Schmechel D, Parma AM, Clark RL, Goodwin FK (1979) Measurement of neuronal and non-neuronal enolase of rat, monkey and human tissues. J Neurochem 33:319

Masuzawa T, Sato F (1983) The enzyme histochemistry of the choroid plexus. Brain 106:55–99

McComb RD, Bigner DD (1985) Immunolocalization of laminin in neoplasms of the central and peripheral nervous systems. J Neuropathol Exp Neurol 44:242–253

McComb RD, Burger PC (1983) Choroid plexus carcinoma. Report of a case with immunohistochemical and ultrastructural observations. Cancer 51:470–475

McComb RD, Jones TR, Pizzo SV, Bigner DD (1986) Localization of factor VIII/von Willebrand factor and glial fibrillary acidic protein in the hemangioblastoma: Implications for stromal cell histogenesis. Acta Neuropathol 56:207–213

McKeever PE, Fligiel SEG, Varani J, Hudson JL, Smith D, Castle R, McCoy JP (1986) Products of cells cultured from gliomas. IV. Extracellular matrix proteins of gliomas. Int J Cancer 37:867–874

McLendon RE, Burger PC, Pegram CN, Eng LF, Bigner DD (1986) The immunohistochemical application of three anti-GFAP monoclonal antibodies to formalin-fixed, paraffin-embedded, normal and neoplastic brain tissue. J Neuropathol Exp Neurol 45:692–703

Memoly VA, Brown EF, Gould VE (1984) Glial fibrillary acidic protein (GFAP) immunoreactivity in peripheral nerve sheath tumors. Ultrastruct Pathol 7:269–275

Mepham BL, Frater W, Mitchell BS (1979) The use of proteolytic enzymes to improve immunoglobulin staining by the P.A.P. technique. Histochem J 11:345–357

Michetti F, DeRenzis G, Donato R, Miani N (1976) Brain specific effect of the S-100 protein on the RNA-polymerase activity in isolated nuclei. Brain Res 105:372–375

Miettinen M, Clark R, Virtanen I (1986) Intermediate filament proteins in choroid plexus and ependyma and their tumors. Am J Pathol 123:231–240

Moll R, Cowin P, Kapprell H-P, Franke WW (1986) Biology of disease Desmosomal proteins: New markers for identification and classification of tumors. Lab Invest 54:4–25

Molnar ML, Stefansson K, Molnar GK, Tripathi RC, Marton LS (1985) Species variations in distribution of S-100 in retina. Demonstration with a monoclonal antibody and a polyclonal antiserum. Invest Ophthalmol Vis Sci 26:283–288

Moore BW (1965) A soluble protein characteristic of the nervous system. Biochem Biophys Res Commun 19:739–744

Moore BW (1975) Brain-specific proteins: S-100 protein, 14.3.2. protein, and glial fibrillary protein. Adv Neurochem 1:137–155

Motoi Y, Yoshino T, Hayashi K, Nose S, Horie Y, Ogawa K (1985) Immunohistochemical studies on human brain tumors using anti-Leu 7 monoclonal antibody in paraffin-embedded specimens. Acta Neuropathol (Berl) 66:75–77

Mukai M, Torikata C, Iri H, Morikawa Y, Shimizu K, Shimoda T, Nukina N, Ihara Y, Kageyama K (1986) Expression of neurofilament triplet proteins in human neural tumors. An immunohistochemical study of paraganglioma, ganglioneuroma, ganglioneuroblastoma, and neuroblastoma. Am J Pathol 122:28–35

Nakagawa Y, Perentes E, Rubinstein LJ (1986) Immunohistochemical characterization of oligodendrogliomas: An analysis of multiple markers. Acta Neuropathol (Berl) 72:15–22

Nakajima T, Kameya T, Tsumuraya M, Shimosato Y, Isobe T, Ishioka N, Okuyama T (1983) Immunohistochemical demonstration of neuron-specific enolase in normal and neoplastic tissues. Biomed Res 4(5):495–504

Nakajima T, Kameya T, Tsumuraya M, Shimosato Y, Kato K (1984) Enolase distribution in human brain tumors, retinoblastomas and pituitary adenomas. Brain Res 308:215–222

Nakamura Y, Becker LE (1983) Subependymal giant-cell tumor: Astrocytic or neuronal? Acta Neuropathol (Berl) 60:271–277

Nakamura Y, Becker LE, Marks A (1983) Distribution of immunoreactive S-100 in pediatric brain tumors. J Neuropathol Exp Neurol 42:136–145

Nakazato Y, Ishizeki J, Takahashi K, Yamaguchi H, Kamei T, Mori T (1982) Localization of S-100 protein and glial fibrillary acidic protein-related antigen in pleomorphic adenoma of the salivary gland. Lab Invest 46:621–626

Nesland JM, Holm R, Johannessen JV, Gould VE (1986) Neurone specific enolase immunostaining in the diagnosis of breast carcinomas with neuroendocrine differentiation. Its usefulness and limitations. J Pathol 148:35–43

Odelstal L, Phalman S, Nilsson K, Larsson E, Lackgreu G, Johansson K-E, Hjerten S, Grotte G (1981) Neuron-specific enolase in relation to differentiation in human neuroblastoma. Brain Res 224:69–82

Osborn M, Weber K (1983) Tumor diagnosis by intermediate filament typing: A novel tool for surgical pathology. Lab Invest 48:372–394

Osborn M, Dirk T, Käser H, Weber K, Altmannsberger M (1986) Immunohistochemical localization of neurofilaments and neuron-specific enolase in 29 cases of neuroblastoma. Am J Pathol 122:433–442

Paetau A, Mellström K, Vaheri A, Haltia M (1980) Distribution of a major connective tissue protein, fibronectin, in normal and neoplastic human nervous tissue. Acta Neuropathol (Berl) 51:47–51

Påhlman S, Esscher T, Nilsson K (1986) Expression of gamma-subunit of enolase, neuron-specific enolase, in human non-neuroendocrine tumors and derived cell lines. Lab Invest 54:554–560

Palmer JQ, Kasselberg AG, Netsky MG (1981) Differentiation of medulloblastoma – Studies including immunohistochemical localization of glial fibrillary acidic protein. J Neurosurg 55:161–169

Pasquier B, Lachard A, Pasquier D, Couderc P, Delpech B, Courel MN (1983) Protéine gliofibrillaire acide (GFA) et tumeurs nerveuses centrales. Etude immunohistochimique d'une série de 107 cas. Ann Pathol 3:203–211

Pearce JM, Edwards YH, Harris H (1976) Human enolase isozymes: Electrophoretic and biochemical evidence for three loci. Ann Hum Genet 39:263–276

Perentes E, Rubinstein LJ (1985) Immunohistochemical recognition of human nerve sheath tumors by anti-Leu 7 (HNK-1) monoclonal antibody. Acta Neuropathol (Berl) 68:319–324

Perentes E, Rubinstein LJ (1986) Immunohistochemical recognition of human neuroepithelial tumors by anti-Leu 7 (HNK-1) monoclonal antibody. Acta Neuropathol (Berl) 69:227–233

Perentes E, Rubinstein LJ, Herman MM, Donoso LA (1986) S-Antigen immunoreactivity in human pineal glands and pineal parenchymal tumors. A monoclonal antibody study. Acta Neuropathol (Berl) 71:224–227

Pilkington GJ, Lantos PL (1982) The role of glutamine synthetase in the diagnosis of cerebral tumors. Neuropathol Appl Neurobiol 8:227–236

Pinkus GS, Kurtin PJ (1985) Epithelial membrane antigen – A diagnostic discriminant in surgical pathology: Immunohistochemical profile in epithelial, mesenchymal, and hematopoietic neoplasms using paraffin sections and monoclonal antibodies. Hum Pathol 16:929–940

Pixley SK, DeVellis J (1984) Transition between immature radial glia and mature astrocytes studied with a monoclonal antibody to vimentin. Dev Brain Res 15:201–209

Pruss R, Mirsky R, Raff MC, Thorpe R, Dowding AJ, Anderton BH (1981) All classes of intermediate filaments share a common antigenic determinant defined by a monoclonal antibody. Cell 27:419–428

Raff M, Mirsky R, Fields K, Lisak R, Dortman S, Silberberg DH, Gregson N, Leibowitz S, Kennedy M (1978) Galactocerebroside is a specific cell-surface antigenic marker for oligodendrocytes in culture. Nature 274:813–816

Raff MC, Fields K, Hakomori SI, Mirsky R, Pruss RM, Winter J (1979) Cell-type-specific markers for distinguishing and studying neurons and the major classes of glial cells in culture. Brain Res 174:283–308

Raff MC, Miller RH, Noble M (1983) A glial progenitor cell that develops in vitro into an astrocyte or an oligodendrocyte depending on culture medium. Nature 303:390–396

Ramaekers FCS, Puts JJG, Moesker O, Kant A, Huysmans A, Haag D, Jap PHK, Herman CJ,

Vooijs GP (1983) Antibodies to intermediate filament proteins in the immunohistochemical identification of human tumors: An overview. Histochem J 15:691–713
Rehm H, Wiedenmann B, Betz H (1986) Molecular characterization of synaptophysin, a major calcium-binding protein of the synaptic vesicle membrane. EMBO J 5:535–541
Roessmann U, Velasco ME, Sindley SD, Gambetti P (1980) Glial fibrillary acidic protein (GFAP) in ependymal cells during development. An immunoperoxidase study. Brain Res 200:13–21
Roessmann U, Velasco ME, Gambetti P, Autilio-Gambetti L (1983) Neuronal and astrocytic differentiation in human neuroepithelial neoplasm. An immunohistochemical study. J Neuropathol Exp Neurol 42:113–121
Rorke LB (1983) The cerebral medulloblastoma and its relationship to primitive neuroectodermal tumors. J Neuropathol Exp Neurol 42:1–15
Roussel G, Delaunoy JP, Nussbaum JL, Mandel P (1979) Demonstration of a specific localization of carbonic anhydrase C in the glial cells of rat CNS by an immunohistochemical method. Brain Res 160:47–55
Royds JA, Parsons MA, Taylor CB, Timperley WR (1982) Enolase isoenzyme distribution in the human brain and its tumors. J Pathol 137:37–49
Rubinstein LJ (1975) The cerebellar medulloblastoma. Its origin, differentiation, morphological variants, and biological behavior. In: Vinken PJ, Bruyn GW (eds) Handbook of clinical neurology, vol 18. North-Holland, Amsterdam, pp 167–193
Rubinstein LJ, Brucher JM (1981) Focal ependymal differentiation in choroid plexus papillomas. An immunoperoxidase study. Acta Neuropathol (Berl) 53:29–33
Rueger DC, Huston JS, Dahl D, Bignami A (1979) Formation of 100 Å filaments from purified glial fibrillary acidic protein in vitro. J Mol Biol 135:53–68
Salisbury JR, Isaacson PG (1985) Synovial sarcoma: An immunohistochemical study. J Pathol 147:49–57
Schachner M, Hedley-Whyte ET, Hsu DW, Schoonmaker G, Bignami A (1977) Ultrastructural characterization of glial fibrillary acidic protein in mouse cerebellum by immunoperoxidase labelling. J Cell Biol 75:67–73
Schiffer D, Giordana MT, Mauro A, Migheli A (1984) GFAP, F VIII/RAg, laminin and fibronectin in gliosarcomas: An immunohistological study. Acta Neuropathol (Berl) 63:108–116
Schiffer D, Giordana MT, Mauro A, Migheli A, Germano I, Giaccone G (1986) Immunohistochemical demonstration of vimentin in human cerebral tumors. Acta Neuropathol (Berl) 70:209–219
Schindler E, Gullotta F (1983) Glial fibrillary acidic protein in medulloblastomas and other embryonic CNS tumors of children. Virchows Arch (Pathol Anat) 398:263–275
Schlegel R, Banks-Schlegel S, McLeod JA, Pinkus GS (1980) Immunoperoxidase localization of keratin in human neoplasms. A preliminary survey. Am J Pathol 101:41–50
Schmechel DE (1985) Gamma-subunit of the glycolytic enzyme enolase: Non-specific or neuronspecific. Lab Invest 52:239–242
Schmechel DE, Marangos PJ, Brightman MW (1976) Neuron-specific enolase is a molecular marker for peripheral and central neuroendocrine cells. Nature (Lond) 276:834
Schmechel D, Marangos PJ, Zis AP, Brightman M, Goodwin FK (1978) Brain enolases as specific markers of neuronal and glial cells. Science 199:313
Schmechel DE, Brightman MW, Marangos PJ (1980) Neurons switch from non-neuronal enolase to neuron-specific enolase during differentiation. Brain Res 190:195
Schmitt HP (1983a) Rapid anaplastic transformation in gliomas of adulthood. "Selection" in neurooncogenesis. Pathol Res Pract 176:313–323
Schmitt HP (1983b) Rapid anaplastic transformation of gliomas in childhood. Neuropediatrics 14:137–143
Schnegg JF, Diserens AC, Carell S, Accolla RS, deTribolet N (1981) Human glioma-associated antigens detected by monoclonal antibodies. Cancer Res 41:1209–1213
Schuller-Petrovic S, Gebhart W, Lassmann H, Rumpoldt H, Kraft D (1983) A shared antigenic determinant between natural killer cells and nervous tissue. Nature 306:179–181
Schwechheimer K (1986) Nervale Tumormarker. Verh Dtsch Ges Pathol 70:82–103
Schwechheimer K, Kartenbeck J, Moll R, Franke WW (1984) The vimentin filament-desmosome cytoskeleton of diverse types of human meningiomas: A distinctive diagnostic feature. Lab Invest 51:584–591

Schwechheimer K, Wiedenmann B, Franke WW (1987) Synaptophysin: A reliable marker for medulloblastomas. Virchows Arch 411:53–59

Sensenbrenner M, Devilliers G, Bock E, Porte A (1980) Biochemical and ultrastructural studies of cultured rat astroglial cells. Effect of brain extract and dibutyryl cyclic AMP on glial fibrillary acidic protein and glial filaments. Differentiation 17:51–61

Shaw G, Weber K (1982) Differential expression of neurofilament triplet proteins in brain development. Nature 298:277–279

Shaw G, Osborn M, Weber K (1981) An immunofluorescence microscopical study of the neurofilament triplet proteins, vimentin and glial fibrillary acidic protein within the adult rat brain. Eur J Cell Biol 26:68–82

Shaw GE, Debus E, Weber K (1984) The immunological relatedness of neurofilament proteins of higher vertebrales. Eur J Cell Biol 34:130–138

Shimada H, Aoyama C, Chiba T, Newton WA (1985) Prognostic subgroups for undifferentiated neuroblastoma: Immunohistochemical study with anti-S-100 protein antibody. Hum Pathol 16:471–476

Slowik F, Jellinger K, Gaszo L, Fischer J (1985) Gliosarcomas: Histological, immunohistological, ultrastructural and tissue culture studies. Acta Neuropathol (Berl) 67:201–210

Smith DA, Lantos PL (1985) Immunocytochemistry of cerebellar astrocytomas: With special note on Rosenthal fibers. Acta Neuropathol (Berl) 66:155–159

Spaar FW, Ahyai A, Spaar U, Gazso L, Zimmermann A (1986) Flow-cytophotometry of nuclear DNA in biopsies of 45 human gliomas and after primary culture in vitro. Clin Neuropathol 5:157–175

Stefansson K, Wollmann R (1980) Distribution of glial fibrillary acidic protein in central nervous system lesions of tuberous sclerosis. Acta Neuropathol (Berl) 52:135–140

Stefansson K, Wollmann R (1981) Distribution of the neuronal specific protein, 14-3-2, in central nervous system lesions of tuberous sclerosis. Acta Neuropathol (Berl) 53:113–117

Sternberger LA, Sternberger NH (1983) Monoclonal antibodies distinguish phosphorylated and nonphosphorylated forms of neurofilaments in situ. Proc Natl Acad Sci USA 80:6126–6130

Suzuki F, Umeda Y, Kato K (1980) Rat brain enolase isozymes: Purification of three forms of enolase. J Biochem 87:1587–1594

Takahashi K, Isobe T, Ohtsuki Y, Akagi T, Sonobe H, Okuyama T (1984) Immunohistochemical study on the distribution of α and β subunits of S-100 protein in human neoplasm and normal tissues. Virchows Arch (Cell Pathol) 45:385–396

Tapia FJ, Polak JM, Barbosa AJA, Bloom SR, Marangos PJ, Dermody C, Pearse AGE (1981) Neuron-specific enolase is produced by neuroendocrine tumors. Lancet 808–811

Taratuto AL, Molina H, Monges J (1983) Choroid plexus tumors in infancy and childhood. Focal ependymal differentiation. An immunoperoxidase study. Acta Neuropathol (Berl) 59:304–308

Tascos NA, Parr J, Gonatas NK (1982) Immunocytochemical study of the glial fibrillary acidic protein in human neoplasms of the central nervous system. Hum Pathol 13:454–458

Taylor CB, Royds JA, Timperley WR (1986) Alpha and gamma enolase in the assessment of tumors of neuroectodermal origin. In: Staal EJ, van Veelen CWM (eds) Markers of human neuroectodermal tumors, chapter 9. CRC Press, Boca Raton, Florida, pp 119–154

Theaker JM, Gatter KC, Esiri MM, Fleming KA (1986) Epithelial membrane antigen and cytokeratin expression by meningeomas: An immunohistological study. J Clin Pathol 39:435–439

Trapp BD, Itoyama Y, MacIntosh TD, Quarles RH (1983) P^2 protein in oligodendrocytes and myelin of the rabbit central nervous system. J Neurochem 40:47–54

Traub P (1985) Intermediate filaments. A review. Springer, Berlin Heidelberg New York Tokyo, pp 1–256

Trojanowski JQ, Lee V, Pillsbury N, Lee S (1982) Neuronal origin of human esthesioneuroblastoma demonstrated with anti-neurofilament monoclonal antibodies. N Engl J Med 307:159–161

Trojanowski JQ, Lee VM-J, Schlaepfer WW (1984) An immunohistochemical study of human central and peripheral nervous system tumors, using monoclonal antibodies against neurofilaments and glial filaments. Hum Pathol 15:248–257

van Eldick LJ, Jensen RA, Ehrenfried BA, Whetsell WO Jr (1986) Immunohistochemical localization of S-100 β in human nervous system tumors by using monoclonal antibodies with specificity for the S-100 β polypeptide. J Histochem Cytochem 34:977–982

Vanstapel MJ, Peeters B, Cordell J, Heyns W, DeWolf-Peeters C, Despet V, Mason D (1985) Production of monoclonal antibodies directed against antigenic determinants common to the α- and β-chain of bovine brain S-100 protein. Lab Invest 52:232–238

Velasco ME, Dahl D, Roessmann U, Gambetti P (1980) Immunohistochemical localization of glial fibrillary acidic protein in human glial neoplasms. Cancer 45:484–494

Velasco ME, Roessmann U, Gambetti P (1982) The presence of glial fibrillary acidic protein in the human pituitary gland. J Neuropathol Exp Neurol 41:150–163

Velasco ME, Ghobrial MW, Ross ER (1985) Neuron-specific enolase and neurofilament protein as markers of differentiation in medulloblastoma. Surg Neurol 23:177–182

Vinores SA, Rubinstein LJ (1985) Simultaneous expression of GFAP and NSE by the same reactive or neoplastic astrocytes. Neuropathol Appl Neurobiol 11:349–359

Vinores SA, Bonnin JM, Rubinstein LJ, Marangos PJ (1984) Immunohistochemical demonstration of neuron-specific enolase in neoplasms of the CNS and other tissues. Arch Pathol Lab Med 108:536–540

Wai-Kwan AY, Luna M, Borit A (1985) Vimentin and glial fibrillary acidic protein in human brain tumors. J Neurooncol 3:35–38

Wang E, Cairncross JG, Liem RKH (1984) Identification of glial filament protein and vimentin in the same intermediate filament system in human glioma cells. Proc Natl Acad Sci USA 81:2102–2106

Weber K, Shaw G, Osborn M, Debus E, Geisler N (1983) Neurofilaments, a subclass of intermediate filaments: Structure and expression. Cold Spring Harbor Symp Quant Biol 47:717–729

Weidenheim KM, Campbell WG Jr (1986) Perineurial cell tumor. Immunocytochemical and ultrastructural characterization. Relationship to other peripheral nerve tumors with a review of the literature. Virchows Arch (Pathol Anat) 408:375–383

Weiss SW, Langloss JM, Enzinger FM (1983) The value of S-100 protein in the diagnosis of soft tissue tumors with particular reference to benign and malignant Schwann cell tumors. Lab Invest 49:299–308

Weller RO, Steart PV, Moore IE (1986) Carbonic anhydrase C as a marker antigen in the diagnosis of choroid plexus papillomas and other tumors: An immunoperoxidase study. In: Walker MD, Thomas DGT (eds) Biology of brain tumors, chapter 16. Nijhoff, Boston, pp 115–120

Wiedenmann B, Franke WW (1985) Identification and localization of synaptophysin, an integral membrane glycoprotein of M^r 38,000 characteristic of presynaptic vesicles. Cell 41:1017–1028

Wiedenmann B, Franke WW, Kuhn C, Moll R, Gould VE (1986) Synaptophysin: A marker protein for neuroendocrine cells and neoplasms. Proc Natl Acad Sci USA 83:3500–3504

Wikstrand CJ, Bourdon MA, Pegram CN, Bigner DD (1982) Human fetal brain antigen expression common to tumors of neuroectodermal origin: Gliomas, neuroblastomas, and melanomas. J Neuroimmunol 3:43–62

Wikstrand CJ, Grahmann FC, McComb RD, Bigner DD (1985) Antigenic heterogeneity of human anaplastic gliomas and glioma-derived cell lines defined by monoclonal antibodies. J Neuropathol Exp Neurol 44:229–241

Yamaguchi Y (1980) Studies on immunohistochemical localization of S-100 and GFA proteins in the rat nervous system and in human brain tumors. Brain Nerve 32:1055–1064

Yoshii Y, Maki Y, Tsuboi K, Tomono Y, Nakagawa K, Hoshino T (1986) Estimation of growth fraction with bromodeoxyuridine in human central nervous system tumors. J Neurosurg 65:659–663

Yung WKA, Borit A, Dahl D, Wang E (1984) Keratin and vimentin in meningiomas. J Neuropathol Exp Neurol 43:299

Yung WKA, Luna M, Borit A (1985) Vimentin and glial fibrillary acidic protein in human brain tumors. J Neurooncol 3:35–38

Zheng J, Ivarsson B, Collins VP (1986) Monoclonal antibodies to GFAP epitopes available in formaldehyde fixed tissue. Acta Pathol Microbiol Immunol Scand Sect A 94:353–361

Zuber P, Hamou M-F, de Tribolet N (1987) Identification of proliferating cells in human gliomas using the monoclonal antibody Ki-67. Neurosurgery (to be published)

Zülch KJ (1979) Histologic Classification of Tumors of the Central Nervous System. International Histological Classification of Tumors, No 21. World Health Organization, Geneva

Viral Tumor Markers

TH. LÖNING and K. MILDE

1 Introduction

In diagnostic virology, isolation of viruses by culturing infectious material and serological immunoassays have usually met the clinical needs for:
1. accurate identification of the infectious agent,
2. definition of the state of the disease (acute, convalescence, recovery),
3. information on the infectiousness of patients,
4. monitoring the immune status of patients,
5. estimation of the prognosis (KINGSBURY and FALKOW 1985).

Traditional morphological methods (light microscopical recognition of cytopathic effects, ultrastructural detection of virus particles, SEIFERT et al. 1984) usually fail to fulfill these clinical demands. Very recently, DNA hybridization probes have been applied in order to detect viral genomes in cells, tissues and biological fluids (MCDOUGALL et al. 1985, 1986; HAASE 1986; SCHUSTER et al. 1986; LÖNING et al. 1986a, b). This methodology – initially carried out on filters after extraction and dissection of nucleic acids (HAMES and HIGGINS 1985) – has gained particular importance for in situ recognition of infectious cells. These techniques are even more relevant in tumor virology, since for some oncogenic viruses (e.g. hepatitis B, papillomaviruses) cell culture studies are

Table 1. Virus infection and human oncogenesis

Virus	Tumor	Cofactor	Putative oncogenic events
HTLV 1	Adult T-cell leukemia	?	tat gene activation
EBV	Burkitt lymphoma	Malaria	chromosomal rearrangement
	Nasopharyngeal carcinoma	plant extracts	?
HBV	Hepatocellular carcinoma	aflatoxins, alcohol, malnutrition	?
HPV	Squamous cell carcinomas	UV, tobacco, alcohol, HSV?	E 6-gene activation

not feasible, and for the particular tumor patient under study, antibody titres may not be detectable because of supressed Ig production or virus infections hidden in the tissues in a silent form without release of appropriate amounts of antigens. From the practical diagnostic viewpoint, two categories of viruses and virus-associated tumors may be distinguished:

1. Tumors with a high degree of virus replication (e.g. human T lymphotropic viruses (HTLV) – Epstein-Barr virus (EBV) – associated tumors) and formation of viral antigens, which can be maintained in vitro and which elicit measurable antibodies in vivo.
2. Low(non-)productive tumors (e.g. papillomavirus-/hepatitisB-associated carcinomas), which can not be grown in vitro and which do not provoke high titres of antibodies against viral antigens in vivo (Table 1).

Of course, for a given disease, demonstration of infectious organisms does not necessarily give the answer as to the pathogenicity of these findings. Interpretation needs to be even more cautious in the complicated case of the initiation, maintenance and progress of neoplasia. However, in this chapter, the primary concern is not directed at the current evidence of oncogenicity of viruses, which is rather circumstantial in most instances, but to the diagnostic and prognostic value of the assessment of virus infection.

2 Methods

In the field of viral pathology, conventional light and electron microscopy and in particular immunocytochemistry represent interesting tools for the recognition of productive virus infections (ALMEIDA 1980; CHERNESKY and MAHONY 1984). For tumors, however, major technical advances rely upon the introduction of the molecular virological detection of foreign DNA/RNA by means of hybridization with cloned nucleic acid probes containing the whole genetic information or subgenomic segments (SHKLAR 1985; PFISTER 1986).

At the very beginning of this process, viral probes have to be labeled. Radioactive probes labeled with tritium, sulphur, phosphorus, remain the first choice because of their high sensitivity (EDBERG 1985; KULSKI and NORVALL 1985). Biotin, fluorochromes, haptens are further labels, which are now used with great success, and which certainly better meet the demands of most pathological laboratories, since these probes can be stored and handled without the hazard of radioactivity (MYERSON et al. 1984; EDBERG 1985; MCDOUGALL et al. 1985). Viral probes are now even commercialized in the biotinylated form (Enzo, N.Y., see SEIFERT et al. 1984). Thus, my technical comments are strictly confined to the handling of tissues, the hybridization itself and the detection systems, which – at least in our experience – can be rapidly and reliably conducted in the routine situation of surgical pathology. Under routine conditions, pathologists can probably not afford the more sophisticated methods of DNA (Southern blot) and RNA (Northern blot) detection, but they are able to establish in situ, and *filter* techniques in the simplified form of dot (slot) blot hybridization (KULSKI and NORVALL 1985; HAMES and HIGGINS 1985).

Since these techniques attract the attention of more and more cytologists, our working formulae are described here.

2.1 DNA Extraction

In our experience, dot blot hybridization works better with DNA extracted and purified before hand from surgical samples or cytological scrapings/washings, since the very small channels of the manifold in our vacuum device (Schleicher and Schuell, Dassel, FRG) block up very easily with tissue fragments or cells. For DNA extractions, tissue samples of 0,2–1 g are homogenized with 3 ml TE buffer (10 mM Tris-HCl, pH 8.1, 1 mM EDTA) by an ultraturrax (IKA, Janke + Kunkel, Staufen, FRG). After addition of 1 ml lysis buffer (2% sodium dodecyl sulfate, 70 mM Tris-HCl, pH 8.1, 25 mM EDTA), the mixture is incubated overnight at 37° C with 100 µg/ml proteinase K (Merck, Darmstadt, FRG). Then the nucleic acids are purified by extraction with phenol (2 ×) and chloroform/isoamylalcohol (24:1, 3 ×), and RNA is degraded with 5 µg/ml ribonuclease B (Sigma, St. Louis, Mo, USA). DNA is precipitated with 0.1 vol 3 M Na-acetate and 2.5 vols ethanol for 1 h at −70° C, harvested by centrifugation for 60 min at 15000 rpm (−15° C), lyophilized and redissolved in 1 ml 0.1 mM EDTA (Fig. 1).

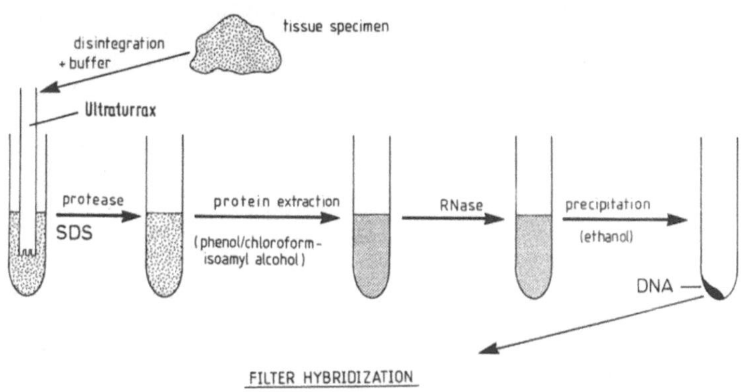

Fig. 1. Tissue processing for DNA extraction and filter hybridization

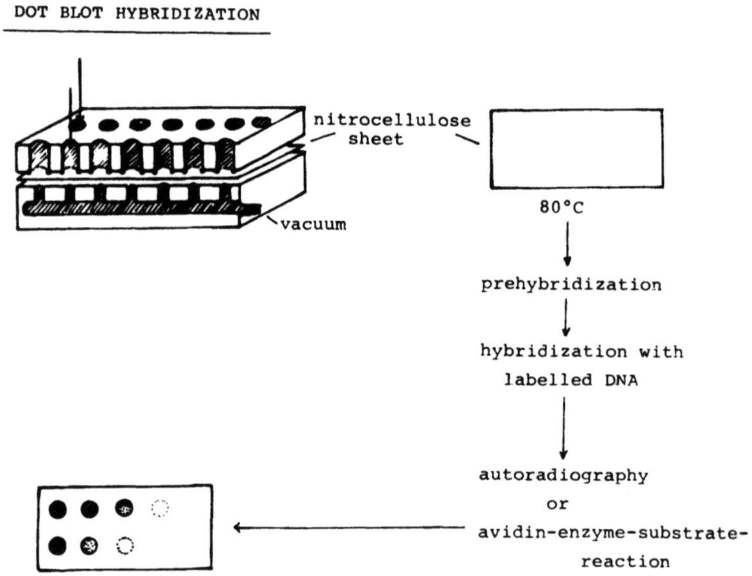

Fig. 2. Spotting and hybridization of cellular DNA

2.2 Dot Blot Hybridization

Aliquots containing 1–20 µg DNA are placed into the wells of a minifold apparatus (Schleicher and Schuell) (Fig. 2) and absorbed to the nitrocellulose membrane (BA 85, Schleicher and Schuell) by suction. The DNA is denatured by placing the membrane for 5 min on a filter paper saturated with 1.5 M NaCl, 0.5 M NaOH, and neutralizing it for 5 min on another filter saturated with 3 M Na-acetate, pH 5.5. Then the air-dried nitrocellulose membrane is baked for 1 h at 80° C.

Prehybridization is carried out for 1–3 h at 42° C in the following solution: 5 × SSC (1 × SSC = 0,15 M NaCl, 0.015 M Na-citrate, pH 7.0), 50 mM Na-phosphate, 50% formamide, 0.1% sodium dodecyl sulfate (SDS), 0.1% bovine serum albumin (BSA), 0.1% ficoll, 0.1% polyvinyl-pyrollidone, 250 µg/ml sheared and denatured herring sperm DNA. The hybridization is conducted overnight in the same solution containing 50–100 ng/ml of the respective cDNA probe labeled with biotin (e.g. deoxy Bio-probe, Enzo, New York, USA), which was denatured beforehand for 10 min at 100° C and chilled on ice. After hybridization, the filter is washed in several changes of 2 × SSC, 0.1% SDS at room temperature, then for 30 min at 55° C in two changes of 0.1 × SSC, 0.1% SDS.

For visualization of the hybridized biotinylated DNA, the Blue Gene-kit (BRL, Eggenstein, FRG) containing a streptavidin-alkaline phosphatase conjugate and the substrates nitroblue tetrazolium (NBT) and 5-bromo-4-chloro-3-indolylphosphate (BCIP) is used.

2.3 In Situ Hybridization

Of course histopathologists usually prefer to look at cells and tissues rather than at spots. In recent years, hybridization protocols were created which allow detection of viral nucleic acids even in paraffin-embedded material (BRIGATI et al. 1983; MYERSON et al. 1984; LÖNING et al. 1986b). Our present protocol is described here briefly for frozen as well as paraffin sections:

Frozen sections are air-dried, fixed in methanol/acetic acid (3:1) for 3 min, immersed in boiling phosphate-buffered saline (PBS, pH 7.4) for 15 s, and cooled in ice water. RNAs are degraded by treatment with 100 µg/ml ribonuclease B from pancreas (Sigma) and 100 units/ml ribonuclease T 1 from Aspergillus oryzae (Sigma) for 1 h at room temperature. After a second fixation in methanol for 3 min, the sections are air-dried.

Paraffin sections are adhered on poly-D-lysin-coated glass slides. Sections are dewaxed in xylene, rehydrated and treated sequentially with 0.02 N HCl, 0.01% Triton X-100 in PBS, 1 mg/ml Pronase E (Calbiochem, Frankfurt, FRG) in 0.05 M Tris-HCl, pH 7.6, 5 mM EDTA, 100 µg/ml ribonuclease B from pancreas (Sigma), and 100 units/ml ribonuclease T 1 (Sigma) in PBS, 100 µg/ml avidin (Sigma) in PBS, and 4% paraformaldehyde in PBS. After washing in PBS, the sections are dehydrated and air-dried.

For *hybridization*, each section is covered with 20 µl of the following hybridization solutions: 2 × SSC, 45% (v/v) formamide, 10% (w/v) dextran sulfate, 0.1 mg/ml herring sperm DNA, and 1.5–2.5 µg/ml biotinylated cDNA (HPV probes were kindly provided by Professor Dr. H. ZUR HAUSEN, Heidelberg, FRG, HIV probes were a generous gift of Dr. H.J. BUHK, Robert Koch Institute, Berlin, Bio-Probes TM for HSV, CMV, EBV and HBV were purchased from Enzo, New York, USA). Section and DNA probes are denatured together in a 90° C water bath for 10 min, then hybridized overnight at 37° C. After hybridization, the sections are washed 2 × 10 min in 1 × SSC, 20% formamide at 36° C (non-stringent conditions) or in 1 × SSC, 45% formamide, 42° C (stringent conditions). Then, a further washing is conducted for 3 × 5 min in 2 × SSC at room temperature.

For the *detection of the biotinylated* DNA, the sections are incubated for 1 h at 37° C with rabbit anti-biotin-antibodies (Enzo) diluted 1:100 in 2 × SSC + 1% BSA. After extensive washing in 2 × SSC, they are incubated with a 1:200 dilution of biotinylated anti-rabbit-IgG (Vector, Burlingame, CA, USA) for 1 h at 37° C, followed by washing and a 1:500 dilution of avidin-alkaline phosphatase (Bresa, Adelaide, Australia) in 100 mM Tris-HCl, pH 7.5, 1 M NaCl, 2 mM $MgCl_2$, 0.05% Triton X-100 for 1 h at 37° C. After further thorough washings in 100 mM Tris-HCl, pH 9.5, 1 M NaCl, 5 mM $MgCl_2$, the sections are incubated with 0.3 mg/ml nitro blue tetrazolium and 0.2 mg/ml 5-bromo-4-chloro-3-indolyl phosphate in 100 mM Tris-HCl, pH 9.5, 100 mM NaCl, 5 mM $MgCl_2$ for 15 min in the dark. In order to inhibit endogeneous alkaline phosphatase activity, 1 µl/ml levamisole (Sigma) is added to the substrate solution. The colour reaction is blocked with 10 mM Tris-HCl, pH 7.5, 1 mM EDTA, and the sections are photographed without and with counterstaining.

3 Non-Hodgkin Lymphoma (NHL) and Its Association with DNA and RNA Viruses: EBV (Epstein-Barr Virus) and HTLV (Human T Cell Lymphotropic Viruses)

3.1 Oncogenic DNA Viruses (EBV)

Although a common etiological denominator of NHL is not known as yet, and arguments are found for all the major concepts of human oncogenesis (spontaneous somatic mutations, physical (radiation), chemical or viral agents), this tumor group has especially attracted the interest of virologists since members of the herpes virus group were discovered to be associated not only with lympho-proliferative diseases in primates, but also in humans (KLEIN 1982). The historical hallmark was the epidemiological observation of high grade lymphomas as endemic in children of Central Africa (BURKITT 1958), and the later recognition that EBV infection is also usually found in those children and their tumors in 80–90% of cases (HENLE and HENLE 1967; KLEIN 1982; ZUR HAUSEN 1985). The early enthusiasm later declined with the finding that EBV is only involved in 20–30% of similar lymphomas of American or European origin (ZUR HAUSEN 1985). It is the present belief that even in African Burkitt lymphomas, EBV is only one actor on the whole stage of B lymphocyte transformation and clonal outgrowth. Recent findings point to three steps of tumor development (KLEIN 1982; KLEIN and KLEIN 1986):

1. Primary EBV infection.

2. Enhanced cell proliferations via exogeneous cofactors (e.g. malaria), which – putatively – suppress or switch off immunological growth control mechanisms.

3. Specific chromosomal events: translocation of the distal arm of chromosome 8 to the chromosomes 14, 2 or 22. 8:14 translocation leads to a close spatial relationship of the c-myc oncogene (located on the chromosome 8 fragment) and the genes for the heavy chain Igs, which are situated at the breakpoint region of chromosome 14. It is tempting to speculate that c-myc activation emerges from this special type of translocation (FUKUHARA and UCHINO 1983; HAMLYN and RABBITTS 1983; MCINTOSH et al. 1983). Again, the impact of EBV infection on chromosomal instability and malignant transformation is far from being understood. In patients with sporadic Burkitt lymphomas, 8:14 translocation was seen without any sign of EBV infection, and even in NHL other than lymphoblastic lymphomas of the Burkitt type (e.g. follicular centrocytic-centroblastic lymphomas) chromosome 14 was seen to be a major acceptor site for genomic material from chromosome 8 or other chromosomes (YUNIS et al. 1982; BLOOMFIELD et al. 1983; GAHRTON et al. 1984).

The biological scenario of diseases of the lymphoid system associated with EBV is outlined in the following issues:

1. In young adults, an overt, self-healing EBV infection of B-lymphocytes is well known and called infectious mononucleosis. This benign lymphoproliferative disease takes the usual course of a virus infection including a prodromal phase of virus entrance (via C3d receptors, WEIS et al. 1986), the acute phase with massive virus production and release from B-lymphocytes eliciting high IgM titres against viral antigens, and the convalescent phase with increasing and, later on, persistent IgG titres for EA (early antigen), VCA (viral capsid antigen), MA (membrane antigen), and EBNA (Epstein Barr nuclear antigen). Extrachromosomal persistence and random integration of viral DNA takes place and leads to a life-long latency state in B lymphocytes of all affected individuals (Fig. 3). In West-Germany, 80% of all adults carry antibodies as a sign of previous EBV infections (ZUR HAUSEN 1981, 1985). After long observation periods of up to 30 years, EBV-infected individuals were reported to carry a slightly increased risk of developing malignant lymphomas (CONNELLY and CHRISTINE 1974; ROSDAHL et al. 1974; MUNOZ et al. 1978).

2. Patients with Burkitt lymphomas follow a rapid and fatal course of tumor progress without treatment. High IgG titres are found for all major EBV antigens. The most reliable signs of virus persistence are the demonstration of nuclear antigens (EBNA) and/or viral DNA in tumor cells. Integrated and extrachromosomal viral DNA can be observed within the tumor (ZUR HAUSEN 1981).

3. Further associations of EBV infections were reported on for primary lymphomas of the central nervous system (ZIEGLER 1982; HOCHBERG et al. 1983) and for lymphomas of North American homosexuals (MAGRATH 1984). Again, persistence of the viral genome, appearance of EBNA in tumor cells, and the

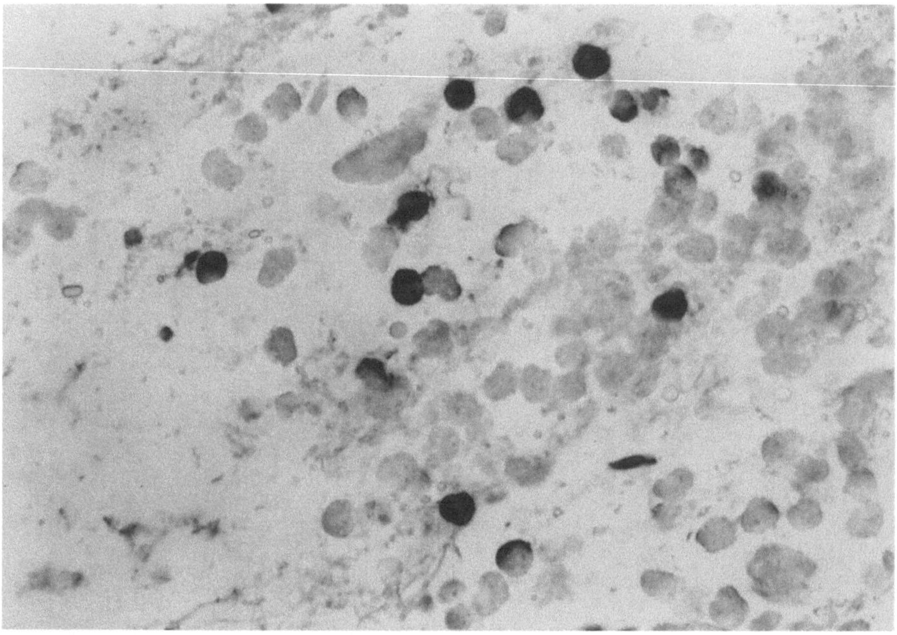

Fig. 3. EBV-DNA harbouring peripheral blood cells (renal transplant recipient). In situ hybridization with a biotinylated EBV probe. × 500

detection of high serum antibody titres for EBV antigens gave rise to the assumption of EBV being one etiological candidate.

3.2 Oncogenic RNA Viruses (HTLV)

Since the discoveries of ROUS (1911) RNA viruses have turned out to be the "classical" source of oncogenic infectious agents, which cause a wide spectrum of sarcomas and numerous neoplasias of the hematopoetic system (including lymphomas), an etiological linkage which even follows the rigorous rules of Koch's postulates – at least for the animal system – (BISHOP 1982; LACEY 1986; GALLO 1986). Those *animal* studies led to major insights into at least some of the principles of virus-associated tumor formation (GALLO 1982; GALLO and WONG-STAAL 1982; WONG-STAAL and GALLO 1985):

1. In acute leucemias, viruses carry their own oncogenes into the host genome. Those oncogenes are the transforming counterparts of normal cellular genes, which – in part – are responsible for the control of normal growth and differentiation (SLAMON et al. 1984; WEINBERG 1985; LACEY 1986). Tumor induction will follow independent of the integration site of the oncogene-harbouring viral DNA.

2. In chronic leucemias, viral promotor sequences (probably located within the long terminal repeats) activate cellular oncogenes. Tumor induction needs

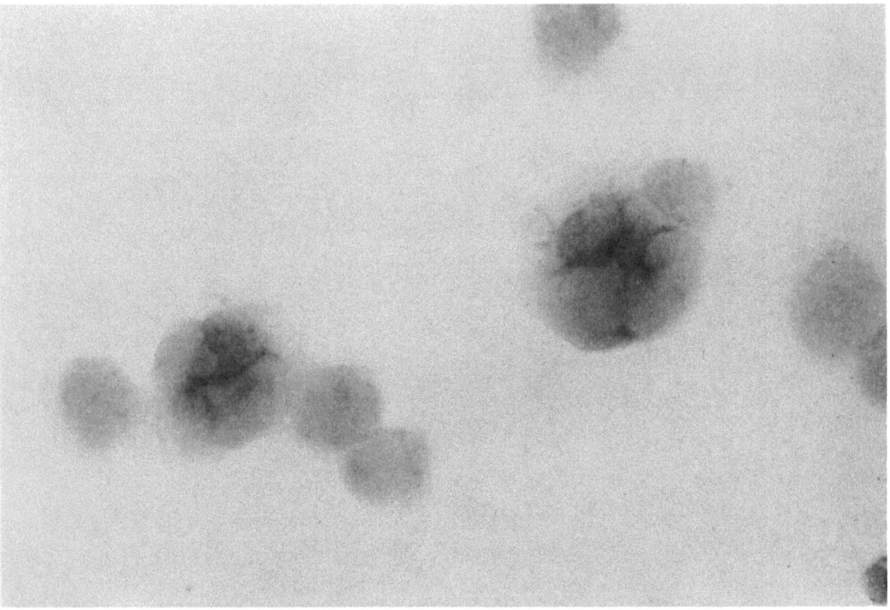

Fig. 4. HIV infected T cell line (H9). In situ hybridization with a biotinylated HTLV III probe reveals positive syncytia. × 500

conserved integration sites in order to bring the viral promotor region next to the cellular oncogene and to influence its function.

3. In some lymphomas, the inserted viral genome codes for regulatory proteins, which are responsible for persistent activation of at least parts of the viral genetic information, and which – in addition – influence the expression of growth conditioning proteins of the host.

In the last five years, the studies of GALLO and associates (see for review: WONG-STAAL and GALLO 1985) have attracted particular attention, because – for the first time – distinct types of retroviruses (HTLV-human T cell lymphotropic viruses) were found to be closely linked to human T cell leukemias/lymphomas. HTLV isolates were obtained from tumor cells of patients with Mycosis fungoides/Sézary syndrome (BUNN et al. 1983), of individuals with adult T cell leukemia (ATL) in endemic regions of South-Western Japan, the Caribbean Islands, and Southern Italy (isolates commonly called HTLV 1, GALLO 1986), of patients with hairy cell leukemia (called HTLV 2, KALYANARAMAN et al. 1982) and from peripheral blood cells of patients with the acquired immune deficiency syndrome (AIDS, isolates commonly termed (HIV/HTLV 3, BARRÉ-SINOUSSI et al. 1983; POPOVIC et al. 1984) (Fig. 4). These HTLV's (HTLV 1,2,3,4) are now defined according to size and structure, nucleic acid homology, the supposed pathogenic principles (e.g. *t*rans*a*cting *t*ranscriptional activity controlled by the viral tat-gene: transactivation of cellular genes for interleukin 2 and its receptor, WONG-STAAL and GALLO 1985), the cellular tropism (helper

T lymphocytes), and action (HTLV 1,2 infections leading to unlimited cell growth, HIV infections causing cell death) (GALLO 1986; GELDERBLOM and PAULI 1986). The final outcome of HIV infected patients with AIDS is not only governed by recurrent opportunistic infections, but also by the development of peculiar sarcomas (Kaposi sarcomas: MOSKOWITZ et al. 1985), and high grade malignant lymphomas (Lancet Editorial 1986). Cytomegalovirus DNA (in Kaposi sarcomas: FENOGLIO et al. 1982; MCDOUGALL et al. 1985; LÖNING and MILDE 1986) and Epstein Barr virus DNA (in malignant lymphomas: Lancet Editorial 1986) have been seen to occur sporadically in those tumors, whereas the presence of HIV has not been unequivocally proved as yet (WONG-STAAL and GALLO 1985) (Fig. 5).

Further studies must also be awaited before the etiological relevance of serological findings of antibodies against HTLV antigens in other NHL including even chronic lymphocytic leukemia can be ascertained (GALLO 1982; BLATTNER et al. 1983).

The same reservations must be attached to the observation of retrovirus-related proteins in a cell line derived from diffuse histiocytic (large cell) lymphoma (GOODENOW et al. 1982). Of course, the discovery of the HTLV's throw some light on older epidemiological studies, which stimulated speculations as to an infectious etiology of NHL. Family clustering has been noted in particular for cases of chronic lymphocytic leukemias, and also for other lymphomas/leukemias (GUNZ et al. 1975; BLATTNER et al. 1976; WINTROBE et al. 1981; BOWEN et al. 1982). KIM and associates (1982) reported increased frequencies of lymphomas in couples. However, it is again EBV which remains a potential etiological candidate, too. Thus, in some cases of lymphomas linked to the X-chromosomal recessive lymphoproliferative syndrome, EBV-related DNA was described to be present (PURTILO et al. 1977; PURTILO 1980).

4 Hepatocellular Carcinoma and Its Association with Hepadnaviruses

The hepadna virus family comprises the human hepatitis B virus (HBV), similar viruses of woodchucks and Beechey ground squirrels in North America, and the duck hepatitis virus of Pekin ducks in China, Japan, and the United States (SUMMERS et al. 1978; MARION et al. 1980; MASON et al. 1980; ROBINSON 1980).

Persistent infections with these viruses are common and may result in chronic hepatitis, cirrhosis, and primary hepatocellular carcinomas (PHC, SZMUNESS 1978). In woodchucks and humans, epidemiologic evidence of the association of HBV infection and PHC is very strong (BEASLEY et al. 1981; POPPER et al. 1981), and for humans, this association is more striking than for any other virus-related cancer.

Immunocytochemical studies directed to tumor-burden liver tissues of sero-positive patients (HBs-Ag^{+}) revealed non-tumorous liver cells to be positive for HBsAg and for HBcAg, but most often failed to detect viral antigens in the tumor itself (KEW 1978). In some studies, scattered HBsAg-positive tumor

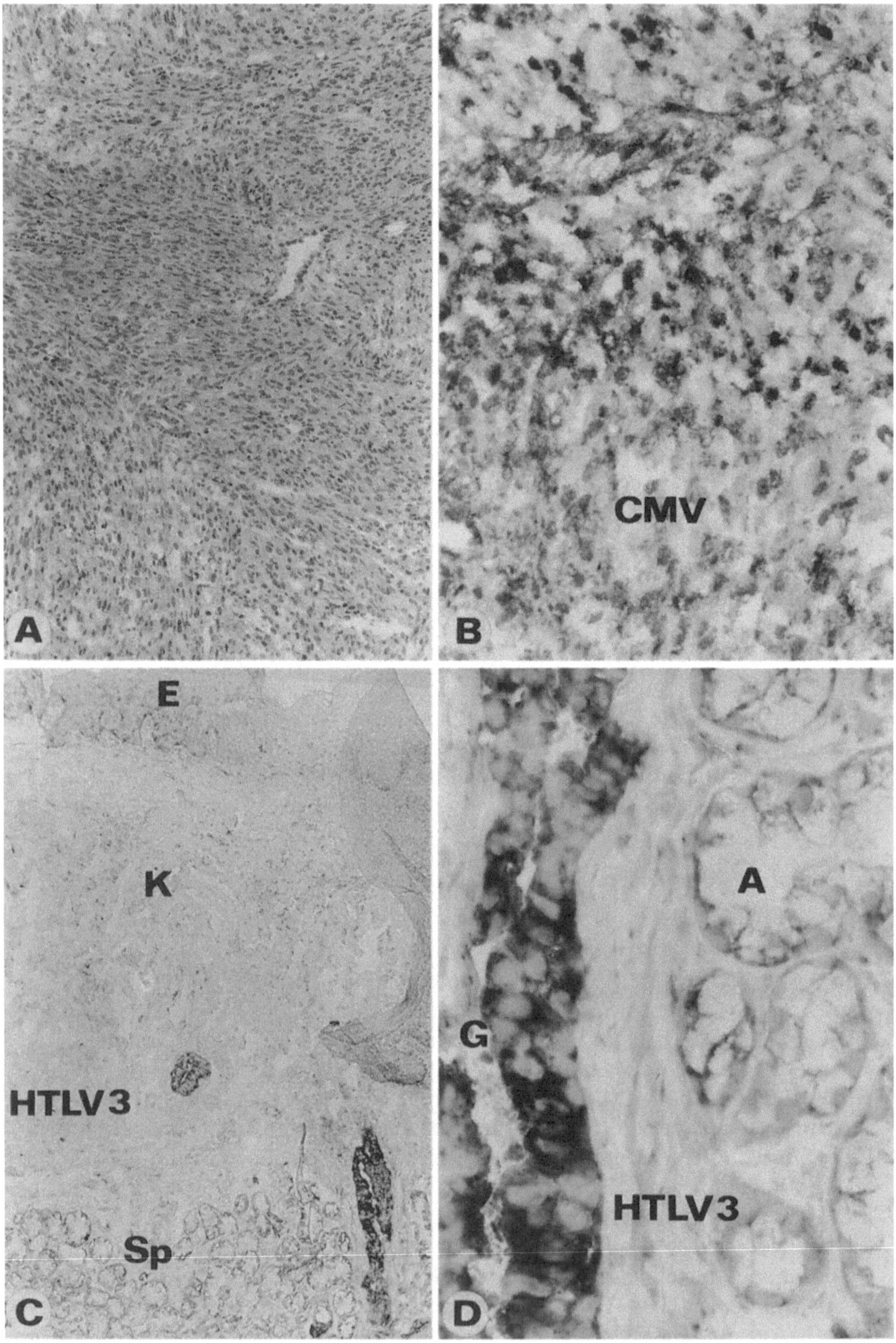
A
B
CMV
E
K
HTLV3
Sp
C
G
A
HTLV3
D

cells were observed, while HBcAg was found only exceptionally (SUZUKI et al. 1985).

These reports should be read with caution, since single positive cells are hard to define as true tumor cells (POPPER 1986). In our own studies, we never unequivocally observed tumor cells or complexes to be positive for HBs/HBc, and even in the non-tumorous liver tissue of seropositive patients, we found viral antigens only to the low extent of 20–30% of over 20 PHC (Figs. 6, 7). In contrast to these disappointing immunological approaches, hybridization methods offered the answer as to the frequency of HBV infections in tumor tissues and tumor cell lines (LUTWICK and ROBINSON 1977; SUMMERS et al. 1978; BRÉCHOT et al. 1980; CHAKRABORTY et al. 1980; EDMAN et al. 1980; TWIST et al. 1981; BRÉCHOT et al. 1982; CHEN et al. 1982; SHAFRITZ 1982; MILLER and ROBINSON 1983). From these experiments, the following findings emerged:

1. The majority of PHC of seropositive individuals contain integrated viral DNA.
2. Tumor cells are clonal with respect to the site of DNA integration in a given tumor.
3. For different PHC's, however, integration sites vary to a large extent. Inserts are generally highly disordered.
4. Multiple insertion sites occur and further obscure the etiological relevance of any of these integrations.
5. As yet, integration has not been regularly linked to oncogenes. Very recently, however, integration was seen next to liver cell DNA homologous to v-erb-A and steroid receptor genes (DEJEAN et al. 1986).
6. The state of integration in PHC must not be different from that in infected non-tumoros tissue.

From the practical diagnostic viewpoint, HBV-DNA appears to be currently the most reliable viral tumor marker in PHC. In our series of PHC examined by dot blot hybridization, we found HBV-DNA in nearly all cases (Fig. 8).

In contrast – at least to our experience – in situ hybridization is very difficult to conduct, which may be due to the insufficient sensitivity of the technique (LÖNING and MILDE 1986). It was of further interest that we found HBV-DNA in some carcinomas other than PHC's, too (e.g. a low differentiated penile carcinoma). Although HBV-DNA has been reported in cells other than hepatocytes (e.g. pancreas, kidney, skin, lymphocytes: BRÉCHOT et al. 1984) our own very preliminary results have to be further confirmed prior to announce a broader spectrum of HBV infections in human tumors.

◁ **Fig. 5. A** H & E section of an oral Kaposi sarcoma shows densely packed mesenchymal cells and slit-like spaces in between. ×67. **B** Same case. Several hybridizing nuclei of mesenchymal cells. In situ hybridization with a biotinylated CMV probe. **C** Oral Kaposi sarcoma (*K*). In situ hybridization with a biotinylated HIV probe. Note the negative epithelium (*E*) and the negative tumor, while some ductal epithelia of underlying minor salivary gland (*Sp*) are positive. ×67. **D** Higher magnification of the salivary gland already shown in **C**. Positive duct (*G*). Negative acini (*A*). In situ hybridization with a biotinylated HIV probe. ×420

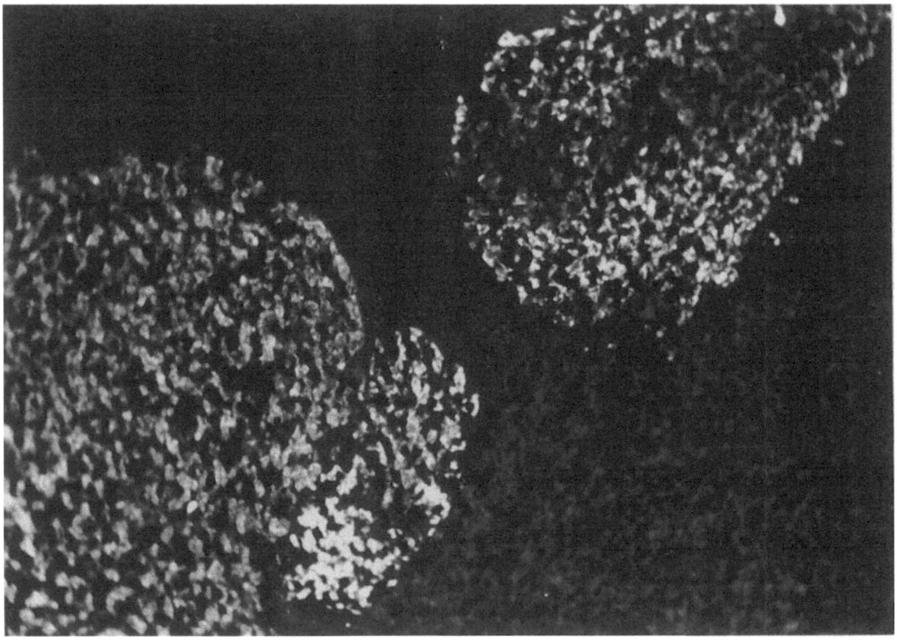

Fig. 6. Non-tumorous liver in the presence of a hepatocellular carcinoma. HBs antigen detected with heterologous antibodies (DAKO). Almost all hepatocytes of cirrhotic nodules are positive. Indirect immunofluorescence. × 160

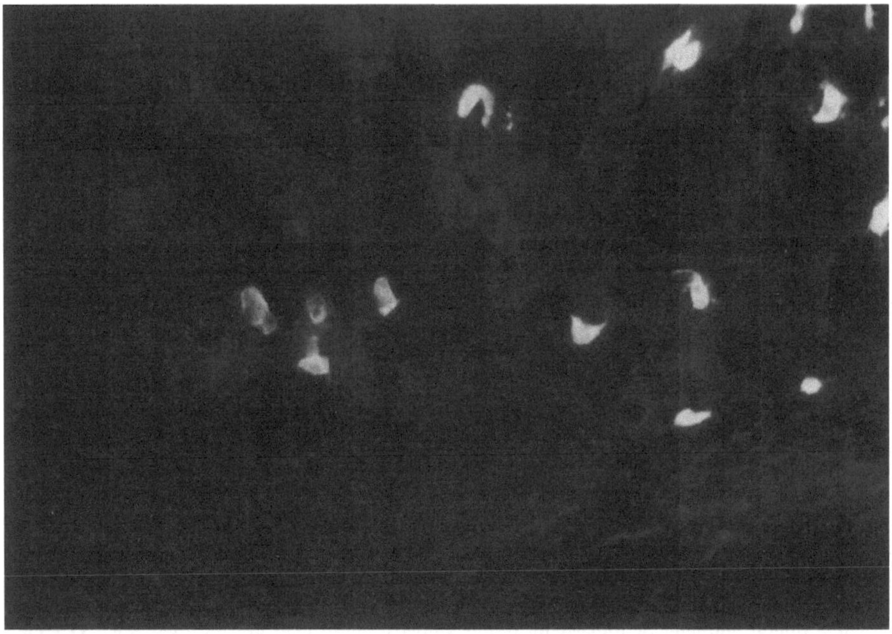

Fig. 7. Hepatocellular carcinoma. Few HBs-positive cells within tumor nodules. Indirect immunofluorescence. × 500

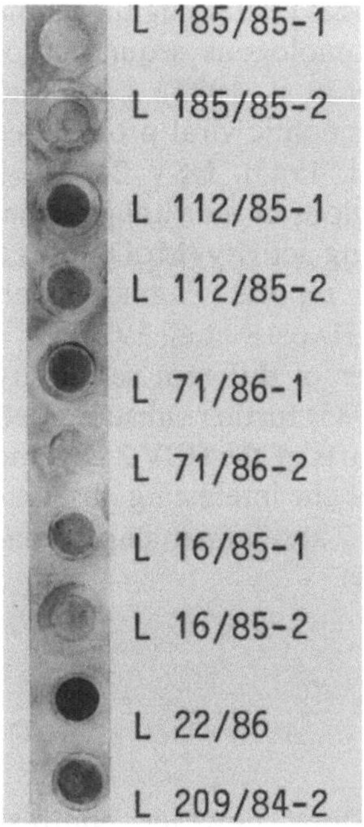

Fig. 8. Dot blot hybridization of six liver carcinomas. Two of the cases (L-209/84-2, L 185/85) represent liver transplants (surgery done because of a hepatocellular carcinoma) anew infiltrated by a cancer, which was discussed to be again hepatocellular carcinoma. This assumption is supported by positive DNA hybridization in tumor tissues in case L 209/84-2, while highly questionable in the negative case L 185/85-1 + 2. The remaining four hepatocellular carcinomas (L 112/85, L 71/85, L 16/85, L 22/86) are also positive for HBV DNA with some variations in different tissue fragments (see the difference of staining intensity in 1/2)

5 Genital Cancer and Its Association with Herpes Simplex Type 2 and with Papillomaviruses

The insights and illusions of science in dealing with the etiology of cancer are best highlighted by the history of infectious organisms accused of being associated with genital cancer. This spectrum ranges from protozoa to viruses (Koss 1979).

5.1 Herpes Simplex Type 2

Herpes simplex type 2 (HSV 2) represents the traditional viral agent assumed to play an oncogenic role, since seroepidemiological studies (Nahmias et al. 1974) revealed elevated titres of serum antibodies against viral antigens in females with cervical cancer. Later on, however, it became clear that this observation was largely overemphasized, taking into account the high prevalence of positive HSV serology in non-selected populations (Corey and Spear 1986). Moreover, although controversially dicussed, viral proteins and viral nucleic acids were claimed to occur in cervical cancer tissues (see for reviews: McDougall et al. 1985; Zur Hausen 1985). Up to 30% of cervical intraepithelial neoplasias (CIN) and cervical carcinomas are said to harbour HSV-specific RNA (McDougall et al. 1980; Maitland et al. 1981; Wilkie et al. 1981).

These studies were first done with hybridization probes containing the whole genomic DNA, and putative cross-reactions with homologous sequences of mammalian DNA were not entirely excluded (Peden et al. 1982). Later on, however, these experiments were confirmed with subgenomic viral probes free of homologous cellular sequences (McDougall et al. 1985). HSV 2-derived RNA species were shown to be read from only three regions of the viral genome, two of them representing DNA regions with transforming activity (McDougall et al. 1985). These observations are in line with earlier reports (Frenkel et al. 1972), but contradict the studies of other groups (Zur Hausen et al. 1974).

The controversy, of course, could well be a matter of different sensitivity of the applied methods. McDougall and associates (1985) further substantiated their results by restriction enzyme analysis and found parts of the HSV 2 genome in cervical as well as in anal carcinomas. They also made the interesting observations that, in some anogenital tumors, genomes of HSV 2 and human papillomaviruses (HPV 6b) may coexist (McDougall et al. 1985).

5.2 Human Papillomaviruses

The, as yet, sporadical observations of HSV 2 infections, however, contrast with the overall high incidence of *human papillomavirus (HPV)* infections in genital cancer. It is the merit of the molecular biological studies of Zur Hausen and Gissmann (Zur Hausen 1986; Gissmann and Schneider 1986) that these viruses are now becoming more and more attractive as causative agents of anogenital tumors. The increasing knowledge on HPV infections of those tumors is partly described in the following list (see for review: Gissmann and Schwarz 1986):

1. HPV-DNA is regularly found in cervical precancer and cancer, and in a high proportion of carcinomas of the lower uro-ano-genital tracts (vaginal, vulvar, penile, urethral, anal).

2. Two types of the heterogeneous HPV group are frequently seen in carcinomas. HPV 16 is present in approximately 50%, and HPV 18 in about 20% of samples. In contrast, about 90% of benign condylomatous lesions harbour HPV 6 DNA.

3. Viral DNA persists in an extrachromosomal state in benign and preinvasive lesions, while becoming integrated in cancer and tumor cell lines. Integration takes place in different regions and different chromosomes of the cellular genome of tumors, although, for a given tumor, tumor cells contain only a single integration site in about 2/3 of cases.

4. In HPV 18 harbouring cell lines, viral DNA is active and RNA's of similar origin are produced (early genes E6, E7, and parts of E1). However, in cancer biopsies, continued expression of the viral genome is found in only some of the specimens. It is assumed, that transcription of viral DNA is mostly required at an early stage of malignant transformation, but not for the maintenance of the neoplastic state. Of course, whether the virus is responsible for tumor induction and/or maintenance, or whether the virus is only one of several

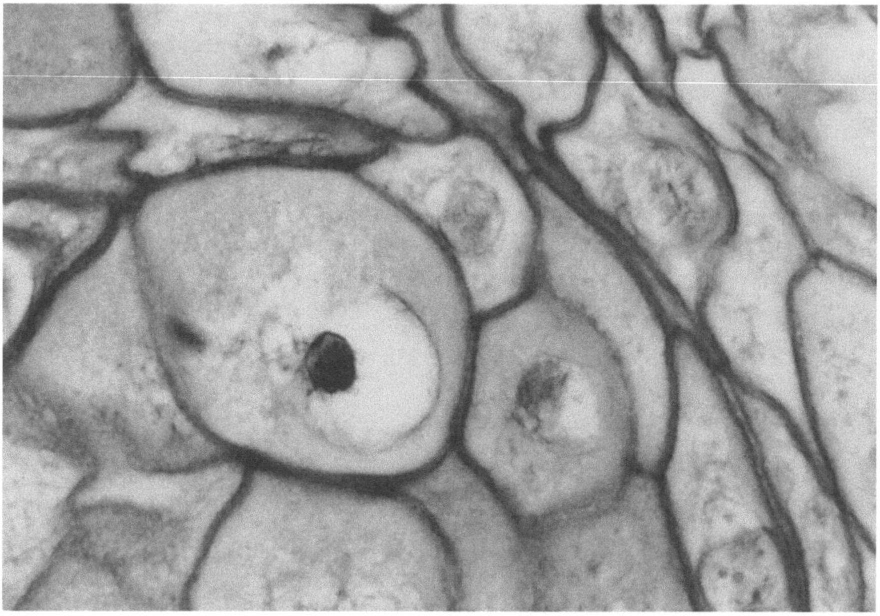

Fig. 9. H & E section showing a typical koilocytic cell with hyperchromatic nucleus and perinuclear halo. ×1250

factors, which simply raise the rate of cell proliferation, makes little difference in a practical diagnostic sense.

5. Among cellular oncogenes, c-myc and c-Ha-ras genes are amplified in a proportion of HPV 16-harbouring cervical cancer. C-myc amplification and activation of transcription is mainly seen in advanced cervical cancer of stage 2–4 (Riou et al. 1985; Orth 1986).

In recent years, increasing attention has been paid to the more practical questions of the prevalence of HPV infections in *non-selected populations*, and of the diagnostic and prognostic implications. Percentages of the subclinical persistence of HPV now pass 20%, when calculated irrespective of HPV types (Lorincz et al. 1985; Pfister 1986). When focussed on HPV 16, the numbers come to 10% (Macnab et al. 1986). For West Germany it has been calculated from this percentage that only about one in 300 women with cervical HPV 16 infection will develop a carcinoma (Pfister 1986).

Although there is a general consensus that HPV DNA occurs in perfectly normal-looking epithelia (McDougall et al. 1986), and, thus, precedes the appearance of the characteristic cytopathic reactions (koilocytosis, dyskeratosis, binucleation, Löning et al. 1986a), cytology and histology remain of prime importance in the diagnosis of the early stages of cervical premalignancy. The histological continuum between CIN and HPV infection is now agreed upon by most pathologists and, particularly from the studies of Reid and associates (1984), it has become clear that no fundamental qualitative differences exist morphologically between subclinical HPV infection and dysplasia (Figs. 9, 10).

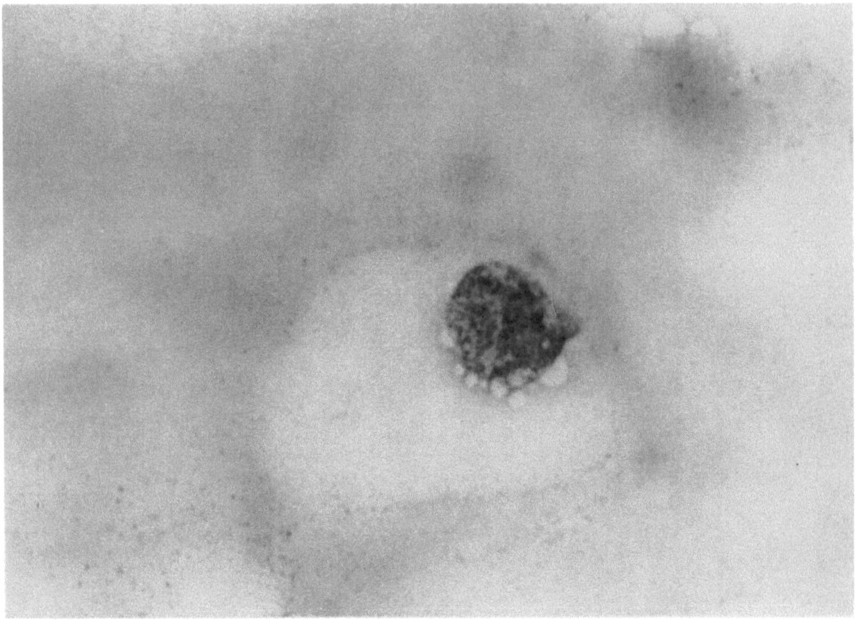

Fig. 10. Koilocytic cell hybridized with HPV 16/18 probe under conditions of high stringency. × 1250

Our own studies included electron microscopy, immunhistochemistry, DNA hybridization (Fig. 11), and DNA cytophotometry, and the following conclusions are drawn from these investigations:

1. Mild cervical dysplasias (CIN 1) show prominent fields of koilocytic cells, contain viral particles and viral antigens in up to 60% of cases, are euploid or sometimes polyploid, and harbour HPV 6/11 or HPV 16/18 DNA.

2. In moderate dysplasias (CIN 2), koilocytic cells, viral particles and viral proteins are found to a minor extent. Lesions show polyploid or even aneuploid patterns, and HPV 16 infections prevail in the latter cases.

3. High grade dysplasias and invasive carcinomas show close morphological relationships, in that the epithelium is highly disordered, and atypia is found throughout the epithelium. This relationship also holds true for the pattern of HPV infection: Koilocytosis is rarely – if ever – seen. Viral particles and viral proteins are usually missed. HPV 16/18 are the predominant virus types. The DNA pattern is almost always aneuploid.

Few prospective studies have seriously scrutinized the biological outcome of HPV 16 infections. The first follow-up study of HPV-associated CIN were recently published by SYRJÄNEN and associates (1986a, b), and showed a striking tendency of HPV 16 lesions to progress (80% of cases), while HPV 6/11 harbouring lesions were seen to regress in up to 45%. Although the reported series of patients (a total of 32 HPV associated cases) is still small and the follow-up period (five years) is probably rather short, investigations of this type will certainly answer the question of whether or not particular HPV infections prime the cervix for neoplastic change.

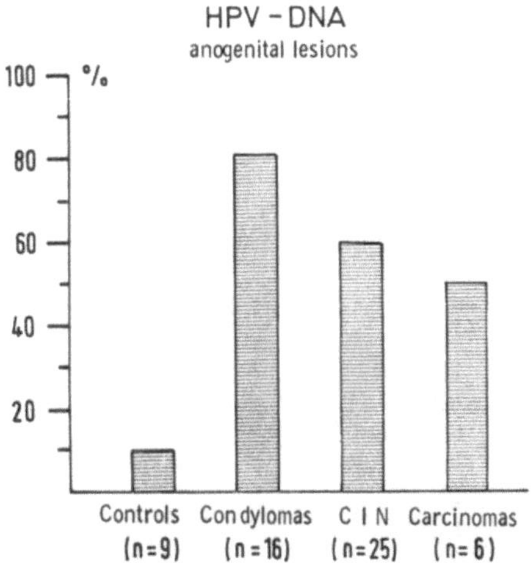

Fig. 11. HPV DNA in anogenital lesions as detected by in situ hybridization (all cases positive for any HPV type, conditions of high and low stringency)

6 Other Carcinomas of the Skin and Mucous Surfaces and Their Association with Viruses (HPV, HSV, EBV)

6.1 Human Papillomaviruses

Among the more than 40 HPV types currently known (Table 2) 11 groups and subgroups are now distinguished on the basis of less than 50% cross-hybridization (COGGIN and ZUR HAUSEN 1979; PFISTER et al. 1986). One of these groups comprise 14 HPV types, which are more or less closely related, and which have all been isolated from patients with *Epidermodysplasia verruciformis* (PFISTER et al. 1986; ORTH 1986). This hereditary generalized verrucosis of the skin represents the first, and still one of the most exciting, papillomavirus-associated human cancer models. It is a particular HPV type (HPV 5) which is most frequently found in primary cancer and metastases of the more than 60 individuals studied as yet (ORTH 1986). Malignant transformation of formerly unsuspicious warts takes place at sunexposed sites in about 1/3 of individuals (ZUR HAUSEN 1985). As yet only casual reports exist to support the association of certain HPV types with other cutaneous preneoplasias and neoplasias (solar keratosis, basal cell carcinoma, keratoacanthoma, even malignant melanoma: PFISTER 1986; ORTH 1986). Moreover, genital types of HPV (HPV 16) were observed in Morbus Bowen in addition to the genital bowenoid papulosis (IKENBERG et al. 1983). Of course, these reports need to be substantiated in molecular biological, pathological and clinical settings. Another source of HPV is being discovered in mucosal tissues of the *upper digestive and respiratory tract* and their benign and malignant tumors (LÖNING et al. 1985; DE VILLIERS et al. 1986; MILDE and LÖNING 1986; COX et al. 1986). Among the benign lesions, oral focal epithelial hyperplasias seem to be peculiar with respect to the supposed genetic background (LÖNING 1984), and the particular HPV type most frequently found (HPV 13: PFISTER et al. 1983; SYRJÄNEN et al. 1984; HENKE et al. 1986).

Table 2. Human papillomavirus types in benign and malignant tumors

Tumor types	Papillomavirus types
Palmoplantar wart	1
Verruca vulgaris	1,2,4,7,26,29
Verruca plana	3,10,27,28
Epidermodysplasia verruciformis (warts)	5,8,9,12,14,15,17,19–25,36,38
Solar keratosis	36
Keratoacanthoma	9,25,37
Morbus Bowen	16
Bowenoid papulosis	16,34,39
Condyloma acuminatum	1,2,6,10,11,16,30
Cervical condylomatous lesion	6,11,16,18,31
Conjunctival papilloma	11
Oral leukoplakia	16
Focal epithelial hyperplasia (Morbus Heck)	13,32
Laryngeal papilloma	6,11
Epidermodysplasia verruciformis (carcinomas)	5,8
Basal cell carcinoma	20
Malignant melanoma	38
Cervical carcinoma	10,11,16,18,31,33,35
Vulvar carcinoma	6,11,16,18
Penile carcinoma	6,11,16,18
Oral carcinoma	2,11,16
Laryngeal carcinoma	30,40
Bronchial carcinoma	16

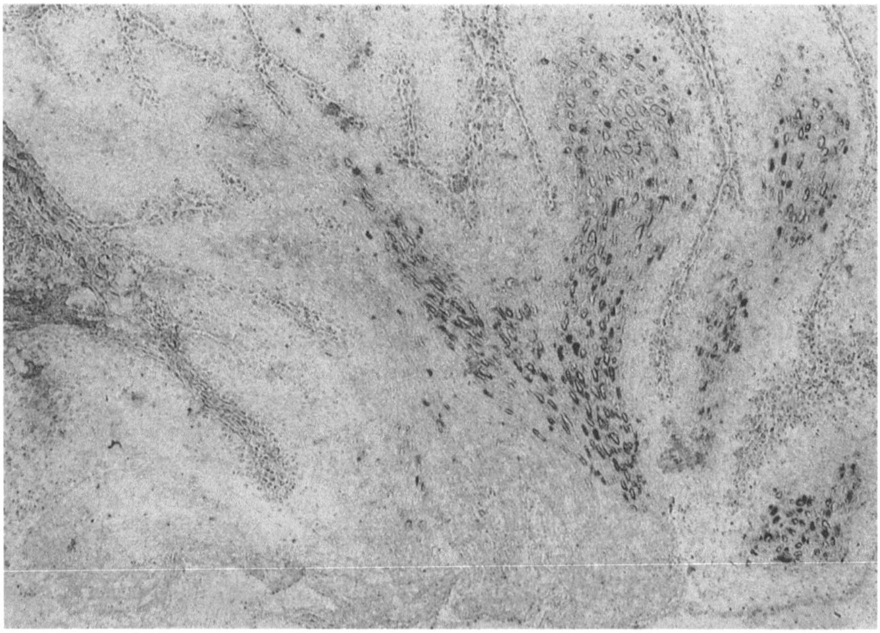

Fig. 12. Oral papilloma. Numerous hybridizing keratinocytes in keratinized parts of the tumor. In situ hybridization with HPV 1,2,6,11,16,18 cocktail under conditions of reduced stringency. × 80

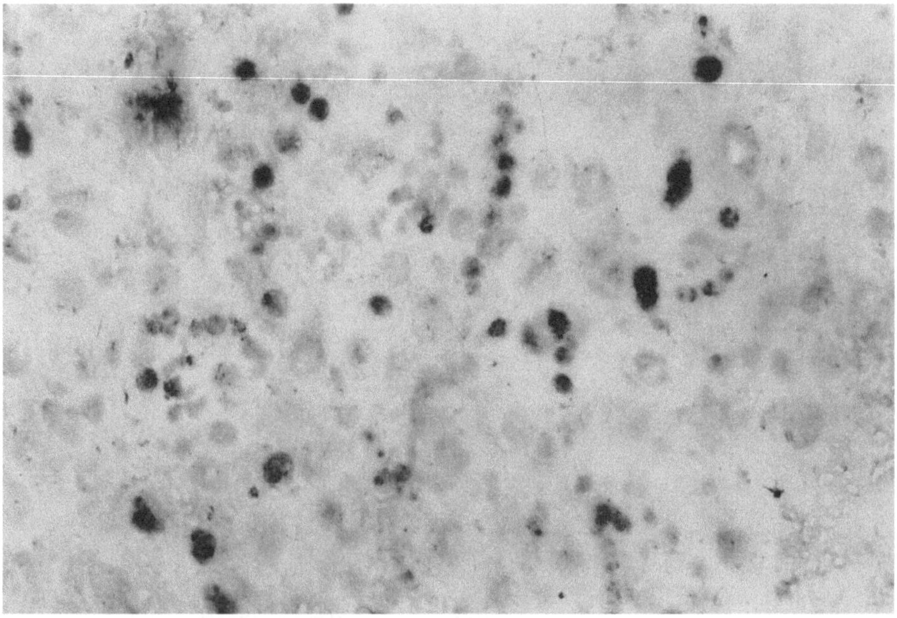

Fig. 13. Oral squamous cell carcinoma. In situ hybridization with HPV 16/18 probe under conditions of high stringency. Scattered hybridizing cells. × 500

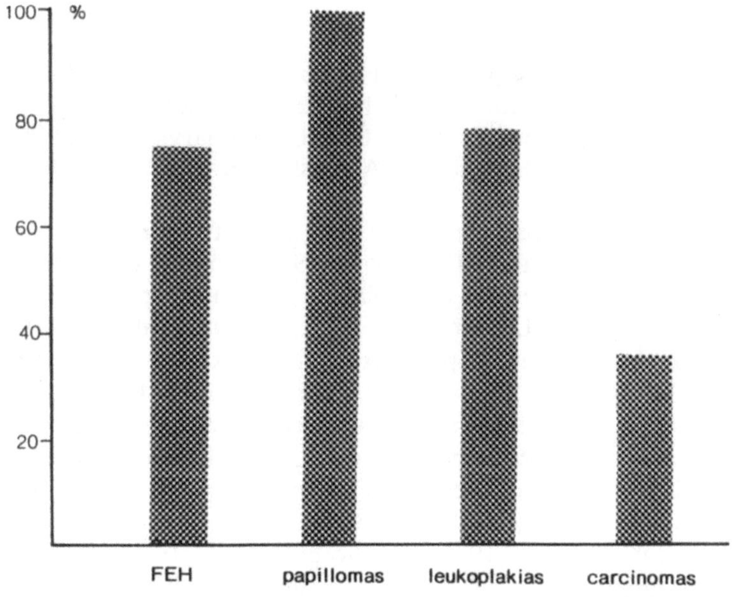

Fig. 14. HPV DNA in oropharyngeal lesions as detected by in situ hybridization (*FEH* = focal epithelial hyperplasia/Morbus Heck). All cases, positive for any HPV type, conditions of high and low stringency

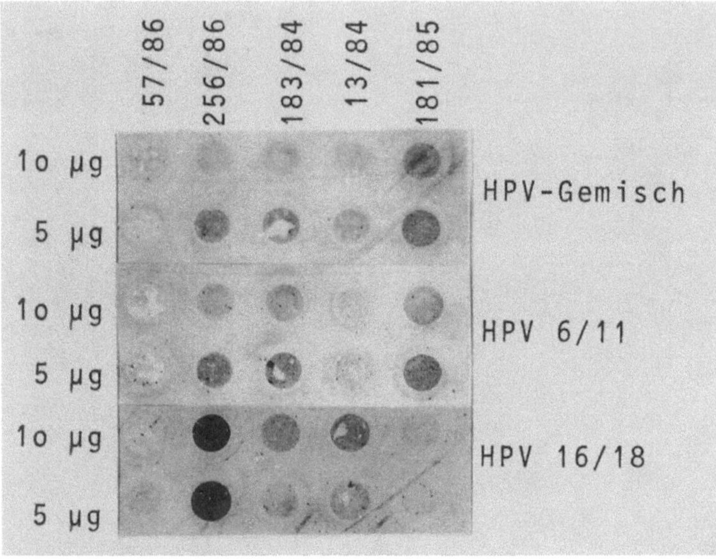

Fig. 15. Dot blot hybridization under conditions of low stringency (HPV-Gemisch = cocktail of HPV 1,2,6,11,16,18-DNA) and high stringency (for HPV 6/11 and HPV 16/18). Cases include two oral squamous cell carcinomas (L 57/86 = negative for any HPV, L 183/84 = positive for both HPV 6/11 and HPV 16/18), one oral leukoplakia (L 13/84 = positive for HPV 16/18), one penile condyloma (L 181/85 = positive for HPV 6/11), and uterine cervix, which was clinically and morphologically unsuspicious (L 256/86 = positive for HPV 6/11 and especially for HPV 16/18). At the left border amount of spotted cellular DNA

Warts and condylomas of the oropharyngeal mucosa are very rare and contain the HPV types of their cutaneous (HPV 1, 2, 3, 4) or genital (HPV 6) counterparts (PFISTER 1986). True papillomas of the mouth, pharynx and larynx were most often seen to harbour HPV 11-DNA (GISSMANN et al. 1982, 1983) (Fig. 12).

In these benign lesions, virus particles, or at least viral structural proteins, were usually observed (JENSEN et al. 1982; LÖNING 1984; LÖNING et al. 1984; LÖNING et al. 1985). Even in some oral leukoplakias without dysplasia, HPV antigens were reported to occur (LÖNING et al. 1984; LÖNING et al. 1985). In our series of cancer of the oral cavity, pharynx, and larynx, we found HPV DNA in 40% of over 20 cases (LÖNING and MILDE 1986) (Figs. 13–15).

We found HPV 6/11 and HPV 16/18 infections to occur and even to coexist in some lesions. In contrast to genital cancer, however, HPV 16 infections were not seen to predominate in malignant lesions. Few mainly casual reports are concerned with the association of HPV infections with esophageal and bronchial carcinomas (HPV 16: STREMLAU et al. 1985; OSTROW et al. 1985). In an extensive analysis of over 200 samples of normal tissues and tumors of different origin and histology, OSTROW and associates (1985) found HPV DNA in a very small percentage of cases (about 2%). Our own data on normal tissues and tumors outside the anogenital and oropharyngeal and laryngeal mucosa are consistent with these findings.

Table 3. EBV-DNA in lymphoepithelial carcinomas

	WOLF and SEIBL (1984)	LÖNING and MILDE (1986)[a]
Epipharynx	100%	60%(10/17 cases)
Tonsil	25%	20%(1/5 cases)

[a] This study was done on formalin-fixed, paraffin-embedded material

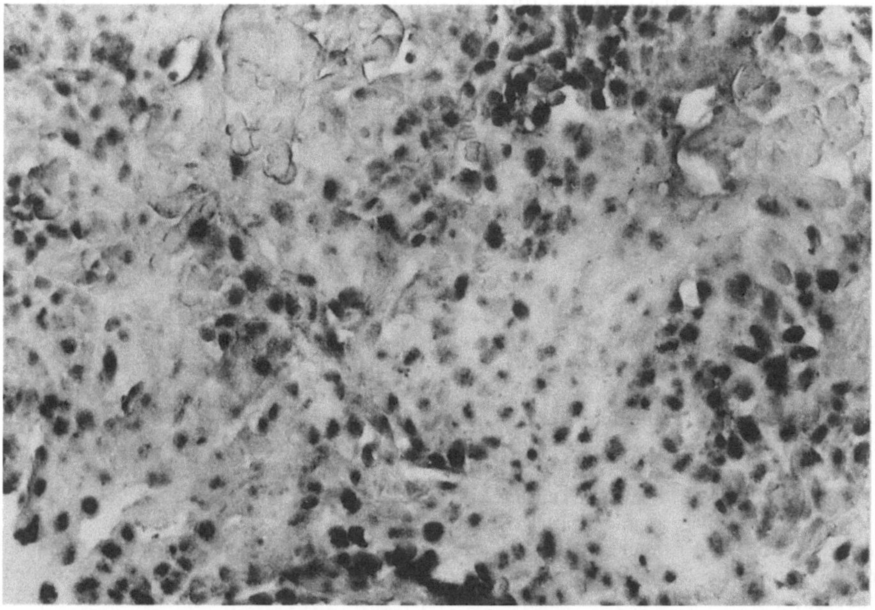

Fig. 16. Epipharynx-carcinoma. In situ hybridization with biotinylated EBV DNA. Hybridizing epithelial and lymphoid cells. × 500

6.2 Herpes Viruses

In addition to HPV, members of the *herpes virus family* must always be kept in mind as etiological (co-)factors for oropharyngeal tumourogenesis. Coexistence has recently been demonstrated for HPV and EBV in AIDS-associated hairy leukoplakia (GREENSPAN et al. 1984, 1985). Coexistence may also occur for HPV and *HSV 1* in oral cancer, since HSV-related RNA was recently claimed to be present (EGLIN et al. 1983). As already discussed for herpes simplex viruses in association with genital cancer, HSV and its relevance for human cancerogenesis (genital and oral cancerogenesis) is very controversially treated in the literature (see for review: SCULLY 1983).

There exists, however, general consensus about another of the herpes viruses *(EBV)* and its very strong association with cancer of the epipharynx (see for review: WOLF and SEIBL 1984). These carcinomas, formerly called lymphoepithelial tumors (KRUEGER and BERTRAM 1983), almost always contain EBV DNA and the EB nuclear antigen (ZUR HAUSEN 1985; LÖNING and MILDE 1986). In countries with high incidence of nasopharyngeal carcinomas (e.g. Southern China, ZENG et al. 1985), hybridization is hoped to become a diagnostic supplement to usual serological surveys, which include monitoring of IgA titres for VCA and EA. Nasopharyngeal carcinomas together with Burkitt lymphomas are the only human tumors which consistently contain EBV DNA. Recently, we were even able to detect EBV DNA in paraffin-embedded specimens (Table 3, Fig. 16). In contrast to carcinomas of the epipharynx, results on undifferentiated carcinomas of tonsils are not as convincing so far (WOLF and SEIBL 1984; LÖNING and MILDE 1986).

Acknowledgements. This study was supported by grants from the *Deutsche Forschungsgemeinschaft* (Lo 285/2-3) and the *Hamburger Stiftung zur Förderung der Krebsbekämpfung* (I 208). The authors thank Ms. I. ORLT and MRS. H. HINZE for invaluable technical assistance.

References

Almeida JD (1980) Practical aspects of diagnostic electron microscopy. Yale J Biol Med 53:5–18

Barré-Sinoussi F, Chermann JC, Rey F, Nugeyre MT, Chamaret S, Gruest J, Daugeut C, Axler-Blin C, Vezinet-Brun F, Rouzioux C, Rosenbaum W, Montagnier L (1983) Isolation of a T-lymphotropic retrovirus from a patient at risk for acquired immune deficiency syndrome (AIDS). Science 220:868–871

Beasley RP, Lin CC, Hwang LY et al. (1981) Hepatocellular carcinoma and hepatitis B virus: a prospective study of 22.707 men in Taiwan. Lancet 2:1129–1133

Bishop JM (1982) Retroviruses and cancer genes. Adv Cancer Res 37:1–32

Blattner WA, Strober AV, Muchmore RM, Blaese S, Broder JF, Fraumeni JF Jr (1976) Familial chronic lymphocytic leukemia. Immunologic and cellular characterization. Ann Intern Med 84:554–557

Blattner WA, Gibbs WN, Saxinger C, Robert-Guroff M, Clark J, Lofters W, Hanchard B, Campbell M, Gallo RC (1983) Human T-cell leukaemia/lymphoma virus-associated lymphoreticular neoplasia in Jamaica. Lancet II:61–64

Bloomfield CD, Arthur DC, Frizzera G, Levine EG, Peterson BA, Gajl-Peczalska KJ (1983) Nonrandom chromosome abnormalities in lymphoma. Cancer Res 43:2975–2984

Bowen LM, Grafton WD, Gelder FB, Williams DM (1982) Malignant lymphoma occurring in a family. Am J Clin Pathol 78:821–824

Bréchot C, Pourcel C, Louise A, Rain B, Tiollais P (1980) Presence of integrated hepatitis B virus DNA sequences in cellular DNA of human hepatocellular carcinoma. Nature 286:533–535

Bréchot C, Pourcel C, Hadchouel M, Dejean A, Louise A, Scotto J, Tiollais P (1982) State of hepatitis B virus DNA in liver diseases. Hepatology 2:27S–34S

Bréchot C, Lugassy C, Dejean A, Pontisso P, Thiers V, Berthelot P, Tiollais P (1984) Hepatitis B virus DNA in infected human tissues. In: Alter HJ (ed) Viral hepatitis and liver disease. Grune & Stratton, pp 395–409

Brigati DJ, Myerson D, Leary JJ, Spalholz B, Travis SZ, Fong CKY, Hsiung GD, Ward DC (1983) Detection of viral genomes in cultured cells and paraffin-embedded tissue sections using biotin-labeled hybridization probes. Virology 126:32–50

Bunn PA Jr, Schechter GP, Jaffe E, Blayney D, Young RC, Matthews MJ, Blattner W, Broder S, Robert-Guroff M, Gallo RC (1983) Clinical course of retrovirus-associated adult T-cell lymphoma in the United States. N Engl J Med 309:257–264

Burkitt D (1958) A sarcoma involving the jaws in African children. Br J Surg 46:218–223

Chakraborty PR, Ruiz-Opazo N, Shouval D, Shafritz DA (1980) Identification of integrated hepatitis

B virus DNA and expression of viral RNA in an HBsAg-producing human hepatocellular carcinoma cell line. Nature 286:531–533
Chen DS, Hoyer BH, Nelson J, Purcell RH, Gerin JL (1982) Detection and properties of hepatitis B virus DNA in liver tissues from patients with hepatocellular carcinomas. Hepatology 2:42S–46S
Chernesky MA, Mahony JB (1984) Detection of viral antigens, particles and early antibodies in diagnosis. Yale J Biol Med 57:757–776
Coggin JR Jr, Zur Hausen H (1979) Workshop on papillomaviruses and cancer. Cancer Res 39:545–546
Connelly RR, Christine BW (1974) A cohort study of cancer following infectious mononucleosis. Cancer Res 34:1172–1178
Corey L, Spear PG (1986) Infections with herpes simplex viruses. N Engl J Med 314:686–691, 749–757
Cox M, Maitland N, Lynas C, Crane J, Prime S, Scully C (1986) HPV Identification in human oral diseases. J Oral Pathol (to be published)
Dejean A, Bougueleret L, Grzeschik KH, Tiollais P (1986) Hepatitis B virus DNA integration in a sequence homologous to v-erb-A and steroid receptor genes in a hepatocellular carcinoma. Nature 322:70–72
De Villiers EM, Weidauer H, Le JY, Neumann C, Zur Hausen H (1986) Papillomviren in benignen und malignen Tumoren des Mundes und des oberen Respirationstrakt. Laryngol Rhinol Otol 65:177–179
Edberg SC (1985) Principles of nucleic acid hybridization and comparison with monoclonal antibody technology for the diagnosis of infectious diseases. Yale J Biol Med 58:425–442
Edman JC, Gray P, Valenzuela P, Rall LB, Rutter WJ (1980) Integration of hepatitis B virus sequences and their expression in a human hepatoma cell. Nature 286:535–538
Eglin RP, Scully C, Lehner T, Ward-Booth P, McGregor IA (1983) Detection of RNA complementary to herpes simplex virus in human oral squamous cell carcinoma. Lancet 2:766–768
Fenoglio CM, Oster MW, Logerfo P, Reynolds T, Edelson R, Patterson JAK, Medeiros E, McDougall JK (1982) Kaposi's sarcoma following chemotherapy for testicular cancer in a homosexual man. Hum Pathol 13:955–959
Frenkel N, Roizmann B, Cassai E, Nahmias A (1972) DNA fragment of herpes simplex 2 and its transcription in human cervical cancer tissue. Proc Natl Acad Sci USA 69:3784–3789
Fukuhara S, Uchino H (1983) Subclasses of 14q+marker-positive lymphoid cancer. N Engl J Med 308:1603–1604
Gahrton G, Juliusson G, Robert KH, Zech L (1984) Chromosomal aberrations in low-grade malignant B-cell lymphoproliferative neoplasias. Second International Conference on Malignant Lymphoma, Lugano, June 13–16
Gallo RC (1982) Regulation of human T-cell proliferation: T-cell growth factor, T-cell leukemias and lymphomas, and isolation of a new c-type retrovirus. In: Rosenberg SA, Kaplan HS (eds) Malignant lymphomas: Etiology, immunology, pathology, treatment. Academic Press, New York, pp 201–218
Gallo RC (1986) Personal reflections on the origin of human leukemia. In: Neth R, Gallo RC (eds) Modern trends in human leukemia VII. Wilsede, June 22–27 (to be published)
Gallo RC, Wong-Staal F (1982) Retroviruses as etiologic agents of some animal and human leukemias and lymphomas and as tools for elucidating the molecular mechanism of leukemogenesis. Blood 60:545–557
Gelderblom H, Pauli G (1986) LAV/HTLV-III: Vergleich mit anderen Retroviren und Einordnung in die Subfamilie der Lentivirinae. AIFO 2:61–72
Gissmann L, Schneider A (1986) Human papillomavirus DNA in preneoplastic and neoplastic genital lesions. In: Peto R, Zur Hausen H (eds) Viral etiology of cervical cancer, Banbury Report, vol 21. Cold Spring Harbor Lab, pp 217–224
Gissmann L, Schwarz E (1986) Persistence and expression of human papillomavirus DNA in genital cancer. In: Papillomaviruses. Ciba Foundation Symposium 120. Wiley, New York, pp 190–207
Gissmann L, Diehl V, Schultz-Coulon H, Zur Hausen H (1982) Molecular cloning and characterization of human papilloma virus DNA derived from a laryngeal papilloma. J Virol 44:393–400
Gissmann L, Wolnik L, Ikenberg H, Koldovsky U, Schnürch HG, Zur Hausen H (1983) Human papillomavirus types 6 and 11 DNA sequences in genital and laryngeal papillomas and in some cervical cancers. Proc Natl Acad Sci USA 80:560–563

Goodenow RS, Liu SL, Fry KE, Levy R, Kaplan HS (1982) Expression of c-type RNA viral proteins by a human lymphoma cell line. In: Rosenberg SA, Kaplan HS (eds) Malignant lymphomas: Etiology, immunology, pathology, treatment. Academic Press, New York, pp 185–200

Greenspan D, Greenspan JS, Conant M, Petersen V, Silverman S Jr, de Souza Y (1984) Oral "hairy" leukoplakia in male homosexuals: evidence of association with both papillomavirus and a herpes-group virus. Lancet 2:831–834

Greenspan JS, Greenspan D, Lennette ET, Abrams DI, Conant MA, Petersen V, Freese UK (1985) Replication of Epstein-Barr Virus within the epithelial cells of oral "hairy" leukoplakia, an AIDS-associated lesion. N Engl J Med 313:1564–1571

Gunz FW, Gunz JP, Veale AMO, Chapmann CJ, Houston IB (1975) Familial leukaemia: A study of 909 families. Scand J Haematol 15:117–131

Haase AT (1986) Analysis of viral infections by in situ hybridisations. J Histochem Cytochem 34:27–32

Hames BD, Higgins SJ (1985) Nucleic acid hybridisation. A practical approach. In: Rickwood D, Hames BD (eds) Practical approach series. IRL Press, Oxford Washington

Hamlyn PH, Rabbitts TH (1983) Translocation joins c-myc and immunoglobulin γ 1 genes in a Burkitt lymphoma revealing a third exon in the c-myc oncogene. Nature 304:135–139

Henke P, Milde K, Löning Th, Strømme-Koppang H (1986) HPV 13 and focal epithelial hyperplasia: DNA hybridisation of paraffin-embedded specimens. Virchows Arch (Pathol Anat) 411:193–198

Henle G, Henle W (1967) Immunofluorescence, interference and complement fixation techniques in the detection of the herpes type virus in Burkitt tumor cell lines. Cancer Res 27:2442–2446

Hochberg FH, Miller G, Schooley RT, Hirsch MS, Feorino P, Henle W (1983) Central nervous system lymphoma related to Epstein-Barr virus. N Engl J Med 309:745–748

Ikenberg H, Gissmann L, Gross G, Grußendorf-Conen EI, Zur Hausen H (1983) Human papillomavirus type 16 related DNA in genital Bowen's disease and in bowenoid papulosis. Int J Cancer 32:563–566

Jenson AB, Lancaster WD, Hartmann DP, Shaffer EL Jr (1982) Frequency and distribution of papillomavirus structural antigens in verrucae, multiple papillomas and condylomata of the oral cavity. Am J Pathol 107:212–218

Kalyanaraman VS, Sarngadharan MG, Robert-Guroff MR, Miyoshi I, Blayney D, Golde D, Gallo RC (1982) A new subtype of human T-cell leukemia Virus (HTLV-II) associated with a T-cell variant of hairy cell leukemia. Science 218:571–573

Kew MD (1978) Hepatoma and HBV. In: Vyas GN, Cohen SN, Schmid R (eds) Viral hepatitis, Franklin Institute Press, Philadelphia, p 439

Kim TH, Michel RP, Melnychuk D, Freeman CR (1982) Non-Hodgkin's lymphoma in couples. Cancer 50:2728–2730

Kingsbury DT, Falkow S (1985) Rapid detection and identification of infectious agents. Academic Press, New York

Klein G (1982) The role of Epstein-Barr virus in the etiology of Burkitt's lymphoma and nasopharyngeal carcinoma. In: Rosenberg SA, Kaplan HS (eds) Malignant lymphomas: Etiology, immunology, pathology, treatment. Academic Press, New York, pp 155–173

Klein G, Klein E (1986) Conditioned tumorigenicity of activated oncogenes. Cancer Res 46:3211–3224

Koss LG (1979) Diagnostic cytology and its histopathologic bases, 3rd edn. Lippincott, Philadelphia

Krueger GRF, Bertram G (1983) Klassifikation und klinisch-pathologische Korrelation des Nasopharynxkarzinoms. Arch Oto-Rhino-Laryngol Suppl I:8

Kulski JK, Norval M (1985) Nucleic acid probes in diagnosis of viral diseases of man. Brief review. Arch Virol 83:3–15

Lacey SW (1986) Review: Oncogenes in retroviruses, malignancy, and normal tissues. Am J Med Sci 291:39–46

Lancet Editorial (1986) Malignant lymphomas in homosexuals. Lancet 1:193–194

Löning Th (1984) Immunpathologie der Mundschleimhaut. Orales Immunsystem-Entzündungsreaktionen-Tumor-"Marker"-Virusnachweis. In: Veröffentlichungen aus der Pathologie, vol 121. Fischer, Stuttgart New York, p 92

Löning Th, Milde K (1986) Virale Tumormarker. Verh Dtsch Ges Pathol 70:104–115

Löning Th, Reichart P, Staquet MJ, Becker J, Thivolet J (1984) Occurrence of papillomavirus structural antigens in oral papillomas and leukoplakias. J Oral Pathol 13:155–165

Löning Th, Ikenberg H, Becker J, Gissmann L, Hoepfner I, Zur Hausen H (1985) Analysis of oral papillomas, leukoplakias and invasive carcinomas for human papillomavirus type related DNA. J Invest Dermatol 84:417–420
Löning Th, Hoepfner I, Milde K (1986a) Zur Typisierung von Schleimhautpapillomen und -carcinomen. Pathologe 7:187–191
Löning Th, Milde K, Foss HD (1986b) In situ hybridization for the detection of cytomegalovirus (CMV) infection. Virchows Arch (Pathol Anat) 409:777–790
Lorincz A, Lancaster WD, Jenson B, Kurmann R, Delgado G, Sanz L, Reid R, Temple G (1985) A study of anogenital HPV infections in the United States and South America. Workshop on Papillomaviruses. Molecular and pathogenetic mechanisms, Kuopio, Finland, August 25–29
Lutwick LI, Robinson WS (1977) DNA synthesis in the hepatitis B Dane particle DNA polymerase reaction. J Virol 21:96–104
Macnab JCM, Walkinshaw SA, Cordiner JW, Clements JB (1986) Human papillomaviruses in clinically and histologically normal tissue of patients with genital cancer. N Engl J Med 315:1052–1058
Magrath IT (1984) Burkitt's lymphoma: Clinical aspects and treatment. In: Molander DW (ed) Diseases of the lymphatic system. Diagnosis and therapy. Springer, Berlin Heidelberg New York, pp 103–139
Maitland NJ, Kinrose JH, Busuttil A, Ludgate SM, Smart GE, Jones KW (1981) The detection of DNA tumor virus-specific RNA sequences in abnormal human cervical biopsies by in situ hybridization. J Gen Virol 55:123–137
Marion PL, Salazar FH, Alexander JF, Robinson WS (1980) State of hepatitis B viral DNA in a human hepatoma cell line. J Virol 33:795–806
Mason WS, Seal G, Summers J (1980) Virus of Pekin ducks with structural and biological relatedness to human hepatitis B virus. J Virol 36:829–836
McDougall JK, Galloway DA, Fenoglio CM (1980) Cervical carcinoma detection of herpes simplex virus RNA in cells undergoing neoplastic changes. Int J Cancer 25:1–8
McDougall JK, Beckmann AM, Galloway DA, Nelson JA, Myerson D (1985) Defined viral probes for the detection of HSV, CMV and HPV. In: Kingsbury DT, Falkow S (eds) Rapid detection and identification of infectious agents. Academic Press, New York, pp 89–107
McDougall JK, Beckmann AM, Kiviat NB (1986) Methods for diagnosing papillomavirus infection. In: Papillomaviruses, Ciba Foundation Symposium 120. Wiley, New York, pp 86–103
McIntosh RV, Cohen BB, Steel CM, Read H, Moxley M, Evans HJ (1983) Evidence for involvement of the immunoglobulin heavy-chain gene locus in the 8:14 translocation of human B lymphomas. Int J Cancer 31:275–279
Milde K, Löning Th (1986) Detection of papilloma virus DNA in oral papillomas and carcinomas: application of in situ hybridization with biotinylated HPV 16 probes. J Oral Pathol 15:292–296
Miller RH, Robinson WS (1983) Integrated hepatitis B virus DNA sequences specifying the major viral core polypeptide are methylated in PLC/PRF/5 cells. Proc Natl Acad Sci USA 80:2534–2538
Moskowitz LB, Hensley GT, Gould EW, Weiss SD (1985) Frequency and anatomic distribution of lymphadenopathic Kaposi's sarcoma in the acquired immunodeficiency syndrome. An autopsy series. Hum Pathol 16:447–456
Munoz N, Davidson RJL, Witthoff B, Ericson JE, De-Thé G (1978) Infectious mononucleosis and Hodgkin's disease. Int J Cancer 22:10–13
Myerson D, Hackman RC, Myerson JD (1984) Diagnosis of cytomegaloviral pneumonia by in situ hybridization. J Infect Dis 150:272–277
Nahmias AJ, Naib ZM, Josey WE (1974) Epidemiological studies relating genital herpetic infection to cervical carcinoma. Cancer Res 34:1111–1117
Orth G (1986) Epidermodysplasia verruciformis: a model for understanding the oncogenicity of human papillomaviruses. In: Papillomaviruses, Ciba Foundation Symposium 120. Wiley, New York, pp 157–174
Ostrow RS, Zachow K, Weber D, Okagaki T, Fukushima M, Clark BA, Twiggs LB, Faras AJ (1985) Presence and possible involvement of HPV DNA in premalignant and malignant tumors. In: Papillomaviruses: Molecular and clinical aspects. UCLA Symp Mol Cell Biol. Alan R Liss, pp 101–122
Peden K, Mounts P, Hayward GS (1982) Homology between mammalian cell DNA sequences and human herpesvirus genomes detected by a hybridization procedure with high-complexity probe. Cell 31:71–80

Pfister H, Hettich I, Runne U, Gissmann L, Chilf GN (1983) Characterization of human papillomavirus type 13 from focal epithelial hyperplasia Heck lesion. J Virol 47:363–366

Pfister H (1986) Papillomviren und Tumorkrankheiten des Menschen. Ber Pathol 103:177–186

Pfister H, Krubke J, Dietrich W, Iftner T, Fuchs PG (1986) Classification of the papillomaviruses – mapping the genome. In: Papillomaviruses, Ciba Foundation Symposium 120. Wiley, New York, pp 3–22

Popovic M, Sarngadharan MG, Read E, Gallo RC (1984) Detection, isolation, and continuous production of cytopathic retroviruses (HTLV III) from patients with AIDS and pre-AIDS. Science 224:497–500

Popper H (1986) The relation between hepatitis B virus infection and hepatocellular carcinoma. Hepato-gastroenterol 33:2–5

Popper H, Shih JWK, Gerin JL, Wong DC, Hoyer BH, London WT, Sly DL, Purcell RH (1981) Woodchuck hepatitis and hepatocellular carcinoma: correlation of histologic with virologic observations. Hepatology 1:91–98

Purtilo DT (1980) Epstein-Barr-virus-induced oncogenesis in immune-deficient individuals. Lancet I:300–303

Purtilo DT, DeFlorio D Jr, Hutt LM, Bhawan J, Yang JPS, Otto R, Edwards W (1977) Variable phenotypic expression of an X-linked recessive lymphoproliferative syndrome. N Engl J Med 297:1077–1081

Reid R, Crum CP, Herschman BR, Fu YS, Braun L, Shah KV, Agronow SJ, Stanhope CR (1984) Genital warts and cervical cancer. III. Subclinical papillomaviral infection and cervical neoplasia are linked by a spectrum of continuous morphologic and biologic change. Cancer 53:943–953

Riou GF, Barrois M, Dutronquay V, Orth G (1985) Presence of papillomavirus DNA sequences, amplification of c-myc and c-Ha-ras oncogenes and enhanced expression of c-myc in carcinomas of the uterine cervix. In: Papillomaviruses: Molecular and clinical aspects, UCLA Symp Mol Cell Biol. Alan R Liss, pp 47–56

Robinson WS (1980) Genetic variation among hepatitis B and related viruses. Ann NY Acad Sci 354:371–378

Rosdahl N, Larsen SO, Clemmesen J (1974) Hodgkin's disease in patients with previous infectious mononucleosis: 30 years' experience. Br Med J II:253–256

Rous P (1911) A sarcoma of the fowl transmissible by an agent separable from the tumor cells. J Exp Med 13:397–409

Schuster V, Matz B, Wiegand H, Traub B, Kampa D, Neumann-Haefelin D (1986) Detection of human cytomegalovirus in urine by DNA-DNA and RNA-DNA hybridization. J Infect Dis 154:309–314

Scully C (1983) Viruses and Cancer: Herpesviruses and tumors in the head and neck. A review. Oral Surg 56:285–292

Seifert G, Löning Th, Hoepfner I (1984) Morphologische Diagnostik bei Virusinfektionen – Pathohistologie, Immunhistologie, Hybridisierungstechnik, Elektronenmikroskopie. Pathologe 5:326–342

Shafritz DA (1982) Hepatitis B virus DNA molecules in the liver of HBsAg carriers: mechanistic considerations in the pathogenesis of hepatocellular carcinoma. Hepatology 2:35S–41S

Shklar J (1985) DNA hybridization in diagnostic pathology. Hum Pathol 16:654–658

Slamon DJ, DeKernion JB, Verma IM, Cline MJ (1984) Expression of cellular oncogenes in human malignancies. Science 224:256–262

Stremlau A, Gissmann L, Ikenberg H, Stark M, Bannasch P, Zur Hausen H (1985) Human papillomavirus type 16 related DNA in an anaplastic carcinoma of the lung. Cancer 55:1737–1740

Summers J, Smolec JM, Snyder R (1978) A virus similar to human hepatitis B virus associated with hepatitis and hepatoma in woodchucks. Proc Natl Acad Sci USA 75:4533–4537

Suzuki K, Uchida T, Horiuchi R, Shikata T (1985) Localization of hepatitis B surface and core antigens in human hepatocellular carcinoma by immunoperoxidase methods. Replication of complete virions of carcinoma cells. Cancer 56:321–327

Syrjänen K, Mäntyjärvi R, Parkkinen S, Väyrynen M, Saarikoski S, Syrjänen S, Castren O (1986a) Prospective follow-up in assessment of the biological behavior of cervical HPV-associated dysplastic lesions. In: Peto R, Zur Hausen H (eds) Viral etiology of cervical cancer. Banbury Rep, vol 21. Cold Spring Harbor Lab, pp 167–178

Syrjänen S, Syrjänen K, Ikenberg H, Gissmann L, Lamberg M (1984) A human papillomavirus closely related to HPV 13 found in a focal epithelial hyperplasia lesion (Heck disease). Arch Dermatol Res 276:199–200

Syrjänen S, Syrjänen K, Mäntyjärvi R, Parkkinen S, Väyrynen M, Saarikoski S, Castren O (1986b) Human papillomavirus (HPV) DNA sequences demonstrated by in situ DNA hybridization in serial paraffin-embedded cervical biopsies. Arch Gynecol 239:39–48

Szmuness W (1978) Hepatocellular carcinoma and the hepatitis B virus: Evidence for a causal association. Prog Med Virol 24:40–69

Twist EM, Clark HF, Aden DP, Knowles BB, Plotkin SA (1981) Integration pattern of hepatitis B virus DNA sequences in human hepatoma cell lines. J Virol 37:239–243

Weinberg RA (1985) The action of oncogenes in the cytoplasm and nucleus. Science 230:770–776

Weis JJ, Fearon DT, Klickstein LB, Wong WW, Richards SA, De Bruyn Kops A, Smith JA, Weis JH (1986) Identification of a partial cDNA clone for the C3d/Epstein-Barr virus receptor of human B lymphocytes: Homology with the receptor for fragments C3b and C4b of the third and fourth components of complement. Proc Natl Acad Sci USA 83:5639–5643

Wilkie NM, Eglin RP, Sanders PG, Clements J (1981) Interactions between virus and host molecules. Academic Press, New York

Wintrobe MM, Lee GR, Boggs DR, Bithell TC, Foerster J, Athens JW, Lukens JN (1981) Clinical hematology, 8th edn. Lea & Febiger, Philadelphia, pp 1681–1725

Wolf H, Seibl R (1984) Benign and malignant disease caused by EBV. J Invest Dermatol 83:88s–95s

Wong-Staal F, Gallo RC (1985) Human T-lymphotropic retroviruses. Nature 317:395–403

Yunis JJ, Oken MM, Kaplan ME, Ensrud KM, Howe RR, Theologides A (1982) Distinctive chromosomal abnormalities in histologic subtypes of non-Hodgkin's lymphoma. N Engl J Med 307:1231–1236

Zeng Y, Zhang LG, Wu YC, Huang NQ, Li JY, Wang YB, Jiang MK, Fang Z, Meng NN (1985) Prospective studies on nasopharyngeal carcinoma in Epstein-Barr virus IgA/VCA antibody-positive persons in Wuzhou City, China. Int J Cancer 36:545

Ziegler J (1982) Burkitt's lymphoma. S Afr Cancer Bull 26:122–141

Zur Hausen H (1981) Virusinfektionen des hämatopoietischen Systems: das Epstein-Barr-Virus und andere bisher wenig charakterisierte Virusinfektionen. Verh Dtsch Ges Pathol 65:146–151

Zur Hausen H (1985) Onkoviren. In: Gross G, Schmidt CG (eds) Klinische Onkologie. Thieme, Stuttgart New York, pp 4.1–4.7

Zur Hausen H (1986) Papillomaviruses in human urogenital cancer-Established results and prospects for the future. In: Peto R, Zur Hausen H (eds) Viral etiology of cervical cancer. Banbury Report, vol 21. Cold Spring Harbor Lab, p 327

Zur Hausen H, Schulte-Holthausen H, Wolf H, Dörries K, Egger HC (1974) Attempts to detect virus-specific DNA in human tumors. II. Nucleic acid hybridizations with complementary RNA of human herpes group viruses. Int J Cancer 13:657–664

Cell, Tissue and Organ Specific Tumor Markers: An Overview

I. DAMJANOV

1 Introduction

Immunohistochemistry, it can be said without any hesitation, represents the most significant methodological and conceptual breakthrough and contribution to diagnostic tumor pathology that this discipline has seen in the entire second half of the twentieth century. Although there are many other new techniques and biomedical innovations that are being applied to pathology (SEIFERT 1986; GOULD 1986) only immunohistochemistry has become readily available to pathologists outside of the elite research laboratories, and has firmly become part of the daily practice of histopathology of the eighties (GATTER et al. 1985).

Like all other innovations, diagnostic immunohistopathology has passed through several phases which roughly correspond to the phase of discovery, formulation of general principles and solution of critical technical problems; gradual acceptance by the research community; transition from research into the routine laboratory; and finally acceptance by the community of pathologists and clinicians at large. Each of these phases was marked, as with any other development in science, by a predictable response: incredulity, cautious exploration by those who thought that the technique had some merit and gradual acceptance by those who became personally convinced that the new approach was indeed useful. Thereafter followed a phase of more or less uncritical overenthusiasm, which is now replaced by the phase of critical evaluation with weighing of positive and negative aspects and *real* contributions. We are, I believe, now entering that final phase and the stage is being set for the ultimate assessment of the contributions of this method to pathology.

Many aspects of immunohistochemistry have been reviewed and highlighted in this volume. I will thus only add a personal overview of the contributions of immunohistochemistry in an attempt to outline how it has influenced our daily practice of pathology, list briefly some of the major problems that have or have not been solved, and also mention at least some of the disappointments that were inevitable due to the high expectations and overenthusiasm of the original proselytizers.

2 General Approaches and Strategies for Identification of Tumor Markers

Search for tumor markers has concentrated on several basic questions which in essence reflect the major concerns of diagnostic pathologists. Most prominent among these questions are:

1. Are there markers that would allow distinction between normal and neoplastic cells?
2. How to recognize and distinguish benign from borderline and clinically malignant tumors?
3. How to distinguish low grade malignant tumors from their more malignant counterparts?
4. Since tumor diagnosis is primarily based on subjective evaluation of histologic sections, would it be possible to improve the diagnostic precision and make it more reliable, reproducible and objective i.e. quantifiable?

2.1 Markers of Early Neoplasia

Malignant transformation is a complex, unpredictable and poorly understood process that occurs in several successive stages (FARBER 1984a, b; RUBIN 1985; BALMAIN 1985). The phenotype of transformed cells is highly variable. Studies of early experimental neoplasia have reported numerous morphological and biochemical changes associated with malignant transformation (FARBER 1984a). However, these markers of early neoplasia vary depending on the organ involved, the carcinogen applied, the stage of neoplasia and many other poorly defined determinants. Thus, it has been shown that the cells in early neoplastic foci of the liver accumulate glycogen, express high levels of glucose-6-phosphate dehydrogenase and differ from adjacent livers with regards to their ability to accumulate exogenous iron (FARBER 1984a, b). On the other hand, papillomas of the skin express an entirely different spectrum of histochemical markers (FOULDS 1969). In most cases experimental data are not directly applicable to human pathology, and there are few cases in which the markers of early experimentally induced animal neoplasia have found application in clinical pathology. Alpha fetoprotein (AFP), discovered originally in rats, is probably the most prominent example of such transfer of knowledge from experimental to diagnostic laboratory (ABELEV 1971).

The basic premise underlying the search for markers of early neoplasia is that the tumor cells differ from normal cells. Theoretically, one could thus hypothesize that cancer cells either lack some of the characteristics of fully differentiated mature cells or that they have acquired some new features. Along these lines it has been shown that some tumor cells, in contrast to their differentiated counterparts, may lack antigens (LIMAS and LANGE 1980), express aberrantly or miss antigens of the major histocompatibility complex (MOMBURG et al. 1986; GUY et al. 1986), or express only the backbone of the oligosaccharide complex that carries the blood group determinants (VOWDEN et al. 1986). These markers of incomplete differentiation or maturation belong in general to the group of carbohydrate antigens (HAKOMORI 1984, 1985). Although there is ample evidence that complex cell surface or secreted glycoconjugates are altered in neoplastic cells, and although there are numerous monoclonal antibodies to these antigens, it is questionable whether any "atypical" or altered glycoconjugates could serve as diagnostic markers of early neoplasia. Hence, antibodies to putative markers of incipient neoplasia still remain to be developed.

Acquisition of new characteristics on evolving cells is even more difficult to document. Most of the "new" cell surface tumor markers described in literature have been shown not to be unique to tumors and are thus of limited diagnostic value (FEIZI 1985). With considerable hope, it was thought that the demonstration of potentially oncogenic viruses such as papilloma virus (KIRCHNER 1986) in the cervical canal or Epstein Barr virus in lymphoid cells (KLEIN and KLEIN 1984; PURTILO 1984) could facilitate the morphologic diagnosis of neoplasia. However, soon after antibodies or molecular probes to these viruses were developed it became apparent that the presence of viruses in tissues does not contribute significantly to the diagnosis of preneoplasia or even incipient and overt neoplasia (PILOTTI et al. 1984; FALSER 1984; SINGER et al. 1985; COLEMAN and RICHMAN 1985). It remains to be seen whether the immunohistochemical demonstrability of well documented oncogen products could provide more reliable evidence of neoplasia (HORAN HAND et al. 1984; THOR et al. 1984). However, recent reports indicate that antibodies to oncogen related proteins react not only with malignant cells but also with benign tumors and normal cells, questioning the validity of this diagnostic approach (GHOSH et al. 1986).

2.2 Benign Versus Malignant Tumors

The criteria for the distinction of benign from corresponding malignant tumors are not always precise. Thus, for example, renal cell adenomas are arbitrarily separated from carcinomas on the basis of their size – tumors measuring less than two centimeters in diameter being labeled as benign in contrast to the larger tumors which could have the same histologic appearance but prove to be clinically malignant. Leiomyosarcoma are most reliably distinguished from their benign counterparts on the basis of mitotic counts rather than on the morphology of tumor cells (KEMPSON and BARI 1970). Indistinct borders between cystadenoma and cystadenocarcinoma of the ovary led to the concept of so-called "borderline-tumors" (for recent studies see e.g. CHAITIN 1985; ERHARDT et al. 1985). In view of these examples, it would be most desirable to have reagents that would distinguish benign from malignant tumors.

Unfortunately, all the efforts in this respect have failed, and most of the studies performed so far have reported only quantitative rather than qualitative differences between malignant and benign tumors. Although such data could still be used in selected cases it is doubtful that quantitative differences detected immunohistochemically could ever be translated into precise diagnostic terms needed for chemotherapeutic or surgical treatment of neoplasia.

Monoclonal antibodies like Ki67 (GERDES et al. 1983), which react with nuclear antigens selectively expressed during mitosis could theoretically be applied to better assess the proliferative tumor cell compartment. Although Ki67 provides more complete and reliable data (BURGER et al. 1986), its application to tumor pathology is only an amplification of the time honored practice of a semiquantitative assessment of the number of mitoses.

Expression of Thomsen-Friedenreich antigen on neoplastic urothelium was found to correlate with the malignancy of the tumors (LIMAS and LANGE 1986). However, the same antigen is expressed on 10% of non-invasive tumors. Accordingly, such histochemical data could contribute to the overall assessment of the tumor but cannot stand on their own without conventional light microscopy.

Reagents that recognize qualitative differences between neoplastic and normal cells have not yet been developed. Judging from the experience of the last fifteen years it appears that such markers will not become readily available in the near future.

2.3 Histochemical Grading of Malignancy

In certain organs, like the urinary bladder, histologic grading of tumors is of paramount clinical importance. Criteria for histologic grading are based on semi-quantitative assessment of several features in tissue sections or cytologic smears (see e.g. HIGHMAN 1986). Thus there is a great need for more objective and less observer-dependent evaluation. Study of the expression of blood group antigens carries considerable promise for objective assessment of malignancy and grading of urinary bladder tumors (LIMAS et al. 1979; LIMAS and LANGE 1980). Cumbersome erythocyte binding techniques have been replaced by immunohistochemical methods based on polyclonal and monoclonal antibodies (JUHL et al. 1986). Nevertheless despite a considerable body of data accumulated so far this approach has not received wide acceptance and is used only in highly specialized laboratories.

Histologic grading of tumors can be supplemented by several immunohistochemical methods. Thus semi-quantitative assessment of laminin, based on the assumption that malignant tumors produce less basement membrane material than corresponding normal tissues or equivalent benign tumors led to suggestions that basement membrane components be used in overall histochemical evaluation of malignancy (Birembaut et al. 1985; d'Ardenne et al. 1986; Faber et al. 1986). More malignant, i.e. less differentiated, tumors produce presumably less "luxury proteins" typical of the mature cells in the organ of their origin. However, like all other methods that evaluate the loss of certain characteristics, rather than the appearance of unique new features this approach has obvious disadvantages and is of limited diagnostic value for routine pathology.

Tumor growth rate assessment is of potential clinical significance, and with time this approach could replace the standard semi-quantitative and subjective grading of tumors. Several methods have been tested including autoradiography with tritiated thymidine on freshly dissected tumors (Camplejohn and Aherne 1974); labeling of tumor cells with bromodeoxyuridine and immunohistochemical detection of mitotic cells with antibodies to this analogue of uridine (Hoshino et al. 1986); cytofluorimetry, (Hall and Fu 1985; Goh and Jass 1986). Recent reports on the use of antibodies selectively reactive with nuclei of mitotic cells promise to make this approach more reproducible and technically less complex (Gerdes 1985). However, it should be remarked that all these techniques are still in experimental phases, often require fresh tumor tissue and are usually not applicable to paraffin embedded material. Thus, there is no indication whether they will ever become routine.

2.4 Monoclonal Antibodies to Tumor Markers

The search for immunohistochemical tumor markers started on a large scale following the discovery by Gold and Friedman (1965) of the so-called carcinoembryonic antigen (CEA). The search received a new impetus following the introduction of monoclonal antibody technique to tumor pathology (Foster 1982; Neville et al. 1982; Damjanov and Knowles 1983). However, more than ten years after the seminal technical discovery of Kohler and Milstein (1975) there are still major problems pertaining to the application of monoclonal antibodies in tumor histochemistry:

a) Some antibodies, specifically reactive with sera of tumor bearing patients, react with many normal tissues. Thus they are of little value to pathologists, although they might be useful in the clinics for serologic detection and follow-up of malignancy (Itzkowitz and Kim 1986; Sack and Kim 1986).

b) Some antibodies, when injected into the tumor patients, localize specifically in tumors (Shen et al. 1984; Allum et al. 1986). However, when used histochemically many of these antibodies do not discriminate between normal and neoplastic cells (Bara et al. 1986). Obviously the antigens expressed in vivo are different on tumor cells than on normal cells and are either inaccessible to the intravenously injected antibody or do not react with it for unknown reasons.

c) Some antibodies whose biochemical specificity was determined by immunoprecipitation or immunoelectrophoresis on Western blots, when applied to tissues may react not only with the biochemically determined epitope but also show cross reactivity with other substances. For example, some antibodies against vimentin react also with other intermediate filaments in tissue sections, although biochemically their reactivity may be limited to vimentin.

d) Antibodies produced against a highly purified molecule may be specific for that molecule in various biochemical tests, but could also be reactive with other macromolecules in which the specific epitope is expressed. Thus it was shown that antibodies to carbohydrate components of the epidermal growth factor receptor crossreact with blood group determinants which obviously had the same carbohydrate sequence (Picard et al. 1985).

Accordingly a few notes of caution seem in order for future work with monoclonal antibodies:

1. Specificity of polyclonal or monoclonal antibodies should be tested both biochemically and histochemically before any antibody is recommended for practical use in immunopathology or serology laboratory.

2. Antibodies useful for one form of immunodiagnosis are not necessarily adequate for others. An antibody highly sensitive in ELISA may not be a good immunohistochemical reagent and vice-versa.

3. There are no short-cuts and foolproof approaches in the search for tumor associated and specific antigens. A targeted effort, supported by multiphasic testing will produce more reliable results than random "fishing expeditions" although there is no good reason for a priori dismissal of the latter approach as well. New avenues of research should be explored. The full potential of human monoclonal antibodies has not yet been explored. For example, antibody "class-switch" leading to change of immunoglobulin isotype (Steplewski et al. 1985) could make many antibodies more useful. Idiotype-antiidiotype techniques are being used to combine advantages of monoclonal antibody specificity with polyclonal antibody reactivity (Herlyn et al. 1986). Other new modalities remain to be tested and fully evaluated.

3 Contribution of Histochemistry to Histogenetic and Phenotypic Classification of Tumors

Since the earliest days of microscopic pathology, histogenesis of tumors has been one of the most attractive enigmas of oncology. Advent of histochemistry opened some new avenues to this problem (Damjanov 1983; Gould 1986). Classical concepts of histogenesis of many tumors were placed under scrutiny, questioned, or reinterpreted (e.g. Mazur and Clark 1983; Miettinen and Virtanen 1984). However, as soon became obvious, only a few equivocal conclusions were reached and the histogenesis of most tumors encountered in daily practice of diagnostic pathology remains obscure. Pragmatic pattern recognition based on phenotypic identification of specific cell types thus still represents the mainstay of histopathology.

3.1 Cell Lineage Studies

According to the postulates of classical embryology, all human tissues could be developmentally traced back to one of three primordial germ layers-ectoderm, mesoderm and endoderm – and cell lineages derived from these germ layers (GARDNER 1985). Although these tenets, initially formulated in the study of ascidian and amphibian embryology are not entirely applicable to mammalian embryos (SVAJGER et al. 1986) they have had a profound influence on tumor pathology. The teaching that carcinomas represent tumors of ectodermal/endodermal origin and sarcomas tumors originating from mesodermal tissues is just one of the salient examples of this concept.

Experimental pathologists and embryologists have extensively used immunohistochemistry for the study of cell lineages. To universal disappointment these studies have shown that there are no immutable markers. Most cells in the developing embryo and fetus do not retain the markers of their predecessors and their phenotype, although influenced by the cell lineage, is in most instances finally shaped by the influences operating within developmental fields (LEHTONEN et al. 1985).

Most embryological cell lineage studies performed with immunohistochemical techniques indicate that markers can be used for tracing cell development over short periods. However, even those anterograde prospective studies cannot be performed in all morphogenetic periods, and it is most unwise to use markers retrospectively on differentiated cells to identify their progenitors. Accordingly, it is also risky to make histogenetic assumptions about the progenitors of tumor cells on the basis of their phenotype, even if light or electron microscopic observations are underpinned with immunohistochemical data.

Like the embryologic studies, cell surface markers of tumor cells proved to be developmentally unstable. Markers considered to be unique to endodermal derivatives, were subsequently found on ectodermal or mesodermal cells and vice versa (e.g. LEHTONEN et al. 1985). As in embryogenesis, intermediate filament polypeptides still appear to be the most stable cell lineage markers, although even these show considerable "switching". Keratin, which is one of the best markers for epithelial cells (OSBORN and WEBER 1983) is expressed in some mesodermal tumors, such as synovial cell sarcoma (MIETTINEN and VIRTANEN 1984; FISHER 1986) or the rhabdoid tumor of the kidney and soft parts (SOTELO-AVILA et al. 1986). Expression of keratin in the normal murine ependymal cells and some human tumors originating in the ventricle (MIETTINEN et al. 1986) cannot be fully explained from present data on the histogenesis of brain and is another example of intermediate filament type switching in development of both normal and neoplastic structures. Likewise, the expression of neurofilament proteins in some endocrine tumors, such as insulinomas of pancreas or parathyroid adenomas (MIETTINEN et al. 1985) originating from neurofilament negative precursors, remains to be fully explained.

3.2 Cell Phenotype Markers

The classification of tumors based on phenotypic recognition of neoplastic cells rather than on uncertain histogenetic principles appears to be the least controversial approach of diagnostic histopathology. Immunohistochemistry has contributed considerably to improvements in the precision with which one can identify various cell types. Numerous cell type markers have been identified. Overall they can be classified as broader *cell class specific* and more restricted *cell type* or *tissue specific* markers.

3.2.1 Cell Class Specific Markers

Immunohistochemistry is the technique of choice for broad classification of cells. Among various antibodies that belong to this category, those recognizing intermediate filament polypeptides are still top of the list of diagnostically useful reagents. Despite many notable exceptions, keratin is still considered to be one of the best markers for epithelial cells. Neurofilaments are good markers for neural cell tumors (SASAKI et al. 1985), although they appear in some neuroectodermal cell tumors (MIETTINEN et al. 1985a, b) and may be barely detectable or not expressed at all in some primitive neural tumors such as neuroblastomas (OSBORN and WEBER 1983). Desmin has a restricted expression and is found almost exclusively in muscle cell tumors (OSBORN et al. 1986a). The only other cell positive for desmin are endometrial stromal cells.

Glial acidic fibrillar protein is a reliable marker of glial cells (OSBORN and WEBER 1983) and is found outside of the central nervous system only in peripheral nerve tumors (GOULD et al. 1986). Vimentin, a marker of connective tissue cells, is not such a good cell class marker. Thus it is expressed as the sole intermediate filament protein in some neuroectodermal tumors such as melanomas (RAEMAKERS et al. 1983) and is coexpressed with keratin in some normal epithelial cells and carcinomas such as those originating from endometrium, thyroid, kidney, lung and possibly other organs (MIETTINEN et al. 1984; GOWN and VOGEL 1985; MCNUTT et al. 1985; DABBS et al. 1986).

Coexpression of vimentin and GFAP has also recorded in gliomas (SCHIFFER et al. 1986; HERPERS et al. 1986). Vimentin expression has been recorded in epithelial cells adapted to in vitro growth and in metastatic lesions. Thus, vimentin cannot be taken with certainty as a consistent marker or exclusive connective tissue cells, which up until now still do not have a class specific unique marker.

Many epithelial cells and tumors originating from them can be identified with antibodies to the so-called milk fat globule or epithelial membrane antigen (EMA) (ORMEROD et al. 1983; PINKUS et al. 1985). Antibodies to EMA originally raised against human milk fat globule membrane in rabbits were reproduced in monoclonal form in mice (HILKENS et al. 1984) and are readily applicable to paraffin embedded formalin fixed pathology specimens. It should, however, be remembered that they are not tumor specific and may occur in many normal tissues as well (HELLE and KROHN 1986). Unfortunately, poorly differentiated carcinomas often lack this antigen (WILSON et al. 1986).

White blood cells express a high molecular weight glycoprotein known as leukocyte common antigen (CLA) (for recent reviews see e.g. TUBBS and SHEIBANI 1984). Leukocyte common antigen may be partially destroyed or lost during paraffin embedding, although it survives postmortem autolysis of at least 72 hours (PALLESEN and KNUDSEN 1985) and could be of considerable importance in autopsy pathology. In conjunction with antibodies to keratin antibodies to CLA may help in the analysis of large cell tumors and the distribution of carcinomas from lymphomas (LAUDER et al. 1984).

Neural and neuroectodermal cells and tumors contain synaptophysin, a glycoprotein of restricted anatomic distribution, which thus could be considered a class marker for neural and closely related neuroectodermal and neural crest derived cells (WIEDENMANN et al. 1986). In the same category are chromogranins, a group of proteins found in the neuroendocrine granules (WILSON and LLOYD 1984; ANGELETTI 1986). In contrast to synaptophysin which does not occur outside of neural and neuroendocrine or neuroectodermal cells, chromogranins occur in some other cells. However, since these cells are in quite uncommon locations (ANGELETTI 1986) and rarely give rise to tumors, antibodies to chromogranin have become a useful probe for identification of neuroendocrine tumors and even neuroendocrine components in other tumors (INOUE et al. 1986).

3.2.2 Specialized Cell and Tissue Specific Markers

Search for specialized cell specific markers has been constantly marred by major disappointments. Most of the antigens originally considered to be specific for a specialized cell or tissue turned out in final analysis to be less restricted in their distribution than originally thought. Chorionic gonadotropin, originally thought to be trophoblast specific was found not only in germ cell tumors but in many tumor cells (BRAUNSTEIN et al. 1986; FUKAYAMA et al. 1986; MEREDITH et al. 1986). S-100, initially considered neural tissue specific, was detected in numerous normal and neoplastic cells (VANSTAPEL et al. 1986). Antibodies to the so-called neuron-specific enolase react with numerous non-neural tissues and tumors (HAIMOTO et al. 1985). Prostatic acid phosphatase may be found in carcinoid tumors (SOBIN et al. 1986). Anti-Leu 7 monoclonal antibody, a reliable marker of human natural killer cells (TRINCHIERI and PERUSSIA 1984), was found subsequently to react with normal human nervous system, neurogenic tissue, peripheral nerve sheath tumors, and neuroepithelial tumors (PERENTES and RUBINSTEIN 1986). Even alpha feto-protein, one of the earliest tissue specific markers, originally thought to be an exclusive marker of yolk sac tumors and hepatocellular carcinomas, was later reported in other tumors such as e.g. gastric carcinomas (ISHIKURA et al. 1986) and pancreatoblastoma (ISEKI et al. 1986).

Polypeptide hormones, normally secreted by specific cells in distinct anatomic locations, have for a long time been considered as markers of specific cell types. While this assumption holds in general true for most normal tissues, many hormones or hormone like substances are ectopically synthesized by aberrant tumor cells. More extensive immunohistochemical studies could on the other hand disclose that some of the previously held assumptions on the restricted distribution of cells synthesizing a certain hormone in normal tissues may be incorrect. As the best example one could take calcitonin. This calcium regulating hormone, originally extracted from the thyroid was subsequently described in bronchial and bronchiolar mucosa, in the thymus, parathyroids and even the prostate and the epithelium of the anal canal (FETISSOF et al. 1986a, b). Nevertheless, antibodies to hormones, if properly used, are still excellent markers for endocrine tissues, and some like thyroglobulin (WILSON et al. 1986) are highly specific for certain organs.

New antibodies to specialized cell specific antigens are still being reported and only time will show whether they are as specific as claimed by the scientists who have discovered them. The specificity of such new antibodies as those to melanoma (GOWN et al. 1986), renal tumors (OOSTERWIJK et al. 1986) or the diagnostic value of antibody to retinal S-antigen in the diagnosis of pineocytomas (KORF et al. 1986) and many others remain to be determined.

Despite the fact that most claims about "exclusive" or "unique" tissue specific markers for specialized cells or antigens proved to be over-optimistic, there is no reason for complacency. On the other hand, even if the original claims do not withstand more critical evaluations, antibodies produced to these antigens could still be useful diagnostic reagents if applied critically and under well-defined conditions. Thus, antibody to S-100 and natural killer cells are still useful reagents for the diagnosis of melanoma (HAGEN et al. 1986) and antibodies to calcitonin, if applied to thyroid tumors, are still most valuable for distinction of medullary carcinoma from other thyroid tumors (SOBRINHO-SIMOES et al. 1985). The value of antibody to a leukocyte antigen (Leu M1) in the study of solid tumors has also been demonstrated (SHEIBANI et al. 1986).

3.2.3 Differentiation Markers

Differentiation and maturation of cells in certain constantly proliferating tissue compartments, such as skin, intestine or bone marrow occurs in a well regulated predictable manner. Immunohistochemistry has made major contributions to the understanding of normal differentiation of lymphoid and myeloid cells, skin or intestine, and other organs. Since the malignant transformation can occur at any stage of the normal proliferation-differentiation sequence, it was possible histochemically to phenotype tumors and compare them with cell stage in normal histogenesis at which the neoplasia has arisen. Most remarkable progress was made in the lymphoma research (TUBBS and SHEIBANI 1984; KNOWLES et al. 1986; WEISS et al. 1986; ANDREESEN et al. 1986). However, similar phenotyping of tumors is in order for all other organ systems and it is a matter of time before this becomes a reality.

4 Diagnostic Immunohistopathology

The extent of future use of immunohistochemistry in diagnostic pathology will depend on many factors. This volume contains many examples of how this most valuable addition to the armoury of the contemporary pathologist has increased our understanding of tumor biology and increased the precision of morphologic diagnosis. Hence, only selected aspects of practical tumor histopathology will be addressed here.

4.1 Histochemical Differential Diagnosis of Tumors

"Is it a carcinoma, sarcoma, melanoma, neuroepithelial tumor or lymphoma?" This question, usually asked by pathologists confronted with a tumor, is usually answered in 85–90% of routine cases upon microscopic study of hematoxylin and eosin stained slides. The remaining 10–15% of the cases, representing the usual "problem" cases, fall into several categories:

A. Poorly differentiated tumors, showing limited histiotypic or organotypic differentiation;
B. Tumors presenting an unexpected or highly unusual pattern;
C. Rare tumors, newly recognized pathologic entities, or sundry conditions that are not commonly encountered in everyday practice;
D. Metastatic tumors with no obvious primary.

Morphologically, most of the "problem" cases or tumors that cannot be definitively or unequivocally diagnosed by light microscopy occur in four histologic patterns:

A. Small blue cell tumors,
B. Spindle cell tumors,
C. Large or medium sized polyhedral cell tumors, and
D. Pleomorphic tumors.

The immunohistochemical approach to these tumors will primarily depend on the initial histopathologic impression, the experience and the routine of the pathologist, the anatomic location of the tumor and the differential diagnostic possibilities considered from the clinical information available at the time of examination. In practical terms the approach will also depend on the technical capabilities of the laboratory, availability of antibodies and also on the nature of material that is still available for examination. Certain antibodies are not applicable to routine paraffin embedded formalin fixed tissues and for their optimal usage frozen tissues are a sine qua non. Necrotic tumors, or inadequately fixed and autolyzed tumors are not suitable for all immunohistochemical studies, although many antigens apparently survive various forms of maceration and are demonstrable even in inadequately preserved tissues (Pallesen and Knudsen 1985).

To obtain maximum benefit from immunohistochemical studies it is essential to ask the right questions and select the antibodies only after the differential diagnosis of the lesion has been defined as narrowly as possible by standard light microscopy, and the conventional histochemical techniques used in routine pathology laboratories. Standard criteria of tumor histopathology should not only serve as general guidelines, but should also serve as the principal basis for classification of tumors to which all the other ancillary results should be referred and correlated for the formulation of the final diagnosis. If the immunohistochemical studies are inconclusive, it is still best to base the final decision on the light microscopic diagnosis and the overall evaluation of the case.

The approach to each problem should be individually designed and the empirical criteria best suited for the laboratory and the pathologist in charge. On the basis of empirical data from many pathology laboratories algorithms

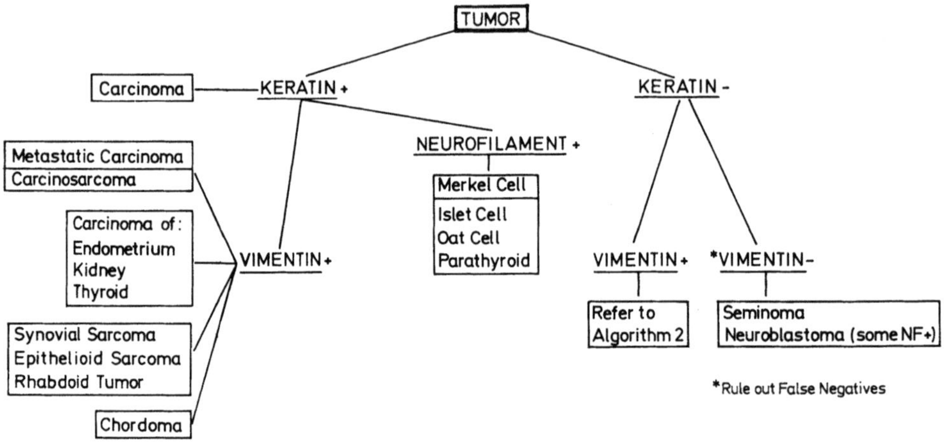

Fig. 1

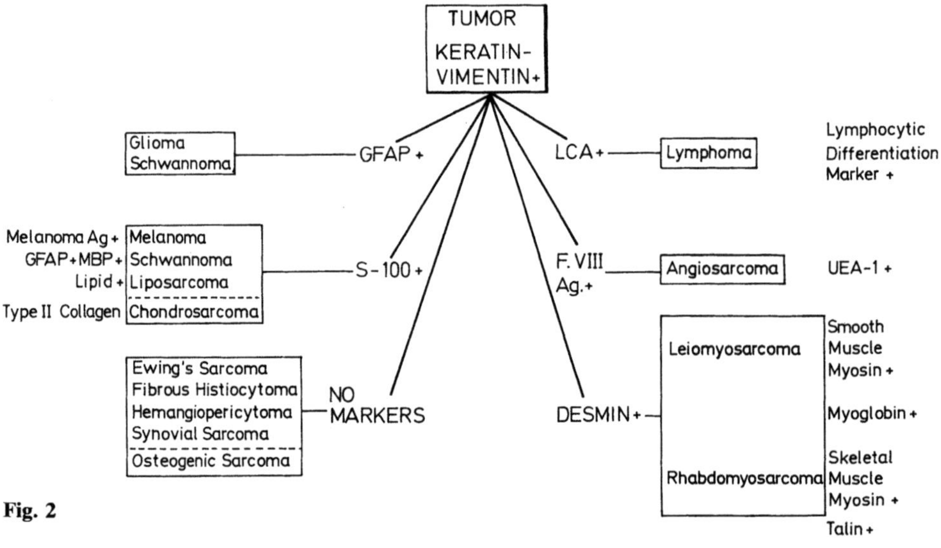

Fig. 2

have been proposed for a step by step approach and these, with some modifications, could be applied to most, if not all, solid tissue tumors. Such a flow chart, modified from ROHOLL et al. (1985) is present in Figure 1 and Figure 2 and is based on the premise that most, if not all, epithelial tumors express keratin polypeptides as major components of their intermediate filament cytoskeleton (OSBORN and WEBER 1983). Epithelial membrane antigen (EMA) could be used instead of keratin as the marker for epithelial cells in the initial screen and the combined application of antibodies to keratin and EMA seems to be even more productive (BATTIFORA 1984).

Keratin positive tumors represent in most instances carcinomas. Appropriately fixed or freshly frozen specimens can be further analyzed for the expression of specific keratin polypeptides (OSBORN et al. 1986b) or vimentin. Vimentin, a typical marker of mesenchymal cells, if expressed as the only intermediate filament protein is a strong indication that the tumor is not a carcinoma and is thus either sarcoma, melanoma or a hematopoietic neoplasm. In that case a battery of tests with antibodies to specific mesenchymal cell markers should be performed as outlined in Figure 2. Keratin positive, vimentin positive tumors could be carcinomas, and there are several organs, such as endometrium, kidney and thyroid that give rise to such tumors (GOWN et al. 1985; MIETTINEN et al. 1985). Metastatic carcinomas, like the carcinoma cells explanted in vitro may express both keratin and vimentin positive cells. Sarcomatoid carcinomas of the esophagus are another example, although in these tumors one could find both coexpression of vimentin and keratin in the same cells and distinct vimentin and keratin positive cell populations.

It is important to note that certain tumors of soft tissues, most notably biphasic synovial cell sarcomas, (MIETTINEN and VIRTANEN 1984), epithelioid sarcomas (MIETTINEN et al. 1985c) and rhabdoid tumors (SOTELLO-AVILA et al. 1986) coexpress vimentin and keratin in the same cells.

Keratin positive cells may in some tumors coexpress neurofilament proteins. Thus, it has been shown that keratin and neurofilaments are coexpressed in some oat cell carcinomas and carcinoids of the lung (LEHTO et al. 1985) islet cell carcinomas of pancreas, especially insulomas (MIETTINEN et al. 1985b), parathyroid adenomas (MIETTINEN et al. 1985a). Merkel cell carcinomas of the skin coexpress keratin and neurofilaments with great regularity (GOULD et al. 1985). The coexpression of several types of intermediate filament proteins in the same cells is biologically a most intriguing phenomenon which so far has not received full explanation (GOULD 1985). However, such aberrant expression of markers should be kept in mind to avoid overdogmatic interpretation of immunohistochemical findings. Keratin negative, vimentin negative tumors should be evaluated most carefully to exclude spuriously negative results. Vimentin is poorly preserved in routinely processed tissues embedded in paraffin and negative staining with antibodies to vimentin should thus be interpreted with utmost caution. Nevertheless some tumors, such as seminoma or those that have very little cytoplasm like some neuroblastomas appear unreactive with antibodies to all intermediate filaments (OSBORN and WEBER 1983).

Vimentin positive, keratin negative tumors comprise mostly benign and malignant soft tissue neoplasms, melanomas and neuroectodermal neoplasms, gliomas and schwannomas. Other intermediate filament proteins, such as glial fibrillary acidic protein and desmin may be used to further characterize these tumors (ROHOLL et al. 1985). Other markers for these tumors are reviewed elsewhere in this volume. New markers, such as titin for rhabdomyosarcomas (OSBORN et al. 1986) or smooth muscle specific myosin (LONGTINE et al. 1985) could be of additional use.

There are still no universally accepted markers for several sarcomas. This group includes most prominently fibrous histiocytomas, Ewing's sarcoma, hemangiopericytoma, synovial cell sarcoma. The nature of stem cells of these tumors has not yet been defined and it is not known whether the polymorphism in some of these represents abortive differentiation or progression into a more anaplastic form of neoplasia (BROOKS 1986).

5 Perspective

This volume contains numerous examples illustrating the diagnostic value of immunohistochemistry in routine pathology. Antibodies, such as those against intermediate filaments, epithelial membrane antigen, hormones and organ specific polypeptides and many others are already widely used by pathologists worldwide. New antibodies are being developed and tested. Diagnostic pathology has not yet reached its peak, and it is reassuring to note that immunohistochemistry is here to stay and that it is on the best way to realize its full potential.

References

Abelev GI (1971) Alpha fetoprotein in oncogenesis and its association with malignant tumors. Adv Cancer Res 14:295–358

Allum WH, MacDonald F, Anderson P, Fielding JWL (1986) Localization of gastrointestinal cancer with a ^{131}I labeled monoclonal antibody to CEA. Br J Cancer 53:203–210

Andreesen R, Bross KY, Osterholz J, Emmrich F (1986) Human macrophage maturation and heterogeneity: analysis with a newly generated set of monoclonal antibodies to differentiation antigens. Blood 67:1257–1264

Angeletti RH (1986) Chromogranins and neuroendocrine secretion. Lab Invest 55:387–390

Balmain A (1985) Transforming ras oncogenes and multistage carcinogenesis. Br J Cancer 51:1–7

Bara J, Zabaleta EH, Mollicone R, Nap M, Burtin P (1986) Distribution of GICA in normal gastrointestinal and endocervical mucosae and mucinous ovarian cysts using antibody NS 19-9. Am J Clin Pathol 85:152–159

Battifora H (1984) Recent progress in the immunohistochemistry of solid tumors. Semin Diagn Pathol 1:251–271

Birembaut P, Caron Y, Adnet JJ, Foidart JM (1985) Usefulness of basement membrane markers in tumoral pathology. J Pathol 145:283–296

Braunstein GD, Thompson R, Princler GL, McIntire KR (1986) Trophoblastic proteins as tumor markers in nonseminomatous germ cell tumors. Cancer 57:1842–1845

Brooks JJ (1986) The significance of double phenotypic patterns and markers in human sarcomas: A new model of mesenchymal differentiation. Am J Pathol 125:113–123

Burger PC, Shibata T, Kleihues P (1986) The use of the monoclonal antibody Ki-67 in the identification of proliferating cells: Application to surgical neuropathology. Am J Surg Pathol 10:611–617

Camplejohn RS, Aherne WA (1974) In vitro labeling of childhood cancers with tritiated thymidine. Br J Cancer 29:487–489

Chaitin BA (1985) Mucinous tumors of the ovary. A clinicopathologic study of 70 cases. Cancer 55:1958–1962

Coleman DV, Richman PI (1985) Human papillomavirus infection and cancer of the uterine cervix. J Pathol 145:2007–2012

Dabbs DJ, Geisinger KR, Norris HT (1986) Intermediate filaments in endometrial and endocervical carcinoma. The diagnostic utility of vimentin patterns. Am J Surg Pathol 10:568–576

Damjanov I (1982) Antibodies to intermediate filaments and histogenesis. Lab Invest 47:215–217

Damjanov I, Knowles BB (1983) Monoclonal antibodies and tumor associated antigens. Lab Invest 48:510–523

D'Ardenne AJ, Kirkpatrick P, Wells CA, Davies JD (1986) Laminin and fibronectin in adenoid cystic carcinoma. J Clin Pathol 39:138–144

Erhardt K, Auer G, Bjorkholm E, Forsslund G, Moberger B, Silversward C, Wicksell G, Zetterberg A (1985) Combined morphologic and cytochemical grading of serous ovarian tumors. Am J Obstet Gynecol 151:356–361

Faber M, Wewer UM, Berthelsen JG, Liotta LA, Albrechtsen R (1986) Laminin production by human endometrial stromal cells relates to the cyclic and pathologic state of the endometrium. Am J Pathol 124:384–398

Falser N (1984) The capsid antigen of Epstein-Barr virus in nasopharyngeal carcinomas: comparison of serum antibody titre, cellular antigen content and histopathology. Arch Otorhinolaryngol 239:15–23

Farber E (1984a) Pre-cancerous steps in carcinogenesis: their physiological adaptive nature. Biochim Biophys Acta 738:171–180

Farber E (1984b) The multistep nature of cancer development. Cancer Res 44:4217–4223

Feizi T (1985) Demonstration by monoclonal antibodies that carbohydrate structures of glycoproteins and glycolipids are onco-developmental antigens. Nature 314:53–57

Fetissof F, Bertrand G, Guilloteau D, DuBois MP, Lanson Y, Arbeille B (1986a) Calcitonin immunoreactive cells in prostate gland and cloacal derived tissues. Virchows Arch Pathol Anat 409:523–533

Fetissof F, Bruandet P, Arbeille B, Penot J, Marboeuf Y, Le Roux J, Guilloteau D, Beaulieu J-L (1986b) Calcitonin-secreting carcinomas of the prostate. An immunohistochemical and ultrastructural analysis. Am J Surg Pathol 10:702–710

Fisher C (1986) Synovial sarcoma: ultrastructural and immunohistochemical features of epithelial differentiation in monophasic and biphasic tumors. Hum Pathol 17:996–1008

Foster CS (1982) Lymphocyte hybridomas. Cancer Treatment Rev 9:59–84

Foulds L (1969) Neoplastic development, vol I. Academic Press, New York

Fukayama M, Hayashi Y, Koike M, Hajikano H, Endo S, Okumura H (1986) Human chorionic gonadotropin in lung and lung tumors. Immunohistochemical study on imbalanced distribution of subunits. Lab Invest 55:433–443

Gardner RL (1985) Clonal analysis of early mammalian development. Phil Trans R Soc Lond B 312:163–178

Gardner RL (1986) Cell mingling during mammalian embryogenesis. J Cell Sci Suppl 4:337–356

Gatter KC, Alcock C, Heryet A, Mason DY (1985) Clinical importance of analysing malignant tumors of uncertain origin with immunohistologic techniques. Lancet 1:1302–1305

Gerdes J (1985) An immunohistological method for estimating cell growth fractions in rapid histopathological diagnosis during surgery. Int J Cancer 35:169–171

Gerdes J, Schwab U, Lemke H, Stein H (1983) Production of a mouse monoclonal antibody reactive with a human nuclear antigen associated with cell proliferation. Int J Cancer 31:13–20

Ghosh AK, Moore M, Harris M (1986) Immunohistochemical detection of ras oncogene p21 product in benign and malignant mammary tissue in man. J Clin Pathol 39:428–435

Goh HS, Jass JR (1986) DNA content and the adenoma-carcinoma sequence in the colorectum. J Clin Pathol 39:387–392

Gold P, Friedman SO (1965) Demonstration of tumor-specific antigens in human colonic carcinomata by immunological tolerance and absorption techniques. J Exp Med 121:439–462

Gould VE (1985) The coexpression of distinct classes of intermediate filaments in human neoplasms. Arch Pathol Lab Med 109:984–985

Gould VE (1986) Histogenesis and differentiation: a reevaluation of these concepts as criteria for the classification of tumors. Hum Pathol 17:212–215

Gould VE, Moll R, Moll I, Lee I, Franke WW (1985) Neuroendocrine (Merkel) cells of the skin – hyperplasias, dysplasias, and neoplasms. Lab Invest 52:334–353

Gould VE, Moll R, Moll I, Lee I, Schwechheimer K, Franke WW (1986) The intermediate filament complement of the spectrum of nerve sheath neoplasms. Lab Invest 55:463–474

Gown AM, Vogel AM (1985) Monoclonal antibodies of human intermediate filament proteins. III. Analysis of tumors. Am J Clin Pathol 84:413–424

Gown AM, Vogel AM, Hoak D, Gough F, McNutt MA (1986) Monoclonal antibodies specific for melanocytic tumors distinguish subpopulations of melanocytes. Am J Pathol 123:195–203

Guy K, Krajewski AS, Dewar EA (1986) Expression of MHC Class II antigens in human B-cell leukemia and non-Hodgkin's lymphoma. Br J Cancer 53:161–173

Hagen EC, Vennegoor C, Schlingemann RO, Van Der Velde EA, Ruiter DJ (1986) Correlation of histopathological characteristics with staining patterns in human melanomas assessed by (monoclonal) antibodies reactive on paraffin sections. Histopathology 10:689–700

Haimoto H, Takahashi Y, Koshikawa T, Nagura H, Kato K (1985) Immunohistochemical localization of γ-enolase in normal human tissue other than nervous and neuroendocrine tissues. Lab Invest 52:257–263

Hakomori S (1984) Tumor associated carbohydrate antigens. Ann Rev Immunol 2:103–126

Hakomori S (1985) Aberrant glycosylation in cancer cell membranes as focused on glycolipids – overview and perspectives. Cancer Res 45:2405–2414

Hall TL, Fu YS (1985) Applications of quantitative microscopy in tumor pathology. Lab Invest 53:5–21

Helle M, Krohn K (1986) Immunohistochemical reactivity of monoclonal antibodies to human milk fat globule with breast carcinomas and with other normal and neoplastic tissues. Acta Pathol Microbiol Scand A 94:43–51

Herlyn D, Ross AH, Koprowski H (1986) Anti-idiotypic antibodies bear the internal image of a human-tumor antigen. Science 232:100–102

Herpers MJHM, Ramaekers FCS, Aldeweireldt J, Moesker O, Slooff J (1986) Co-expression of glial fibrillary acidic protein – and vimentin – type, intermediate filaments in human astrocytomas. Acta Neuropathol 70:333–339

Highman WJ (1986) Transitional carcinoma of the upper urinary tract: a histological and cytopathological study. J Clin Pathol 39:297–305

Hilkens J, Buijs F, Hilgers J, Hageman P, Calafat J, Vandervalk M (1984) Monoclonal antibodies against human milk-fat globule membranes detecting differentiation antigens of the mammary gland and its tumors. Int J Cancer 34:197–206

Horan-Hand P, Thor A, Wunderlich D, Muraro R, Caruso A, Schlom J (1984) Monoclonal antibodies of predefined specificity detect activated ras gene expression in human mammary and colon carcinomas. Proc Natl Acad Sci USA 81:5227–5231

Hoshino H, Nagashima T, Murovic JA (1986) In situ cell kinetics studies on human neuroectodermal tumors with bromodeoxyuridine labeling. J Neurosurg 64:453–459

Inoue M, DeLellis RA, Scully RE (1986) Immunohistochemical demonstration of chromogranin in endometrial carcinomas with argyrophil cells. Hum Pathol 17:841–847

Iseki M, Suzuki T, Koizumi Y, Hirose M, Laskin WB, Nakazawa S, Ohaki Y (1986) Alpha-fetoprotein-producing pancreatoblastoma. A case report. Cancer 57:1833–1835

Ishikura H, Kirimoto K, Shamoto M, Miyamoto Y, Yamagiwa H, Itoh T, Aizawa M (1986) Hepatoid adenocarcinomas of the stomach. An analysis of seven cases. Cancer 58:119–2126

Itzkowitz SH, Kim YS (1986) New carbohydrate tumor markers. Gastroenterology 90:491–493

Juhl BR, Hartzen SH, Hainau B (1986) A, B, H antigen expression in transitional cell carcinomas of the urinary bladder. Cancer 57:1768–1775

Kempson RL, Bari W (1970) Uterine sarcomas: classifications, diagnosis, and prognosis. Hum Pathol 1:331–369

Kirchner H (1986) Immunobiology of human papilloma virus infection. Prog Med Virol 33:1–41

Klein G, Klein E (1984) The changing faces of EBV research. Prog Med Virol 30:87–106

Knowles DM (1986) The human T-cell leukemias: Clinical cytomorphologic, immunophenotypic and genotypic characteristics. Hum Pathol 17:14–33

Knowles DM, Pelicci PG, Dalla-Favera R (1986) T-cell receptor beta chain gene rearrangements: genetic markers of T-cell lineage and clonality. Hum Pathol 17:546–551

Kohler G, Milstein C (1975) Continuous cultures of fused cells secreting antibody of predefined specificity. Nature 256:495–497

Korf HW, Klein DC, Zigler JS, Gery I, Schachenmayr W (1986) S-antigen-like immunoreactivity in a human pineocytoma. Acta Neuropathol 69:165–167

Lauder I, Holland D, Mason DY, Gowland G, Cunliffe WJ (1984) Identification of large cell undifferentiated tumors in lymph nodes using leucocyte common and keratin antibodies. Histopathology 8:259–272

Lehto V-P, Miettinen M, Virtanen I (1985) Dual expression of cytokeratin and neurofilaments in bronchial carcinoid cells. Int J Cancer 35:421–425

Lehtonen E, Virtanen I, Saxen L (1985) Reorganization of intermediate filament cytoskeleton in induced metanephic mesenchymal cells is independent of tubule morphogenesis. Dev Biol 108:481–490

Limas C, Lange P (1980) Altered reactivity for A, B, H antigen in transitional cell carcinomas of the urinary bladder: A study of the mechanisms involved. Cancer 46:1366–1373

Limas C, Lange P (1986) T-antigen in normal and neoplastic urothelium. Cancer 58:1236–1245

Limas C, Lange P, Fraley EE (1979) A, B, H, antigens in transitional cell tumors of the urinary bladder: correlation with the clinical course. Cancer 44:2099–2107

Longtine JA, Pinkus GS, Fujiwara K, Corson JM (1985) Immunohistochemical localization of smooth muscle myosin in normal human tissues. J Histochem Cytochem 33:179–184

Mazur MT, Clark HB (1983) Gastric stromal tumors. Reappraisal of histogenesis. Am J Surg Pathol 7:507–519

Meredith RF, Wagman LD, Piper JA, Mills AS, Neifeld JP (1986) Beta-chain human chorionic gonadotropin-producing leiomyosarcoma of the small intestine. Cancer 58:131–135

McNutt M, Bolen J, Gown AM, Hammar S, Vogel AM (1985) Coexpression of intermediate filaments in human epithelial neoplasms. Ultrastruct Pathol 9:35–37

Miettinen M, Virtanen I (1984) Synovial sarcoma – a misnomer. Am J Pathol 117:18–23

Miettinen M, Lehto V-P, Virtanen L (1982) Keratin in the epithelial-like cells of classical biphasic synovial sarcoma. Virchows Arch (Cell Pathol) 40:157–161

Miettinen M, Franssila K, Lehto V-P, Paasivuo R, Virtanen I (1984) Expression of intermediate filament proteins in thyroid gland and thyroid tumors. Lab Invest 50:262–270

Miettinen M, Clark R, Lehto V-P, Virtanen I, Damjanov I (1985a) Intermediate filament protein in parathyroid glands and parathyroid adenomas. Arch Pathol Lab Med 109:986–989

Miettinen M, Lehto V-P, Dahl D, Virtanen I (1985b) Varying expression of cytokeratin and neurofilaments in neuroendocrine tumors of human gastrointestinal tract. Lab Invest 52:429–436

Miettinen M, Virtanen I, Damjanov I (1985c) Coexpression of keratin and vimentin in epithelioid sarcoma. Am J Surg Pathol 9:460

Miettinen M, Clark R, Virtanen I (1986) Intermediate filament proteins in choroid plexus and ependyma and their tumors. Am J Pathol 123:231–240

Momburg F, Degener T, Bacchus E, Moldenhauer G, Hammerling G, Moller P (1986) Loss of HLA-A, B, C and de novo expression of HLA-D in colorectal cancer. Int J Cancer 37:179–184

Neville AM, Foster CS, Moshakis V, Gore M (1982) Monoclonal antibodies and human tumor pathology. Hum Pathol 13:1067–1081

Oosterwijk E, Ruiter DJ, Wakka JC, Huiskens VD, Meij JW, Jonas U, Fleuren G-J, Zwartendijk J, Hoedemaker P, Warnaar SO (1986) Immunohistochemical analysis of monoclonal antibodies in the diagnosis of renal cell carcinomas. Am J Pathol 123:301–309

Ormerod MG, Steele K, Westwood JH, Mazzini MN (1983) Epithelial membrane antigen: partial purification, assay and properties. Br J Cancer 48:533–541

Osborn M, Weber K (1983) Tumor diagnosis by intermediate filament typing: A novel tool for surgical pathology. Lab Invest 48:371–394

Osborn M, Hill C, Altmannsberger M, Weber K (1986a) Monoclonal antibodies to titin in conjunction with antibodies to desmin separate rhabdomyosarcomas from other tumor types. Lab Invest 55:101–108

Osborn M, Van Lessen G, Weber K, Kloppel G, Altmannsberger M (1986b) Differential diagnosis of gastrointestinal carcinomas by using monoclonal antibodies specific for individual keratin polypeptides. Lab Invest 55:497–504

Perentes E, Rubinstein LJ (1986) Immunohistochemical recognition of human neuroepithelial tumors by anti-Leu 7 (HNK-1) monoclonal antibody. Acta Neuropathol 69:227–233

Picard JK, Hounsell EF, Gregorio M, Rees AR, Feiz IT (1985) Monoclonal antibody (EGR/G49) reactive with the epidermal growth factor receptor of A431 cells recognizes the blood group ALe^b and ALe^y structures. Mol Immunol 22:689–693

Pilotti S, Rilke F, Shah KV, Delle Torre G, De Palo G (1984) Immunohistochemical and ultrastructural evidence of papilloma virus infection associated with in situ and microinvasive squamous cell carcinomas of vulva. Am J Surg Pathol 8:751–761

Pinkus GS, Etheridge CL, O'Connor EM (1986) Are keratin proteins a better tumor marker than epithelial membrane antigen? A comparative immunohistochemical study of various paraffin-embedded neoplasms using monoclonal and polyclonal antibodies. Am J Clin Pathol 85:269–277

Poston RN, Sidhu YS (1986) Diagnosing tumors on routine surgical sections by immunohistochemistry: use of cytokeratin, common leucocyte and other markers. J Clin Pathol 39:514–523

Purtilo DT (1984) Defective immune surveillance in viral carcinogenesis. Lab Invest 51:373–385

Ramaekers FCS, Puts JJG, Moesker O, Kant A, Vooijs GP (1983) Intermediate filaments in malignant melanomas. Identification and use as markers in surgical pathology. J Clin Invest 71:635–643

Rubin H (1985) Cancer as a dynamic developmental disorder. Cancer Res 45:2935–2942

Sack TL, Kim YS (1986) Pancreatic cancer-associated carbohydrate antigens. Front Gastroent Res 12:48–59

Sasaki A, Ogawa A, Nakazato Y, Ishida Y (1985) Distribution of neurofilament protein and neuron-specific enolase in peripheral neuronal tumors. Virchows Arch (Pathol Anat) 407:33–41

Schiffer D, Giordana MT, Mauro A, Migheli A, Germano I, Giaccone G (1986) Immunohistochemical demonstration of vimentin in human cerebral tumors. Acta Neuropathol 70:209–210

Seifert G (1986) Molekularbiologische Aspekte der Tumorpathologie: Onkogene-Zellrezeptoren-Tumormarker. Pathologe 7:69–74

Sheibani K, Battifora H, Burke JS (1986) Antigenic phenotype of malignant mesotheliomas and pulmonary adenocarcinomas: An immunohistologic analysis demonstrating the value of Leu M1 antigen. Am J Pathol 123:212–219

Shen J-W, Atkinson B, Koprowski H, Sears HF (1984) Binding of murine immunoglobin to human tissues after immunotherapy with anticolorectal carcinoma monoclonal antibodies. Int J Cancer 33:465–468

Singer A, Wilters J, Walker P, Jenkins D, Slavin G, Cowdell H, To A, Husain O (1985) Comparison of prevalence of human papilloma virus antigen in biopsies from women with cervical intraepithelial neoplasia. J Clin Pathol 38:855–857

Sobin LH, Hjermstad BM, Sesterhenn IA, Helwig EB (1986) Prostatic acid phosphatase activity in carcinoid tumors. Cancer 58:136–138

Sobrinho-Simoes MA, Nesland JM, Holm R, Johannesen JV (1985) Transmission electron microscopy and immunocytochemistry in the diagnosis of thyroid tumors. Ultrastruct Pathol 9:255–275

Sotelo-Avila C, Gonzalez-Crussi F, deMello D, Vogler C, Gouch WM, Gale G, Pena R (1986) Renal and extrarenal rhabdoid tumors in children: a clinicopathologic study of 14 patients. Semin Diagn Pathol 3:151–163

Steplewski Z, Spira G, Blaszczyk M, Lubeck MD, Radbruch A, Illges H, Herlyn D, Rajewsky K, Scharff M (1985) Isolation and characterization of anti-monosialoganglioside monoclonal antibody 19-9 class switch variants. Proc Natl Acad Sci USA 82:8653–8657

Svajger A, Levak-Svajger B, Skreb N (1986) Rat embryonic ectoderm as renal isograft. J Embryol Exp Morphol 94:1–27

Thor A, Horan Hand P, Wunderlich D, Caniso A, Muraro R, Schlom J (1984) Monoclonal antibodies define differential ras expression in malignant and benign colonic diseases. Nature 311:562–565

Trinchieri G, Perussia B (1984) Human natural killer cells: biologic and pathologic aspects. Lab Invest 50:489–501

Tubbs R, Sheibani K (1984) Immunohistology of lymphoproliferative disorders. Semin Diagn Pathol 1:272–284

Vanstapel M-J, Gatter KC, deWolf-Peeters C, Mason DY, Desmet VD (1986) New sites of human S-100 immunoreactivity detected with monoclonal antibodies. Am J Clin Pathol 85:160–168

Vowden P, Lowe AD, Lennox ES, Bleehen NM (1986) Are blood group isoantigens lost from malignant prostatic epithelium? Immunohistochemical support for the preservation of the H isoantigen. Br J Cancer 53:307–312

Weiss LM, Trela MJ, Cleary ML, Turner RR, Warnke RA, Sklar J (1985) Frequent immunoglobulin and T-cell receptor gene rearrangements in "histiocytic" neoplasms. Am J Pathol 121:369–373

Weiss LM, Strickler JG, Hu E, Warnke RA, Sklar J (1986) Immunoglobulin gene rearrangements in Hodgkin's disease. Hum Pathol 17:1009–1014

Wiedenmann B, Franke WW, Kuhn C, Gould VE (1986) Synaptophysin: a marker protein for neuroendocrine cells and neoplasms. Proc Natl Acad Sci USA 83:3500–3506

Wilson BS, Lloyd RV (1984) Detection of chromogranin in neuroendocrine cells with a monoclonal antibody. Am J Pathol 115:458–468

Wilson NW, Pambakian H, Richardson TC, Stokoe MR, Makin CA, Heyderman E (1986) Epithelial markers in thyroid carcinomas: an immunoperoxidase study. Histopathology 10:815–829

Subject Index

Index of Volumes 71–76 Current Topics in Pathology

H.-E. SCHAEFER, Inflammatory Disease of the Human Lung of Definite or Presumed Viral Origin. Cytologic and Histologic Topics
W. KISSLER, Formal Genesis of Pulmonary Fibrosis: Experimental Investigations
K.-M. MÜLLER, G. MÜLLER, The Ultrastructure of Preneoplastic Changes in the Bronchial Mucosa
D. FRANCIS, M. JACOBSEN, Pulmonary Blastoma

Volume 74: Dermatopathology. Edited by C.L. BERRY

D. WEEDON, Melanoma and Other Melanocytic Skin Lesions
T.J. RYAN, S.M. BURGE, Cutaneous Vasculitis
R.B. GOUDIE, A.S. JACK, B.M. GOUDIE, Genetic and Development Aspects of Pathological Pigmentation Patterns
N.A. WRIGHT, Changes in Epidermal Cell Proliferation in Proliferative Skin Diseases
W. STERRY, Mycosis fungoides
A.C. CHU, Bullous Dermatoses

Volume 75: The Human Thymus. Edited by H.K. MÜLLER-HERMELINK

B. VON GAUDECKER, The Development of the Human Thymus Microenvironment
G.G. STEINMANN, Changes in the Human Thymus During Aging
G. JANOSSY, M. BOFILL, L.K. TREJDOSIEWICZ, H.N.A. WILLCOX, M. CHILOSI, Cellular Differentiation of Lymphoid Subpopulations and Their Microenvironments in the Human Thymus
N. DOUROV, Thymic Atrophy and Immune Deficiency in Malnutrition
CH. NEZELOF, Pathology of the Thymus in Immunodeficiency States
H. WEKERLE, H.K. MÜLLER-HERMELINK, The Thymus in Myasthenia Gravis
H.K. MÜLLER-HERMELINK, M. MARINO, G. PALESTRO, Pathology of Thymic Epithelial Tumors

Volume 76: Neuropathology. Edited by C.L. BERRY (in preparation)

J.H. ADAMS, The Autopsy in Fatal Non-Missile Head Injuries
J.R. ANDERSON, Viral Encephalitis and Its Pathology
R.O. WELLER, A General Approach to Neuropathological Problems
C.L. SCHOLTZ, Dementia in Middle and Late Life

MIX
Papier aus verantwortungsvollen Quellen
Paper from responsible sources
FSC® C105338

If you have any concerns about our products,
you can contact us on
ProductSafety@springernature.com

In case Publisher is established outside the EU,
the EU authorized representative is:
Springer Nature Customer Service Center GmbH
Europaplatz 3, 69115 Heidelberg, Germany

Printed by Libri Plureos GmbH
in Hamburg, Germany